CATALOG OF SOLAR PARTICLE EVENTS
1955–1969

ASTROPHYSICS AND SPACE SCIENCE LIBRARY

A SERIES OF BOOKS ON THE RECENT DEVELOPMENTS
OF SPACE SCIENCE AND OF GENERAL GEOPHYSICS AND ASTROPHYSICS
PUBLISHED IN CONNECTION WITH THE JOURNAL
SPACE SCIENCE REVIEWS

VOLUME 49

CATALOG
OF SOLAR PARTICLE EVENTS
1955–1969

*Prepared under the Auspices of Working Group 2 of the Inter-Union Commission
on Solar-Terrestrial Physics by*

H. W. DODSON and E. R. HEDEMAN
The McMath-Hulbert Observatory of the University of Michigan, Pontiac, Mich., U.S.A.,

R. W. KREPLIN
The Naval Research Laboratory, Washington, D.C., U.S.A.,

M. J. MARTRES
Observatoire de Paris, Meudon, France

V. N. OBRIDKO
IZMIRAN Moscow, U.S.S.R.

M. A. SHEA and D. F. SMART
Air Force Cambridge Research Laboratories, Bedford, Mass., U.S.A.

H. TANAKA
Research Institute of Atmospherics, Toyokawa, Japan

edited by

Z. ŠVESTKA
*Fraunhofer Institute, Freiburg im Breisgau, F.R.G.,
presently at American Science and Engineering, Cambridge, Mass., U.S.A.*

P. SIMON
Observatoire de Paris, Meudon, France

with the assistance of L. FRITZOVÁ-ŠVESTKOVÁ (Freiburg and Cambridge) and A. GUITART (Meudon)

D. REIDEL PUBLISHING COMPANY
DORDRECHT-HOLLAND / BOSTON-U.S.A.

Library of Congress Catalog Card Number 74–81944

ISBN-13:978-94-010-1744-2   e-ISBN-13:978-94-010-1742-8
DOI: 10.1007/978-94-010-1742-8

Published by D. Reidel Publishing Company
P.O. Box 17, Dordrecht, Holland

Sold and distributed in the U.S.A., Canada, and Mexico
by D. Reidel Publishing Company, Inc.
306 Dartmouth Street, Boston,
Mass. 02116, U.S.A.

TABLE OF CONTENTS

PREFACE

This Catalog originated as a common enterprise of solar physicists and space scientists under the auspices of the Second Working Group of the Inter-Union Commission of Solar Terrestrial Physics (IUCSTP). It is a pleasure to acknowledge the support we received from the IUCSTP president, Dr H. Friedman, and the IUCSTP Secretary, Dr E. R. Dyer during the several years we have spent on this project.

The aim of our work has been to assemble all observations of particle events from the first PCA observation in 1955 through two solar cycle maxima up to the end of 1969, in order to preserve these data from the first epoch of the space era in a concise form for use in the future. Because the techniques of observation have very much improved in the present solar cycle, there is a tendency to consider the observations before 1964 as incomplete and out-of-date; one must not forget, however, that the 19th solar cycle greatly differed from cycle No. 20 in the level of its activity, and also may have been the last cycle of strong activity for many decades to come. Therefore, the particle event observations before 1964 should be compiled in a consistent manner for comparison with later observations, and we believe that the Catalog achieves this.

The rapid development of the exploration techniques necessarily results in a significant amount of inhomogeneity in the Catalog, as increasingly smaller events were detected. This lack of homogeneity cannot be avoided and the reader should be well aware of its existence. *Before using the Catalog, a careful reading of the Introduction is essential.*

The Catalog consists of three Parts. For *Part 1* and *Appendix A*, M. A. Shea and D. F. Smart have summarized observations of particle increases during the 15 years studied. In addition to the excerption of published data many *unpublished observations* were included through the cooperation of K. A. Anderson, R. P. Bukata, J. R. Burrows, C. Y. Fan, M. Gros, M. van Hollebeke, S. W. Kahler, H. Leinbach, R. P. Lin, K. G. McCraken, F. B. McDonald, A. J. Masley, U. R. Rao, J. A. Simpson, S. Singer, S. N. Vernov, and J. R. Wang, to all of whom our particular thanks are due. A significant contribution to the list of ground-based data was made by E. R. Hedeman, and the list of PCA events prior to 1964 was checked by G. C. Reid.

In a very detailed and careful study H. W. Dodson and E. R. Hedeman have tried to find the source, or alternative sources, of each particle event in Part 1 of the Catalog, or in Appendix A, using criteria explained in the Introduction. Their work has clearly demonstrated how difficult it is to find the actual sources of the particles on the Sun, as soon as the study is extended to particle events of relatively low energy.

The final selection of the events to be included in Part 1 of the Catalog or Appendix A, respectively, the choice of abbreviated selected data to be published on each event, and the final association with solar and interplanetary sources, has been made by the Editors (by Švestka for 1955–1966, and by Simon for 1967–1969). The opinions of H. Tanaka (radio emission) and R. W. Kreplin (X-ray bursts) were a helpful tool when checking the associations suggested by H. W. Dodson and E. R. Hedeman.

All the flares which have been considered to be obvious or probable sources of particle events are listed in *Part 2*, with a brief description of their characteristic features. This part was prepared by the Editors (for the same periods as above), using a summary of radio events prepared by H. Tanaka, and a list of X-ray bursts prepared by R. W. Kreplin. Unpublished X-ray data have been made available by J. P. Conner, S. R. Kane, S. Singer, and Ch. Wende. Extensive use has been made of data published in the *Quarterly Bulletin on Solar Activity* (Zürich) and the *Solar-Geophysical Data* (Boulder). The final arrangement of Parts 1 and 2 was made by L. Fritzová-Švestková and A. Guitart.

Finally, *Part 3* contains a description of the active regions on the Sun, in which the flares of Part 2 occurred. M. J. Martres has prepared the basic section of this part, having received considerable support from many solar observatories: Athens (Greece), Boulder and Carnarvon (WDC-A Boulder, U.S.A.), Culgoora (CSIRO, Australia), Debrecen (Hungary), Capri F (Fraunhofer Institute, Freiburg, F.R.G.), Honolulu and Haleakala (Hawaii, U.S.A.), Lockheed (U.S.A.), McMath-Hulbert (Pontiac, U.S.A.), Meudon and St. Michel de Haute Provence (France), Mitaka (Japan), Mt. Wilson (U.S.A.), Capri S (Saltsjöbaden, Sweden), Tashkent and Voroshilov (U.S.S.R.). *Appendix B* and the references to Part 3 were prepared by V. N. Obridko, with the assistance of Y. N. Dolginova and E. V. Ivanov. Since this section differed in some aspects from the style of the other parts of the Catalog, it has been rearranged, renumbered, and transliterated by the Meudon librarian, M. Guidoni. Nevertheless, this Appendix remains a somewhat less homogeneous constituent of the Catalog, since the latest changes and improvements of Part 1 could not be reflected fast enough in Obridko's laborious system of references.

Finally I must say that I greatly underestimated the extent of this work when I first proposed it in 1969. Therefore, I deeply admire all the authors and my co-editor, since they never hesitated to continue, even when the work continuously expanded. Let me express my deep gratitude to all of them for their enthusiastic approach to the enormous amount of work we had to do. P. Simon's and my particular thanks are extended to M. A. Shea and D. F. Smart of Bedford, H. W. Dodson and E. R. Hedeman of Pontiac, and M. J. Martres of Meudon, who carried the main burden of this enterprise.

Freiburg, 31 December, 1973   ZDENĚK ŠVESTKA
                                                                Chairman
                                                                IUCSTP Working Group 2

ACKNOWLEDGEMENTS

This Catalog could not have been accomplished without the assistance of many scientists throughout the world. Our particular thanks are extended to:

J. A. SIMPSON for unpublished data from IMP 1, 2, 3, and 4;

J. A. SIMPSON and C. Y. FAN for unpublished data from Pioneers 6 and 7;

S. N. VERNOV for a special list of the U.S.S.R. observations of proton increases in the stratosphere and interplanetary space;

K. A. ANDERSON and R. P. LIN for unpublished data on solar electron events;

J. H. KING and the U.S. National Space Science Data Center for their cooperation in providing considerable satellite data;

L. DESZÖ for white-light solar photographs specially prepared for the Catalog by the staff of the Debrecen Observatory;

J. RAYROLE and I. SORU ESCAUT for the magnetic field data from Meudon;

We also express our gratitude to

G. C. REID for checking the list of the PCA events;

M. VAN HOLLEBEKE, J. R. WANG, and F. B. McDONALD for unpublished data from IMP 4;

S. KAHLER, S. R. KANE, and K. A. ANDERSON for unpublished OGO data;

K. G. McCRACKEN, U. R. RAO, and R. P. BUKATA for unpublished data from Pioneers 6 and 7;

J. R. BURROWS for unpublished Alouette 1 data;

J. P. CONNER for unpublished X-ray data from the VELA 3A and 3B satellites;

M. GROS for unpublished HEOS data;

Ch. WENDE for unpublished X-ray data from Explorer 33;

S. SINGER for unpublished data from VELA 4; and

L. M. ADKINS, G. BANOS, A. BRUZEK, V. EVANS, G. HOSINSKY, J. V. LINCOLN, M. McCABE, S. MARTIN, and F. MORYAMA for supplying optical and magnetic field solar data.

Finally, we acknowledge with thanks private communications we received from L. E. ANDERSON, J. R. ASBRIDGE, C. D. BOSTROM, F. B. McDONALD, K. C. HSIEH, H. S. HUDSON, D. H. JELLY, A. J. MASLEY, H. LEINBACH, and D. J. WILLIAMS.

The aid we received from Y. N. DOLGINOVA and E. V. IVANOV in Moscow and M. GUIDONI, J. GAPIHAN, and G. SERVAJEAN in Meudon when preparing the third part of the Catalog is highly appreciated.

THE EDITORS

INTRODUCTION

Part 1

This part of the Catalog contains only confirmed particle events with > 10 MeV proton flux in excess of 0.1 proton $(cm^2 s ster)^{-1}$ before December 1965, and in excess of 0.01 proton $(cm^2 s ster)^{-1}$ since December 1965. 'Confirmed' means that at least two independent measurements indicate the existence of the event.

In the HEADING of each event we give its *serial number*, the *date* of the earliest onset reported, an approximate *onset time*, and the *importance* of the event, according to the classification system of Smart and Shea (1971), reproduced for reader's convenience in Table I.

TABLE I

Solar proton event classification system (Smart and Shea, 1971)

	First digit	Second digit	Third digit
Digit	>10 MeV proton flux $(cm^2 s ster)^{-1}$	Daylight polar cap absorption at 30 MHz	Sea level neutron monitor increase
-2	$10^{-2}-<10^{-1}$	$-$	$-$
-1	$10^{-1}-<10^{0}$	$-$	$-$
0	$10^{0}-<10^{1}$	No increase	No increase
1	$10^{1}-<10^{2}$	<1.5 dB	$<3\%$
2	$10^{2}-<10^{3}$	$1.5-<4.6$ dB	$3-<10\%$
3	$10^{3}-<10^{4}$	$4.6-<15$ dB	$10-<100\%$
4	$\geqslant 10^{4}$	$\geqslant 15$ dB	$\geqslant 100\%$
X	measurements not available		
( )	the digit is uncertain or implied		

Observed ground-based and/or space data on each particle event are arranged according to the onset times below each heading. A complete line (all data is not always available) comprises the following information:

FIRST COLUMN: Kind of observation or name of the spacecraft. The abbreviations used:

VHF   very high frequency ionospheric scatter data for the polar cap absorptions (Bailey, 1964);

f-min   polar cap absorption measurements by means of ionosondes;

RIOM ~ 30 MHz riometer measurements of polar cap absorptions;

f-fix   polar cap absorption measurements by means of high-sensitivity backscatter sounding of the lower ionosphere at 2.3 MHz (Gregory, 1963);

VLF   polar cap absorption as measured from modifications of the propagation of very low frequency waves;

Pr > 2.3 GeV	neutron monitor record of an increase corresponding to the presence of protons with energies in excess of 2.3 GeV;
Balloon	particle flux measurements carried out during a balloon flight;
Rocket	particle flux measurements carried out during a rocket flight;
ALOU	Alouette
EXPL	Explorer
MAR	Mariner
PION	Pioneer
VEN	Venus

SECOND COLUMN: Kind of particles and the energy range recorded. The abbreviations used:

GLE	ground level effect, implying > 500 MeV protons in the particle flux;
PCA	polar cap absorption, implying protons of the order of 10 MeV in the particle flux;
Pr > 55 MeV	specifies the energy range of protons. Most of the time, this is an abbreviation for an energy 'window' listed in Table II (for a given spacecraft and reference). The absence of any information in Table II means that the upper limit for the recorded energy is not specified by the experimenters.
El > 40 keV	electrons with energies in excess of 40 keV;
Pr 14 MeV	differential flux measurement at the mean energy of the proton channel. The lower edge of the channel is always given.
Pr (10–15 gr)	protons penetrating to the balloon floating at altitudes corresponding to 10–15 grams of atmospheric depth.

THIRD COLUMN: Onset time of the particle event.

FOURTH COLUMN: Time of the maximum particle flux.

If not indicated otherwise, the time in these two columns refers to the day given in the heading. Note that 25^h10^m = next day 01^h10^m.

Whenever the time refers to a day different from the date in the heading, the day is added, e.g. $29^d14^h25^m$.

In these two columns we also indicate the accuracy of the time given, according to the following scheme:

$22^h15^m(3) - 45^m$	the stations given in references give the time within these limits (the earliest time being given in ref. (3));
$22^h40^m \pm 02^m$	an exact time between 22^h38^m and 42^m;
22^h00^m	the accuracy is not exactly known, but it is expected to be significantly smaller than $\pm 30^m$;
22^h	the time accuracy is not known.

In all other cases we give a *code*:

Δ1	time accurate to within one hour;
Δ2	time uncertainty greater than one hour;
Δ3	start superimposed on enhancement of previous event;
Δ4	estimate because of magnetospheric perigee interference;
Δ5	missing data; or estimate because of missing data;
Δ6	estimate because of broad maximum;

Δ7   saturation of counter during event maximum; the exact time of the maximum flux and its actual value are not known; the data reported give the onset time and the level of saturation;

Δ8   a questionable time (for low-energy protons), due to electron contamination (cf. Appendix 1).

The numbers in brackets give references.

FIFTH COLUMN: The duration in hours. This is only an approximate value, since (a) many authors give the duration in days (i.e. $\pm 12^h$), (b) many events overlap, and (c) generally the duration depends upon the noise level of the detector (cf. Appendix 3).

SIXTH COLUMN: Magnitude of the event. The abbreviations used:

2.0 dB	2.0 decibels at 30 MHz riometer;
3.3 ± 1.0%	3.3% increase in the neutron monitor flux;
0.3 Pr	0.3 protons $(cm^2 \, s \, ster)^{-1}$ (measurement of integrated flux);
0.3* Pr	0.3 protons $(cm^2 \, s \, ster \, MeV)^{-1}$ (measurement of differential flux);
20 El	20 electrons $(cm^2 \, s \, ster)^{-1}$;
190 c	190 counts s^{-1}.

SEVENTH COLUMN: References (in brackets).

The *ground level effects* (GLE) are presented in a composite abbreviated form:

earliest onset				station which reported the earliest onset
	maximum		highest flux	sea level station which reported the highest flux
		longest duration		station which reported the longest duration

In order to save space, PCA observations often have been combined together in one line. In such cases we give the earliest and latest time in Columns 3–5. At space observations we sometimes write two sets of data in one line; they are separated by '/', e.g. Pr > 40/>23 MeV, 0.8/4.3 Pr, etc. When only one value (one time) is given, it means that this value (this time) is the same for both measurements. $09^h/12^h \Delta 2$ means $09^h \Delta 2/12^h \Delta 2$; $09^h \Delta 2/\Delta 1$ means $09^h \Delta 2/09^h \Delta 1$.

For *spaceprobes* the position also is given, in brackets below the data:

PION 6 at 0.98 AU, ES $-5°$ means: Pioneer 6 was at 0.98 astronomical unit from the Sun, 5 deg east from the Sun-Earth line (ES $+5°$ means 5 deg west). This entry is not repeated for subsequent events as long as it does not change.

IMPORTANT NOTE ON THE SELECTION OF DATA

The Catalog does not report all the observations available, since this would at least double its contents. Those who are interested in the full set of data (in computer form) should contact M. A. Shea and D. F. Smart. We have omitted:

(a) Measurements which strongly deviated from another reasonably homogeneous set of measurements (e.g. late PCA onsets);

TABLE II
List of the energy windows at different spacecrafts

Spacecraft	Abbreviation	Energy window	References
AIMP 2	Pr > 0.32 MeV	0.32–6.3 MeV	(193)
AIMP 2	Pr > 0.3 MeV	0.3–6.3 MeV	(210)
ALOU 1	Pr > 1.3 MeV	1.3–7 MeV	(156, 157)
ATS 1	Pr > 35 MeV	35–53 MeV	(167)
ATS 1	Pr > 17 MeV	16.6–21 MeV	(167)
EXPL 12	Pr > 330 MeV	330–600 MeV	(45)
EXPL 12	Pr > 130 MeV	130–330 MeV	(45)
EXPL 12	Pr > 55 MeV	55–118 MeV	(63)
EXPL 12	Pr > 9 MeV	9–14 MeV	(45)
EXPL 12	Pr > 7 MeV	7–14.5 MeV	(63)
EXPL 12	Pr > 0.5 MeV	0.5–4.2 MeV	(87)
EXPL 33	Pr > 0.5 MeV	0.5–4.2 MeV	(87)
HEOS	Pr > 63 MeV	63–200 MeV	(208)
HEOS	Pr > 41 MeV	41–63 MeV	(206)
HEOS	Pr > 24 MeV	24–64 MeV	(202, 208)
HEOS	Pr > 6 MeV	6.7–25 MeV	(201, 202)
HEOS	Pr > 5 MeV	5–7 MeV	(202)
HEOS	El > 0.5 MeV	0.5–0.6 MeV	(202)
HEOS 1	Pr > 200 MeV	200–1500 MeV	(208)
HEOS 1	Pr > 1 MeV	1–13 MeV	(218)
IMP 1	Pr > 6.5 MeV	6.5–190 MeV	(68)
IMP 1	Pr > 0.9 MeV	0.9–190 MeV	(68)
IMP 2	Pr > 90 MeV	90–190 MeV	(76)
IMP 2	Pr > 19 MeV	19–90 MeV	(76)
IMP 2	Pr > 6.5 MeV	6.5–19 MeV	(76)
IMP 2	Pr > 0.9 MeV	0.9–190 MeV	(76)
IMP 3	Pr > 90 MeV	90–190 MeV	(80)
IMP 3	Pr > 19 MeV	19–90 MeV	(80)
IMP 3	Pr > 6.5 MeV	6.5–19 MeV	(80)
IMP 3	Pr > 0.9 MeV	0.9–190 MeV	(80, 84)
IMP 3	El > 3 MeV	3–10 MeV	(67)
IMP 4	Pr > 29 MeV	29.5–94.2 MeV	(107)
IMP 4	Pr 19 MeV	19–80 MeV	(182)
IMP 4	Pr 15 MeV	15–18.7 MeV	(185)
IMP 4	Pr > 9 MeV	9.6–18.8 MeV	(107)
IMP 4	Pr 7 MeV	7–7.4 MeV	(225)
IMP 4	Pr > 7 MeV	7–55 MeV	(174)
IMP 4	Pr > 6 MeV	6–55 MeV	(176)
IMP 4	Pr 5 MeV	5–9.4 MeV	(196)
IMP 4	Pr 4 MeV	4.4–5 MeV	(225)
IMP 4	Pr 4 MeV	4–19 MeV	(182)
IMP 4	Pr > 3.5 MeV	3.5–7.6 MeV	(176)
IMP 4	Pr > 1.8 MeV	1.8–3.3 MeV	(166, 179)
IMP 4	Pr > 1 MeV	1–10 MeV	(181)
IMP 4	Pr > 0.8 MeV	0.8–9.6 MeV	(107, 173)
IMP 4	Pr > 0.7 MeV	0.7–7.6 MeV	(174)
IMP 4	Pr > 0.7 MeV	0.7–55 MeV	(176)
IMP 4	Pr 0.5 MeV	0.56–0.6 MeV	(175)
IMP 4	El > 3 MeV	3–12 MeV	(190)
IMP 4	El > 3 MeV	3–10 MeV	(118, 179)
IMP 4	El > 0.5 MeV	0.5–1 MeV	(178)
IMP 4	El > 0.3 MeV	0.3–0.9 MeV	(118, 166, 190, 191)
IMP 4	El > 0.17 MeV	0.17–1 MeV	(178)
IMP 5	Pr 17 MeV	17–19.7 MeV	(215)
IMP 5	Pr 2.5 MeV	2.5–4 MeV	(215)
IMP 5	Pr > 1 MeV	1–10 MeV	(212)

Table II (continued)

Spacecraft	Abbreviation	Energy window	References
INJUN 1	Pr > 1.5 MeV	1.5–15 MeV	(61, 143)
INJUN 1	Pr > 1 MeV	1–15 MeV	(59, 153)
INJUN 1	Pr > 1.4 MeV	1.4–17 MeV	(64)
MAR 2	Pr > 0.5 MeV	0.5–10 MeV	(147, 149)
MAR 4	Pr > 1 MeV	1–170 MeV	(77, 79)
MAR 4	Pr > 0.5 MeV	0.5–11 MeV	(79, 87)
OGO 4	Pr > 1.2 MeV	1.2–40 MeV	(180)
OGO 5	Pr > 90 MeV	90–110 MeV	(186)
OGO 5	El > 12 MeV	12–45 MeV	(186)
OGO 5	El > 0.3 MeV	0.3–0.9 MeV	(166)
OGO 6	Pr > 5 MeV	5–11 MeV	(184)
OGO 6	Pr 5 MeV	5–6 MeV	(211)
OGO 6	Pr 2 MeV	2–2.1 MeV	(211)
OGO 6	Pr 1.1 MeV	1.17–1.2 MeV	(211)
PION 6	Pr > 73 MeV	73.2–175 MeV	(70, 104)
PION 6	Pr > 44 MeV	44–77 MeV	(84, 125)
PION 6	Pr > 13 MeV	13.9–73.2 MeV	(104, 154)
PION 6	Pr > 13 MeV	13–70 MeV	(70)
PION 6	Pr > 7.5 MeV	7.5–45 MeV	(207)
PION 6	Pr > 7.5 MeV	7.5–44 MeV	(84, 125)
PION 6	Pr > 0.6 MeV	0.6–13.9 MeV	(104, 154)
PION 6	Pr > 0.6 MeV	0.6–13 MeV	(70)
PION 7	Pr > 73 MeV	73–165 MeV	(119)
PION 7	Pr > 64 MeV	64.5–81.2 MeV	(84, 125, 165)
PION 7	Pr > 47 MeV	47.4–64.5 MeV	(84, 125, 165)
PION 7	Pr 15 MeV	15–18.7 MeV	(185)
PION 7	Pr > 12 MeV	12.7–73 MeV	(84, 119)
PION 7	Pr > 7.2 MeV	7.2–47.4 MeV	(84)
PION 7	Pr > 0.6 MeV	0.6–12.7 MeV	(107, 119)
PION 8	Pr > 42 MeV	42–59 MeV	(207)
PION 8	Pr > 11 MeV	11.1–25.7 MeV	(183)
PION 8	Pr > 7 MeV	7.5–45 MeV	(199)
PION 9	Pr > 32 MeV	32–41 MeV	(207)
PION 9	Pr > 10 MeV	10.5–24.6 MeV	(217)
PION 9	Pr > 7 MeV	7.5–45 MeV	(199)
VELA 4	Pr 10 MeV	10–13.6 MeV	(168)
VELA 4	Pr 5 MeV	5–20 MeV	(168)
VELA 4	Pr 4.5 MeV	4.5–6.3 MeV	(168)
VELA 4	Pr 1.4 MeV	1.4–2.05 MeV	(168)
VELA 4	Pr 0.6 MeV	0.68–0.95 MeV	(168)
VELA 4	Pr 0.55 MeV	0.55–0.68 MeV	(168)
VEN 2	Pr > 1 MeV	1–5 MeV	(103)
VEN 4	Pr > 1 MeV	1–5 MeV	(43, 177)
VEN 6	Pr > 1 MeV	1–5 MeV	(43, 209)
ZOND	Pr > 1 MeV	1–5 MeV	(43, 103)

(b) Incomplete measurements (with codes $\Delta 3$ or higher, or without time or flux data), when a similar observation could be obtained elsewhere;

(c) Observations which duplicated other measurements without adding any new information to them. When doing this, we have always selected the most complete set of data, and we always included data for the highest and lowest energies observed and the earliest onset time.

(d) Measurements carried out aboard spaceprobes too far from the Earth-Sun line.

At the BOTTOM of all the data we suggest the *source* of the particle event. The criteria used for these associations are briefly described in Appendix 2. For the benefit of an easy quick look we use the following markings:

○ a flare-associated acceleration process, the flare being known;
□ a flare-associated process, but the flare was on the invisible hemisphere (Appendix 2);
△ a flux increase associated with a sudden commencement;
◇ a modulation effect; this includes gradual increases of the particle flux during geomagnetic disturbance, sc magnetic storms in progress (i.e. with an onset preceding the
   particle flux), recurrent particle events, as well as possible flux modulations in space.

The degree of certainty of these associations is expressed in the following way (an example for flares):

● the association is certain;
⊙ the association is probable;
○ the association is possible, but for some reason open to doubt;
⊘ this flare could not (or probably did not) give rise to the particle event, but it might
   have contributed to it.

For some events, mostly listed in Appendix A, but in the last years also in Part 1, it has been extremely difficult to find the source of particles on the Sun. Therefore, we had to use two additional categories of associations:

Origin unknown means that one cannot find any source which might be reasonably associated with the event; whilst

origin uncertain means that there may be possible sources on the Sun, but we want to emphasize our doubts about their association with the event discussed. In some cases this uncertainty is due to unsatisfactory observations of the particle event itself.

The *flares* are characterized by four data (details on them can be found in Part 2):
Onset time (2050 = $20^h 50^m$ on the day in the heading);
position (N16 W27 = $16°$ north, $27°$ west from the central solar meridian);
Hα importance (1+, 2, 2+, etc. before 1966; 1N, 1B, 2F, 2N, 2B, etc. since 1966);
active region (McM = McMath plage number).

Other abbreviations used:

GMS	geomagnetic storm;
sc	sudden commencement starting a geomagnetic storm;
DS	dynamic radio spectrograph;
SWF	short-wave fade-out;
μ-burst	microwave radio burst;
Type II, III, or IV	radio bursts of type II, III, or IV, respectively, in decimetric, metric, or dekametric range.

When there are several associations suggested, their order is according to the probability we ascribe to them. There are events, of course, when the order is arbitrary, since it is essentially impossible to decide which association should be preferred (all being 'possible', but doubtful). In some cases two or more different associations may be correct, e.g. several flares contribute, a sudden commencement modifies a flare-associated flux, a distinct flare event sets in a 'permanent particle flux from the active region', or, quite randomly, a flare particle event occurs during a magnetically disturbed period when a 'modulated particle increase' is in progress.

There are many particle events which show two components in their development: a flare-associated (prompt) and an sc-associated (delayed) increase. They are listed as one event only when we feel sure that the second increase did not have another origin. However, if there is any ambiguity in the source of the probable sc-increase, this increase is numbered as a new event.

Throughout the Catalog references are made to the particle events listed in Appendix A.

Appendix A

In order to list any reported particle event without making the Catalog extensively long, we have decided to take out from Part 1 three kinds of events and present them in an abbreviated form in Appendix A:

(1) Unconfirmed events (UN), which are based on a single measurement, or (in later years) on a few measurements of contradictory character. Many of them are real, as one can verify from their recurrence. However some (and we do not know which of them) may represent fictitious increases.

(2) Low-energy events (L), the number of which generally increases from one year to another, as the sensitivity of receptors is improving. Therefore, the definition of a 'low-energy event' is time-dependent. We had to consider as such all increases with less than 0.1 proton (cm^2 s ster)$^{-1}$ above 10 MeV before December 1965, but only those with less than 0.01 proton (cm^2 s ster)$^{-1}$ above 10 MeV since December 1965 (launch of Pioneer 6).

(3) Pure electron events (PE), for which no protons have been recorded.

Each entry gives the date, the event category (UN, L, or PE), reference(s) in brackets, and a very brief description of the particle measurements in the following sequence:

spacecraft or method of detection,
kind of particles and energy range,
onset time,
time of maximum (if known),
duration (if known),
flux or counts (if known),
remarks.

For flare associations we give

the Hα flare importance (imp.),
flare onset time ($1720 = 17^h 20^m$ on the day of the event; otherwise the flare day is added: $19^d 1720$ if the particle event is on the 20th),
flare heliographic longitude (e.g. W53 = 53° west), and
the active region (McM = the McMath plage number).

All abbreviations are the same as in the Part 1 of the Catalog.

We have included in Part 1 and Appendix A all particle increases, with only two exceptions:

(1) An unconfirmed increase has been omitted, when this has been recommended by the observer himself.

(2) Increases in the >4 MeV electron flux (Simnett et al., 1970) have been omitted, since they anticorrelate with solar activity, obviously being of non-solar origin.

Some unconfirmed events (UN) appear to be clearly associated with a 'typical proton flare', and the association is hence marked as 'certain' (●). In this connection it is important to emphasize that the signs ●, ⊙, and ○ only estimate the degree of certainty of the association, not that of the production of particles. Thus the sign ● does not mean "This flare surely produced particles", but only "In the case that this particle event was real, it was certainly caused by this flare".

There are some events for which the low-energy proton flux arrived too early when compared with electrons, higher energy protons, or the source occurrence on the Sun. These cases are obviously due to electron contamination, as is explained in more detail in Appendix 1. There are also contradictory observations when low-energy protons were recorded by one spacecraft but not by another in a similarly favourable position. Before using this as a reason for omission of the event in question, one must first consider the background levels of the detectors, which vary within wide limits. Differences in these levels may also cause significant discrepancies between the apparent onset times. These problems are explained in more detail in Appendix 3.

Part 2

This part contains data on all flare sources of the particle events in Part 1 and Appendix A, which have been marked as certain (●), probable (⊙), or contributing (⊘) sources.

In the HEADING of each flare we give the *date* of the flare, the *degree of certainty* in the association (●, ⊙, or ⊘), the *active region* (McMath plage number), *number* of the particle event associated in Part 1 with the *event classification* in brackets (cf. Table I), or a reference to Appendix A, and the *comprehensive flare index* (CFI) after Dodson and Hedeman (1971).

The comprehensive flare index

$$CFI = A + B + C + D + E,$$

where,

A (= 1 to 3) = short wave fade (SWF) or other sudden ionospheric disturbance (SID) importance,
B (= 1 to 3) = importance of Hα flare,
C = characteristic of log of ∼10 cm flux in units of 10^{-22} W $(m^2 Hz)^{-1}$,
D = dynamic spectrum events: type II = 1, continuum = 2, type IV = 3,
E = characteristic of log of ∼200 MHz flux in units of 10^{-22} W $(m^2 Hz)^{-1}$.

The comprehensive flare index ranges from 0 for subflares without significant ionizing or radio frequency emission to 15, 16, and 17 for flares that were outstanding in all aspects of electromagnetic flare radiation. When some kind of observations (e.g. the dynamic spectrum) is not available, only a lower limit for the comprehensive flare index can be given.

In the FIRST LINE below the heading we give *observations of the Hα flare*: time of its beginning, maximum (sometimes not available) and end, position on the disk, importance, reference(s) in brackets, and the approximate time delay between the flash phase of the flare and the onset of the particle event, Δt. No reference is given (throughout all lines in Part 2) if all the data in the line were obtained from the *Quarterly Bulletin on Solar Activity* (ref. (4)).

In all other lines the time data (onset − maximum − end), the importance, and the references are arranged in exactly the same way.

The SECOND LINE (sometimes two or three lines) gives information on the X-ray flux as reflected in *sudden ionospheric disturbances* (SID). When available, we always give data on the short wave fade (SWF), since this is the SID type most commonly used in the past. When no SWF was observed, we give another type of SID: sudden enhancement of atmospherics (SEA), sudden cosmic noise absorption (SCNA), sudden phase anomaly (SPA), or sudden frequency deviation (SFD). Both SWF and another type(s) of SID are presented only when the other SID gives significant additional information. The most powerful SID effects have an importance of 3+.

Since 1966 the NEXT TWO LINES report direct measurements of *soft and hard X-rays* aboard satellites. Soft X-rays are taken from 0.5−15 Å VELA 3A and 3B detectors, 1−8 Å SOLRAD 8 (Kreplin, 1966), 9 (Kreplin and Moran, 1969), and OGO 4 (Kreplin *et al.*, 1969) detectors, 2−12 Å EXPLORER 33 (Drake *et al.*, 1969) and 35 detectors and 0.5−3 Å OSO 5 detectors. The time data are accurate within ± 1 or 2 min. Hard X-rays come from 10−50 keV data aboard OGO 1 and 3, 10−30 keV (NRL experiment), 10−130 keV (Kane and Anderson, 1970) OGO 5 detectors, 20−150 keV OGO 6 and 7.7−210 keV OSO 3 (Hudson *et al.*, 1969) detectors. Amongst several soft X-ray data, we report the most precisely documented. Out of several hard X-ray data, we report the highest energy event and if 'no event' is reported, we refer to the satellite working at that time.

Most of the time, the X-ray burst importance has been established subjectively by R. W. Kreplin according to the following scale:

VSM very small   LG   large,
SM   small,   VLG very large.
SIG   significant (moderate),

In 1968 and 1969 the OGO 5 hard X-ray data are scaled by S. R. Kane according to their flux above 20 keV: 1 for a flux between 10^{-8} and 10^{-7} erg cm^{-2} s^{-1}, 2 for 10^{-7} to 10^{-6}, and 3 for 10^{-6} to 10^{-5}. From March 1967 to July 1968 several OSO 3 hard X-ray bursts are reported as 'extraordinary burst' (EB) by H. Hudson.

The data available before 1966 are not reported in the Catalog, since they were very scarce.

The NEXT TWO LINES give information on the *radio burst* as measured at single frequencies. Due to the developments in observational techniques we have found it useful to change the presentation since the beginning of 1966.

Before 1966 the upper line gives information about the burst at the highest frequency recorded, while the lower line always shows the burst at, or near, 200 MHz. In some cases, when these two entries did not characterize the burst properly, a third frequency also has been added. The number given in the importance column is log of the flux in units of 10^{-22} W (m^2 Hz)$^{-1}$.

Since 1966 the upper line always gives information about the burst at, or near, 3.0 GHz. Instead of presenting the highest frequency recorded, we give a spectral type of the burst in the fifth column:

P5   means that the spectrum shows a peak at 5 GHz; P5(2.3) means that log of the maximum flux at 5 GHz was 2.3.

U1/9   means that the flux is minimum at 1 GHz and rises up to 9 GHz; no measure-
         ments are available at higher frequencies;
/9   has the same meaning, but the frequency of the minimum flux is not known;
0.6\   means that the flux falls toward high frequency, without any maximum in the
         microwave range;
U2P7   means that the flux is minimum at 2 GHz and peaks at 7 GHz;
3—9   means a flat spectrum between 3 and 9 GHz.

All the other entries are the same as before 1966.

The NEXT LINES describe the *dynamic spectrum records*, if available. However, in order to save space, 'No DS data' is often reported as a remark in the single-frequency lines. 'Type IV, no DS data' means that no DS observations were carried out, but a spectral diagram composed of single-frequency records shows the occurrence of a type IV burst.

Finally, in the LAST LINES remarks are sometimes added, referring to *white-light emission, spray ejection,* or *loop system prominences* (abbreviated as 'loops').

Part 3

This part contains maps and a description of selected active regions in which the 142 'certain' (●) flares of Part 1 occurred. Each of the 93 sets of data illustrates the disk passage of a proton center and Hα pictures of 58 flares are reproduced. Each set of data is identified in the HEADING by the serial number and the importance of the particle event (as given in Part 1), the date, the onset time, the heliographic coordinates and the importance of the solar flare (according to Part 2), and (in the next line) the Carrington coordinates and the day of the central meridian passage of the relevant active region.

The UPPERMOST GRAPH reports the chronology of the flare activity in the active region, the time scale being that of the synoptic map below: going from the right (east limb) to the left (west limb). The lengths of the bars are proportional to the importance of the flares as given in the *Quarterly Bulletin on Solar Activity*; before 1966 we use the highest reported importance as the final evaluation of the flare. The 'certain' proton events (●) are marked as black circles. Dotted baseline means that the active region was behind the limb at that particular time.

The corresponding strip of the SYNOPTIC MAP below it (180° in longitude) always has 60° in latitude: the proton region is at the center of this strip and a broad horizontal line represents the solar equator. The plages, spots and filaments are sketched according to the conventions used in the *Cartes Synoptiques de la Chromosphere* (Meudon): Plages are hatched; the dashing is more dense when the plage is brighter. Sunspots are marked as circles of different size, the size being proportional to the sunspot area. Filament positions are represented by lengthy stripes, filled in when the filament was visible for most of the time.

Below we give several COMPOSITE DRAWINGS of the main spots, faculae and filaments for selected days around the particle event(s); when available, also magnetic field data is reported. In these graphs, full lines are used for marking spots as they are visible in the K_1 spectroheliograms at Meudon, dashed lines surround plages (as in K_1 or K_3), and hatched lines represent filaments or vortices (in Hα light). In each graph we give the day and two time data: the first time is related to the Hα material and the second one to the K-line data. Depending on the material available, magnetic field polarities (N, S) are reported either for spots only (according to the Mt. Wilson data or to *Solnechnye Dannye*), or for both spots and plage (full line for the north polarity and broken line for the south polarity, according to the Mt. Wilson, Sacramento Peak, or Meudon data). The field of black crosses in some of these graphs shows the site of the proton flare for those events which cannot be illustrated by flare photographs.

In many cases, negative Hα PICTURES OF THE FLARE(S) and positive photospheric images of the sunspot group in white light could be added to the set of data. We have decided to use negative prints in the flare reports, since they are the best presentation for seeing details; the selected flare pictures mostly show the beginning and the maximum of the event.

In the composite drawings and the flare and photospheric images a uniform scale and orientation has been adopted. It is shown on the first set of data (particle event No. 2). All the composite drawings are from Meudon and all photospheric photographs from Debrecen. The observatories which supplied other data are identified by abbreviations.

Appendix B

This Appendix contains a list of selected active regions which produced a proton event associated with a flare classified as ● or ⊙ until the end of 1965, and ● only since 1966 (in both Part 1 and Appendix A; thus Appendix B describes more regions than Part 3, in which only active regions producing a ● flare in Part 1 have been considered). Most of the data on the active regions in this Appendix have been taken from *Solar-Geophysical Data* (Boulder). However, some information had to be supplemented from *Solnechnye Dannye* (Pulkovo), *Heliographic Maps of the Photosphere* (Zürich), and *Quarterly Bulletin on Solar Activity* (Zürich). For the period before 1965 the spot area and number data were often taken from R. S. Gnevysheva's *Catalog of Solar Activity* (Proceedings of GAO Pulkovo). These data are marked with an asterisk. Mt. Wilson magnetic data were also taken from the lists published in *Publ. Astron. Soc. Pacific*.

In the HEADING we give the McMath plage number (McM), from 1964 also the Meudon active region number (M), the heliographic latitude, and the date of the central meridian passage (CMP) in tenths of a day. (14.5 Nov. means 12^h UT on 14th November.) In the SECOND LINE reference is given to the most important particle events which are presumed to have had their origin in the active region being described.

The arrangement of data BELOW THE HEADING is as follows:

Age of the active region (Age 3 means three solar rotations), description of its *development*, and *characteristics* of the active region *at the time of its CMP* (this data is in brackets if no measurements were available exactly on the day of the CMP). Here the following abbreviations are used:

Ca 6100/3.5   area of the calcium plage (in Ca II K line) was 6100 millionths of the solar hemisphere, and its intensity was 3.5 (on a scale 1 to 5);

spots 640/19   there were 19 spots in the active region, with an area equal to 640 millionths of the solar hemisphere;

Zürich classification of the sunspot group: A-B-C-D-E-F (maximum development)-G-H-J;

Mt. Wilson classification of the sunspot group:

α   unipolar spot,
β   bipolar group (βp, βf — the preceding or following spot, respectively, is more
       developed),
βγ   bipolar group with magnetic irregularities,
γ   magnetically complex group,
δ   magnetically complex group with a common penumbra to both polarities.

On 1968 and 1969, the 3.2 cm flux and the 3.2 to 7.5 cm flux density ratio are added after H. Tanaka.

Finally, the last lines give references to literature in which the active region, the particle-producing flares in it, or the particle events associated with it, have been described. Here the number without brackets is the number of the reference, while the number in brackets is a code indicating which type of data can be found in it. The meaning of the code is as follows:

(1) Structure and evolution of the active region;
(2) magnetic field measurements;
(3) flare data;
(4) radio emission;
(5) X-ray emission;
(6) particle emission and geophysical effects.

27-Day Index

This index (16 tables at the end of the book) plots all particle events of Part 1 and Appendix A in dependence on the 27-day Bartels period. Each entry contains the day of the event (the month is given in the heading) and event's importance (according to Table I or using the abbreviations of Appendix A — UN, L, PE; '2 PE' means two pure-electron events on that particular day). PCA events in excess of 1.5 dB are in bold type.

Appendix 1 to the Introduction

Electron Contamination

One serious problem encountered during the compilation of the proton events was the determination of the correct onset of an event. From our current knowledge we realize this problem has existed from essentially the first space borne instruments. The gas discharge counters carried aboard the early U.S.A. Explorer satellites and Soviet Sputniks would respond to any ionizing radiation that penetrated into their sensitive volume, but they were not capable of differentiating between electrons or protons. For example, the Geiger counters on Explorer 7 would respond to protons with energy >30 MeV or electrons >2.5 MeV or Bremsstrahlung with a low efficiency, but could not distinguish which type of radiation actually caused the devise to count (Line and Van Allen, 1964). The first convincing measurements of solar electrons were obtained by comparing the response of a system of detectors involving three thin window Geiger counters, an ionization chamber, and a solid state detector (Van Allen and Krimigis, 1965). The studies of solar electrons that have been accomplished during the 20th solar cycle demonstrate that low energy electrons (>40 keV) often arrive at the Earth before the energetic protons. As the sensor technology developed so that very small events could be recognized in the data, the early electron arrival became a significant problem in determining proton event onsets. Advanced proton sensors were designed so that they would be insensitive to 40 keV electrons; however, it has developed that many of the very sensitive proton detectors were responding to fast electrons. For example it was deduced that proton detectors on IMP 1 were responding to electrons as the satellite passed through the magnetosheath and bow shock (Fan *et al.*, 1966). Subsequent work both of an experimental and theoretical nature (Lupton and Stone, 1972) has helped resolve the electron sensitivity of these instruments.

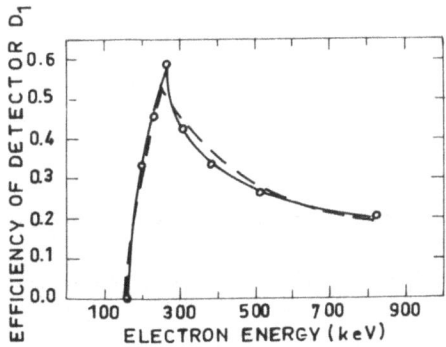

Fig. 1.   The detection efficiency to electrons for IMP 1.

Figure 1 shows the detection sensitivity to electrons for IMP 1. The low energy channels ($0.9 \leqslant Pr < 190$ MeV) of IMP 1, IMP 2, and IMP 3 (Simpson, private communication) are similar in their electron response. Figure 2 illustrates the measured sensitivity of an IMP 4 instrument to electrons. Inspection of Figure 1 indicates that the low energy sensor element will respond efficiently to electrons with energies greater than 200 keV. The response of the low energy proton channels to electrons can be roughly estimated by extrapolating the >40 keV electron flux to 200 keV using the typical electron spectra given by Lin (1970). As an example consider a >40 keV electron flux of 200 $(cm^2 s ster)^{-1}$ extrapolated to 200 keV using a spectral slope of -2.5. This results in an estimated electron flux of ~1.6 at 200 keV. This extrapolated electron flux, combined with the relative electron detection efficiency indicated in Figure 1 would give an increase of the order of 0.5 flux unit in the channel response. This is well within the sensitivity limits of the low energy proton detectors on the IMP spacecraft.

The electron contamination problem is even more serious in the more sensitive low-energy proton channels on Pioneer 6 and 7 (0.6 to 13 MeV) and on IMP 4 (0.8 to 9.6 MeV). The very low background on these sensors makes them extremely susceptible to electron contamination. The response of the Pioneer 7 (0.6 to 12.7 MeV) and the IMP 4 (0.8 to 9.6 MeV) low-energy proton channel to an electron event on 26 August 1967 is illustrated in Figure 3. Normally measurements by two independent spacecraft would be quite convincing; however, in this case for this period there was an electron insensitive low energy proton channel operative on IMP 4 at this time. The comparison of the 0.8 to 9.6 MeV sensor (electron sensitive) with the 1.0 to 10 MeV sensor (electron insensitive),

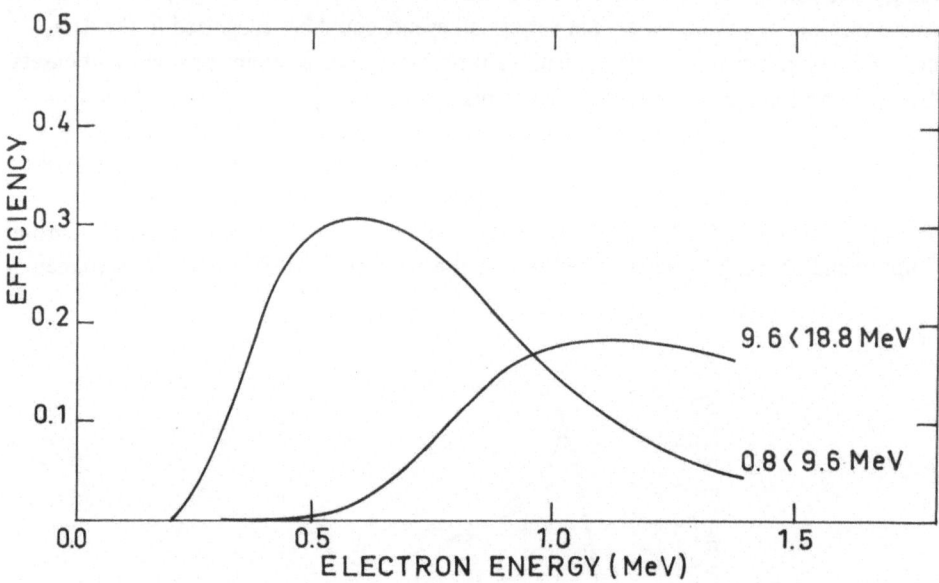

Fig. 2.   The measured sensitivity of an IMP 4 instrument to electrons.

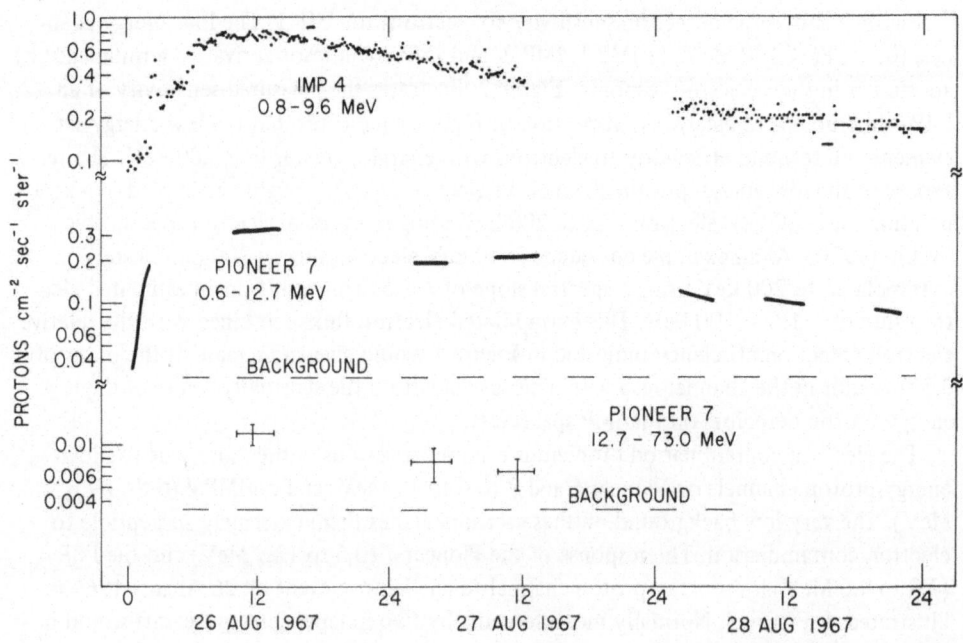

Fig. 3.   The response of the Pioneer 7 and IMP 4 low-energy proton channels to an electron event on
26 August 1967.

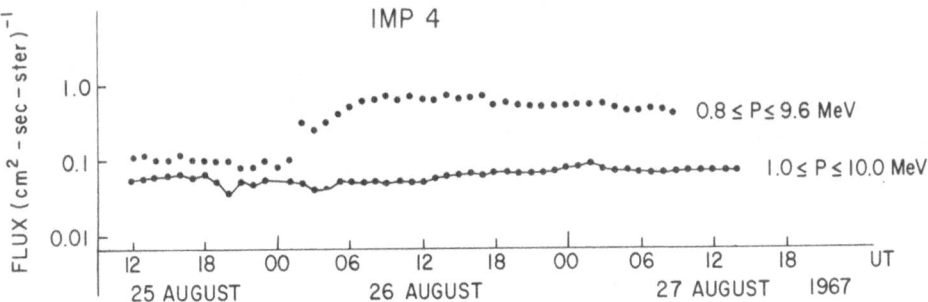

Fig. 4.   Comparison of the electron sensitive (0.8–9.6 MeV) and electron insensitive (1.0–10.0 MeV) sensor on IMP 4 for the event illustrated in Figure 3.

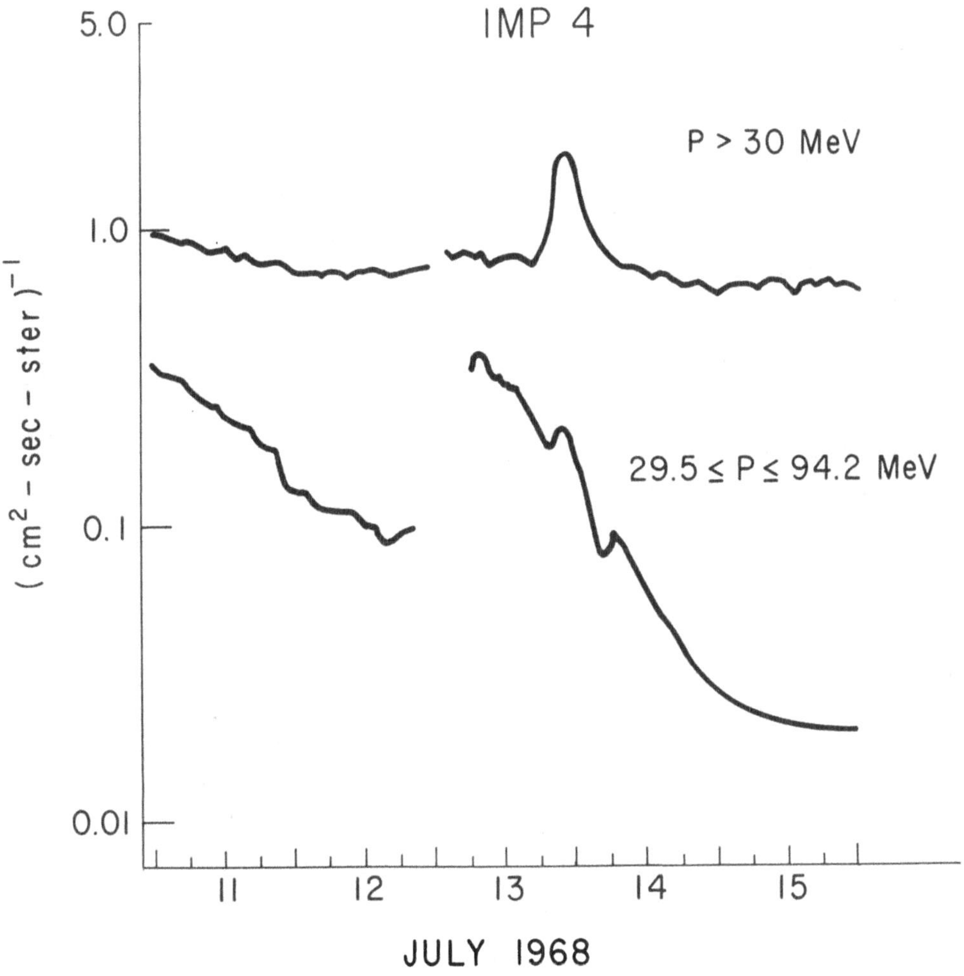

IMP 4

$P > 30$ MeV

$29.5 \leq P \leq 94.2$ MeV

JULY 1968

Fig. 5.   Comparison of another electron sensitive (> 30 MeV) sensor on IMP 4 with 29.5–94.2 MeV sensor which should be electron insensitive. The peak on 13 July is due to relativistic electrons.

illustrated in Figure 4, casts severe doubt as to the reality of any proton increase at this time.

Figure 5 illustrates the response of two similar energy proton detectors on IMP 4 between 11–15 July 1968. This figure shows an apparent event on 13 July on the >30 MeV proton detector which is not obvious on the 29.5 to 94.2 MeV proton detector. This event has been identified (Simnett *et al.*, 1970) as containing a significant flux of relativistic (0.3–0.9 MeV) electrons. The 29.5 to 94.2 MeV channel on IMP 4 (Simpson, private communication) requires a triple coincidence of three detector elements (the last element

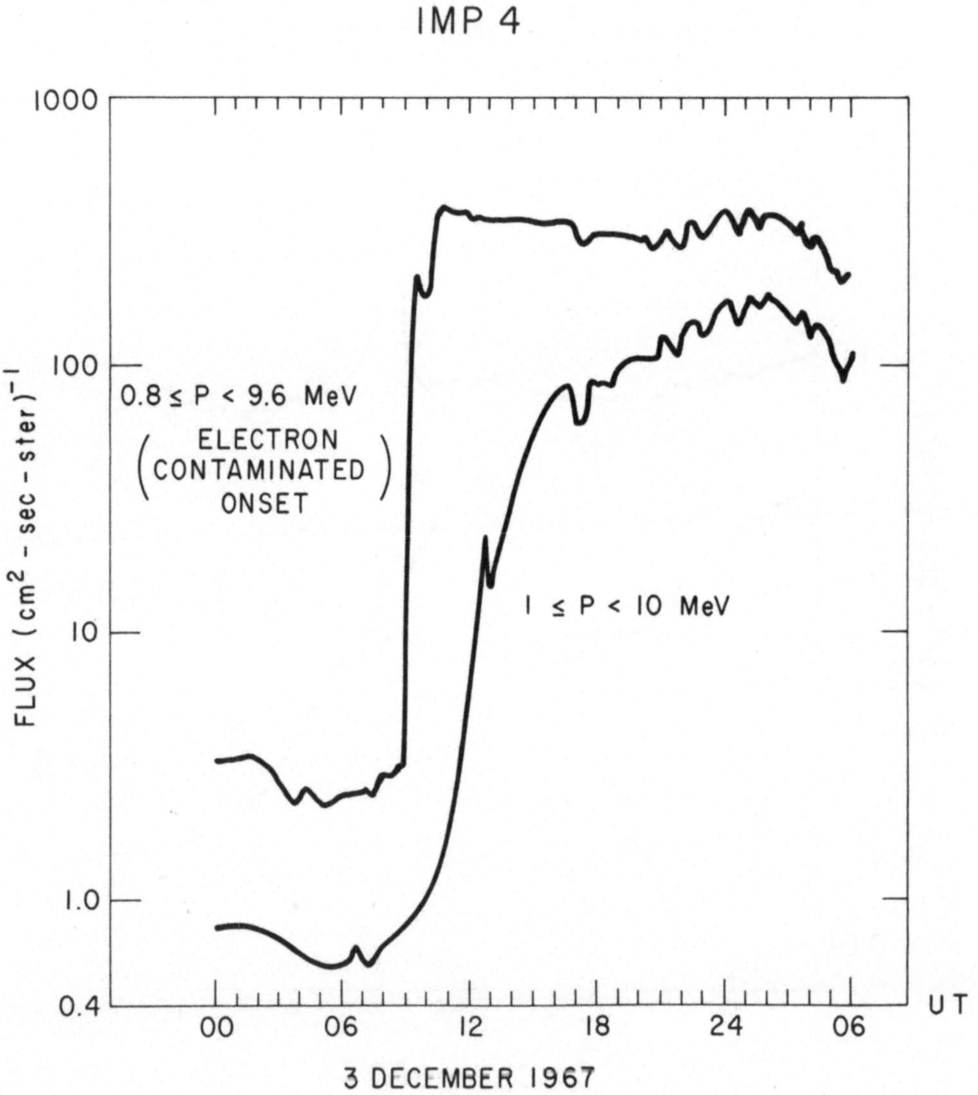

IMP 4

Fig. 6. Comparison of the electron sensitive (0.8–9.6 MeV) and electron insensitive (1.0–10.0 MeV) sensor on IMP 4 for the event on 3 December, 1967.

having a threshold of 15 MeV energy deposition) and so is relatively insensitive to electron contamination. The > 30 MeV solar proton monitoring experiment channel will record any particle that penetrates the 1.6 mm copper shield and deposits more than 150 keV in a thick detector and so is quite sensitive to this type of contamination.

The problem of electron contamination can become quite serious when trying to select accurate onset times for various proton energy channels. Figure 6 illustrates the response of the 0.8 to 9.6 MeV and the 1 to 10 MeV proton detectors on IMP 4 for the event on 3 December 1967. In this case the 0.8 to 9.6 MeV proton channel is contaminated by electrons whereas the 1 to 10 MeV channel has a very thin detector such that is is unlikely that the electrons will deposit sufficient energy (200 keV) to trigger the counter.

TABLE III

The electron sensitivity of some of the sensors used in this Catalog

Spacecraft	Energy range (MeV)	Electron detection efficiency	Reference
Mariner 4	$1.0 \leqslant Pr < 170$	Assumed to be similar to IMP 3	(77) in Part 1
IMP 1, 2, and 3	$0.9 \leqslant Pr < 190$	Insensitive below 160 keV; 60% at 200 keV	Fan *et al.* (1966)
IMP 4	$0.8 \leqslant Pr < 9.6$	Insensitive below 200 keV; 30% at 500 keV	K.C. Hsieh, private communication
IMP 4	$9.6 \leqslant Pr < 18.8$	Insensitive below 0.5 MeV; 17% efficiency at 1.1 MeV	
IMP 4	$1.0 \leqslant Pr < 10.0$	Insensitive to electrons	(181) and (212) in Part 1
IMP 4	$Pr > 10.0$	Insensitive to electrons < 300 keV; 30% efficiency to electrons 700 keV	C.O. Bostrom, private communication
Pioneer 8 and 9	$Pr > 14.0$	60% efficiency to electrons > 600 keV	W.R. Webber, private communication
Pioneer 8	$11.1 \leqslant Pr < 25.7$	Insensitive to electrons	
Pioneer 9	$10.5 \leqslant Pr < 24.6$	Insensitive to electrons	
Pioneer 6	$0.6 \leqslant Pr < 12$	Insensitive below 150 keV; $\sim 60\%$ efficiency at 200 keV	J.A. Simpson and C.Y. Fan, private communication
Pioneer 7	$9.6 \leqslant Pr < 13$	Insensitive below 150 keV; $\sim 60\%$ efficiency at 200 keV	
IMP 4	$Pr > 0.7$	Corrected for electron contamination	(176) in Part 1

Some solar proton instruments are designed so that the electron contamination can be resolved. Sophisticated instruments which perform pulse height analysis can determine dE/dx (energy loss per unit thickness of the detector) and E (total energy deposition) measurements that enable electron contamination of the proton channels and proton contamination of electron channels to be eliminated. Measurements given by references (182) and (185) in Part 1 are obtained in this manner. Some of the more recent measurements (references (181) and (212) in Part 1) have sensors specifically designed to avoid electron contamination. However, many of the low energy measurements in the Catalog have serious electron contamination problems. In Table III we have attempted to list the known electron sensitivity of some of the sensors used in this Catalog.

Appendix 2 to the Introduction

Criteria for the Association of Solar Phenomena with the Particle Events

For each reported particle enhancement a review was made of solar and geophysical circumstances on the day, or days, prior to the event. Consideration was given not only to flares, ionospheric, and radio frequency events, but also to the central meridian passage of active centers and the formation of new regions on the disk. Geomagnetic disturbances, sc effects, and 27-day recurrences also were noted. In addition, for each event, an attempt was made to evaluate the 'certainty' of specific solar or geophysical circumstances as the source of an observed particle enhancement. In these evaluations we tried to adhere to certain criteria or guidelines. All flares, SID's and major radio frequency events within the preceding ~24 hr were considered. (Systematic surveys of X-ray data were not made by us.) Flares which occurred tens of minutes or a small number of hours prior to the particle events were given greater certainty than apparently equivalent flares with longer time delays. Flares that were above average in ionizing, radio frequency, or Hα emission were considered to be more probable sources of enhanced particles than flares without such attributes. The emission characteristics of the flares were codified through the derivation of the experimental comprehensive flare index for each flare considered (Dodson and Hedemann, 1971). Although a western predominance in the location of particle-associated flares was known prior to the start of this project, an effort was made to include eastern flares whenever time-associations or flare-characteristics indicated the suitability of their consideration.

In general, 'flare on the invisible hemisphere' was suggested as a possible or probable source of a particle enhancement only when (1) a center of activity with a history of flare production was known to be a relatively small number of days beyond the west limb, or (2) there was good time association between observed prominence activity at the limb (eruptive prominence, surges, loops) and SID's or radio frequency events, but no flare was reported. In some instances an event was assigned to this category when a flare was reported as at or near 90° from the central meridian, but the center of activity in which it probably was based was, at the time, well beyond the limb. In most cases, the records of particle events ascribed to flares on the invisible hemisphere were required to show the shape or characteristics usually observed in enhancements confidently associated with known flares.

When possible, data for the Hα flares were taken from the *Quarterly Bulletin on Solar Activity*. For the years 1964–1967, importance estimates are those in published reevaluations of *Quarterly Bulletin Data* (Dodson and Hedeman, 1968 and 1972). For certain events, subflares and unconfirmed flares reported only in SGD bulletins, were suggested as possible solar sources.

The report of several flares as the possible source of a particle event suggests that the event probably was flare-associated, but that the flare identification remains ambiguous.

Many particle enhancements occurred at times for which reasonable flare-associations could not be made. For some of these events there was close time-association with geomagnetic disturbance, sudden commencements, the central meridian passage of regions or zones of activity, or the formation of new regions on the disk. For other events, a 27-day recurrence pattern seemed to be present. Relationships of these types were duly indicated in the source evaluations. In a number of instances, for particle events in both Part I and

Appendix A, there were several possible sources, and the comment 'source uncertain' frequently was made. Finally it was necessary to report 'source unknown' for a large number of weak particle events.

Data and evaluations of the above types were summarized and presented to the editors for their use in the preparation of the final form of the catalogue. The user of the catalogue should remember that efforts to identify probable sources for the observed particle enhancements reflect only 'best judgments' and apparently 'reasonable evaluations' made in the absence of firm knowledge of the mechanism, or mechanisms, associated with the acceleration of solar particles.

Appendix 3 to the Introduction

Sensitivity Levels of the Various Detectors

The sensitivity of detectors to solar particle events has increased by many orders of magnitude during the period covered by this catalog. This is principally the result of the development of new types of sensors with increased sensitivity.

Early in the 19th solar cycle, the Earth's ionosphere acted as the medium for a number of the more sensitive techniques available. This is demonstrated by the number of reports using VHF measurements, f-min variations, fixed frequency techniques or riometer observations as the means of detecting solar particle influxes. When spacecraft first enabled us to carry instrumentation above the atmosphere, the early particle detectors had sensitivities of about the same threshold as the ionospheric measurement techniques. The gas discharge counters recorded any ionizing radiation capable of penetrating the relatively thick counter walls, and scintillators could measure the energy deposition which enabled detailed spectra to be resolved. However, limited data coverage restricted the usefulness of the early spacecraft. Later spacecraft such as the interplanetary monitoring platforms (IMP satellites) and spacecraft launched into heliocentric orbits greatly improved the data coverage and carried instruments that were orders of magnitude more sensitive than either the early Earth orbiting spacecraft of Earth-based sensors.

The criterion used for the selection of particle data included in the catalog varies considerably during the 15 yr covered. For the 19th solar cycle, we have tended to use almost every available type of sensor and sensing technique. During the 20th solar cycle we have been rather selective in the data we have entered into the particle catalog, preferring measurements obtained beyond the magnetosphere in the Earth-Moon system to those obtained in the magnetosphere (altitudes less than ~ 10 Earth radii). This selection criterion results in a bias against data obtained at synchronous orbit (6.6 Earth radii) or by polar orbiting satellites. There are times when the interplanetary measurements are missing or scarce, and for these periods we have used satellite data obtained in the magnetosphere.

Quantum jumps in detector sensitivity are marked by the launch of IMP 1 in 1964 and again with the launch of Pioneer 6 in December 1965. These large increases in sensitivity are the result of advances in sensor technology. Consequently we have simultaneous observations of instruments of varying degrees of sensitivity on various spacecraft (such as comparison of the Pioneer 6 data with the IMP 3 data) or between instruments on the same spacecraft. These differences in sensitivity generate the situation where an event that is detected by one type of sensor may not be discernible by another sensor. An example of this is given in Figure 7 where the relative sensitivity of the 0.9 to 190 MeV channel on

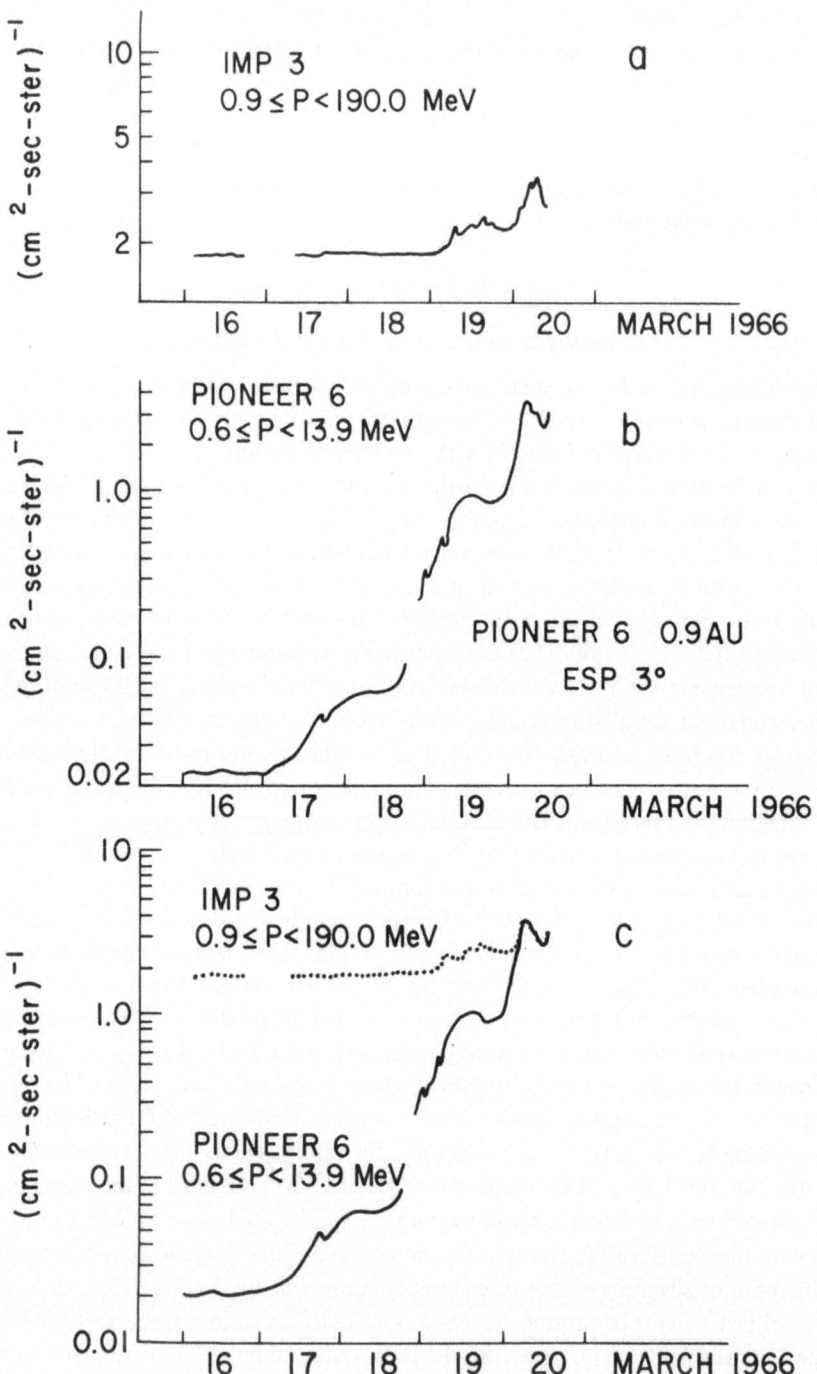

Fig. 7. (c) compares the relative sensitivity of (a) the 0.9 to 190 MeV channel on IMP 3 with that of (b) the 0.6 to 13.9 MeV channel on Pioneer 6, showing how events clearly recorded on Pioneer 6 are not seen (or begin much later) on IMP 3.

IMP 3 is compared with the 0.6 to 13.9 MeV channel on Pioneer 6. This figure illustrates the situation often encountered by the editors in compiling the catalog. It was the policy of the editors to place those events observed on only one sensor in Appendix A. Many of the low energy events in Appendix A are unquestionably valid measurements, but due to the differences in sensitivity of the various instruments, they were not confirmed by other sensors.

The galactic cosmic radiation generates the background in charged particle detectors so that events that are significantly smaller than the sensor background are not discernible. An example of this is shown in Figure 8 where two equivalent instruments on the same spacecraft are compared. In this figure the 29.5 to 94.2 MeV channel has a very low background $[\sim 0.02 \ (cm^2 \ s \ ster)^{-1}]$ due to guard counters inhibiting particles from all directions other than through the narrow acceptance aperture, while the > 30 MeV channel records all particles above the threshold level from any direction and has a relatively high background of about 0.7 $(cm^2 \ s \ ster)^{-1}$. The small discrete proton event at 1620 UT observable on the low background channel as a maximum flux of 0.04 $(cm^2 \ s \ ster)^{-1}$ is an order of magnitude below the background of the > 30 MeV channel and is not discernible in the data from this sensor. A listing of some of the various sensor backgrounds is given in Table IV.

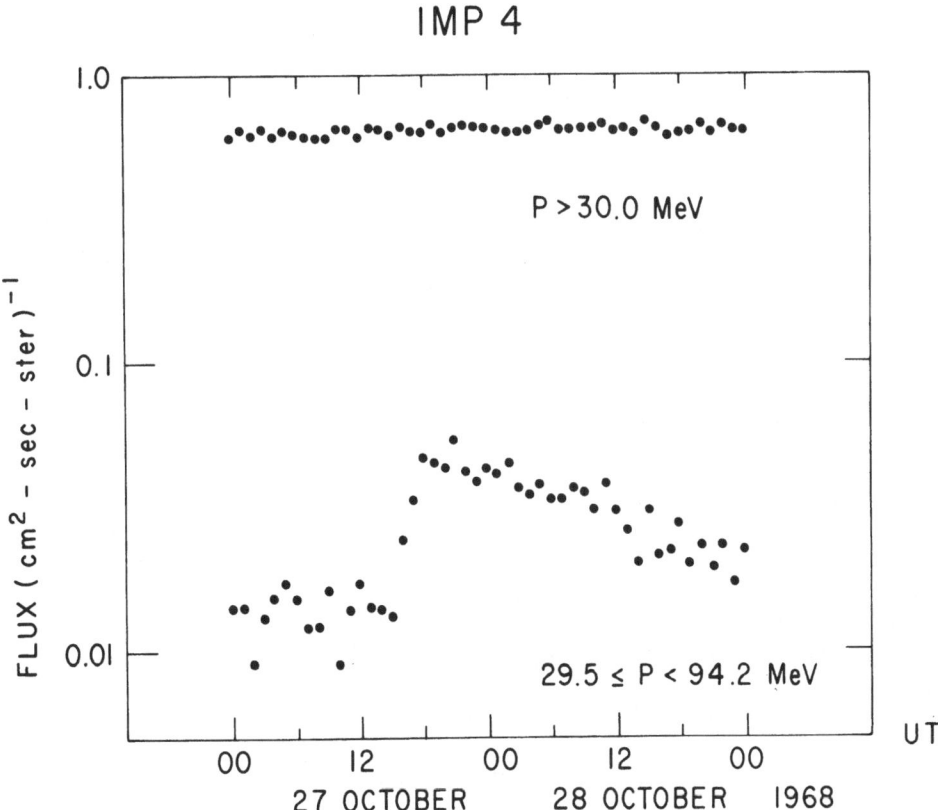

Fig. 8. Difference in the background level, and in consequence of it in the discernibility of proton events, at two instruments on the same spacecraft (IMP 4).

TABLE IV

The sensor backgrounds

Satellite	Energy range (MeV)	Background (cm^2 s ster)$^{-1}$
IMP 1	$0.9 \leqslant Pr < 190$	0.4
	$6.5 \leqslant Pr < 190$	0.2
	$19 \leqslant Pr < 190$	0.2
	$90 \leqslant Pr < 190$	0.8
IMP 2	$0.9 \leqslant Pr < 190$	0.6
	$6.5 \leqslant Pr < 190$	0.3
	$19 \leqslant Pr < 190$	0.1
	$90 \leqslant Pr < 190$	0.2
IMP 3	$0.9 \leqslant Pr < 190$	1.5
	$6.5 \leqslant Pr < 19$	0.02
	$19 \leqslant Pr < 90$	0.02
	$90 \leqslant Pr < 190$	0.3
PIONEER 6	$0.6 \leqslant Pr < 13.9$	0.002
	$7.4 \leqslant Pr < 44.0$	0.01[a]
	$13.9 \leqslant Pr < 73.2$	0.006
	$73.2 \leqslant Pr < 175$	0.5
	$Pr \geqslant 175$	0.8
PIONEER 7	$0.6 \leqslant Pr < 12.7$	0.002
	$7.2 \leqslant Pr < 47.4$	0.01[a]
	$12.7 \leqslant Pr < 73$	0.006
	$73 \leqslant Pr < 165$	0.1
	$Pr \geqslant 165$	0.5
IMP 4	$0.8 \leqslant Pr < 9.6$	0.1
	$9.6 \leqslant Pr < 18.8$	0.08
	$29.5 \leqslant Pr < 94.2$	0.001
	$Pr \geqslant 94.2$	0.2
	$1.0 \leqslant Pr < 10.0$	0.05
	$Pr \geqslant 10.0$	0.3
	$Pr \geqslant 30.0$	0.6
	$Pr \geqslant 60.0$	0.6
IMP 5	$1.0 \leqslant Pr < 10.0$	0.05
	$Pr \geqslant 10.0$	0.6
	$Pr \geqslant 30.0$	1.0
	$Pr \geqslant 60.0$	1.0
AIMP 2	$El > 0.04$	10[b]
IMP 3	$El > 0.04$	10
IMP 4	$El > 0.04$	10[b]

[a] Sectored data.

[b] 25 for 1968 and 1969.

References

Bailey, D.K.: 1964, *Planetary Space Sci.* **12**, 495.

Dodson, H.W. and Hedeman, E.R.: 1968, WDC-A Report UAG-2, ESSA, Boulder.

Dodson, H.W. and Hedeman, E.R.: 1971, WDC-A Report UAG-14, NOAA, Boulder.

Dodson, H.W., Hedeman, E.R., and de Miceli, M.R.: 1972, WDC-A Report UAG-19, NOAA, Boulder.

Drake, J.F., Gibson, J., and Van Allen, J.A.: 1969, *Solar Phys.* **10**, 433.

Fan, C.Y., Gloeckler, G., and Simpson, J.A.: 1966, *J. Geophys. Res.* **71**, 1937.

Gregory, J.B.: 1963, *J. Geophys. Res.* **68**, 3097.

Hudson, H.S., Peterson, L.E., and Schwartz, D.A.: 1969, *Astrophys. J.* **157**, 389.

Kane, S.R. and Anderson, K.A.: 1970, *Astrophys. J.* **162**, 1003.

Kreplin, R.W.: 1966, 'Final Data and Calibrations for the Explorer 30 (NRL Solrad 8) X-ray
    Monitoring Experiment (1965–93A)', E.O. Hulburt Center for Space Research, U.S. Naval
    Research Laboratory, Washington D.C.
Kreplin, R.W. and Horan, D.M.: 1969, 'The NRL Solrad 9 Satellite, Solar Explorer B, 1968–17A',
    NRL Report 6800.
Kreplin, R.W., Horan, D.M., Chubb, T.A., and Friedman, H.: 1969, in C. de Jager and Z. Švestka
    (eds.), *Solar Flares and Space Research*, North-Holland, Amsterdam, p. 121.
Lin, R.P.: 1970, *Solar Phys.* **12**, 266.
Lin, W.C. and Van Allen, J.A.: 1964, *Proc. of the International School of Physics 'Enrico Fermi'*,
    Course 24, Space Exploration and the Solar System, Academic Press, New York, p. 194.
Lupton, J.E. and Stone, E.C.: 1972, *Nuclear Instr. Methods* **98**, 189.
Simnett, G.M., Cline, T.L., Holt, S.S., and McDonald, F.B.: 1970, *Acta Phys. Acad. Sci. Hungaricae*
    29, *Suppl. 2*, 649.
Smart, D.F. and Shea, M.A.: 1971, *Solar Phys.* **16**, 484.
Van Allen, J.A. and Krimigis, S.M.: 1965, *J. Geophys. Res.* **70**, 5737.

PART 1

CATALOG OF SOLAR PARTICLE EVENTS, 1955–1969

1	1955	January 16	$\sim 22^h$				X 2 0
f-min	PCA		$\sim 20^h(1)$		48^h		(1, 3, 6)
VHF	PCA		22^h30^m		48^h	~ 2.0 dB	(2)

Source: ⊙ flare < 2130   N33   W41   3   McM 3065   (onset in (1) earlier)

	1955	Feb. 1, Nov. 19, Dec. 6		events in Appendix A			
2	1956	February 23	03^h45^m				X 3 4
Pr > 16.5 GeV	GLE (141)		$03^h45^m \pm 05^m$				several
				$03^h52^m \pm 07^m$		$4554 \pm 12\%$	Leeds
					24^h		Leeds
VHF	PCA		04^h00^m	22^h	123^h	13 dB	(2)
f-min	PCA		$04^h(30)$–06^h		48–96^h	strong	(1, 3, 6, 30)

Source: ● flare ≦ 0334   N23   W80   3   McM 3400

3	1956	March 10	09^h				X 2 0
VHF	PCA		09^h00^m	11^d23^h	160^h	3.5 dB	(2)
f-min	PCA		$14^h(3)$		72–168^h		(1, 3, 6)

Source: ● flare < 0515   N16   E88   2   McM 3432

	1956	Apr. 15		event in Appendix A			
4	1956	April 27	$\leqq 22^h$				X (1) 0
f-min	PCA		$20^h(10)$–22^h		24–48^h	weak	(1, 3, 6, 10,
			(1, 3)				30)

Source: ⊙ flare 2050   N16   W27   1+   McM 3474
      ◇ severe GMS began with an sc 26^d2112

	1956	May 14, Aug. 28		events in Appendix A			
5	1956	August 31	12^h50^m				X 2 1
Pr > 2.3 GeV	GLE (141)		$12^h50^m \pm 15^m$				several
				$13^h07^m \pm 05^m$		$3.3 \pm 1.0\%$	Chicago
					4^h30^m		Climax
VHF	PCA		14^h30^m	$1^d04^h30^m$	69^h	4.9 dB	(2)
f-min	PCA		$15^h00^m(3)$–18^h		48–96^h	strong	(1, 3, 6, 30)

Source: ● flare 1226   N15   E15   3   McM 3643

	1956	Nov. 8		event in Appendix A			

6	1956	November 13	$\geqq 14^h$				X 3 0
f-min	PCA	14^h		48^h		(6)
VHF	PCA	20^h00^m	14^d23^h	63^h	5.4 dB	(2)
f-min	PCA	$14^d < 00^h(3)-16^h$		$48-72^h$		(1, 3)

Source: ⊙ flare < 1430 N16 W10 2   McM 3753
     ⊘ flare 14^d1037 S20 W55 3   McM 3751
     ○ flare < 0157 N28 W50 2+   McM 3747

	1956	Dec. 25	event in Appendix A				
7	1957	January 20	15^h				X 2 0
VHF	PCA	15^h00^m	21^d07^h	86^h	4.1 dB	(2)
f-min	PCA	$18^h(1)-21^d15^h$		$48-96^h$		(1, 3, 6, 16, 30)

GLE (highly doubtful) reported in (126)

Source: ● flare < 1100 S30 W18 3 McM 3820
     △ sc $21^d12^h56^m$ explains some delayed PCA onsets

8	1957	February 21	$\sim 18^h$			X (1) 0
f-min	PCA	$18^h00^m(3)$		72^h	(3, 10)
f-min	PCA	$22^d05^h-16^h$		$62-84^h$	(1, 16, 30)

Source: ⊙ flare 1605 N20 W33 3+ McM 3856

	1957	March 28, Apr. 2	events in Appendix A				
9	1957	April 3	$\sim 10^h$				X 2 0
f-min	PCA	$10^h15^m(3)-11^h$		168^h	strong	(3, 39)
VHF	PCA	13^h30^m	$4^d03^h30^m$	65^h	3.9 dB	(2)

Source: ● flare 0825 S14 W60 3 McM 3907

Note in Appendix A that (1) gives PCA onset as soon as 2^d23^h.

	1957	Apr. 6	event in Appendix A				
10	1957	April 11	$\sim 13^h$				X (1) 0
f-min	PCA	13^h00^m		$144-168^h$		(1, 3)
f-min	PCA	12^d17^h		$24^h(10)$		(10, 39)
	(possibly a second event)					

Source: ⊘ flare 1722   S23 E05 2+ McM 3923
     ⊘ flare 12^d1850 S25 W73 2   McM 3916

No suitable sources known prior to the onset times.

11	1957 April 19	02^h				X (1) 0
f-min	PCA	02^{h}00^m(3)–03^h		72–96^h	weak	(1, 3)

Source: ⊙ flare 17^{d}2000   N20   E69   3+   McM 3941
       ⊙ flare 18^d < 1310   S16   E64   2   McM 3944
       ○ flare 18^{d}2025   N32   E56   2   McM 3941

	1957 May 5	event in Appendix A				
12	1957 May 8–9	?				X (1) 0
f-min	PCA	8^{d}01^h(3)–9^{d}05^h		48–96^h	weak	(1, 3)

Source: ○ flare 8^{d}0455   S26   E43   1+   McM 3972
       ○ flare 7^{d}1016   S27   E40   2   McM 3972
       ◇ a weak GMS starts ~8^{d}21^h; the event is sequential with Event 10

	1957 May 19, May 30	events in Appendix A				
13	1957 June 19	22^h				X (1) 0
f-min	PCA	22^{h}15^m(3, 30)–23^h		105–190^h	strong	(1, 3, 30, 39)
VHF	PCA	20^d			weak	(2)
f-min	PCA	20^{d}18^h				(6)

Source: ⊙ flare 1609   N20   E45   2   McM 4024
       ○ flare 0610   S38   E22   2   McM 4021

14	1957 June 22	05^h				X 3 0
RIOM, VHF, f-min	PCA	05^{h}00^m(2, 3)	24^{d}01^h(2)	115–144^h	5.0 dB	(2, 3, 6, 12)

Source: ○ flare 21^{d}2210   N12   W01   1+   McM 4024 (DS continuum)
       ○ flare 0236   N23   E12   2   McM 4024 (microwaves only)

	1957 June 28	event in Appendix A				
15	1957 July 1	00^h–12^h				X (1) 0
f-min	PCA	~00^h		>24^h		(14)
f-min	PCA	12^h(6)–14^h		33–>48^h		(6, 15)

Source: ⊙ flare 30^{d}0924   N10   W02   2+   McM 4039
       ◇ a severe GMS is in progress, increasing at ~15^h

16	1957 July 3	08^h				X 3 0
f-min	PCA	08^{h}15^m				(16)
f-min	PCA	09^h–11^h		54–96^h	strong	(1, 3, 14, 19, 23, 30)
RIOM + VHF	PCA	10^{h}00^m(2)	22^h	52^h	6–9.2 dB	(2, 12)

Source: ● flare 0712   N12   W41   3+   McM 4039

	1957 July 19	event in Appendix A

17	1957 July 24	20^h?				X (1) 0
f-min	PCA	10^h			weak	(23)
f-min	PCA	20^{h}15^m(3)–45^m		24^h	weak	(3, 30)
RIOM + VHF	PCA	20^{h}15^m–30^m		~12^h	2 dB	(12, 88)
f-min	PCA	21^h–25^{d}01^h (second onset after (23))		6–21^h		(14, 19, 23)

Source: ● flare 1712   S24   W27   3   McM 4070 (onset in (23) earlier)

18	1957 July 28	15^h				X (1) 0
f-min	PCA	15^h(3, 10, 14)–21^h		24^h	weak	(3, 10, 14, 23)

Source: ○ flare 1348   S23   W85   2   McM 4070 (□ greater part of McM 4070 beyond W limb)
        ◇ recurrent particle stream related to McM 4039

19	1957 August 9	~15^h				X 2 0
f-min	PCA	15^{h}00^m		72^h		(3, 10)
VHF	PCA	16^{h}00^m	10^{d}02^h	50^h	3.1 dB	(2)
RIOM + f-min	PCA	20^h–24^h		24–72^h		(1, 12, 14, 23)

Source: ⊙ flare 1330   S33   W77   1   McM 4082
        ○ flare 0619   S09   E75   2   McM 4099
        ⊡ most of McM 4082 beyond W limb, loops 1437-2357

20	1957 August 27–28	≦ 28^{d}13^h				X (1) 0
f-min	PCA	28^{d}04^h		>12^h	weak	(23)
f-min	PCA	28^{d}13^h(19)–15^h		166–216^h		(1, 6, 16, 18, 19)

(The long duration in (1), (6) and (19) includes the forthcoming events. The date 27^{d}14^h in (6) was a misprint (9).)

Source: ◇ possible onset of long-enduring particle stream associated with McM 4125, located at about E50 on 27^d
        ⊙ flare 28^d < 0913   S31   E33   3   McM 4125

21	1957	August 28	22^h					X 2 0
f-min	PCA		$21^h(14)–22^h30^m$		$>72^h–192^h$	strong	(3, 14)
RIOM + VHF	PCA		$\sim22^h(21)–24^h00^m$	$29^d07^h(2)$	$\sim27^h$	3.2 dB	(2, 21)
f-min	PCA		$29^d04^h–06^h$		$47^h(15)$	strong	(15, 23, 30)

Source: ⊙ flare 2010   S28   E30   2+   McM 4125
       ○ flare causing Event 21

22	1957	August 29	$11^h–13^h$					X 3 0
f-min	PCA		11^h				(18)
RIOM + VHF	PCA		$13^h00^m(12)–14^h$	$30^d02^h(2)$	$58–>77^h$	8.2–9 dB	(2, 12)
Balloon	Pr ≳ 100 MeV?		$\sim30^d00^h$		(interpretation unclear)	20% increase	(21)

Source: △ particles preceding sc 1921, probably associated with one of the flare sources of Events 21 and 22
      ○ flare <0545   N25   E34   2   McM 4124
      ○ flare 1031   S25   E20   2   McM 4125

23	1957	August 31	14^h					X 3 0
Balloon	Pr > 100 MeV		$13^h40^m(\pm03^m)$			0.03 Pr	(21)
f-min	PCA		$14^h15^m(3,30)–15^h$		$>72^h$	strong	(3, 14, 30)
RIOM + VHF	PCA		$15^h00^m–30^m$	1^d03^h	$46–48^h$	4.9–5 dB	(2, 12)

Source: ● flare 1257   N25   W02   3   McM 4124
    ⊘ ⎡ flare 0521   S32   W04   2+   McM 4125
      ⎣ flare 0544   N13   E02   2   McM 4124
    ◇ gradual onset of a major GMS at 12^h

24	1957	September 2	15^h					X 3 0
RIOM + VHF	PCA		⎡$15^h00^m(18)–$ ⎣$17^h00^m(2)$	$3^d02^h(2)$	$32–46^h$	7.2–9 dB	(2, 12, 18)
f-min	PCA		$15^h00^m(3)–19^h$		$24–72^h$	strong	(3, 14, 19)

Source: ⊙ flare 1257   N10   W26   2   McM 4124
      ○ flare 1313   S34   W36   2+   McM 4125
      ⊘ flare 3^d1412   N24   W30   3   McM 4124, also see Appendix A

	1957	Sept. 3, Sept. 10	events in Appendix A

25		1957	September 12	02^h–09^h				X 1 0
f-min		PCA		$02^h(23)$–04^h		8–33^h		$(1, 19, 23)$
f-min		PCA		$08^h(3, 19)$–09^h	13^d02^h	<24–48^h		$(3, 14, 16, 19)$
RIOM + VHF		PCA		$09^h00^m(88)$		>18–24^h	0.5 dB	$(12, 88)$

(In (19) the onset time 04^h was observed on the northern, while 08^h on the southern hemisphere.)

Source: ☉ flare $11^d < 0236$ N13 W02 3 McM 4134
      ○ flare 0703   N09 W15 2 McM 4134
      ⊘ flare 1510   N11 W18 2 McM 4134
      △ sharp PCA increase at the time of sc 13^d0047

26		1957	September 18	$\sim20^h$				X (1) 0
f-min		PCA		20^h–21^h		144^h (incl. Event 28)		$(1, 16)$
f-min		PCA		$19^d00^h45^m(23)$–08^h		48^h		$(3, 14, 23)$

Sources: ● flare <1722 N23 E08 3+ McM 4151
        ○ flare <1303 N23 E10 3 McM 4151
        ⊘ flare $19^d<0350$ N23 E02 3 McM 4151 (also see Appendix A)

	1957	Sept. 20	event in Appendix A

27		1957	September 21	$\geqq11^h$				X 3 0
f-min		PCA		$11^h(18)$–12^h		48^h		$(6, 18)$
f-min		PCA		$13^h(16)$–15^h				$(14, 16)$
RIOM + VHF		PCA		$17^h00^m(2)$	22^d11^h	48^h	5–5.1 dB	$(2, 12)$

Source: △ sc 1005 may be responsible for the early onsets
      ○ flare 1330 N10 W06 3 McM 4152

f-min		PCA		$22^d10^h(23)$–12^h		$>24^h$	strong	$(3, 14, 23)$

Source: ▲ sc 22^d1345 starts a great GMS

28		1957	September 26	21^h				X 2 0
f-min + VHF		PCA		21^h00^m		24–72^h	weak, 1.9 dB	$(1, 2, 14, 19, 23)$
f-min + RIOM		PCA		$\leqq23^h15^m$		29–48^h	2 dB	$(3, 12)$

Source: ● flare 1907 N22 E15 3 McM 4159

	1957	Oct. 5, Oct. 20	events in Appendix A

29		1957	October 20	$\sim 18^h$				X 3 0
f-min	PCA			$17^h(20)–19^h$		$35–72^h$		$(1, 19, 20)$
f-min + VHF	PCA			$21^h(2)–22^h$	$21^d19^h(2)$	$62–64^h$	7.8 dB	$(2, 10, 23)$
f-min + RIOM	PCA			$21^d00^h(14)–11^h$		48^h	5 dB	$(3, 12, 14,$ $16, 30)$

Source: ● flare 1637   S26   W45   3+   McM 4189

Note in Appendix A that (18) gives the PCA onset as early as 13^h.

30	1957	November 4	$\geqslant 23^h$				X 2 0
f-min	PCA		$4^d23^h(1)–5^d07^h(3)$		$24–52^h$		$(1, 3, 10,$ $14, 16, 23)$
VHF	PCA		$5^d02^h00^m$	5^d12^h	46^h	2.6 dB	(2)

Source: ? 0840 oustanding meter burst ($\sim$9000 su at 200 MHz for about an hour) (22)
   □ 2240–2242 type II, no flare associated
   ⊘ flare $5^d < 1205$ S24   W54   2   McM 4207

	1957	Nov. 24, Nov. 26	events in Appendix A			
31	1957	December 17	$03^h?$			X (1) 0
f-min	PCA		03^h	$>24^h$		(14)
f-min	PCA		$12^h(3)–16^h$	$24–43^h$	weak	$(3, 23)$

Source: ⊙ flare 16^d1125 N17   E50   2   McM 4314
   ⊘ flare <0734 N20   E41   2   McM 4314

32	1957	December 28	23^h			X (1) 0
f-min	PCA		$23^h(10,23)–24^h$	$<12–36^h$		$(10, 14, 23)$

Source: ● flare 2229   N25   W50   2   McM 4321

	1958	Jan. 25	event in Appendix A			
33	1958	February 10	05^h			X 2 0
f-min	PCA		$05^h00^m(19)–30^m$		$48–88^h$		$(19, 23)$
VHF	PCA		06^h00^m	20^h	37^h	3.2 dB	(2)
f-min + RIOM	PCA		$06^h–07^h30^m$		48^h	>12 dB	$(1, 3, 12,$ $30)$

Source: ● flare 9^d2108 S12   W14   2+   McM 4400
   △ sc 11^d0126 starts a severe GMS which might have modified the particle flux

	1958	March 3	event in Appendix A

34	1958	March 11	03^h				X (1) 0
f-min	PCA		03^{h}00^m		<21^h	weak	(15, 23)
f-min	PCA		04^h(10)–05^h(3)		48^h		(1, 3, 10)

Source: ⊙ flare <0030   N11   E12   1   McM 4449

35	1958	March 14	15^h				X (1) 0
f-min	PCA		15^{h}00^m		31^h	weak	(15, 23)
f-min	PCA		16^{h}00^m(1, 3)–45^m		27–72^h		(1, 3, 30)

Sources: ⊙ flare 1454   S21   W85   2   McM 4445
          ◇ GMS begins with an sc 1212

36	1958	March 17	<11^h				X X 0
Balloon	Pr > 140 MeV		<11^h			1.0 Pr	(43, 97)
f-min	PCA		17^d(1), 18^d(3)		192–380^h		(1, 3)

Source: ◇ GMS 17^d–22^d with gradual onset ∼17^{d}06^h; this is the first balloon measurement in the U.S.S.R. (43) and
          the cutoff energy might have been misinterpreted due to the occurrence of the GMS
          ○ flare 0438   N10   E18   1+   McM 4456
          ○ flare 1008   N10   E15   1+   McM 4456
          □ beyond E limb?

	1958	March 21	event in Appendix A				
37	1958	March 23	∼15^h				X 2 0
VHF	PCA		15^h	25^{d}16^h	≧ 53^h	3.2 dB	(2)

(Also reported in (12, 18, 19, 40); onset originally given as 18^{h}30^m in (88).)

f-min		PCA cf. Event 36 and Appendix A for onset on earlier days					(1, 3, 10)

Source: ● flare 0947   S14   E78   3+   McM 4476

38	1958	March 25	≦ 15^h				X 3 0
f-min	PCA		08^h(3)–10^h		87–192^h	strong	(1, 3, 30)
			(03^h originally given in (20) corrected in (3))				
f-min	PCA		12^h–15^{h}45^m(23)		72–96^h	strong	(6, 16, 23)
VHF + RIOM	PCA		15^{h}30^m(2)	26^{d}04^{h}30^m	122^h	10–12 dB	(2, 12, rec. 95)
Balloon	Pr > 220 MeV				>26^{d}18^h	0.11 Pr	(89, 90, 91)

Source: △ sc 1540, probably related to ● responsible for Event 37
          ○ flare <0529   N17   E25   2   McM 4474
          ○ flare <0557   S15   E50   2   McM 4476

39	1958	March 31	~16^h				X (1) 0
f-min		PCA	16^h(39)		60^h(10)		(10, 39)

Source: ○ flare 30^{d}0944   S16   W20   2+   McM 4476 (possibly type IV)
        ○ flare 0038   S08   W23   2   McM 4476 (SWF 3+, μ-burst, type III)

40	1958	April 10	~08^h				X 2 0
f-min		PCA	06^h(3)–08^h		72^h		(1, 3, 10, 39)
VHF		PCA	09^{h}00^m	18^h	68^h	4.4 dB	(2)
RIOM + *f*-min		PCA	11^{h}30^m–45^m		40–60^h	4.5 dB	(12, 23, 30, rec. 95)

Source: □ McM 4483 and 4484 one day beyond W limb
        McM 4476 five days beyond W limb

41	1958	June 4	≧23^h				X (1) 0
f-min		PCA	23^h(1)–5^{d}08^{h}15^m(23)		48–72^h		(1, 3, 10, 16, 23, 30)

Source: ● flare <2147   N14   W58   2   McM 4578

42	1958	June 6	14^h				X (1) 0
f-min		PCA	13^{h}45^m(3)		35–48^h	weak	(3, 19)

Source: ⊙ flare 0436   N16   W78   2   McM 4578

43	1958	July 7	01^h				X 4 0
f-min		PCA	01^{h}00^m(18)–03^h		104–144^h	strong	(1, 3, 16, 18, 19, 30)
VHF + RIOM		PCA	[01^{h}30^m(12)– 03^{h}30^m	8^{d}01^{h}30^m	96–>120^h	>15– 23.7 dB	(2, 12, rec. 95)
Balloon		Pr > 130 MeV	<8^{d}09^h			1.4 Pr	(43, 97)

Source: ● flare 0020   N25   W08   3+   McM 4634
        △ sc 8^{d}0748 starts a great GMS

44	1958	July 29	04^h				X 2 0
f-min		PCA	04^h–04^{h}15^m		9–48^h	weak	(1, 3, 16, 19, 30)
RIOM + VHF		PCA	04^{h}05^m(12)–30^m		24–>30^h	~1.6 dB	(2, 12)

Source: ● flare <0259   S14   W44   3   McM 4659

45   1958   August 16 06^h X 3 0

VHF + RIOM	PCA	$06^h00^m(2, 12)-30^m$	$22^h(2)$	$48-71^h$	$12.1-$ >15 dB	(2, 12, 30, rec. 95)
f-min	PCA	$06^h(1)-08^h(23)$		$50-72^h$	strong	(1, 3, 16, 19, 23, 30)
EXPL 4	Pr > 90 MeV	$<11^h$	$<17^d11^h$		34c	(48)

Source: ● flare 0433   S14   W50   3+   McM 4686

---

46   1958   August 21 $\sim15^h$ X 2 0

f-min	PCA	14^h	192^h(incl. Events 47, 48)		(1)
f-min	PCA	$14^h45^m(3)-15^h45^m$	$16->24^h$		(3, 30)
RIOM + VHF	PCA	$15^h00^m(12)-17^h30^m$		$1.5-3$ dB	(2, 12, rec. 95)

Source: ⊙ flare 19^d2118 N18   E26   2   McM 4708
        ⊙ flare 20^d0042 N16   E18   2+   McM 4708

A great GMS starts with an sc 22^d0228

---

47   1958   August 22 15^h X 3 0

f-min	PCA	$14^h(19)-15^h00^m$		$54-83^h$	strong	(19, 23)
Balloon	Pr >100 MeV	15^h30^m	$<23^d04^h30^m$		0.4 Pr	(92)
VHF, RIOM	PCA	$15^h30^m(2)$	$23^d02^h30^m(2)$	84^h	$>10-$ 10.6 dB	(2, 12, rec. 95)
f-min	PCA	$15^h30^m(3)-16^h$		$59->96^h$		(3, 6, 30)
EXPL 4	Pr > 90 MeV	$<23^d09^h$			84c	(48)

Source: ● flare 1428   N18   W10   3   McM 4708

---

48   1958   August 26 01^h X 4 0

RIOM + VHF	PCA	$01^h00^m(12)-03^h30^m$	20^h30^m	70^h	$>13-$ 16.6 dB	(2, 12, rec. 95)
f-min	PCA	$02^h00^m(30)-03^h00^m(23)$		$50-60^h$	strong	(3, 16, 23, 30)
EXPL 4	Pr > 90 MeV	$<08^h25^m$			190c	(48)
Balloon	Pr > 100 MeV	$<09^h$			0.2 Pr	(43, 97)

Source: ● flare 0005   N20   W54   3   McM 4708

1958   Sept. 14, Sept. 22   events in Appendix A

49	1958	September 22	14^h				**X 3 0**
VHF + RIOM	PCA		$14^h00^m(2)$–30^m	23^d12^h	80^h	4.0–5.0 dB	(2, 12)
f-min	PCA		$14^h30^m(3)$–16^h		72–96^h		(1, 3, 19, 30)

Source: ⊙ flare 0738　S19　W42　2　McM 4765　(Note in Appendix A that (23) gives the PCA onset as early as 0530.)

　　　○ flare 1009　N17　W65　2　McM 4756

	1958	Oct. 3	event in Appendix A				
50	1959	January 26	$\sim 14^h$				**X (1) 0**
f-min		PCA	$14^h(3)$–$15^h(39)$		48^h	weak	(3, 10, 39)

Sources: ⊙ flare 0842　N16　W61　3　McM 4969
　　　　　⊙ flare 1027　N16　W61　3　McM 4969
　　　　　◇ a GMS reaches maximum intensity $\sim 15^h$

51	1959	February 13	08^h				**X 2 0**
VHF	PCA		08^h00^m	20^h	74^h	2.6 dB	(2)
f-min	PCA		09^h–$<10^h$		72–96^h	weak	(1, 3)

Source: ⊙ flare $12^d < 2301$　N13　E48　3　McM 5009
　　　　◇ increase in GMS $\sim 10^h$, duration about the same as of the PCA

52	1959	May 10	23^h				**X 4 0**
RIOM	PCA		23^h00^m		221^h	>17 dB	(26, rec. 95)
RIOM + VHF	PCA		⎡$11^d00^h30^m(2)$– ⎣01^h30^m	$12^d02^h30^m(2)$	170–200^h	>15–22 dB	(2, 12)
f-min	PCA		$11^d01^h30^m(3)$–03^h		160–312^h	strong	(1, 3, 6, 16, 30)
Balloon	Pr $>$ 130 MeV		$<11^d09^h$		50^h	6 Pr	(43, 97)
Balloon	Pr $>$ 105 MeV		$<12^d04^h$			$\leqq 52$ Pr	(27, 91)

Sources: ● flare 2102　　　N18　E47　3+　McM 5148
　　　　　⊘ flare 11^d2006　N10　E41　3　McM 5148
　　　　　⊘ flare 13^d0509　N22　E26　2+　McM 5148
　　　　　△ sc 11^d2330 followed by a great GMS may explain the increased proton flux in (27) and it also probably determined the time of maximum flux in (2)

53	1959	June 9	$\sim 24^h$				**X (1) 0**
f-min	PCA		9^d (time unknown)		$\sim 84^h$		(10)
f-min	PCA		$10^d00^h45^m$		24^h	weak	(3)

Source: ● flare 1707　N17　E90　2　McM 5204
　　　　○ flare 0430　S31　E20　1　McM 5194 (with metric cont.)

54	1959 June 13	$\geq 08^h$				X 1 0
f-min	PCA	08^h		96^h		(1)
f-min + RIOM	PCA	13^h–13^h30^m		96^h	weak, 1.5 dB	(3, 10, 12)

Source: ⊙ flare 12^d0735 N21   E65   2+   McM 5204
          ○ flare 0357   N17   E58   1   McM 5204

	1959 July 9	events in Appendix A				
55	1959 July 10	04^h?				X 4 0
RIOM	PCA	04^h			>17 dB	(26, rec. 95)
f-min	PCA	06^h15^m(3)–10^h		>96^h	strong	(3, 6, 16, 30)
VHF + RIOM	PCA	07^h00^m	11^d12^h(2)	360^h	>15–20 dB	(2, 12)
Balloon	Pr 50–300 MeV	<12^h40^m			240 Pr	(93, 94)
Balloon	Pr > 140 MeV	<13^h	11^d11^h	70^h	9 Pr	(43, 97)
Balloon	Pr (10–15 gr)	<11^d05^h	at $11^d20^h15^m$ count rate 62 times background			(29)

Source: ● flare 0206   N20   E60   3+   McM 5265
          ⊙ flare 9^d1930 N18   E67   2   McM 5265

Note in Appendix A that (43) recorded >100 MeV protons as early as 9^d09^h and (1) gives PCA onset at 9^d20^h.
A great GMS starts with an sc 11^d1625.

56	1959 July 14	$\leq 07^h$				X 4 0
RIOM	PCA	04^h45^m				(18, rec. 95)
RIOM + VHF	PCA	[07^h00^m(12, 26)–30^m	$15^d03^h30^m$(2)	72^h	>15–23.7 dB	(2, 12, 26)
f-min	PCA	continues			strong	(1, 3, 6, 16, 30)
Balloon	Pr > 140 MeV	$08^h\triangle1$	$15^d09^h\triangle2$	55^h	1.2 Pr	(43, 97)
Balloon	Pr 50–300 MeV	<23^h			1600 Pr	(93, 94)
Balloon	Pr > 200 MeV	<15^d01^h	$15^d05^h\triangle2$		1.0 Pr	(29)
Balloon	Pr > 88 MeV		$15^d11^h\triangle2$		241 Pr	(29)

Source: ● flare <0325   N17   E04   3+   McM 5265
          △ sc 15^d0803 followed by a great GMS is associated with the flux maxima observed

57	1959 July 16	22^h				X 4 2
RIOM	PCA	22^h00^m			16 dB	(26, rec. 95)
Pr > 1.5 GeV	GLE (141)	$24^h \pm 02^h$				several
			$16^h \pm 02^h$		$10 \pm 2\%$	[Resolute
						Bay
				20^h		several
RIOM + VHF	PCA	[$\leq 22^h50^m(12)-$	$17^d10^h(2)$	67^h	>15–21.2 dB	(2, 12)
		24^h00^m				
f-min	PCA	continues			strong	(1, 3, 6, 16,
						30)
Balloon	Pr (18gr)	$<17^d01^h30^m$	evidence of excess radiation			(29)
Balloon	Pr > 230 MeV	$<17^d08^h30^m$		80^h	15 Pr	(43, 97)
Balloon	Pr (7 gr)	$<17^d12^h30^m$	$>17^d18^h$			(29)
Balloon	Pr 85–400 MeV	$<18^d03^h45^m$	18^d06^h	$\sim 200^h$	1500 c	(49)
Balloon	Pr (8 gr)	$<18^d07^h30^m$	$18^d08^h30^m$			(29)

Source: ● flare 2114   N16   W31   3+   McM 5265
△ sc 17^d 1638 followed by a great GMS is associated with the flux maxima observed

	1959 Aug. 2	event in Appendix A				
58	1959 August 18	11^h				X 2 0
f-min + VHF	PCA	$10^h45^m(18)-11^h30^m(2)$		72^h	~ 1.9 dB	(1, 2, 18)
f-min	PCA	12^h-13^h		48^h		(3, 10, 16)

Source: ● flare 1014   N12   W33   3   McM 5323

59	1959 August 19	$\sim 09^h$			X (1) 0
f-min	PCA	$09^h(39)-10^h$	23^h		(30, 39)
f-min	PCA	continues			(1, 3, 16)

Source: ⊙ flare 18^d1654   N05   E16   2+   McM 5329

60	1959 September 2	$\sim 04^h$				X (1) 0
VHF	PCA	04^h		48^h	very weak	(88)
RIOM	PCA	2^d (no time)			very weak	(26, 31)
EXPL 6	Pr > 75 MeV	$3^d04^h45^m$	$3^d05^h20^m$	10^h	factor 2 incr.	(32)

Source: ⊙ flare 1^d1923   N12   E60   2+   McM 5355

	1959 Sept. 12, Nov. 9, Nov. 30, Dec. 2, Dec. 21	events in Appendix A

61 1960 January 11 12^h? (0) (2) 0

f-fix, f-min, VHF	PCA	$22^h(35)$–$12^d03^h(2)$		36–96^h	weak, ~2 dB	(2, 3, 35, 88)
EXPL 7	Pr > 30 MeV	$12^h\Delta2$	$12^d17^h\Delta2$		0.16 Pr	(42)

Source: ● flare < 2040   N22   E02   3   McM 5527
(gradual onset on EXPL 7 prior to this flare)

---

62 1960 January 13 ~16^h X (1) 0

f-min	PCA	$16^h(30)$–$20^h(10)$	13–24^h	(10, 15, 30)
f-min, f-fix	PCA	continues		(3, 35)
EXPL 7	Pr > 30 MeV	continues		(42)

Source: △ sc 1859, onset of GMS
○ flare 12^d1646 S10   W37   1   McM 5525

1960   Jan. 16, Feb. 7, Feb. 15, Feb. 29, March 10   events in Appendix A

63 1960 March 17 ~18^h (−1) (1) 0

f-fix	PCÁ	18^h		72^h		(35)
EXPL 7	Pr > 30 MeV		18^d (no time)	~72^h	0.02 Pr	(42)

Sources: □ McM 5592 + 5593 one day beyond W limb
◇ recurrent gradual GMS 15^d12^h–18^d12^h

1960   March 28 (twice)   events in Appendix A

64 1960 March 29 07^h (−1) 2 0

PION 5	Pr > 25 MeV	$07^h\Delta1(38)$	$23^h\Delta2$		0.032 Pr	(37, 38)
VHF + f-min	PCA	08^h00^m	$31^d10^h(2)$	73^h (incl. Ev. 65, 66)	2.6 dB	(2, 15)
f-min + f-fix	PCA	$11^h(35)$–14^h		> 24^h		(8, 35)

Sources: ● flare 0650   N12   E30   2+   McM 5615
⊘ flare 28^d2042 N14   E37   2   McM 5615 (see Appendix A)

---

65 1960 March 30 ~09^h X (1) 0

f-fix	PCA	$04^h30^m\Delta1$			(35)
PION 5	Pr > 25 MeV	$06^h\Delta3$	$15^h\Delta2$	0.0010 c	(37)
f-min	PCA	$09^h30^m(18)$–11^h	17^h	weak	(3, 18, 30)
VHF	PCA	continue			(2)

Source: ● flare 0216   N09   E15   1+   McM 5615

66		1960 March 30	$\geq 15^h$				X 3 0
f-min + f-fix	PCA	$15^h(15, 30)$–16^h				(8, 15, 35, 39)
PION 5	Pr > 25 MeV	$15^h\Delta3$	$20^h\Delta2$		0.0015 c	(37)
VHF	PCA	20^h (superposed on Ev. (64))		$>36^h$	very weak	(88)
VHF	PCA	continues (Event 64)				(2)
f-min	PCA	continues		$>48^h$	strong	(3)

Source: ● flare 1455   N12   E11   2   McM 5615

PION 5	Pr > 25 MeV	$31^d03^h\Delta3$	$31^d08^h\Delta2$		0.002 c	(37)
f-min	PCA	$\begin{bmatrix}31^d07^h15^m(18)-\\09^h15^m\end{bmatrix}$				(8, 18, 30)
RIOM	PCA	$\begin{bmatrix}31^d<07^h30^m(12)-\\14^h(26)\end{bmatrix}$		$5->14^h$	3–7 dB	(22, 26, 38)

Source: ◊ GMS starts gradually at ~31^d09^h; also compare the maximum time given in Event 64 (2)

67		1960 April 1	09^h30^m				1 2 0
EXPL 7	Pr > 30 MeV	09^h33^m	$10^h\Delta2$	70^h	32.8 Pr	(36)
Balloon	$\begin{bmatrix}Pr>160/\\>130\,MeV\end{bmatrix}$	$09^h45^m/10^h\Delta1$	~11^h		0.16/0.3 Pr	(38/43)
RIOM	PCA	$\begin{bmatrix}09^h35^m(18)-\\10^h05^m(26)\end{bmatrix}$		$46->86^h$	3–4.2 dB	(18, 22, 26)
$\begin{bmatrix}VHF, f\text{-min},\\f\text{-fix}\end{bmatrix}$	PCA	$09^h45^m(8)$–10^h	$16^h(2)$	$62->96^h$	$\begin{bmatrix}strong,\\3.6\,dB\end{bmatrix}$	(2, 3, 8, 18, 30, 35, 39)
PION 5	Pr > 25 MeV	$<14^h$	$<14^h$		2.8 Pr	(37)

Source: ● flare 0843   N12   W11   3   McM 5615

PION 5	Pr > 25 MeV	$3^d03^h\Delta3$	$3^d04^h\Delta2$		0.16 Pr	(37)
f-min	PCA	3^d00^h				(8)

Source: △ sc 2^d2313

	1960	Apr. 3, Apr. 4		events in Appendix A		

68		1960 April 5	$\geq 04^h$				1 2 0
f-fix + f-min	PCA	04^h		23–96^h		(30, 35)
EXPL 7	Pr > 30 MeV	$<06^h15^m$	$>12^h30^m$	45	0.4 Pr	(36)
VHF + RIOM	PCA	$\begin{bmatrix}07^h00^m(2)-\\08^h50^m\end{bmatrix}$	23^h	18–55^h	3–3.5 dB	(2, 18, 22, 26, 30)
f-min	PCA	$07^h(8)$–10^h				(8, 18, 39)
PION 5	Pr > 25 MeV	$<15^h$	$20^h\Delta2$		2.5 Pr	(37)

Source: ● flare <0215   N12   W63 $\geqslant2$ McM 5615

	1960	Apr. 15	event in Appendix A				
69	1960	April 28	03^h				(2) 2 0

VHF + RIOM	PCA	$\begin{bmatrix} 02^h30^m(2)- \\ 03^h20^m \end{bmatrix}$	14^{h}30^m	24–30^h	2.5–3 dB	(2, 18, 26)
EXPL 7	Pr > 30 MeV	03^{h}23^m	< 19^h		65.5 Pr	(42, 56)
Balloon	Pr ≳ 160 MeV	03^{h}30^m(±15^m)				(38)
f-min	PCA	04^h		15^h		(3, 18, 30)
f-fix, RIOM	PCA	05^h(22)–06^h		> 24^h	3 dB	(22, 35)
Balloon	Pr > 120 MeV	<08^{h}30^m	< 11^h		1.2 Pr	(43)

Source: ● flare <0130   S05   E34   3   McM 5645
△ sc 27^{d}2001 followed with a GMS

70	1960	April 29	≧02^h				(1) 3 0

f-fix + RIOM	PCA	02^h(35)–04^h		45–>120^h	7.5 dB	(26, 35)
VHF, RIOM, ⎤ f-min ⎦	PCA	$\begin{bmatrix} 05^h00^m(2)- \\ 06^h15^m \end{bmatrix}$	30^{d}08^h	36–>114^h	11.2–14 dB, strong	(2, 3, 18, 22, 38)
EXPL 7	Pr > 30 MeV		30^{d}01^{h}Δ5		1.22 Pr	(42)

Source: ● flare <0107   N14   W21   2+   McM 5642
▲ sc 30^{d}0132 followed with a GMS (related to maximum)

71	1960	May 4	10^{h}30^m				(1) 2 4

Pr > 4.1 GeV	GLE	10^{h}30^m ± 05^m				several
			10^{h}37^m ± 01^m		290 ± 10%	Churchill
				~9^h		Mt Washington
VHF	PCA	10^{h}30^m	12^{h}30^m	8^h	3.4 dB	(2)
RIOM	PCA	10^{h}44^m(22)–50^m		3–>49^h	3.7–5 dB	(22, 26)
f-min + f-fix	PCA	10^{h}45^m(3, 30)–12^h		5–>48^h	weak	(3, 30, 35, 39)
Balloon	Pr > 150 MeV	11^{h}Δ1	12^{h}Δ2	35^h	1 Pr	(43)
EXPL 7	Pr > 30 MeV		< 18^h		3.1 Pr	(42)

Source: ● flare 1000   N13   W90   3   McM 5642

72	1960	May 6	≧ 14^h				(1) 3 0

f-min + f-fix	PCA	$\begin{bmatrix} 14^h(39)- \\ 16^h(15, 35) \end{bmatrix}$		> 48–72^h	strong	(3, 15, 18, 30, 35, 39)
VHF + RIOM	PCA	$\begin{bmatrix} <16^h20^m(18)- \\ 18^h30^m \end{bmatrix}$	8^{d}04^h(2)	48–103^h	8.7–>15 dB	(2, 18, 22, 30, 136)
EXPL 7	Pr > 30 MeV	≤ 18^h	7^{d}19^{h}Δ2		2.0 Pr	(42)

Source: ● flare 1404   S08   E07   3+   McM 5653
▲ sc 8^{d}0421 followed with a great GMS is related to the maximum flux in (2, 12)

73		1960	May 9	~08^h			X (1) 0
f-fix + f-min	PCA			08^h(35)–11^h		13–>72^h	(3, 15, 35)
RIOM	PCA			continues			(18, 22, 30)

Source: ☉ flare <0704   S11   E52   3   McM 5657
　　　　□ McM 5642 and 5645 are 2–5 days beyond W limb

	1960	May 12 (twice)		events in Appendix A			
74	1960	May 13	06^{h}15^m				(1) 2 0

f-min + f-fix	PCA	⌈06^{h}15^m(18)– ⌊08^{h}45^m(3)		23–48^h		(3, 18, 30, 35)
. RIOM	PCA	⌈06^{h}20^m(22)– ⌊08^{h}00^m	13^h(22)	<24–>65^h	4–4.5 dB	(18, 22, 26, 30)
VHF	PCA	07^{h}30^m	15^{h}30^m	36^h	3.6 dB	(2)
Balloon	Pr > 100 MeV	<08^{h}30^m	12^{h}Δ2		0.2 Pr	(43)
EXPL 7	Pr > 30 MeV	<13^{h}30^m	15^{h}Δ2		3.2 Pr	(42)

Source: ● flare 0519　　　　N30　W67　3　McM 5654
　　　　⊘ flare 12^d<1342　N30　W59　1+　McM 5654 (cf. Appendix A)

Note in Appendix A that (43) recorded >100 MeV protons as early as 12^{d}09^h and (3) gives PCA on 12^d. According to (88) some evidence for PCA onset can be traced as early as 13^{d}04^h.

75		1960	May 17	15^h–21^h			(0) (1) 0
f-fix + f-min	PCA			15^h(35)–21^h		24–48^h	weak	(3, 35)
EXPL 7	Pr > 30 MeV			18^{d}12^{h}Δ5		0.07 Pr	(42)

Source: ? type II + IV 1743, no flare, but surge activity 1726–1743, S09, E33, McM 5663
　　　　□ McM 5654 about 3 days beyond W limb

76		1960	May 26	10^h–12^h			(0) (1) 0
f-fix + f-min	PCA			10^h(35)–13^h(15)		62–72^h	weak	(3, 15, 35)
EXPL 7	Pr > 30 MeV			12^{h}Δ5		0.07 Pr	(42)

Source: ● flare 0850　N14　W15　2+　McM 5669

	1960	May 28		events in Appendix A			

77	1960	June 1	$\sim 10^h$				(1) (1) 0

EXPL 7	Pr > 30 MeV	$<10^h21^m$	$12^h\Delta 2$		0.40 Pr	(42)
f-min + f-fix	PCA	$12^h(3)-14^h$		$109-144^h$		(3, 15, 35)

Source: ● flare 0823   N29   E46   3+   McM 5680

f-min	PCA	3^d20^h		72^h		(10)
EXPL 7	Pr > 30 MeV	$<4^d12^h$		$\sim 20^h$	0.09 Pr	(42)
f-min + f-fix	PCA	continues				(3, 35)

Source: △ sc 4^d0250 followed by a moderate GMS probably related to the flare 1^d0823

	1960	June 15	event in Appendix A			
78	1960	June 25	$\sim 17^h$			X (1) 0

f-fix + f-min	PCA	$17^h(35)$		$>48^h$	(35, 53)

Source: ⊙ flare 1136   N21   E06   3   McM 5713
      ⊙ flare 1659   N19   W01   1   McM 5713
      ⊘ flare 2039   N18   W04   2+   McM 5713
      ○ flare <1026   N19   E03   1+   McM 5713 (weak type II + IV)
      ◇ $\sim 12^h$ onset of gradual GMS

(It is strange that no stronger particle event occurred.)

79	1960	June 27	$\sim 23^h$			X (1) 0

f-fix + f-min	PCA	$23^h(35)-28^d04^h(15)$		$36-48^h$	weak	(3, 15, 35)

Source: ⊙ flare 2140   N22   W27   3   McM 5713
      ⊙ flare 0418   N20   W19   1+   McM 5713
      ⊙ flare 26^d2358 S08   E34   3   McM 5719
      ◇ a severe GMS in progress (sc 0146)

	1960	June 28, July 14	events in Appendix A			
80	1960	August 12	$\sim 00^h$			(0) (1) 0

f-fix + f-min	PCA	$00^h(35)-13^d$ (no time)		$120-240^h$	weak	(3, 35)
EXPL 7	Pr > 30 MeV		$07^h\Delta 2$		0.18 Pr	(56)

Sources: ⊙ flare 11^d1916 N22   E26   2+   McM 5794
       ⊙ flare 11^d0233 N21   E35   2   McM 5794
       ◇ The long interval of particle emission is coincident with the interval of strong 10cm emission, 11^d-23^d, and
         with the transit of a wide zone of active plages; McM 5794 leads the way and McM 5811 closes the zone.
       □ active prominence SE 11^d2241, type IV 2248-2308

81	1960 August 14	~13^h				(0) (1) 0
f-min	PCA	13^{h}15^m	40^h			(30)
f-min + f-fix	PCA	continues				(3, 35)
EXPL 7	Pr > 30 MeV	<15^{d}11^{h}30^m		0.065 Pr		(42)

Sources: ◇ see Event 80
    △ sc 1510 followed with a weak and brief disturbance
    ⊙ flare 0511 N22 W06 2+ McM 5794
    ○ flare 1306 N20 E36 2   McM 5799 (microwaves + type III)

82	1960 August 26	~10^h				X (1) 0
f-fix + f-min	PCA	10^h(35)	96–120^h	weak		(3, 35)

Source: □ The active plages mentioned in Event 80 are on or beyond W limb.

	1960 Aug. 29, Sept. 1	events in Appendix A				
83	1960 September 3	02^{h}30^m				2 2 1
Pr > 1.5 GeV	GLE (141)	02^h±01^h				several
			09^h±01^h		4.5%	Murmansk
				~15^h		Ellsworth
Balloon	Pr > 90 MeV	02^{h}30^m ± 5^m	13–14^h		5.8 Pr	(54, 73)
VHF + RIOM	PCA	05^{h}00^m(2)–<12^h	4^{d}12^h(2)	>72–89^h	2.5–4 dB	(2, 22, 135)
f-fix + f-min	PCA	07^h(35)–09^{h}30^m		60–>330^h	strong	(3, 18, 30, 35)
Balloon	Pr > 100 MeV		13^h	60^h	5 Pr	(43)
EXPL 7	Pr > 30 MeV		4^{d}03^h△2	>120^h	20.6 Pr	(42)
Rocket	Pr > 14 MeV	flight 14^{h}08^m			19 Pr	(55)

Source: ● flare 0037 N18 E88 2+ McM 5837

RIOM	PCA	23^h		66^h	4 dB	(26)

Source: ▲ sc 4^{d}0230 followed by a great GMS probably contributes to (26) and PCA max.

	1960 Sept. 25	event in Appendix A				
84	1960 September 26	07^h				(1) (2) 0
RIOM + f-min	PCA	07^h(15)–09^h	20–96^h	~2 dB		(3, 15, 18, 39)
Rocket	Pr > 14 MeV	flight 27^{d}14^{h}44^m		~7 Pr		(55)

Source: ● flare 0525 S22 W64 1+ McM 5858

Note in Appendix A that (35) gives the PCA onset as early as 25^{d}21^h.

85	1960	October 3–4	?			X (1) 0
f-fix	PCA		3^d16^h	240^h		(35)
f-min	PCA		$4^d12^h(15)-16^h$	$15-96^h$		(3, 15)

Source: ◇ gradual GMS begins 4^d14^h; this event is sequential with Events 80+81 and July 14 (Appendix A)

	1960	Oct. 11	event in Appendix A			
86	1960	October 29	$\sim12^h$			X (1) 0
f-fix + f-min	PCA		$12^h(35)$	$72-192^h$	weak	(3, 35)

Source: ● flare 1026   N22   E27   3   McM 5909
　　　　◇ increase in GMS intensity at $\sim12^h$

87	1960	November 10	$\sim18^h$			X (1) 0
f-fix	PCA		18^h	$>24^h$		(35)

(This event is unconfirmed and should be listed in Appendix A. It is included here since it completes the sequence of events related to McM 5925.)

Source: ● flare 1009   N28   E28   3   McM 5925

88	1960	November 11	04^h			X (1) 0
f-fix + f-min	PCA		04^h	$12->24^h$		(3, 15, 35)

Source: ● flare 0305   N28   E12   2+   McM 5925

89	1960	November 12	13^h30^m			4 4 4
Pr > 4.1 GeV	GLE (141)		$13^h30^m\pm05^m$			Climax
				$19^h59^m\pm01^m$	$135\pm4\%$	Thule
				29^h		Mt Washington
f-min + f-fix	PCA		$13^h30^m(15)-$ $15^h15^m(3)$	$56->72^h$	strong	(3, 15, 18, 35, 39, 133)
VHF	PCA		14^h00^m	13^d06^h 73^h	21.2 dB	(2)
RIOM	PCA		$14^h00^m(18)-$ 15^h00^m	$13^d09^h\Delta2(132)$ $42-65^h$	$>14->22$ dB	(18, 22, 26, 30, 132, 134)
EXPL 7	Pr > 30 MeV			$23^h30^m\Delta2$	954 Pr	(42)
Balloon	Pr > 80 MeV			$22^h\Delta2$	200 Pr	(129)
Rocket	Pr > 10 MeV		launched 23^h32^m		5000 Pr	(131)
Balloon	Pr > 100 MeV		$<14^d08^h$		2.5 Pr	(43, 130)

Source: ● flare 1315   N27   W04   3+   McM 5925
　　　　△ sc 1348 followed by a great GMS

90	1960	November 14	~22^h				X (1) 0
f-fix + f-min	PCA	22^h(35)		>24^h			(3, 35)

Source: ● flare 0246   N27   W20   2+   McM 5925
◇ a major GMS in progress

91	1960	November 15	02^{h}30^m				4 4 3
Pr > 4.1 GeV	GLE (141)	02^{h}30^m ± 05^m					several
			03^{h}10^m ± 05^m			88%	Deep River
				24^h			Uppsala
f-min	PCA	02^{h}30^m(15)–03^h		138–>144^h		strong	(3, 15)
				(incl. Event 92)			
f-min	PCA	⌈04^{h}30^m(133)– ⌊06^{h}30^m					(18, 30, 133)
VHF	PCA	04^{h}30^m	19^{h}30^m	79^h		20 dB	(2)
RIOM	PCA	05^{h}00^m(26)–05^m	22^{h}Δ1(132)	84^h		>20 dB	(22, 26, 132)
RIOM + f-fix	PCA	07^{h}20^m(134)–09^h		72–>96^h		>22 dB	(18, 35, 134)
Balloon	Pr > 130 MeV		09^{h}Δ5	90^h		7 Pr	(43, 130)
Balloon	Pr > 200 MeV		12^{h}Δ2				
Balloon	Pr > 95 MeV		23^{h}Δ2			200 Pr	(129)
EXPL 7	Pr > 30 MeV		23^{h}Δ2			875 Pr	(42)

Source: ● flare 0207   N25   W35   3   McM 5925
△ sc 1304 followed by a major GMS

92	1960	November 19	12^h				X (1) 0
f-fix + f-min	PCA	12^h(35)		>48^h			(15, 35)

Source: ▢ McM 5925 at and beyond west limb
⊘ flare 1522   N26   W90   2   McM 5925 – suggested by (15) who gives no onset time

93	1960	November 20	21^h				(3) 3 2
Pr > 2.3 GeV	GLE (141)	21^{h}00^m ± 30^m					several
			22^{h}30^m ± 30^m			8 ± 1%	Mc Murdo
				26^h ± 04^h			⌈Kerguelen ⌊Isl.
EXPL 7	Pr > 30 MeV	~22^h	21^{d}01^{h}Δ2			141 Pr	(42, 56)
f-min + f-fix	PCA	⌈23^{h}00^m(3, 15)– ⌊21^{d}00^h		111–>300^h		strong	(3, 15, 35)
VHF + RIOM	PCA	⌈21^{d}02^{h}00^m(2)– ⌊06^h	21^{d}12^h–17^h	24–51^h		3.0–8 dB	(2, 22, 26, 30, 132, 134)
Balloon	Pr > 100 MeV	<21^{d}14^h				0.5 Pr	(43)

Source: ● flare 2017   N28   W90   2   McM 5925
▲ sc 21^{d}0632 followed by a major GMS contributed to PCA max.

94	1960	December 6	$\sim 05^h$			X (1) 0
f-fix + f-min	PCA		05^h		144^h	(15, 35)
f-min	PCA		08^d		48^h	(3)

Source: ● flare $5^d 1825$ N26 E74 3+ McM 5929
◇ onset of a gradual GMS $\sim 08^h$

	1961	Feb. 13, Feb. 18, March 17		events in Appendix A	

95	1961	April 13	$\sim 12^h$		X (1) 0
f-min	PCA		$12^h(15), 14^d(3)$	$24^h - 36^h$	(3, 15)

Source: ◇ storminess $13^d - 15^d$ starting with an sc 1450; sequential with storminess and unconfirmed events on March 17 and Feb. 18

96	1961	July 11	$19^h 30^m$				X 1 0
Balloon	Pr > 270 MeV	$19^h 25^m \Delta 1$	$22^h \Delta 2$			0.06 Pr	(58)
f-min + RIOM	PCA	$20^h (3, 15) - 22^h (24)$		$13^h - 24^h$		1–1.5 dB	(3, 15, 24, 40)
Balloon	Pr > 80 MeV	$< 20^h 37^m$	$22^h \Delta 2$				(57)

Source: ● flare 1615 S07 E32 3 McM 6171

97	1961	July 12	12^h				(2) 4 0
f-min	PCA	$11^h 15^m - 13^h$		$107 - 140^h$			(3, 15)
Balloon	Pr > 80 MeV	$12^h 15^m$	$< 13^d 00^h$				(57)
RIOM + VHF	PCA	$13^h (2, 52)$	$13^d 11^h - 15^h$	$60 - 80^h$	17–20 dB		(2, 24, 40, 52)
INJUN 1	Pr > 1 MeV	$16^h \Delta 5$	$13^d 14^h \Delta 5$	120^h	33000 Pr		(59)
INJUN 1	Pr > 40 MeV	$< 17^h$	$13^d 00^h \Delta 5$		1.75 Pr		(59)
Balloon	Pr > 100 MeV	$18^h \Delta 1$	25^h		0.5 Pr		(43)
Balloon	Pr 100–200 MeV	$20^h \Delta 2$			0.6 Pr		(58)
Balloon	Pr > 77.5 MeV	13^d			1.7 Pr		(138)

Source: ● flare 1000 S07 E23 3 McM 6171
△ sc $13^d 1113$ followed by a great GMS contributed to PCA maximum

---

98	1961	July 15	$\sim 16^h$			X (2) 0
RIOM	PCA	$15^h 45^m \Delta 3$			3 dB	(13, 40)
f-min	PCA	$16^h \Delta 3 (15)$				(3, 15)
RIOM	PCA	$19^h \Delta 3$	$22^h \Delta 2$	$> 24^h$	3.5 dB	(24)

(A masked event; the original sources in (13) and (40) unknown; at Kiruna RIOM possibly increased absorption by $20^h (106)$; Resolute Bay RIOM no event (52); balloon no event (57). According to (155) the available RIOM data does not confirm the presence of an PCA larger than 1 dB additional absorption superposed on Event 97.)

Source: ● flare 1508 S07 W20 2 McM 6171
⊘ flare 1433 N13 E15 3 McM 6172

	1961	July 17	event in Appendix A				
99	1961	July 18	10^h				3 3 3

f-min	PCA	$\leq 10(3,15)-12^h$		$106-120^h$		$(3,15,39)$
INJUN 1	$Pr > 40$ MeV	$10^h \Delta 2$	$20^h \Delta 2$		72 Pr	(59)
$Pr > 2.3$ GeV	GLE (141)	$10^h 15^m \pm 05^m$				several
			$11^h 37^m \pm 08^m$		$23.5 \pm 1.4\%$	Thule
				$12^h \pm 02^h$		⌈Thule,
						⌊Mc Murdo
Balloon	$Pr > 80$ MeV	$10^h 23^m$	$<18^h \Delta 5$		140 Pr(140)	(57, 140)
Balloon	⌈$Pr > 125/$	$10^h 30^m \Delta 1/$	$15^h 30^m \Delta 1/$		30/40 Pr	(58/138)
	⌊> 100 MeV		$16^h \Delta 2$			
VHF + RIOM	PCA	$11^h 30^m(2)-<12^h$	$19^h 30^m(2)$	$55-60^h$	8.7–10 dB	(2, 13, 24, 40)
Balloon	$Pr > 130$ MeV		$19^d 12^h \Delta 5$		2.6 Pr	(43)

Source: ● flare 0920 S07 W59 3+ McM 6171
   △ sc 1123

100	1961	July 20	$16^h 10^m$				(1) 2 2

INJUN 1	$Pr > 1.0$ MeV	$16^h \Delta 2$	$22^h \Delta 3$	$>80^h$	750 Pr	(153)
$Pr > 2.3$ GeV	GLE (141)	$16^h 10^m \pm 05^m$				several
			$16^h 45^m \pm 15^m$		$7 \pm 2\%$	⌈Ottawa,
						⌊Churchill
				$\sim 8^h$		several
Balloon	$Pr > 80$ MeV	$16^h 20^m$	$18^h \Delta 1$			(57)
Balloon	$Pr > 100$ MeV	$17^h \Delta 1$	$17^h \Delta 1$	25^h	1.2 Pr	(43)
f-min + RIOM	PCA	$17^h(15)-21^d 03^h$	$21^d 14^h(24)$	$24-120^h$	5 dB	(3, 13, 15, 24, 40)
INJUN 1	$Pr > 40$ MeV	$<20^h$	$20^h \Delta 2$		3.9 Pr	(153)

Source: ● flare 1553 S06 W90 3 McM 6171
   △ sc 1550

101	1961	July 24	$\sim 09^h$			(0) (1) 0
f-min	PCA	$09^h(15)$		$86-168^h(3)$		(3, 15)

(the long duration includes July 28 event in Appendix A)

Source: ● flare 0410 N15 E15 2+ McM 6178

INJUN 1	$Pr > 1$ MeV	$<26^d 13^h$	$26^d 19^h \Delta 2$	$>60^h$	650 Pr	(59)

Source: △ sc $26^d 1951$ followed with a GMS, possibly initiated by the flare $24^d 0410$ (or the flare $24^d 1722$)

	1961	July 28, Aug. 10	events in Appendix A

102	1961 September 7	00^h?				(0) 1 0
EXPL 12	Pr > 40/>23 MeV	00^h∆2	15^h∆2		0.8/4.3 Pr	(61)
f-min	PCA	00^h(60, 74)	17^h(106)	48–60^h		(3, 60, 74, 106)
RIOM	PCA	<10^h	17^h	24^h	1 dB	(74, 106)
EXPL 12	Pr 87 MeV		12^h∆2		0.01* Pr	(63)
INJUN 1	Pr > 1.5 MeV		8^{d}02^h∆2	24^h	35 Pr	(61)

Source: ◇ remnants of Event 97 (cf. event on Aug. 10 in Appendix A) (60)
  □ but no active region has gone around W limb since September 1

UNUSUAL long-lived radio emission from McM 6212 on 6^{d}1535–2225 (2.8 GHz) and 1738–2254 (DS continuum) (62)

103	1961 September 10	20^{h}36^m–45^m				(2) 2 0
EXPL 12	Pr > 40/>23 MeV	20^{h}36^m/	11^{d}02^h∆2		0.55/6.8 Pr	(61)
EXPL 12	Pr > 600 MeV	20^{h}45^m	23^h∆1	12^h	0.06 Pr	(63)
EXPL 12	Pr 295/175 MeV	20^{h}50^m	23^h∆1		⌈0.00025 */ ⌊0.0014* Pr	(63)
EXPL 12	Pr 87 MeV	20^{h}50^m	23^h∆1	50^h	0.018* Pr	(63)
RIOM	PCA	⌈21^h(2, 52, 74)– ⌊23^h	11^{d}14^h–21^h	56–72^h	2.9–5.5 dB	(2, 40, 52, 74)
f-min	PCA	21^h(74)–23^{h}15^m(3)		31–48^h		(3, 15, 74)
EXPL 12	Pr 14.5 MeV	22^h∆1	11^{d}20^h∆2	50^h	8.0* Pr	(63)
EXPL 12	Pr 3.8 MeV	11^{d}01^h∆1	12^{d}10^h∆2		40* Pr	(63)
INJUN 1	Pr > 1.5 MeV		12^{d}01^h∆2		940 Pr	(61)

Source: ● flare 1958   N08   W80   1   McM 6212

	1961 Sept. 18	event in Appendix A				
104	1961 September 28	22^{h}40^m				2 2 0
Balloon	(5g)	22^{h}40^m ± 05^m				(46)
EXPL 12	Pr > 600 MeV	22^{h}40^m∆1	23^{h}40^m∆1	50^h	2.3 Pr	(45)
EXPL 12	Pr > 330 MeV	22^{h}40^m∆1	23^{h}40^m∆1	50^h	0.004 Pr	(45)
EXPL 12	Pr 295/175 MeV	22^{h}40^m∆1	⌈23^{h}40^m/ ⌊29^{d}01^h∆1	50^h	0.01*/ 0.032* Pr	(63)
EXPL 12	Pr > 130 MeV	22^{h}40^m∆1	29^{d}02^h∆1	50^h	0.025 Pr	(45)
EXPL 12	Pr 87/14.5 MeV	22^{h}40^m∆1/	⌈29^{d}02^h∆1/ ⌊10^h∆4	50^h/	0.1*/0.8* Pr	(63)
f-min	PCA	23^h–23^{h}15^m		85–168^h		(3, 15, 39, 52)
VHF + RIOM	PCA	23^h–23^{h}37^m		77^h	1.7–1.8 dB	(2, 22, 40, 52)
INJUN 1	Pr > 40 MeV	<23^{h}13^m	29^{d}03^h∆2		8 Pr	(143)
EXPL 12	Pr 7.9/5.7 MeV	∆7	29^{d}10^h/20^h∆4		1.6*/7* Pr	(63)
EXPL 12	Pr 3.8/2.2 MeV	∆7	29^{d}20^h∆4		14*/40* Pr	(63)

Source: ● flare 2202   N13   E29   3   McM 6235

(*continues on next page*)

104 continued

EXPL 12	Pr > 30/>9.0 MeV	$30^d20^h30^m\Delta3$	$30^d21^h30^m\Delta1$	$30^h/92^h$	1.2/175 Pr	(45)
EXPL 12	Pr > 3.0 MeV	$30^d19^h30^m\Delta4$	$30^d21^h30^m\Delta2$		1600 Pr	(142)
INJUN 1	Pr > 1.5 MeV		$30^d22^h\Delta2$	92^h	7700 Pr	(143)
f-min + RIOM	PCA	continues	$30^d22^h(52)$		2.8–3.3 dB	(3, 22, 52)

Source: ▲ sc 30^d1348 followed by a great GMS

105	1961 October 26	$\sim18^h$				$(-1)\,(1)\,0$
f-min	PCA	18^h	27^d18^h	48^h		(60)
EXPL 12	Pr > 3 MeV	$22^h\Delta4$	$27^d18^h\Delta2$	48^h	20 Pr	(142)

Source: ◇ recurrent particle stream, remnants of Event 104; great GMS of unknown origin 26^d–29^d, sc 28^d0820

106	1961 November 10	15^h10^m				1 1 0
EXPL 12	Pr 295 MeV	15^h10^m	15^h45^m	16^h	0.0028* Pr	(63)
EXPL 12	Pr > 200/>60 MeV	15^h10^m	15^h45^m		3.2/5.0 Pr	(63)
f-min	PCA	$15^h(15)$–16^h		31–>48^h		(3, 15, 74)
EXPL 12	Pr 175/87 MeV	15^h20^m	15^h45^m	40^h	$\begin{bmatrix}0.016*/\\0.06* \text{ Pr}\end{bmatrix}$	(63)
EXPL 12	Pr 14.5/3.8 MeV	$16^h30^m/45^m$	$\begin{bmatrix}19^h30^m/\\11^d01^h00^m\Delta1\end{bmatrix}$		2.5*/20* Pr	(63)
VHF + RIOM	PCA	$15^h30^m(145)$–16^h	18^h–20^h	15–50^h	1–2.2 dB	(2, 74, 144, 145)

Source: ● flare 1434   N19   W90   1+   McM 6264

	1961 Dec. 1	event in Appendix A				
107	1962 February 1	$\sim21^h$				$(1)\,(2)\,0$
f-min	PCA	$20^h30^m(3, 74)$–22^h		44–48^h		(3, 15, 74)
RIOM	PCA	22^h	$2^d14^h\Delta1$	25^h	2.8 dB	(146)
INJUN 1	Pr > 1.4 MeV	$<2^d19^h40^m$	$2^d20^h\Delta5$	48^h	2600 Pr	(64)

Source: ○ flare 0901   N10   W36   2   McM 6326⎤
      ○ flare 1634   N09   W38   1   McM 6326⎦ The only active region on the disc.
      □ ? (geomagnetic field extremely quiet)

108	1962 February 4	$\sim12^h$?				$(0)\,(1)\,0$
RIOM	PCA?	12^h	15^h		3 dB	(74, 106)
INJUN 1	Pr > 1.4 MeV	$16^h30^m\Delta2$	$5^d00^h\Delta2$	48^h	490 Pr	(64)
f-min	PCA	$22^h(15)$		12–24^h		(3, 15)

Source: ○ flare 3^d1223 N09   W63   1   McM 6326 (Δt too long, no other suitable flares)
      □ McM 6326 at and beyond west limb
      ◇ a moderate and short-lived GMS starts with an sc 0930

109   1962   February 20 $<12^h$ (0) 1 0

RIOM	PCA	$<12^h$	barely detectable			(74)
INJUN 1	Pr > 1.4 MeV	$15^h\Delta2$	22^d	79^h	100 Pr	(64)
RIOM	PCA	$<22^d07^h$	23^d08^h	$>35^h$	0.7 dB	(75)

Source: ○ flare 0548   S10   E80   2   McM 6351
         ○ flare $19^d < 1245$ S09   E80   2   McM 6351
         ○ flare $19^d < 1304$ N10   E77   2   McM 6352

       1962   Oct. 6   event in Appendix A

110   1962   October 23 17^h30^m 0 1 0

MAR 2	Pr $> 12/>0.5$ MeV	$17^h23^m/30^m$	$21^h\Delta2/$	$40^h/$	1.0/400 Pr	[(148, 149/ 147)
EXPL 14	Pr 278/210 MeV	$17^h30^m\Delta1$	$18^h30^m\Delta2/\Delta1$	$/6^h$	⌈0.00012*/ ⌊0.0003* Pr	(63)
EXPL 14	Pr 130/73 MeV	$17^h30^m\Delta1$	$19^h30^m\Delta1$	8^h	⌈0.0009*/ ⌊0.0028* Pr	(63)
RIOM	PCA	17^h30^m	23^h	16^h	0.9 dB	(75)
EXPL 14	Pr 17/6.0 MeV	$19^h30^m\Delta1/$	$24^d01^h\Delta2/$		0.12*/0.6* Pr	(63)

(MAR 2 at 0.9 AU, ES $-17.3°$.)

Source: ⊙ flare 1642   N03   W70   2   McM 6581
         ◇ a moderate recurrent GMS begins on 22^d, in sequence S1 (first observed in August 1962)

---

111   1963   February 9 $\sim18^h$ -1 1 0

EXPL 14	Pr $> 10/>2.9$ MeV	$18^h\Delta2$	$11^d12^h\Delta2$	96^h	0.2/6 Pr	(44)
f-min	PCA	18^h45^m		192^h	weak	(3)
RIOM	PCA	⌈18^h42^m (74, 75)– ⌊22^h	10^d11^h(146) 11^d18^h(75)	148^h	0.6–2.4 dB	(74, 75, 146)
ALOU 1	Pr > 1.3 MeV	$<10^d03^h$	$11^d12^h\Delta5$	96^h	110 Pr	(156)

Source: ◇ a recurrent GMS of 6^d duration begins 9^d18^h–21^h, in the sequence S1

       1963   Feb. 15/16, March 8, Apr. 4   events in Appendix A

112   1963   April 15 12^h15^m 1 1 0

RIOM + f-min	PCA	$12^h(38)$–13^h	16^d00^h(75)	18–96^h	0.9 dB	(3, 15, 74, 75)
EXPL 14	Pr $> 23/>3.4$ MeV	$12^h15^m/12^h\Delta5$	$14^h30^m\Delta1/$	$20^h/96^h$	2.2/>9.0 Pr	(150/44)
ALOU 1	Pr $> 33/>1.3$ MeV	$<16^h$	$<16^h/16^d05^h\Delta2$		2.3/220 Pr	(156)

Source: ⊙ flare 1034   S11   W06   2   McM 6766 (the only active region on the disc)
         ◇ a minor GMS starts sequence S2 of recurrent storms

---

       1963   Apr. 24   events in Appendix A

113   1963   May 1 $\sim$00^h -1 (1) 0

EXPL 14	⌈Pr > 10/>6.1/ ⌊>3.4 MeV	00^{h}Δ2	2^{d}06^{h}Δ2	144^h	0.9/2.5/ 4.0 Pr	(44)
f-min	PCA	12^h		72^h	weak	(3)

Sources: ◇ a major recurrent GMS April 30–May 4 in sequence S1
        ⊘ flare 0525   N15   E46   2   McM 6790

      1963   May 27, May 29   events in Appendix A

114   1963   June 14   06^h (-1) (0) 0

EXPL 14	Pr > 6.3/>3.5 MeV	06^{h}Δ2	15^{d}04^{h}Δ2	96^h	0.6/1.0 Pr	(44)

Source: ⊙ flare <0225   N09   W34   1   McM 6832

      1963   June 25   event in Appendix A

115   1963   August 6   11^h (0) (1) 0

ALOU 1	Pr > 33 MeV	<11^{h}Δ5	13^{h}Δ5	24^h	1.1 Pr	(156)
ALOU 1	Pr > 1.3 MeV	11^{h}Δ5	07^{d}06^{h}Δ2	>24^h	37 Pr	(156)
f-min	PCA	11^{h}15^m		48^h	weak	(3)

Source: ⊙ flare 0855   N13   W12   2   McM 6909

---

116   1963   August 9   23^h 0 (1) 0

f-min	PCA	23^{h}15^m		48^h	weak	(3)
ALOU 1	Pr > 33/>1.3 MeV		⌈10^d<03^h/ ⌊11^{h}Δ2	/>24^h	1.6/120 Pr	(156)

Source: ● flare 2234   N07   W80   1   McM 6908

      1963   Sept. 14   event in Appendix A

117   1963   September 15 $\sim$10^h X 1 0

RIOM	PCA	<10^h	16^{d}02^{h}Δ5	$\sim$3^h	0.5 dB	(151)
f-min	PCA	10^h(74)–14^h		>36^h	weak	(3, 15, 74)

(No measurable flux from ALOU 1 (156).)

Source: ● flare 0015   N15   E75   2   McM 6964

Note in Appendix A that a very slight riometer absorption was recorded at McMurdo as early as 14$^d\sim$20^h.

       ◇ a major GMS begins 13^{d}19^h in sequence S1

118   1963   September 16 $\leq 16^h$ X 1 0

RIOM	PCA	10^h–20^h	17^d05^h	$>40^h$	0.8 dB	(151, 152)
f-min	PCA	$16^h(3)$–17^d04^h		$>48^h$		(3, 15)

(No measurable flux from ALOU 1 (156).)

Source: ● flare 1430   N12   E48   2   McM 6964
         ○ flare 1300   N10   E50   2   McM 6964 (intense but short burst)
         ○ flare 0325   N11   E57   3   McM 6964 (no DS effects)

1963   Sept. 19   event in Appendix A

119   1963   September 21 00^h 1 2 0

RIOM	PCA	00^h	15^h	54^h	3.1 dB	(2)
ALOU 1	Pr $> 114/>33$ MeV	$/<01^h08^m$		$26^h/52^h$	2.3/23 Pr	(157)
RIOM	PCA	$<03^h(151)$	18^h	$\geq 48^h$	2.5–4 dB	(151, 157)
f-min	PCA	03^h		68–72^h	strong	(3, 15)
ALOU 1	Pr > 1.7 MeV	$04^h\Delta 2$	$20^h\Delta 2$	55^h	300 Pr	(157)
Balloon	Pr > 200 MeV	$<08^h$			0.3 Pr	(43)

Source: ● flare 20^d2314 N10   W09   2   McM 6964

Note in Appendix A that (3) also gives an earlier onset on 19^d.

---

120   1963   September 26 07^h30^m (1) 2 0

RIOM	PCA	$07^h30^m(2)$–08^h	15^h–16^h	48–89^h	3–4.6 dB	(2, 151, 157, 158)
Balloon	Pr > 120 MeV	$<08^h$			1.0 Pr	(43)
f-min	PCA	$08^h(15)$–11^h15^m		52–168^h		(3, 15)
ALOU 1	Pr > 1.3 MeV		$27^d07^h\Delta 2$		3500 Pr	(157)

Source: ● flare <0638 N13   W78   3   McM 6964

1963   Oct. 12   event in Appendix A

121   1963   October 28 08^h $(-1)(1)0$

f-min	PCA	08^h–08^h15^m		48–52^h	weak	(3, 15)
ALOU 1	Pr > 1.3 MeV	$16^h\Delta 2$	$21^h\Delta 5$	48^h	150 Pr	(156)

Source: ● flare <0135 N12   W24   3   McM 7003

1963   Dec. 3,
1964   Jan. 2, Jan. 23, Jan. 28 (twice), Feb. 18, March 3, March 16 (twice)   events in Appendix A

122	1964 March 16	16^h40^m					0 1 0
IMP 1	El > 40 keV	$16^h43^m(\pm03^m)$	$17^h00^m\Delta1$	13^h	750 El	(66)	
IMP 1	El > 3.0 MeV	$17^h\Delta2$			0.05 El	(67)	
IMP 1	Pr > 45/>20 MeV	$17^h20^m/$			/0.7 Pr	(70, 71)	
IMP 1	Pr > 15 MeV	$17^h40^m(\pm15^m)$	$23^h\Delta1$	25^h	0.84 Pr	(69, 72)	
IMP 1	Pr > 6.5/>0.9 MeV	$18^h/17^h\Delta1$	$23^h/17^d07^h\Delta1$	$50/75^h$	3.29/29.6 Pr	(68)	
RIOM	PCA	18^h00^m	17^d00^h	$\sim24^h$	~0.3 dB	(86)	

Source: ● flare 1553   N05   W73   1+   McM 7182

Note in Appendix A that >1 MeV proton flux increased already on 15^d.

	1964	March 19, March 22, March 27, March 31, Apr. 15, May 10, Oct. 4	events in Appendix A
	1965	Jan. 8, Jan. 10, Feb. 2	events in Appendix A

123	1965 February 5	18^h45^m					1 2 0
IMP 2	Pr > 0.9 MeV	$18^h40^m\Delta1$	$6^d01^h\Delta1$	88^h	162 Pr	(76)	
IMP 2	El > 3 MeV	$18^h45^m\Delta1$	$20^h00^m\Delta1$	10^h	0.40 El	(67)	
IMP 2	El > 40 keV	$18^h45^m(\pm05^m)$	$19^h\Delta1$		700 El	(66)	
RIOM + f-min	PCA	$18^h40^m(3)-50^m$	$22^h(86)$	$48-65^h$	1.6 dB	(3, 86)	
IMP 2	Pr > 90/>19 MeV	$19^h\Delta1$	$20^h\Delta1/21^h\Delta7$	$24^h/85^h$	0.91/5.87 Pr	(76)	
IMP 2	Pr > 15 MeV	$19^h05^m(\pm03^m)$	$00^h30^m\Delta1$	38^h	15.0 Pr	(77)	
IMP 2	Pr > 6.5 MeV	$19^h\Delta1$	$6^d01^h\Delta1$	85^h	53 Pr	(76)	

Also recorded on Mariner 4 at 1.2 AU (>55,>15,>0.5 MeV), ES − 2°(77, 78, 79).

Source: ● flare 1750   N08   W25   2   McM 7661

MAR 4	Pr > 0.5 MeV		$7^d06^h\Delta2$		140 Pr	(79)

▲ sc 6^d1414 followed with a major GMS

	1965	Apr. 17, Apr. 20, May 7, May 16, May 20, May 25, May 31, June 1, June 5, June 12, June 13, June 15, June 17, June 28, June 29, July 2, July 4/6, July 10, July 13	events in Appendix A

Mariner 4 data not included after July 15, because of the great difference in position of Mariner 4 and Earth with respect to the Sun.

	1965	July 28, Aug. 16, Sept. 1, Sept. 29	events in Appendix A

124	1965 October 4	11^h					0 (1) 0
OGO 1	Pr > 8 MeV	$10^h40^m\Delta1$	$16^h30^m\Delta4$		24 c	(82)	
IMP 3	Pr > 90/>19 MeV	$11^h\Delta1/\Delta4$	$14^h\Delta1/\Delta4$	$5^h/65^h$	0.62/2.3 Pr	(80)	
IMP 3	Pr > 15 MeV	$11^h17^m(\pm09^m)$	$14^h\Delta5$		2.4 Pr	(69, 72)	
ZOND 3	Pr > 30/>1.0 MeV	$12^h\Delta2/\Delta3$	$16^h\Delta2/$	$40^h/>96^h$	0.39/5.5 Pr	(43)	
	(ZOND 3 at 1.13 AU, ES − 2°)						
IMP 3	Pr >6.5/>0.9 MeV	$12^h\Delta4$		$65^h/81^h$	>1/> 23 Pr	(80)	
IMP 3	El > 3 MeV	probably absent				(67)	
IMP 3	El > 40 keV	absent				(69)	
f-min	PCA	12^h		12^h	very weak	(3, 83)	

Source: ● flare 0937   S21   W30   2   McM 8012

1965　　Oct. 8, Nov. 23, Nov. 25, Dec. 7, Dec. 17, Dec. 24, Dec. 27, Dec. 29　　　events in Appendix A

125	1965	December 30	01^h02^m				$-1\,(0)\,0$
IMP 3	Pr > 0.9 MeV	$01^h\Delta 1$	$03^h\Delta 1$	16^h	5.6 Pr	(80)
IMP 3	El > 40 keV	$01^h02^m(\pm 02^m)$	02^h00^m	20^h	300 El	(66)
IMP 3	Pr > 6.5/> 19 MeV	$02^h/03^h\Delta 2$	$08^h/06^h\Delta 2$	$14^h/11^h$	0.08/0.05 Pr	(80)
PION 6	Pr > 13 MeV	$03^h\Delta 1$	$11^h\Delta 1$	15^h	0.17 Pr	(154)
PION 6	Pr > 44/>7.5 MeV	$04^h\Delta 1$	$12^h/14^h\Delta 1$	$15^h/13^h$	93/1968c	(84)
VEN 2	Pr > 1.0 MeV	event continues (see Appendix A)				
RIOM	PCA					(86)

(PION 6 at 0.98 AU, ES − 1°; VEN − 2 at 0.91 AU, ES − 2°)

Source: ● flare 0006　N09　W70　1?　McM 8105

126	1965	December 30	$\sim 16^h$				$-2\,0\,0$
PION 6	Pr > 0.6 MeV	$16^h\Delta 3$	$1^d03^h\Delta 1$	70^h	55 Pr	(154)
PION 6	Pr > 44/> 13 MeV	$17^h\Delta 3$	$31^d04^h\Delta 1/$ $08^h\Delta 2$	$25^h/100^h$	94c/0.03 Pr	(84/154)
PION 6	Pr > 7.5 MeV	$17^h\Delta 3$	$31^d13^h\Delta 5$		0.17 Pr	(84)
IMP 3	Pr > 6.5/>0.9 MeV	$17^h\Delta 2/\Delta 3$	$31^d12^h\Delta 4$	$20^h/120^h$	0.05/18 Pr	(80)
VEN 2	Pr > 1.0 MeV	continues	$1^d06^h\Delta 7$		5.5 Pr	(103)

Source: ▫ McM 8105 on W limb

1966　　Jan. 2　　　event in Appendix A

127	1966	January 17	$\sim 12^h$				$-2\,0\,0$
PION 6	Pr > 0.6 MeV	$12^h\Delta 2$	$18^d04^h\Delta 1$		15.5 Pr	(104)
PION 6	Pr > 7.5 MeV	$13^h\Delta 2$ gradual $22^h\Delta 2$ fast	$18^d05^h\Delta 2$		0.07 Pr	(84)
PION 6	Pr > 13 MeV	$17^h\Delta 2$	$18^d01^h\Delta 2$	20^h	0.013 Pr	(104)
IMP 3	El > 40 keV	$< 21^h$			20 El	(66)
IMP 3	Pr > 0.9 MeV	$18^d00^h\Delta 2$	$18^d10^h\Delta 2$	19^h	9.4 Pr	(80)
VEN 2	Pr > 1.0 MeV	$18^d01^h\Delta 2$		168^h (incl. Event 131)	5.5 Pr	(103)

(PION 6 at 0.95 AU, ES − 1.5°; VEN 2 at 0.85 AU, ES − 1°)

Source: ● flare < 1029　N19　E27　2B　McM 8131

128	1966	January 18	$\sim 18^h$				$-2\,0\,0$
PION 6	Pr > 0.6 MeV	$18^h\Delta 5$	$19^d03^h\Delta 1$	14^h	100 Pr	(104)
PION 6	Pr > 7.5 MeV	$19^h\Delta 5$	$19^d03^h\Delta 1$			(125)
IMP 3	Pr > 0.9 MeV	$19^h\Delta 3$	$19^d00^h\Delta 1$	9^h	20 Pr	(80)
IMP 3	Pr > 6.5 MeV	no event				
PION 6	Pr > 13 MeV	$20^h\Delta 2$	$23^h\Delta 2$		0.009 **Pr**	**(104)**

Source: ◇ flux modulation of Event 127 (softer spectrum)
　　　　　◇ CMP of the cluster of regions McM 8130+31+33

129 1966 January 19 $\sim 00^h$ -200

IMP 3	El > 40 keV	$00^h \pm 100^m$	$06^h\Delta2$		200 El	(66)
IMP 3	Pr > 0.9 MeV	$04^h\Delta3$	$13^h\Delta1$	29^h	64 Pr	(80)
PION 6	Pr > 13 MeV	$07^h\Delta5$	$10^h\Delta2$	25^h	0.03 Pr	(104)
IMP 3	Pr > 6.5 MeV	$08^h\Delta5$	$10^h\Delta2$	25^h	0.04 Pr	(80)
PION 6	Pr > 7.5/>0.6 MeV	$09^h\Delta5$	$12^h\Delta1/19^h\Delta5$		0.18/138 Pr	(84/104)

Source: ● flare 18^d2253 N20 E07 2B McM 8131

PION 6	Pr > 0.6 MeV	$20^d05^h\Delta3$	$20^d12^h\Delta1$	20^h	586 Pr	(104)
PION 6	Pr > 13 MeV	$20^d06^h\Delta2$	$20^d09^h\Delta1$	14^h	0.013 Pr	(104)
PION 6	Pr > 7.5 MeV	$20^d06^h\Delta1$	$20^d09^h\Delta1$	25^h		(125)
IMP 3	El > 40 keV	$20^d06^h\Delta2$			200 El	(66)
IMP 3	Pr > 6.5/>0.9 MeV	$20^d09^h\Delta3$	$20^d10^h\Delta2$	$6^h/19^h$	0.02/148 Pr	(80)

Source: ▲ sc 20^d0203 starts a GMS

1966 Feb. 3, Feb. 7, Feb. 9, Feb. 12, Feb. 19, Feb. 22, Feb. 27, Feb. 28, March 10, March 17, March 18, March 19 (twice)   events in Appendix A

---

130 1966 March 19 22^h23^m -200

IMP 3	El > 40 keV	$22^h23^m \pm 02^m$	$22^h40^m\Delta1$		35 El	(66)
PION 6	Pr > 7.5 MeV	$23^h10^m\Delta1$	$20^d01^h\Delta1$	9^h	0.045 Pr	(84)
PION 6	Pr > 13 MeV	$23^h26^m \pm 05^m$	$20^d00^h13^m$	13^h	0.026 Pr	(70)
IMP 3	Pr > 0.9 MeV	$20^d00^h\Delta3$	$20^d07^h\Delta1$	10^h	2.0 Pr	(80)
PION 6	Pr > 0.6 MeV	$20^d00^h03^m \pm 05^m$	$20^d06^h08^m$	12^h	3.45 Pr	(70)

Source: ● flare 2131 N15 E15 1B McM 8207

---

131 1966 March 20 12^h -200

IMP 3	El > 40 keV	$>09^h43^m <12^h24^m$			500 El	(66)
PION 6	Pr > 7.5 MeV	$11^h30^m\Delta1$	$18^h\Delta1$	20^h	0.056 Pr	(84)
PION 6	Pr > 13 MeV	$11^h53^m \pm 10^m$	12^h28^m	16^h	0.03 Pr	(70)
PION 6	Pr > 44 MeV	$12^h\Delta1$	$14^h\Delta1$			(125)
IMP 3	Pr > 6.5 MeV	$12^h\Delta2$	$15^h\Delta2$	10^h	0.03 Pr	(80)
PION 6	Pr > 0.6 MeV	$13^h28^m \pm 20^m$	16^h48^m		6.2 Pr	(70)

Source: ● flare 0928 N21 E25 2B McM 8207

1966 Mar. 21   event in Appendix A

---

132 1966 March 21 19^h -200

IMP 3	Pr > 6.5/>0.9 MeV	$19^h\Delta2/\Delta3$	$22^h\Delta1/\Delta2$	$5^h/4^h$	0.03/11.3 Pr (80)
PION 6	Pr > 13/>0.6 MeV	$20^h\Delta5$	$21^h\Delta2/22^d00^h$	$4^h/$	0.017/10.3 Pr (104)
PION 6	Pr > 44 MeV	$21^h\Delta5$	$22^d00^h\Delta2$		(125)

(PION 6 at 0.88 AU, ES + 3.5°)

Source: ○ flare 1820 N21 W08 1B McM 8207
　　　　 ○ flare 1544 N21 W04 1N McM 8207

133 1966 March 22 00^h −2 0 0

IMP 3	El > 40 keV	>21^{d}23^{h}15^m<02^{h}20^m			200 El	(66)
IMP 3	Pr > 6.5 MeV	00^{h}Δ1	01^{h}Δ1	10^h	0.04 Pr	(80)
PION 6	Pr > 13 MeV	00^{h}23^m ± 15^m	03^{h}30^{m}Δ1	20^h	0.03 Pr	(70)
PION 6	Pr > 7.5 MeV	01^{h}Δ2	02^{h}Δ2		0.20 Pr	(125)
IMP 3	Pr > 0.9 MeV	Δ3	02^{h}Δ1	30^h	27.5 Pr	(80)

(PION 6 at 0.88 AU, ES + 4°)

Source: ● flare 21^{d}2138 N19 W10 2B McM 8207

PION 6	Pr > 13 MeV	22^{h}Δ2	23^{d}07^{h}Δ2	25^h	0.05 Pr	(104)
PION 6	Pr > 7.5 MeV	23^{h}Δ2	23^{d}06^{h}Δ2	25^h	0.34 Pr	(84)
IMP 3	Pr > 6.5/>0.9 MeV	23^{d}05^{h}Δ2/Δ3	23^{d}18^{h}Δ1	20^h/19^h	0.06/48 Pr	(80)
PION 6	Pr > 0.6 MeV	23^{d}05^{h}Δ3	23^{d}07^{h}Δ7	19^h	345 Pr	(104)
PION 6	Pr > 44 MeV	23^{d}06^{h}Δ1	23^{d}06^{h}Δ1	5^h		(125)
IMP 3	El > 40 keV	23^{d}12^{h}Δ2			500 El	(66)

Source: ▲ sc 23^{d}0745 starts a major GMS; gradual onset since 23^{d}00^h

---

134 1966 March 24 00^h −1 0 0

IMP 3	Pr > 19/>6.5 MeV	00^{h}Δ1	01^{h}30^m/02^{h}Δ1	3^h	0.16/0.13 Pr	(80)
PION 6	Pr > 7.5 MeV	00^{h}00^{m}Δ1	03^{h}Δ1		0.38 Pr	(84)
PION 6	Pr > 13 MeV	00^{h}18^m ± 05^m	01^{h}06^m	2^{h}43^m	0.6 Pr	(70)
PION 6	Pr > 44 MeV	01^{h}Δ1	02^{h}Δ1	2^h		(125)
IMP 3	Pr > 0.9 MeV	Δ3	01^{h}Δ1	2^h	26.2 Pr	(80)

Source: ⊙ flare 23^{d}2248 N16 W33 1N McM 8207

---

135 1966 March 24 02^{h}50^m 1 2 0

IMP 3	El > 40 keV	>01^{h}53^m<02^{h}41^m	03^{h}00^{m}Δ1	10^h	5700 El	(66)
RIOM	PCA	02^{h}48^m ± 15^m	04^{h}Δ1	38^h	1.8 dB	(83, 86)
OGO 1	P‐75/45 MeV	02^{h}50^m ± 05^m	03^{h}15^m/20^m	3^h/4^h		(105)
OGO 1	Pr 28/12 MeV	03^{h}00^m ± 05^m	03^{h}35^m/04^{h}25^m	5^h/6^h		(105)
PION 6	Pr > 175 MeV	03^{h}Δ1	03^{h}30^m	8^h		(104)
IMP 3	Pr > 90/>19 MeV	03^{h}Δ2/Δ3	04^{h}Δ2/06^{h}Δ7	15^h/57^h	0.26/6.1 Pr	(80)
PION 6	Pr > 13 MeV	03^{h}01^m ± 05^m	03^{h}44^m	70^h	43 Pr	(70)
PION 6	Pr > 73 MeV	03^{h}05^m ± 05^m	03^{h}30^m	20^h	0.78 Pr	(70)
PION 6	Pr > 0.6 MeV	04^{h}58^m ± 02^m	06^{h}08^m		379 Pr	(70)
IMP 3	Pr > 6.5/>0.9 MeV	Δ3	04^{h}Δ1/06^{h}Δ7	72^h/32^h	13.3/373 Pr	(80)

Source: ● flare 0225 N20 W42 2N McM 8207

1966	March 24	event in Appendix A

136	1966	March 25	$\sim 10^h$					−2 0 0

IMP 3	Pr > 0.9 MeV	$10^h\Delta3$	$11^h30^m\Delta1$	80^h	20 Pr	(80)
PION 6	Pr > 44/>7.5 MeV	$<15^h$	$15^h\Delta5$	$35^h/$	/0.05 Pr	(125/84)

(PION 6 at 0.86 AU, ES + 4.5°)

Source: ⊙ flare 0145　N20　W54　2N　McM 8207
　　　　○ flare 0513　N14　W54　2B　McM 8207 (type III)
　　　　△ weak sc 24^d2335 and $\sim25^d12^h$ associated with a minor GMS

	1966	March 26	event in Appendix A				

137	1966	March 27	$\sim16^h$				−2 0 0

PION 6	Pr > 13/>7.5 MeV	$17^h\Delta3/18^h\Delta1$	$20^h\Delta2/\Delta1$	0.017 Pr/	(104/125)
PION 6	Pr > 0.6 MeV	16^h	sudden decrease in intensity		
IMP 3	Pr > 6.5/>0.9 MeV	$18^h/$	very small/noise		(106)

(PION 6 at 0.86 AU, ES + 5°)

Source: △ sc 1935 starts a GMS (delayed event after flare $26^d < 1843$ in Appendix A?)
　　　　○ flare 1218　N18　W85　SN　McM 8207
　　　　○ flare 0805　N18　W82　1N　McM 8207

138	1966	March 28	14^h				−2 0 0

PION 6	Pr > 44/>7.5 MeV	$14^h\Delta2/15^h\Delta5$	$15^h/20^h\Delta2$	10^h	/0.026 Pr	(125/84)
PION 6	Pr > 13 MeV	$15^h25^m \pm 15^m$	17^h30^m	20^h	0.017 Pr	(70)

(IMP 3 in the magnetosphere)

Source: ○ flare 0442　N28　E80　1B　McM 8223 (type II)
　　　　○ flare 1459　N15　W90　1N　McM 8207 (type III)

A major GMS in progress.

	1966	March 31, Apr. 1, Apr. 12 (twice)	events in Appendix A				

139	1966	April 16	00^h				−(2) 0 0

PION 6	Pr > 44 MeV	$00^h\Delta1$	$04^h\Delta2$	10^h		(125)
PION 6	Pr > 0.6 MeV	$00^h\Delta2$	$09^h\Delta5$	7^d	0.45 Pr	(70)
PION 6	Pr > 13/>7.5 MeV	$01^h\Delta1/01^h15^m$	$04^h/05^h\Delta5$	20^h	0.01/0.005 Pr	(104/84)

(PION 6 at 0.84 AU, ES + 11°; no event on IMP 3)

Source: ⊙ flare 15^d0956　N19　E40　2B　McM 8262
　　　　◇ it starts the permanent particle flux associated with McM 8262

	1966	Apr. 25, Apr. 26	events in Appendix A				

140	1966 April 29	07^h				$-1\,0\,0$
IMP 3	$Pr > 0.9$ MeV	07^h slow rise	$20^h\Delta 2$	45^h	3.25 Pr	(80)
IMP 3	$El > 40$ keV	$07^h05^m \pm 10^m$	17^h35^m		35 El	(66)
IMP 3	$Pr > 19/>6.5$ MeV	$09^h\Delta 1/10^h\Delta 2$	$17^h/19^h\Delta 2$	$40^h/30^h$	0.09/0.07 Pr	(80)
PION 6	$Pr > 13/>0.6$ MeV	$10^h/12^h\Delta 5$	$/30^d04^h\Delta 2$	$40^h/70^h$	/14.7 Pr	(104)
PION 6	$Pr > 44/>7.5$ MeV	$<18^h\Delta 5$	$/23^h\Delta 1$		/0.24 Pr	(125/84)

(PION 6 at 0.82 AU, ES $+ 16°$)

Source: ⊡ McM 8272 is going over W limb; McM 8262~4 days beyond W limb

141	1966 May 2	12^h30^m				$0\,(1)\,0$
IMP 3	$Pr > 19$ MeV	$12^h30^m\Delta 1$	$14^h\Delta 4$	70^h	~1.6 Pr	(80)
IMP 3	$Pr > 6.5/>0.9$ MeV	$12^h30^m\Delta 1$	$15^h/13^h\Delta 1$	37^h	1.0/12.5 Pr	(80)
IMP 3	$Pr > 90$ MeV	$13^h\Delta 2$	$15^h\Delta 1$	20^h	0.22 Pr	(80)
VLF	PCA	13^h10^m				(83)

(Also seen at 13^h on Pioneer 6 at 0.82 AU, ES $+ 17°$, up to $Pr > 73$ MeV (84, 104, gaps in the data), and on OGO 1 (107). No >40 keV electrons on IMP 3 (66).)

Source: ■ 1216–1317 type IV (11); 1220–1320 active prominence at NW limb (Sacramento Peak Report); 1223–1237 type II (11)

142	1966 May 4	02^h				$-2\,0\,0$
IMP 3	$Pr > 6.5/>0.9$ MeV	$02^h\Delta 3/\Delta 8$	$03^h\Delta 2/\Delta 1$	$13^h/20^h$	0.09/21 Pr	(80)
IMP 3	$El > 40$ keV	$>02^h15^m < 03^h05^m$			160 El	(66)
PION 6	$Pr > 0.6$ MeV	$<12^h$	$<12^h$	$>41^h$	>6.9 Pr	(104)

(Enhancement slightly indicated at >19 MeV on IMP 3, and >13 MeV on PION 6.)

Source: ⊙ flare 0150 N28 W67 SB McM 8278

143	1966 May 6	01^h				$-1\,0\,0$
IMP 3	$Pr > 19$ MeV	$01^h\Delta 3$	$19^h\Delta 2$	25^h	0.04 Pr	(80)
IMP 3	$Pr > 6.5/>0.9$ MeV	$01^h\Delta 1$	$20^h\Delta 2/\Delta 1$	$37^h/95^h$	0.11/33 Pr	(80)
PION 6	$Pr > 7.5/>0.6$ MeV	$02^h\Delta 5$	$22^h\Delta 5/20^h\Delta 7$	$59^h/8^d$	0.7/60 Pr	(84/104)
PION 6	$Pr > 44/> 13$ MeV	$<06^h/<05^h$	$15^h/17^h\Delta 2$	$/72^h$	/0.02 Pr	(125/104)

(PION 6 at 0.82 AU, ES $+ 18°$; duration at low energies includes Event 144)

Source: uncertain (□?)

	1966 May 7	event in Appendix A				
144	1966 May 8					$-(2)\,0\,0$
PION 6	$Pr > 13/>44$ MeV	$05^h\Delta 3/07^h\Delta 2$	$08^h/07^h\Delta 2$	$30^h/8^h$	0.09 Pr/	(104/125)
IMP 3	$Pr > 6.5$ MeV	$06^h\Delta 2$	$08^h\Delta 2$	25^h	0.02 Pr	(80)

(PION 6 at 0.82 AU, ES $+ 19°$; Event 143 in progress at low energies)

Source: ○ flare 0429 N26 W83 SN McM 8279 (type II)

	1966	May 26, May 27	events in Appendix A				
145	1966	May 28	16^h14^m				$-(2)\,0\,0$
IMP 3	El > 40 keV	$16^h14^m \pm 03^m$	17^h00^m			52 El	(66)
IMP 3	Pr > 0.9 MeV	$\lceil\, 17^h$ slow increase	$30^d05^h\Delta2$	65^h		98 Pr	(80)
		$\lfloor\, 29^d11^h$ sharp rise					
PION 6	Pr $>44/>7.5$ MeV	$18^h\Delta2/18^h10^m$	$<30^d21^h$				(125, 84)
IMP 3	Pr >6.5 MeV	$29^d11^h\Delta2$	$30^d06^h\Delta2$	30^h		0.05 Pr	(80)

(PION 6 at 0.82 AU, ES + 28°)

Source: ● flare 1532   N15   W40   2B   McM 8310

	1966	June 2	event in Appendix A				
146	1966	June 25	15^h36^m				$-2\,0\,0$
IMP 3	El > 40 keV	$15^h36^m \pm 02^m$	17^h30^m			100 El	(66)
IMP 3	Pr $> 19/>6.5$ MeV	$16^h\Delta1$	$22^h/26^d01^h\Delta2$	$30^h/40^h$		0.15/0.11 Pr	(80)
IMP 3	Pr > 0.9 MeV	$16^h\Delta1$	$17^h\Delta1$	95^h		16 Pr	(80)
PION 6	Pr $>44/>7.5$ MeV	$<19^h$	$22^h\Delta2/\Delta1$			/0.01 Pr	(125, 84)

(PION 6 at 0.83 AU, ES + 38°)

Source: ● flare 1525   S24   W09   1B   McM 8348

	1966	June 28, July 3 (twice)	events in Appendix A				
147	1966	July 4	20^h				$-2\,0\,0$
IMP 3	Pr $> 19/>6.5$ MeV	$21^h/22^h\Delta2$	$5^d00^h/03^h\Delta2$	24^h		0.06/0.03 Pr	(80)
IMP 3	Pr > 0.9 MeV	$20^h\Delta2$	$5^d07^h\Delta2$	26^h		0.84 Pr	(80)
PION 6	$\lceil$ Pr $> 13/$	$<5^d11^h$	$<5^d11^h$				(106)
	$\lfloor >7.5/>0.6$ MeV						

Source: ◇ continuation of the 'permanent particle flux' from McM 8362 (see Appendix A)
　　　　○ flare 1925   N34   W16   SN   McM 8362
　　　　□ ?

148	1966	July 7	00^h47^m				1 2 1
Pr > 1.5 GeV	GLE (141)	$00^h42^m \pm 02^m$					Sanae
			$01^h30^m \pm 08^m$			$2.5 \pm 0.3\%$	$\lceil$ Kerguelen
							$\lfloor$ Isl.
				$6^h \pm 1^h$			$\lceil$ Leeds,
							$\lfloor$ McMurdo
IMP 3	El > 3 MeV	$00^h58^m \pm 02^m$	01^h30^m	10^h		3.0 El	(67)
IMP 3	Pr $>90/>19$ MeV	$01^h\Delta1$	$16^h/14^h\Delta7$	50^h		$\sim1.1/15$ Pr	(80/65)
IMP 3	Pr $>6.5/>0.9$ MeV	$01^h\Delta1$	$10^h\Delta1/8^d03^h\Delta7$	50^h		$18.9/\sim750$ Pr	(80)
RIOM	PCA	01^h	12^h	44^h		2.1 dB	(146)
IMP 3	El > 40 keV	$01^h05^m \pm 05^m$	01^h30^m			1500 El	(66)
Balloon	Pr > 100 MeV	01^h10^m	02^h05^m			2.6 Pr	(110)
RIOM	PCA	01^h20^m	$13^h00^m/12^h20^m$	47^h		2.5/2 dB	(111/112)
OGO 3	Pr 32/12 MeV	$01^h30^m\Delta1$	$06^h\Delta2/12^h\Delta4$			$2.5*/10*$ Pr	(108)
OGO 3	Pr 8/5 MeV	$01^h30^m\Delta1$	$12^h\Delta4/8^d04^h\Delta2$			$10*/40*$ Pr	(108)

(continues on next page)

148 continued

EXPL 33	Pr > 0.5 MeV	$03^h\Delta1$	$8^d09^h\Delta1$			(87)

(Summary of the event in (113).)

Source: ● flare 0025   N35   W48   2B   McM 8362

OGO 3	Pr 11/7 MeV	$18^h/19^h\Delta4$	$8^d01^h/03^h\Delta2$	15^h	6*/22*c	(114)
OGO 3	Pr 3 MeV	$18^h\Delta4$	$8^d05^h\Delta2$	30^h	200*c	(114)
IMP 3	El > 40 keV	$8^d02^h\Delta3$	$8^d07^h\Delta2$	24^h	1500 El	(66)
EXPL 33	El > 45 keV	$8^d06^h\Delta1$	$8^d09^h\Delta1$	4^h	1500 El	(114)

(See also maxima at low-energy channels above.)

Source: ♦ 8^d02^h gradual onset of a GMS, due to a 'halo' effect (114) or a sector boundary crossing (115); sc 8^d2102

149	1966 July 9	~05^h				0 0 0
IMP 3	Pr > 19/>6.5 MeV	$05^h\Delta3$	$07^h\Delta1$	60^h	0.65/0.65 Pr	(80)
IMP 3	Pr > 0.9 MeV	$05^h\Delta3$	$17^h\Delta2$	100^h	149 Pr	(80)
PION 6	Pr > 13/>7.5 MeV	$<10^h/09^h50^m\Delta1$	/$12^h\Delta2$		0.87/3.0 Pr	(104/84)

(PION 6 at 0.86 AU, ES + 43°)

Source: ● flare 0310   N35   W75   2B   McM 8362

150	1966 July 14	01^h				−1 0 0
EXPL 33/IMP 3	Pr > 15 MeV	01^h06^m				(69)
IMP 3	El > 40 keV	$<01^h40^m$			15 El	(66)
IMP 3	Pr > 19/>6.5 MeV	$<02^h$	$6^h/10^h\Delta2$	$35^h/25^h$	0.13/0.09 Pr	(80)
IMP 3	Pr > 0.9 MeV	$<02^h$	$10^h\Delta2$	45^h	2.4 Pr	(80)
PION 6	Pr > 13/>0.6 MeV	$<10^h$	$10^h/16^h\Delta1$		0.17/4.3 Pr	(104)

(PION 6 at 0.87 AU, ES + 44°)

Source: ▢ McM 8362 about 3 days beyond W limb

151	1966 July 16					−1 0 0
IMP 3	El > 40 keV	$>22^h40^m<23^h45^m$			60 El	(66)
EXPL 33/IMP 3	Pr > 15 MeV	$23^h00^m\Delta1$				(69)
IMP 3	Pr > 0.9 MeV	$23^h\Delta1$	$17^d12^h\Delta2$	82^h	26 Pr	(80)
IMP 3	Pr > 19/>6.5 MeV	$17^d00^h\Delta1$	$17^d07^h/08^h\Delta2$	$37^h/55^h$	0.20/0.28 Pr	(80)

Source: ▢ 2050 radio continuum + type III, possibly in McM 8362 (116)

	1966 July 25	event in Appendix A	

152	1966 July 29	$\sim 17^h$					$-1\,0\,0$
IMP 3	Pr > 0.9 MeV	$17^h\Delta2$	$31^d20^h\Delta1$	164^h	348 Pr	(80)	
IMP 3	El > 45 keV	$30^d05^h\Delta2$	$31^d20^h\Delta1$	4^d	300 El	(117)	
EXPL 33	Pr > 15 MeV	$30^d06^h\Delta2$	31^d20^h			(117)	
IMP 3	Pr > 19/>6.5 MeV	$30^d17^h\Delta2$	$31^d19^h\Delta2/\Delta1$	$40^h/85^h$	0.05/0.26 Pr	(80)	
IMP 3	Pr > 0.5 MeV	$30^d21^h\Delta2$	$31^d20^h\Delta1$	4^d	1000 Pr	(117)	
IMP 3	Pr > 90 MeV	$31^d00^h\Delta2$	$31^d19^h\Delta2$	35^h	0.05 Pr	(80)	
IMP 3	El > 4 MeV	present				(118)	

Source: ⊙ flare 28^d2214 N36   E33   3B   McM 8413

     ◇ its delayed particles start the 'permanent particle flux' associated with McM 8413 + 8414 (also see preceding event in Appendix A)

	1966 Aug. 6, Aug. 18, Aug. 23, Aug. 27		events in Appendix A				

153	1966 August 28	15^h30^m					$1\,2\,0$
PION 7	Pr > 165 MeV	$15^h30^m\Delta1$	$17^h\Delta1$	12^h	0.69 Pr	(119)	
IMP 3	El > 40 keV	$15^h31^m\pm05^m$	16^h10^m		1200 El	(66)	
		$15^h49^m\pm02^m$					
IMP 3	Pr > 90 MeV	$16^h\Delta1$	$19^h\Delta2$	18^h	0.81 Pr	(80)	
PION 7	Pr > 73 MeV	$16^h\Delta1$	$17^h\Delta1$	35^h	0.95 Pr	(119)	
IMP 3	Pr > 19/>6.5 MeV	$16^h\Delta1$	$21^h\Delta7/19^h\Delta1$	$25^h/13^h$	6.7/12.2 Pr	(80)	
IMP 3	Pr > 0.9 MeV	$16^h\Delta1$	$29^d00^h\Delta7$	15^h	~745 Pr	(80)	
PION 7	Pr > 7.2/>0.6 MeV	$16^h07^m/16^h\Delta1$	$21^h/20^h\Delta5$		15/129 Pr	(84/119)	
RIOM	PCA	16^h10^m			< 1/4.0 dB	(83/152)	
IMP 3	El > 3 MeV	16^h15^m	16^h50^m	7^h	0.25 El	(67)	
PION 7	Pr > 64/>47 MeV	$<17^h$	$18^h/19^h\Delta1$			(125)	
(PION 7 at 1.04 AU, ES = 0°)							
RIOM	PCA	19^h	30^d01^h	56^h	2.4 dB	(146)	

Source: ● flare 1523   N22   E05   3B   McM 8461

PION 7	Pr > 64/>47 MeV	$29^d07^h\Delta5$	$29^d21^h\Delta5/\Delta7$			(125)	
IMP 3	Pr > 6.5/>0.9 MeV	$29^d07^h\Delta3$	⌐$29^d22^h\Delta2/$	85^h	11/745 Pr	(80)	
			⌐$24^h\Delta7$				
PION 7	Pr > 7.2 MeV	$29^d07^h10^m$	$29^d16^h\Delta2$		101 Pr	(84)	
PION 7	Pr > 0.6 MeV	$29^d07^h\Delta5$	$30^d08^h\Delta7$	100^h	1379 Pr	(119)	
IMP 3	Pr > 90 MeV	$29^d10^h\Delta2$	$30^d03^h\Delta3$	45^h	0.67 Pr	(80)	
IMP 3	El > 40 keV	$29^d13^h\Delta2$			3700 El	(66)	
IMP 3	Pr > 0.5 MeV	$29^d19^h\Delta1$	$30^d07^h\Delta1$	35^h	11000 Pr	(66)	
RIOM	PCA	see the maximum above				(146)	

Source: ▲ sc $29^d13^h15^m$ starts a major GMS

     ⊘ flare 29^d1324 N21   W11   1N   8461

     (Summary of the event in (120).)

	1966 Aug. 31	event in Appendix A	

154	1966 September 2						230

IMP 3	El > 40 keV	$>05^h50^m < 09^h00^m$				300 El	(69)
RIOM	PCA	$06^h00^m/08^h$	$12^h30^m/3^d13^h$	108^h		13.0/14.0 dB	(152, 146)
IMP 3	Pr $>90/>19$ MeV	$06^h/06^h30^m\Delta4$	$11^h/08^h\Delta7$	$55^h/51^h$		1.0/30 Pr	(80/65)
IMP 3	Pr $>6.5/>0.9$ MeV	$06^h\Delta4$	$10^h/09^h\Delta7$	$50^h/51^h$		111/744 Pr	(80)
EXPL 33/IMP 3	Pr > 15 MeV	06^h50^m	21^h	8^d			(69)
VLF + RIOM	PCA	$06^h55^m(83)$	$23^h30^m(120)$	7^d		12.0 dB	(83, 120)
PION 7	Pr $>73/>165$ MeV	$<07^h/<11^h$	$11^h\Delta7/15^h\Delta2$	40^h		/14 Pr	(119)
Balloon	Pr > 120 MeV	$08^h\Delta1$	$19^h\Delta1$	120^h		7.0 Pr	(43)

Source: ● flare 0542   N24   W56   3B   McM 8461
△ sc 0823 starts a major GMS
(Summary of the event in (120).)

155	1966 September 4	07^h				(1) (0) 0

PION 7	Pr $>165/>73$ MeV	$07^h\Delta5/\Delta1$	$09^h\Delta1/12^h\Delta7$	$10^h/9^h$	⌈0.009/	(119)
					⌊0.014 Pr	
IMP 3	Pr $>6.5/>19$ MeV	$08^h/09^h\Delta3$	$10^h\Delta1/\Delta7$	$13^h/12^h$	39/6.6 Pr	(80)
IMP 3	Pr >0.9 MeV	$09^h\Delta3$	$11^h\Delta7$	12^h	743 Pr	(80)

(PION 7 at 1.05 AU, ES $= 0°$)
The reality of this event is open to some doubt due to data overflow both on IMP 3 and PION 7. PCA continues.

Source: ● flare 0407   N21   W87   3N   McM 8461

156	1966 September 4	$\sim17^h$				(1) (0) 0

PION 7	Pr $>165/>73$ MeV	$17^h\Delta3/\Delta2$	$<04^h/00^h\Delta2$	$30^h/15^h$	0.17/0.06 Pr	(119)
IMP 3	Pr >19 MeV	$\lesssim21^h$	$5^d05^h\Delta7$	130^h	6.6 Pr	(80)
IMP 3	Pr >6.5 MeV	$\lesssim21^h$	$22^h\Delta1$	140^h	12.2 Pr	(80)
IMP 3	Pr >0.9 MeV	$\lesssim21^h$	⌈22^h and	12^h	12.5 Pr	(80)
			⌊$5^d03^h\Delta1$			

The reality of this event is open to some doubt due to data overflow both on IMP 3 and PION 7. PCA continues.

Source: ☉ flare 1430   N22   W88   SN   McM 8461

	1966 Sept. 11	event in Appendix A		
157	1966 September 12	slow rise		-200

PION 7	Pr $>0.6/>12$ MeV	$00^h\Delta2/04^h\Delta5$		$/54^h$		(119)

(Continuous increase until the onset of Event 166; PION 7 at 1.05 AU, ES $- 1°$.)

IMP 3	Pr >0.9 MeV	13^d08^h	$14^d09^h\Delta2$	27^h	19.4 Pr	(80)

(A very gradual increase for the previous 30 hr.)

IMP 3	Pr $>19/>6.5$ MeV	$13^d12^h/13^h$	$14^d10^h\Delta2$	$23^h/22^h$	0.06/0.08 Pr	(80)

(Continuous increase to Event 166.)

Source: ☉ flare 0925   N12   E90   1N   McM 8505
□ possibly another flare prior to this one beyond E limb
◇ it starts a 'permanent particle flux' associated with McM 8505

158	1966 September 14	10^h30^m					0 0 0
PION 7	Pr > 165 MeV	$10^h30^m\Delta1$	$12^h\Delta1$			0.14 Pr	(119)
PION 7	Pr > 12/>0.6 MeV	$10^h30^m\Delta3/\Delta5$	$13^h/11^h\Delta5$	$<17^h$		2.4/69 Pr	(119)
IMP 3/AIMP 1	Pr > 15 MeV	10^h35^m					(69)
IMP 3	El > 3 MeV	10^h45^m	13^h00^m			0.4 El	(67)
IMP 3	El > 40 keV	$<11^h$				1500 El	(69)
IMP 3	Pr > 90/>19 MeV	$11^h\Delta2/\Delta3$	$13^h\Delta2/15^h\Delta1$	$12^h/11^h$		0.16/1.6 Pr	(80)
PION 7	Pr > 73/>7.2 MeV	$11^h\Delta1$	$12^h\Delta1/13^h\Delta7$			0.12/24 Pr	(119/84)
IMP 3	Pr > 6.5/0.9 MeV	$11^h\Delta3$	$16^h/15^h\Delta1$	11^h		2.4/250 Pr	(80)

Source: ● flare 1014   S21   W90   SN   McM 8484
△ sc 1511 (start of a minor GMS) might be related to the maximum flux

159	1966 September 14	22^h					0 1 0
IMP 3	Pr > 19/>6.5 MeV	$22^h\Delta3$	$15^d02^h/04^h\Delta2$	61^h		0.83/2.8 Pr	(80)
IMP 3	Pr > 0.9 MeV	$22^h\Delta3$	$15^d05^h\Delta7$	60^h		747 Pr	(80)
IMP 3	Pr > 90 MeV	$23^h\Delta2$	$15^d10^h\Delta2$	31^h		0.28 Pr	(80)
PION 7	Pr > 12/>0.6 MeV	$<15^d03^h$	$\lceil15^d04^h\Delta2/$ $\lfloor09^h\Delta7$	$120^h/$		2.7/862 Pr	(119)
RIOM	PCA		15^d02^h			1.2 dB	(152)

Source: ⊡ McM 8484 close behind W limb

160	1966 September 17	10^h08^m					−1 0 0
IMP 3	Pr > 0.9 MeV	$10^h\Delta3$	$11^h\Delta1$	61^h		162 Pr	(80)
IMP 3	El > 40 keV	$10^h08^m\pm03^m$				3000 El	(69)
IMP 3	Pr > 19/>6.5 MeV	$11^h\Delta3$	$12^h\Delta2/\Delta1$	$65^h/42^h$		0.13/0.32 Pr	(80)
PION 7	Pr > 64/>47 MeV	$11^h/12^h\Delta3$	$12^h\Delta1$				(125)

(PION 7 at 1.06 AU, ES − 1°)

Source: ● flare 0940   N24   W63   2N   McM 8496

161	1966 September 20						−1 0 0
IMP 3	Pr > 19 MeV	$19^h\Delta2$	$21^h\Delta2$	20^h		0.11 Pr	(80)
PION 7	Pr > 12 MeV	$19^h\Delta3$	$21^d02^h\Delta1$	90^h		0.28 Pr	(119)
PION 7	Pr > 64/>47 MeV	$20^h\Delta1$	$21^d01^h\Delta1$				(125)
IMP 3	Pr > 6.5 MeV	$20^h\Delta1$	$21^d02^h\Delta1$	25^h		0.18 Pr	(80)
PION 7	Pr > 7.2 MeV	$21^h\Delta3$	$21^d08^h\Delta5$			0.52 Pr	(125)
IMP 3	Pr > 0.9 MeV	$21^h\Delta2$	$22^d08^h\Delta2$	115^h		84 Pr	(80)
PION 7	Pr > 0.6 MeV	$21^h\Delta3$	$22^d21^h\Delta6$	150^h		293 Pr	(119)

Source: ● flare < 1738   N06   W14   2B   McM 8505

162   1966   September 25 18^h $-2\,0\,0$

PION 7	Pr > 12 MeV	$18^h\Delta1$	$26^d16^h\Delta2$	44^h	0.10 Pr	(119)
IMP 3	Pr > 0.9 MeV	$18^h\Delta4$	$23^h\Delta1$	40^h	4.25 Pr	(80)
PION 7	Pr > 64/>47 MeV	$19^h\Delta2$	$26^d04^h/06^h\Delta2$	45^h		(125)
IMP 3	Pr > 6.5/>19 MeV	$19^h\Delta4/20^h\Delta2$	$26^d14^h\Delta2$	40^h	0.08/0.08 Pr	(80)
PION 7	Pr > 0.6 MeV	$26^d00^h\Delta3$	$26^d05^h\Delta2$	40^h	6.9 Pr	(119)

Source: □ 1312–1340 type IV, 1316–1336 type II, 1308–1312 type III, McM 8505 is going over W limb
◇ 'permanent' proton flux from McM 8514 (return of McM 8461), situated near CM; a weak gradual GMS starts at ∼20^h

163   1966   September 27 15^h $-1\,0\,0$

IMP 3/AIMP 3	Pr > 15 MeV	$14^h55^m\pm12^m$				(69)
PION 7	Pr > 12 MeV	$14^h\Delta3$	$19^h\Delta1$	70^h	0.60 Pr	(119)
IMP 3	Pr > 19 MeV	$15^h\Delta1$	15^h30^m	12^h	0.73 Pr	(80)
IMP 3	Pr > 6.5/>0.9 MeV	$15^h\Delta1$	$16^h/19^h\Delta1$	$20^h/45^h$	1.1/50 Pr	(80)
PION 7	Pr > 64/>47 MeV	$16^h\Delta3$	$17^h\Delta1/20^h\Delta5$	$10^h/25^h$		(125)
PION 7	Pr > 7.2/>0.6 MeV	$16^h\Delta3$	$20^h\Delta5/23^h\Delta1$	$30^h/40^h$	1.0/67 Pr	(84/119)

Source: □ 1313–1443 type IV, 1315 X-rays, 1400–2100 loops at NW limb, McM 8505

1966   Oct. 4, Oct. 19, Oct. 20 (twice), Oct. 24, Oct. 30, Nov. 1, Nov. 2, Nov. 3, Nov. 4, Nov. 6, Nov 16, Nov. 20, Nov. 30   events in Appendix A

164   1966   December 11 13^h–18^h $-(2)\,0\,0$

PION 7	Pr > 0.6 MeV	$13^h\Delta2$	$18^h\Delta1$	26^h	14.6 Pr	(119)
PION 7	Pr > 7.2 MeV	$18^h\Delta1$	$19^h\Delta1$	20^h		(125)
IMP 3	Pr > 0.9 MeV	$19^h\Delta1$	$22^h\Delta1$	29^h	2.7 Pr	(80)

(PION 7 at 1.11 AU, ES − 11°)

Source: ⊙ flare 0537   N18   E77   1N   McM 8612
○ flare 10^d2305   N21   E90   SN   McM 8612 (type II)

1966   Dec. 12   event in Appendix A

165   1966   December 13 $\geq00^h$ $-2\,0\,0$

IMP 3	El > 45 keV	$00^h\Delta2$	$10^h\Delta2$	24^h	70 El	(117)
IMP 3	Pr > 0.9 MeV	$02^h\Delta3$	$13^h\Delta1$	140^h	249 Pr	(80)
IMP 3	Pr > 0.5 MeV	$03^h\Delta2$	$20^h\Delta2$	48^h	720 Pr	(117)
IMP 3	Pr > 6.5 MeV	$13^h\Delta2$	$19^h\Delta2$	15^h	0.05 Pr	(80)
EXPL 33	Pr > 15 MeV	$16^h\Delta2$	$17^h\Delta2$	12^h		(117)

Source: ◇ a gradual GMS starts at 01^h (cf. the preceding PION 7 event in Appendix A)
◇ several bright active regions transit the CM on Dec. 10^d–13^d (McM 8606+07+09+10)

1966   Dec. 19 (twice), Dec. 22, Dec. 29, Dec. 31 (four times)   events in Appendix A

	1967	Jan. 3	event in Appendix A				

166	1967	January 5	05^h				$-2\,0\,0$
PION 7	$\lceil$Pr $> 64\,MeV/$ $\lfloor> 47\,MeV$	$05^h\Delta2$	$16^h\Delta1/\Delta2$			35/77 c	(84)
IMP 3	Pr $> 19\,MeV$	$06^h\Delta2$	$22^h\Delta2$	60^h		0.02 Pr	(80)
PION 7	Pr $> 12\,MeV$	06^h	06^d07^h	56^h		0.06 Pr	(119)
IMP 3	Pr $> 6.5/>0.9\,MeV$	$06^h\Delta2$	$13^h/15^h\Delta2$	$60^h/56^h$		0.02/11.5 Pr	(80, 84)

(PION 7 at 1.13 AU, ES $-$ 13.5°)

Source: ○ flare < 0117   S26   E34   2F   McM 8632
          ○ flare 0440   N16   E03   1N?   McM 8631

167	1967	January 7	14^h				$-2\,0\,0$
PION 7	Pr $> 12\,MeV$	$14^h\Delta2$	$8^d00^h\Delta5$	40^h		0.035 Pr	(119)
IMP 3	Pr $> 0.9\,MeV$	$14^h\Delta3$	$8^d02^h\Delta2$	60^h		111.5 Pr	(80)

(PION 7 at 1.13 AU, ES $-$ 14°)

Source: ○ flare 0540   S24   W77   1N?   McM 8629
          ◇ GMS begins with sc, 7^d08^h

	1967	Jan. 8	event in Appendix A				

168	1967	January 11	02^h39^m				$-1\,0\,0$
IMP 3	El $> 40\,keV$	$02^h39^m \pm 1^m$	03^h			350 El	(69)
IMP 3	Pr $> 19\,MeV$	03^h	04^h	19^h		0.44 Pr	(80)
PION 7	Pr $> 12\,MeV$	$<05^h$	15^h	$>17^h$		0.35 Pr	(119)
PION 7	Pr $>64/>47\,MeV$	$06^h\Delta2$	$11^h\Delta2/14^h\Delta5$			43/134 c	(165)
IMP 3	Pr $> 90\,MeV$	$07^h\Delta2$	$20^h\Delta2$	15^h		0.05 Pr	(80)
IMP 3	Pr $>6.5/>0.9\,MeV$	$07^h\Delta3$	$14^h\Delta2/21^h\Delta2$	14^h		0.22/37.5 Pr	(80)

(PION 7 at 1.13 AU, ES $-$ 15°)

Source: ☉ flare 0131   S26   W47   3B?   McM 8632

169	1967	January 11	$<21^h$				$-1\,0\,0$
IMP 3	Pr $>6.5/>0.9\,MeV$	$21^h\Delta3$	$12^d02^h\Delta1/\Delta7$	$40^h/90^h$		2.2/749 Pr	(80)
IMP 3	Pr $>90/>19\,MeV$	$22^h\Delta2/\Delta3$	$\lceil12^d02^h\Delta2/$ $\lfloor01^h\Delta1$	$20^h/30^h$		0.28/0.33 Pr	(80)
PION 7	Pr $> 12\,MeV$	$22^h\Delta3$	$12^d01^h\Delta1$	80^h		0.30 Pr	(119)
PION 6	Pr $> 13\,MeV$	$12^d00^h\Delta3$	$12^d06^h\Delta2$	35^h		0.28 Pr	(104)
IMP 3	Pr $> 0.5\,MeV$	$12^d02^h\Delta2$	$12^d15^h\Delta2$	48^h		16000 Pr	(117)

(PION 7 at 1.13 AU, ES $-$ 15°, PION 6 at 0.91 AU, ES $+$ 55.5°)

Source: ☉ flare 2016   N16   W88   SB   McM 8631
          No geomagnetic disturbance

	1967	Jan. 14	event in Appendix A				

170	1967	January 16	$08^h 30^m$						$-1\,0\,0$
IMP 3		$El > 40\,keV$	$08^h 32^m \pm 07^m$				35 El	(69)	
IMP 3		$Pr > 0.9/> 19\,MeV$	$09^h \Delta 2/10^h$	$22^h/17^d 13^h \Delta 2$	$19^h/30^h$	0.03 Pr	(80)		

(This max. includes next event in Appendix A)

| IMP 3 | | $Pr > 6.5\,MeV$ | $11^h \Delta 1$ | $20^h \Delta 2$ | 18^h | 0.04 Pr | (80) | |
| PION 7 | | $Pr > 12\,MeV$ | $12^h \Delta 5$ | $16^h \Delta 1$ | 40^h | 0.13 Pr | (119) | |

(PION 7 at 1.13 AU, ES $-$ 16°)

Source: □ McM regions 8631 and 8632 respectively 3–2 days behind W limb

	1967	Jan. 17	event in Appendix A					

171	1967	January 18	20^h					$-2\,0\,0$
IMP 3		$Pr > 0.9\,MeV$	$20^h \Delta 2$	$19^d 04^h/22^h \Delta 2$		1.25/2.5 Pr	(80)	
PION 7		$Pr > 12\,MeV$	$21^h 30^m$	$19^d 02^h$	24^h	0.044 Pr	(119)	
PION 7		$Pr > 64/> 47\,MeV$	$22^h \Delta 5$	$23^h \Delta 2$		43.8/103 c	(165)	
IMP 3		$Pr > 19/> 6.5\,MeV$	$23^h \Delta 2/\Delta 1$	$03^h \Delta 2$	12^h	0.028/0.04 Pr	(80)	

Source: ○ flare 0616　N17　E57　1N　McM 8650
　　　　□ McM regions 8631 and 8632 behind W limb 4–5 days

172	1967	January 20	00^h					$-2\,0\,0$
IMP 3		$Pr > 19/> 0.5\,MeV$	$00^h \Delta 1/\Delta 2$	$\begin{bmatrix} 20^d 05^h \Delta 2/ \\ 22^d 00^h \Delta 2 \end{bmatrix}$	$18^h/72^h$	0.04/140 Pr	(80, 117)	
PION 7		$Pr > 12\,MeV$	00^h	$02^h \Delta 2$	48^h	0.05 Pr	(119)	
IMP 3		$Pr > 6.5/> 0.9\,MeV$	$00^h \Delta 1/\Delta 3$	$06^h/05^h \Delta 2$	$20^h/90^h$	0.17/49 Pr	(80)	

(PION 7 at 1.13 AU, ES $-$ 16.5°)

Source: □ active region 8644 two days beyond W limb
　　　　◇ passage at CM of region 8654
　　　　○ flare 19^d 1717　N17　E67　SB　McM 8659

173	1967	January 28	$\leq 02^h$					$0\,1\,1$
IMP 3		$Pr > 0.9\,MeV$	$02^h \Delta 1 \Delta 8$	$06^h \Delta 1$	6^h	25 Pr	(80)	
IMP 3		$El > 40\,keV$	$02^h 08^m \pm 09^m$			450 El	(69)	
MAR 4		$Pr > 40\,MeV$	$02^h 15^m$	$05^h \Delta 2$	$> 6^h$	1 c	(170)	
IMP 3		$Pr > 90/> 19\,MeV$	$03^h \Delta 1$	$04^h/06^h \Delta 1$	$5^h/6^h$	0.11/1.67 Pr	(80)	
IMP 3		$Pr > 6.5\,MeV$	$03^h \Delta 1$	$07^h \Delta 1$	5^h	1.33 Pr	(80)	
$Pr > 3.6\,GeV$		GLE (141)	$03^h 02^m \pm 03^m$				Leeds	

Precursor to major increase (event 174)

ATS 1		Pr 35/16 MeV	$03^h 30^m/04^h \Delta 2$	$06^h \Delta 1$	5^h	$\begin{bmatrix} 0.04*/ \\ 0.14* \end{bmatrix}$ Pr	(167)	
PION 7		$Pr > 64/> 47\,MeV$	$03^h 30^m \pm 30^m$				(165)	
IMP 3		$El > 3\,MeV$	$04^h \Delta 2$	$07^h \Delta 2$	$> 4^h$	0.05 El	(67)	
RIOM		PCA	$04^h \Delta 2$			0.3 dB	(152)	

(PION 7 at 1.13 AU, ES $-$ 18°; MAR 4 at 1.59 AU, ES $+$ 82°)

(continues on next page)

173 continued

Source: □ possibility of region 8654 one day behind W limb or region 8687, four days beyond limb
○ flare 0158   N19   W65   SF   McM 8650

174	1967 January 28	08^h				1 3 3
VELA 2	Pr > 25 MeV	08^{h}10^m ± 10^m	17^{h}15^m	11^d	1300 c	(213)
IMP 3	Pr >90/>19 MeV	08^h/09^{h}Δ3	Δ7	132^h	1.11/6.67 Pr	(80)
IMP 3	Pr >6.5/>0.9 MeV	08^{h}Δ3	20^{h}Δ7/Δ7	132^h/70^h	100/749 Pr	(80)
Pr > 3.6 GeV	GLE (141)	08^{h}10^m ± 10^m				⎡ Mt Welling-
						⎣ ton, Sanae
			11^{h}00^m ± 15^m	~24^h		several
					21 ± 1%	McMurdo
Balloon	⎡ Pr > 300/	/08^{h}26^m ± 02^m	13^{h}Δ1	90^h/80^h	5/60 Pr	(43, 171)
	⎣ > 100 MeV					
IMP 3	⎡ El > 3 MeV/	08^{h}30^{m}Δ3/	10^{h}Δ2/		1/1000 El	(67, 69)
	⎣ > 40 keV	08^{h}25^m ± 03^m				
ATS 1	Pr 35/17 MeV	08^{h}30^m/09^{h}Δ1	18^{h}Δ2	130^h	0.7*/15* Pr	(167)
PION 7	Pr > 165/>73 MeV	09^h/10^{h}Δ5	12^h/13^{h}Δ5	120^h	6.9/4.3 Pr	(119)
RIOM	PCA	09^{h}Δ3	18^{h}Δ2	131^h	7 dB	(152)

Source: □ region 8687 is 60° beyond W limb (219, 220, 221, 222, 223)
○ flare 0743   S23   E19   SB   McM 8667

	1967	Jan. 31	event in Appendix A			
175	1967	February 2	20^h			1 2 0
VELA 2	Pr > 25 MeV	19^{h}37^m ± 05^m	23^{h}45^m	48^h	83 c	(213)
IMP 3	El > 40 keV	20^{h}00^m ± 15^m			250 El	(69)
PION 7	Pr > 165 MeV	20^{h}Δ1	3^{d}01^{h}Δ2	35^h	1.7 Pr	(119)
IMP 3	Pr >90/>19 MeV	20^{h}Δ3	3^{d}07^{h}Δ7/Δ7	40^h/11^d	1.11/6.65 Pr	(80)
IMP 3	Pr >6.5/>0.9 MeV	20^{h}Δ3	3^{d}05^{h}Δ2/Δ7	11^d	31.1/749 Pr	(80)
RIOM	PCA	20^{h}Δ3	3^{d}06^{h}Δ2	130^h	2.6 dB	(152)

(PION 7 at 1.12 AU, ES − 19.5°)

Source: □ region 8659 behind W limb, type II at 1838–1853
○ flare 1945   N27   E47   SN   McM 8680

	1967	Feb. 7	event in Appendix A			

176 1967 February 13 19^h $-1\ 1\ 0$

IMP 3	El > 40 keV	$18^h42^m \pm 02^m$			500 El	(69)
IMP 3	El > 3 MeV	$19^h\Delta2$			0.05 El	(67)
IMP 3	Pr > 19/>6.5 MeV	$19^h\Delta3$	$14^d01^h\Delta5$	$47^h/49^h$	0.44/0.63 Pr	(80)
PION 7	Pr > 12/>73 MeV	$19^h\Delta3/14^d00^h\Delta5$	$14^d11^h/04^h\Delta2$	$44^h/48^h$	2.6/0.05 Pr	(119)
IMP 3	Pr > 0.9 MeV	$19^h\Delta3$	$15^d07^h\Delta2$	48^h	186 Pr	(80)
RIOM	PCA				0.5 dB	(152)

Source: ● flare 1747   N21   W11   3B   McM 8687

PION 7	Pr > 12 MeV	$15^d15^h\Delta3$	$20^h\Delta5$	70^h	1.5 Pr	(119)
IMP 3	Pr > 19/>0.9 MeV	$15^d18^h/19^h\Delta3$	$\lceil16^d01^h\Delta2/$	$40^h/130^h$	0.13/748.6 Pr	(80)
			$\lfloor05^h\Delta7$			
IMP 3	Pr > 6.5/>90 MeV	$15^d20^h\Delta3/23^h\Delta2$	$16^d00^h/01^h\Delta2$	$50^h/10^h$	0.41/0.18 Pr	(80)
RIOM	PCA		16^d02^h		0.5 dB	(169)
(PION 7 at 1.12 AU, ES − 22°)						

Source: ▲ sc storm modulation, sc begins 15^d2348

177 1967 February 27 17^h $0\ 0\ 0$

IMP 3	El > 40 keV	$17^h30^m \pm 05^m$			400 El	(69)
IMP 3	El > 3 MeV	$17^h\Delta1$	$19^h\Delta1$		0.6 El	(67)
PION 7	Pr > 165/>12 MeV	$17^h\Delta1$	$\lceil19^h\Delta1/$	$40^h/6^d$	1.15/6.1 Pr	(119)
			$\lfloor28^d01^h\Delta2$			
IMP 3	Pr > 90/>19 MeV	$17^h\Delta1$	$22^h/28^d01^h\Delta2$	$50^h/150^h$	0.56/2.2 Pr	(80)
IMP 3	Pr > 6.5/>0.9 MeV	$17^h\Delta1/\Delta8$	$21^h\Delta2/20^h\Delta1$	$150^h/160^h$	1.09/117.4 Pr	(80)
(PION 7 at 1.13 AU, ES − 24°)						

Source: ● flare 1637   N27   E02   2N   McM 8704

178 1967 March 9 02^h $-2\ 0\ 0$

IMP 3	El > 40 keV	$01^h19^m \pm 18^m$			55 El	(69)
IMP 3	Pr > 0.9/>19 MeV	$01^h30^m\Delta8/02^h\Delta2$	$09^h\Delta1/06^h\Delta2$	$11^h/25^h$	5/0.04 Pr	(80)
IMP 3	Pr > 6.5 MeV	$02^h\Delta2$	$06^h\Delta2$	10^h	0.04 Pr	(80)
PION 7	Pr > 12 MeV	$05^h\Delta5$	$06^h\Delta5$	48^h	0.1 Pr	(119)
(PION 7 at 1.11 AU, ES − 26°)						

Source: ○ flare 0005   S15   E09   SF   McM 8716
         ○ flare 8^d2305 N16   W42   SB   McM 8714
         □ region 8704 two days beyond W limb

1967   March 9 (twice)   events in Appendix A

179	1967 March 11	19^h					0 1 0
IMP 3	Pr > 0.9 MeV	$19^h\Delta2\Delta8$	$23^h\Delta1$	47^h	261.5 Pr	(80)	
IMP 3	El > 3 MeV	19^h45^m	$22^h\Delta2$		2.5 El	(67)	
IMP 3	El > 40 keV	$19^h48^m\Delta3$			700 El	(69)	
PION 7	Pr > 165/>12 MeV	$20^h\Delta2/\Delta5$	$22^h/12^d05^h\Delta5$	$20^h/120^h$	0.52/1.7 Pr	(119)	
IMP 3	Pr > 6.5 MeV	$20^h\Delta1$	$12^d03^h\Delta1$	47^h	5.53 Pr	(80)	
PION 7	Pr > 0.6 MeV	$20^h\Delta5$	$12^d05^h\Delta5$	$>48^h$	9.5 Pr	(119)	
RIOM	PCA	$19^h\Delta2/$	$\begin{bmatrix}12^d19^h\Delta1/\\02^h\Delta2\end{bmatrix}$	$25^h/$	1/1.6 dB	(146, 152)	

(PION 7 at 1.11 AU, ES − 26.5°)

Source: □ invisible hemisphere ? no flare patrol 11^d1727–1855 and 1925–2040
◇ corotating structure sequential with event of Feb. 13

	1967 March 13 (twice), March 14, March 24, March 27			events in Appendix A			
180	1967 March 27	$>23^h$					(−1) 0 0
IMP 3	El > 45 keV		$29^d00^h\Delta5$	$>72^h$	150 El	(117)	
IMP 3	Pr > 0.9/>6.5 MeV	$21^h30^m\Delta3/23^h\Delta2$	$\begin{bmatrix}28^d10^h\Delta1/\\29^d07^h\Delta2\end{bmatrix}$	$121^h/65^h$	15/0.23 Pr	(80)	
IMP 3	Pr > 19 MeV	$28^d04^h\Delta2$	$29^d07^h\Delta2$	45^h	0.02 Pr	(80)	

Source: ○ flare 2107 N23 W24 1B McM 8740
○ flare 1718 N25 W23 1N McM 8740
◇ corotating structure, CM passage of region 8740

	1967 April 1 (three times), Apr. 3, Apr. 4, Apr. 8			events in Appendix A			
181	1967 April 14	$>19^h$					−2 0 0
IMP 3	El > 45 keV	$19^h\Delta2$	$15^d03^h\Delta2$	48^h	40 El	(117)	
IMP 3	Pr > 0.9 MeV	$19^h/22^h\Delta2$	$15^d10^h\Delta2$	65^h	22 Pr	(80)	
IMP 3	Pr > 19/>6.5 MeV	$15^d05^h\Delta2$	$15^d13^h\Delta2$	$15^h/18^h$	$\begin{bmatrix}0.014/\\0.044\ Pr\end{bmatrix}$	(80)	

Source: ⊙ flare 1703 N24 W71 2N McM 8760

	1967 May 7, May 19	events in Appendix A					
182	1967 May 21	$>21^h$					−2 0 0
VELA 4	Pr 0.55 MeV	$20^h45^m \pm 15^m$	$23^d07^h\Delta1$	47^h	8* Pr	(168)	
PION 7	Pr > 12/>73 MeV	$23^h/22^d00^h\Delta5$	$22^d08^h\Delta5$		6/0.26 Pr	(119)	
VELA 4	Pr 4.5/1.4 MeV	$22^d04^h\Delta2$	$23^d07^h\Delta1$	40^h	0.05*/2.5* Pr	(168)	

(PION 7 at 1.06 AU, ES − 36.5°)

Source: ● flare 1919 N24 E39 2N McM 8818

183	1967 May 23	20^h				3 3 0
f-min	PCA	$20^h\Delta1$ (203, 192)				(192, 224, 203)
AIMP 1	Pr > 15 MeV	20^h03^m	$25^d10^h\Delta1$	80^h	2.5 c	(69)
RIOM	PCA	23^h30^m	$25^d13^h\Delta2$	103^h	11 dB	(152)
VELA 4	Pr 10 MeV	$24^d02^h\Delta2$	$25^d12^h\Delta2$		200* Pr	(168)
VELA 4	Pr 5 MeV	$24^d02^h\Delta2$	$25^d13^h45^m$		1100* Pr	(168)
VELA 4	Pr 1.4 MeV	$24^d04^h\Delta3$	$25^d13^h\Delta1$		10000* Pr	(168)
VELA 4	Pr 0.6 MeV	$24^d05^h\Delta3$	$25^d13^h45^m$		33000* Pr	(168)
IMP 4	Pr > 94 MeV	$\Delta5$	$25^d08^h\Delta1$		0.75 Pr	(107)
IMP 4	Pr > 60/>30 MeV	$\Delta5$	$25^d09^h/08^h30^m$		2.35/32.15 Pr	(173)
IMP 4	Pr 7/4 MeV	$\Delta5$	$25^d08^h/10^h\Delta2$		⌈1500*/ ⌊4000* Pr	(225)

Source: ● flare 1802   N28   E25⌉
         1932   N28   E28⌋   3B   McM 8818
      ▲ major GMS with sc 24^d1726

184	1967 May 28	$>06^h$				2 2 0
f-min	PCA	$03^h\Delta3$ (203)		85^h (203)		(192, 224, 203)
IMP 4	El > 0.3/>2.7 MeV	05^h58^m	$/08^h$	$/72^h$	50/3.7 El	(166)
IMP 4	El > 40 keV	$06^h07^m\pm01^m$	07^h		4500 El	(69)
Balloon	Pr > 100 MeV	$12^h\Delta5$	25^h		2 Pr	(43)
IMP 4	Pr > 94/>29 MeV	⌈$06^h15^m/$ ⌊$06^h20^m\pm20^m$	$07^h15^m/$ 10^h30^m	$45^h/115^h$	3.1/26 Pr	(107)
IMP 4	Pr > 0.8/>9 MeV	⌈$06^h15^m\pm30^m\Delta8/$ ⌊$\pm05^m\Delta8$	$14^h50^m/$ 10^h50^m	$104^h/180^h$	1624/133 Pr	(107)
IMP 4	Pr > 60/>7 MeV	⌈$06^h25^m\pm10^m/$ ⌊$06^h30^m\Delta3$	$08^h\Delta2/10^h\Delta1$	$45^h/>65^h$	9.45/85 Pr	(173, 174)
IMP 4	Pr 4 MeV	$07^h10^m\pm05^m$	$11^h\Delta2$		34.8* Pr	(182)
IMP 4	Pr > 1/0.5 MeV	⌈$08^h\pm10^m\Delta3/$ ⌊$10^h\Delta1$	$20^h/30^d05^h$	$103^h/$	1391/ 4880* Pr	(181, 175)
RIOM	PCA	$06^h\Delta3$	09^h	59^h	4 dB	(146)

Source: ● flare < 0529   N28   W32   3B   McM 8818
      ▲ major GMS with sc 30^d1425

	1967 June 1	event in Appendix A				
185	1967 June 3	$\sim07^h$				0 (1) 0
f-min	PCA	$06^h\Delta2$		09^h		(203)
IMP 4	El > 40 keV	$06^h23^m\pm02^m$	07^h		500 El	(69)
IMP 4	⌈El > 0.3 MeV/ ⌊> 70 keV	$06^h30^m/06^h34^m$	$10^h/$	$/20^h$	170/210 El	(175, 176)
IMP 4	Pr > 9 MeV	$06^h40^m\pm10^m\Delta8$	09^h30^m	52^h	2.8 Pr	(107)
IMP 4	El > 2.7 MeV	$07^h\Delta2$	09^h		2.5 El	(166)
IMP 4	Pr > 30/>6 MeV	$07^h\pm20^m/07^h\Delta1$	$12^h\Delta2/10^h\Delta1$	$15^h/10^h$	0.21/30 Pr	(173, 176)
IMP 4	Pr 19 MeV	$07^h07^m\pm03^m$	$09^h\Delta2$		0.0052* Pr	(182)
IMP 4	Pr > 0.7/>3.5 MeV	$07^h12^m/08^h\Delta1$	$13^h45^m/11^h\Delta1$	$>20^h/09^h$	30/160 Pr	(176)
IMP 4	Pr > 1 MeV	$09^h\pm30^m\Delta3$	13^h40^m	27^h	27.5 Pr	(181)

(continues on next page)

185 continued

Source: ○ flare 0226 N23 E12 1N McM 8831

　　　　□ active region 8818 on invisible hemisphere 2–3 days beyond W limb

	1967　June 4	event in Appendix A				
186	1967　June 5	$\sim 12^h$				$-1\ 0\ 0$
IMP 4	Pr > 9/> 1 MeV	$12^h\Delta 3$	$22^h/18^h\Delta 2$	$15^h/21^h$	0.2/98 Pr	(107, 181)
PION 6	Pr > 13 MeV	$13^h\Delta 5$				(104)
(PION 6 at 0.89 AU, ES + 100°)						

Source: ▲ storm modulation, major magnetic storm $5^d 1915$

187	1967　June 6	$\sim 06^h$				$1\ 2\ 0$
IMP 4	⌈El > 40 keV/ ⌊> 2.7 MeV	$06^h \pm 20^m/06^h06^m$			200/1.9 El	(69, 166)
IMP 4	El > 0.3 MeV	06^h06^m	$13^h\Delta 1$	5^d	22 El	(166)
IMP 4	El > 70 keV	$06^h10^m\Delta 7$	$9^h\Delta 2$	20^h	400 El	(176)
IMP 4	Pr > 0.7 MeV	06^h27^m	$14^h\Delta 2$	12^h	160 Pr	(176)
IMP 4	Pr > 94/> 9 MeV	$06^h35^m \pm 05^m$	⌈$18^h30^m\Delta 4/$ ⌊$7^d01^h\Delta 1$	$42^h/150^h$	0.25/20.5 Pr	(107)
IMP 4	Pr > 30/> 1 MeV	⌈$06^h45^m \pm 15^m/$ ⌊$09^h\Delta 3$	$15^h\Delta 4/$ $7^d03^h30^m$	$60^h/> 120^h$	4.6/108 Pr	(173, 181)
IMP 4	Pr 19/4 MeV	⌈$06^h45^m \pm 30^m/$ ⌊$07^h45^m \pm 15^m$	$\Delta 4$			(182)
RIOM	PCA	09^h	7^d02^h	61^h	1.8 dB	(146)

Source: □ region McM 8824 at W limb; active regions on invisible hemisphere, origin unknown; no flare patrol 0343–0507.

188	1967　June 12	$> 10^h$				$0\ 1\ 0$
IMP 4	Pr > 29/19 MeV	⌈$11^h\Delta 2/11^h30^m \pm$ ⌊60^m	$14^d22^h/18^h\Delta 2$	$120^h/$	0.04/ 0.00426* Pr	(107, 182)
IMP 4	El > 70 keV	$12^h\Delta 2$	$15^d00^h\Delta 2$	7^d	155 El	(176)
IMP 4	Pr 4 MeV	$12^h30^m \pm 60^m$	$14^d22^h30^m$		0.26* Pr	(182)
IMP 4	Pr > 1/> 10 MeV	$16^h/18^h\Delta 2$	⌈$16^d20^h/$ ⌊$14^d18^h\Delta 2$	$> 9^d/120^h$	8.2/1.2 Pr	(181, 173)
RIOM	PCA		15^d16^h		0.3 dB	(169)

Source: ◇ origin unknown; no flare patrol 0701–1144; possibly return to E limb of great region McM 8818

189	1967	June 25	$\sim00^h$					-100
IMP 4	El >45 keV		$26^d09^h\Delta2$			10 El	(117)
IMP 4	Pr $>1/0.5$ MeV	$00^h\Delta2/19^h\Delta1$	$26^d14^h\Delta2/\Delta1$	$>5^d/$		39/820* Pr	(181, 175)
IMP 4	$\lceil$Pr >9 MeV/	$07^h/06^h\Delta2$	$26^d10^h/$	$60^h/72^h$		0.85/230 Pr	(107, 117)
	$\llcorner>300$ keV		$27^d00^h\Delta2$				

Source: $\Diamond$ origin uncertain; particle stream related to regions McM 8854–8863 transit; GMS begins with sc 25^d0221, with another sc 26^d1459

190	1967	July 5	$\sim08^h$					-110
IMP 4	$\lceil$El >2.7 MeV/	$08^h03^m\pm1^m$				0.05/200 El	(166, 69)
	$\llcorner>40$ keV						
IMP 4	$\lceil$El >70 keV/	08^h10^m	$08^h36^m/$	$/2^h$		48/1000 El	(176, 117)
	$\llcorner>20$ keV		08^h45^m				
IMP 4	Pr >30 MeV	$08^h15^m\pm15^m$	08^h45^m	7^h		0.24 Pr	(173)
IMP 4	Pr 19/>9 MeV	$\lceil08^h22^m\pm03^m/$	$12^h\Delta2/\Delta1$	$/50^h$		0.0074*/	(182, 107)
		$\llcorner08^h45^m\pm05^m$				0.75 Pr	
IMP 4	Pr >1 MeV	$09^h45^m\pm15^m$	13^h20^m	42^h		2.8 Pr	(181)
IMP 4	El >0.3 MeV	09^h	10^h			11 El	(175)
IMP 4	Pr 4 MeV	$09^h15^m\pm03^m$	$12^h\Delta2$			0.176* Pr	(182)
RIOM	PCA		19^h			0.3 dB	(169)

Source: $\square$ possibly region McM 8878 at W limb or McM 8863 on invisible hemisphere about 3 days beyond W limb

191	1967	July 7	17^h					X 0 0
PION 6	Pr >175 MeV	17^h	$18^h\Delta2$	10^h		0.16 Pr	(104)
PION 6	Pr$>73/>13$ MeV	17^h	$20^h\Delta2/$	$10^h/$		3 Pr/	(104)
(PION 6 at 0.95 AU, ES + 105°)							

Source: $\circ$ flare 1526   N28   W77   1F   McM 8871

192	1967	July 7	22^h					-100
IMP 4	Pr $>29/>9$ MeV	$22^h/23^h\Delta2$	$8^d13^h/9^d12^h\Delta2$	$48^h/90^h$	0.015/0.93 Pr	(107)
IMP 4	Pr >0.5 MeV	$8^d00^h\Delta2$	$9^d10^h\Delta2$	28^h	7 Pr	(117)
IMP 4	Pr >1 MeV	$8^d13^h\Delta2$	$10^d13^h\Delta2$	72^h	0.7 Pr	(181)

Source: origin uncertain, possibly a particle stream related to CM transit of region McM 8877.

It is also possible that events 191–192 are similar events whose origins are flares beyond W limb

1967	July 9, July 11, July 16, July 23, July 28 (three times), July 29	events in Appendix A

193	1967 July 29	$>15^h44^m$				$-1\,0\,0$
AIMP 2	El >40 keV	$15^h44^m \pm 01^m$	16^h		300 El	(69)
IMP 4	Pr >9 MeV	$18^h\Delta2$	$30^d04^h\Delta2$	12^h	0.2 Pr	(107)
IMP 4	Pr >1 MeV	$18^h\Delta2$	$30^d03^h\Delta5$	12^h	0.8 Pr	(181)
IMP 4	Pr >0.8 MeV	$19^h30^m \pm 30^m\Delta3$	$30^d03^h\Delta2$	12^h	7.3 Pr	(107)

Source: ○ flare 1453　N25　W16　1N　McM 8905
　　　　○ flare 1525　S29　W90　SF　McM 8899

	1967 July 29 (twice), July 30		events in Appendix A			
194	1967 July 30	$>05^h$				$-1\,0\,0$
IMP 4	El >0.3 MeV	05^h16^m	05^h30^m		0.4 El	(166)
IMP 4	Pr >9 MeV	$05^h15^m \pm 05^m\Delta8$	06^h20^m	$>27^h$	0.14 Pr	(107)
AIMP 2	El >40 keV	$05^h20^m \pm 01^m$			850 El	(69)
IMP 4	El >70 keV	05^h22^m	05^h30^m		90 El	(176)
IMP 4	Pr >1 MeV	$06^h30 \pm 30^m\Delta3$	15^h	29^h	8.1 Pr	(181)
IMP 4	Pr >0.8 MeV	$07^h\Delta8$	$09^h30^m\Delta8$	26^h	29.6 Pr	(107)

Source: ● flare 0508　N24　W27　1B　McM 8905

	1967 July 30 (twice)		events in Appendix A			
195	1967 July 31	09^h15^m				$-2\,0\,0$
IMP 4	Pr >9 MeV	$09^h15^m \pm 15^m$	12^h15^m	24^h	0.05 Pr	(107)
IMP 4	Pr >1 MeV	$11^h30^m \pm 30^m\Delta3$	$23^h\Delta2$	24^h	1.9 Pr	(181)
IMP 4	Pr >0.8 MeV	$09^h25^m \pm 15^m\Delta3$	$23^h\Delta2$	24^h	11 Pr	(107)

Source: ⊙ flare 0808　N25　W42　1N　McM 8905
　　　　○ flare 0855　N13　E11　1N　McM 8913

	1967 July 31		event in Appendix A			
196	1967 August 1	$>09^h30^m$				$-1\,0\,0$
IMP 4	Pr >9 MeV	$09^h30^m \pm 60^m\Delta3$	$11^h30^m\Delta4$	20^h	0.65 Pr	(107)
IMP 4	Pr $>1/0.5$ MeV	$\lceil11^h \pm 45^m\Delta3/$ $\lfloor11^h\Delta1$	$15^h\Delta2/$	$8^h/$	1.6 Pr/	(181, 175)

Source: ⊙ flare 0120　N13　W25　SB　McM 8907
　　　　⊙ flare $<$0535　N25　W53　1N　McM 8905

197	1967	August 1	$>17^h43^m$				0 0 0
AIMP 2	El >40 keV	$17^h43^m \pm 02^m$			700 El	(69)	
IMP 4	Pr $>29/>10$ MeV	$\lceil 17^h45^m/$ $\lfloor 18^h55^m \pm 65^m \Delta 4$	$20^h/22^h\Delta 4$	$10^h/18^h$	0.1/6.5 Pr	(107, 173)	
IMP 4	Pr >1 MeV	$18^h45^m \pm 50^m \Delta 3$	$2^d01^h\Delta 1$	$>11^h$	33 Pr	(181)	
VEN 4	Pr >1 MeV	$19^h\Delta 2$	$23^h\Delta 2$		135 Pr	(177)	
(VEN 4 at 0.93 AU, ES $-$ 1°)							

Source: ● flare 1721   N27   W62   2B   McM 8905

198	1967	August 2	05^h15^m				-1 0 0
IMP 4	Pr $>9/>0.8$ MeV	$05^h15^m \pm 15^m \Delta 3$	$06^h/06^h15^m$	28^h	0.26/144 Pr	(107)	
IMP 4	Pr >1 MeV	$05^h15^m \pm 15^m \Delta 3$	$07^h\Delta 2$		47 Pr	(181)	

Source: ○ flare 0505   N28   W64   SB   McM 8905
　　　　 ○ flare 0043   N26   W58   1N   McM 8905

	1967	Aug. 2	event in Appendix A

199	1967	August 3	$>09^h30^m$				-1 0 0
IMP 4	$\lceil$El >0.3 MeV/ $\rfloor > 70$ keV	$09^h30^m/$ $09^h37^m \pm 02^m$	$11^h\Delta 1$	$/17^h$	46/125 El	(175, 176)	
IMP 4	Pr >0.8 MeV	$09^h35^m \pm 05^m \Delta 8$	$11^h45^m\Delta 8$	7^h	85 Pr	(107)	
AIMP 2	El >40 keV	$09^h41^m \pm 01^m$			400 El	(69)	
IMP 4	Pr >10 MeV	$10^h30^m \pm 30^m$	11^h45^m	$>7^h$	0.22 Pr	(173)	
IMP 4	Pr 19/4 MeV	$\lceil 10^h38^m/$ $\lfloor 11^h08^m \pm 10^m$	$12^h\Delta 2$		0.001*/ 0.47* Pr	(182)	

Source: ⊙ flare 0918   N27   W85   1B   McM 8905

200	1967	August 3	$>16^h30^m$				0 0 0
VEN 4	Pr >1 MeV	$16^h\Delta 3$	$4^d01^h\Delta 2$	$>25^h$	30 Pr	(177)	
IMP 4	Pr 0.5 MeV	16^h30^m	4^d06^h		110* Pr	(175)	
IMP 4	Pr $>9/>0.8$ MeV	$\lceil 16^h50^m \pm 10^m/$ $\lfloor \pm 20^m \Delta 3$	$17^h20^m/$ 17^h30^m	$32^h/>23^h$	0.33/66 Pr	(107)	

Source: ○ flare 1628   N28   W70   SB   McM 8905

	1967	Aug. 4	event in Appendix A

201	1967	August 4	23^h				0 0 0
IMP 4	Pr >9 MeV	$23^h\Delta 1$	$5^d02^h30^m$	40^h	1 Pr	(107)	
IMP 4	Pr >1 MeV	$23^h55^m \pm 10^m \Delta 3$	$5^d01^h15^m$		6.1 Pr	(181)	
IMP 4	Pr >0.8 MeV	$5^d00^h\Delta 3$	$03^h\Delta 2$	70^h	13.5 Pr	(107)	

Source: □ flare invisible hemisphere; SWF imp 1, 2213–2238; 10 cm burst; region McM 8905 is just beyond W limb

202   1967   August 8 $>11^h$ −1 0 0

IMP 4	Pr $>0.8/>9$ MeV	$\lceil 11^h \pm 30^m/$ $\lfloor 12^h 30^m \pm 70^m$	$9^d 14^h/13^h \Delta 2$	$56^h/34^h$	2.5/0.4 Pr	(107)
VEN 4	Pr >1 MeV	$9^d 14^h \Delta 2$		144^h (duration includes next event)		(177)

(VEN 4 at 0.91 AU, ES = 0°)

Source: ◇ large and bright regions 8921 and 8926 are near CM 10^d and 12^d

---

203   1967   August 9 $>18^h 30^m$ 0 0 0

IMP 4	Pr >1 MeV	$18^h 30^m \pm 30^m$	$11^d 04^h \Delta 2$	85^h	1521 Pr	(181)
IMP 4	Pr >0.8 MeV	$19^h 30^m \pm 30^m$	$11^d 04^h 30^m$	140^h	4232 Pr	(107)
IMP 4	Pr $>9/>29$ MeV	$\lceil 10^d 03^h 30^m \pm 60^m/$ $\lfloor 06^h \pm 30^m \Delta 4$	$11^d 03^h \Delta 2/$ $10^d 08^h \Delta 4$	$50^h/13^h$	3.5/0.09 Pr	(107)

Source: ⊙ flare 1758   S24   E32   2B   McM 8926
  △ GMS effect with sc $11^d 0555$
  ◇ possible CM transit of region 8921

1967   Aug. 12, Aug. 13 (twice), Aug. 15, Aug. 17, Aug. 18, Aug. 24, Aug. 26, Sept 3, Sept. 9,
Sept. 10 (three times), Sept. 11, Sept. 12 (twice)   events in Appendix A

204   1967   September 17 $<23^h$ −2 1 0

VEN 4	Pr >1 MeV	$16^h \Delta 2$ (max. and duration include next four events)	$19^d 20^h \Delta 2$	140^h	1950 Pr	(177)

(VEN 4 at 0.79 AU, ES + 11°)

IMP 4	Pr $>9/>0.8$ MeV	$23^h \pm 90^m/\pm 10^m$	$\lceil 18^d 16^h/$ $\lfloor 17^h 30^m \Delta 4$	$>35^h$	0.016/ 2035 Pr	(107)
RIOM	PCA		$18^d 18^h$		0.6 dB	(169)

Source: ○ flare 1050   N18   W38   1N(?)   McM 8973
  ○ flare 0353   N15   E61   2B   McM 8985
  ◇ particle stream related to CM transit of zones of plages

1967   Sept. 18   event in Appendix A

205   1967   September 19 02^h −2 (1) 0

f-min	PCA	$01^h \Delta 3$		65^h		(203)
IMP 4	Pr $>9/>1$ MeV	$\lceil 02^h \pm 80^m/$ $\lfloor 03^h \pm 30^m \Delta 3$	$05^h \Delta 2$	$8^h/6^h$	0.05/86 Pr	(107, 181)

Source: ● flare $18^d 2316$ N16   W60   2B   McM 8973
  ▲ GMS with sc $19^d 1957$

206   1967   September 19 10^h20^m $-1\,0\,0$

IMP 4	Pr 19 MeV		$13^h\Delta2$		0.000193* Pr	(182)
IMP 4	Pr > 9/>0.8 MeV	$10^h20^m\pm30^m\Delta3$	$\begin{bmatrix}12^h30^m/\\20^d00^h30^m\end{bmatrix}$	24^h	0.9/695 Pr	(107)
IMP 4	Pr 4 MeV		$14^h\Delta2$		1.25* Pr	(182)
VEN 4	Pr > 1 MeV		$20^h\Delta2$		1950 Pr	(177)

Source: $\triangle$ GMS with sc 19^d 1957

    $\square$ region McM 8972 is at W limb and gives many subflares on 19^d

---

207   1967   September 20 10^h30^m $-1\,1\,0$

IMP 4	Pr > 29 MeV	$10^h30^m\pm30^m$	$12^h\Delta2$	12^h	0.013 Pr	(107)
IMP 4	Pr 19/4 MeV		13^h30^m		$\begin{bmatrix}0.000572*/\\1.93* \text{ Pr}\end{bmatrix}$	(182)
IMP 4	Pr > 9 MeV	$10^h30^m\pm30^m\Delta3$	12^h30^m	65^h	0.8 Pr	(107)
IMP 4	Pr > 1 MeV	$11^h\pm30^m\Delta3$	$13^h\Delta1$	31^h	351 Pr	(181)
RIOM	PCA		$18^h\Delta1$		0.8 dB	(169)

Source: $\triangle$ increase in intensity of previous geomagnetic storm

    $\square$ region 8972 is at W limb on Sept. 20 and gives numerous subflares

---

	1967	Sept. 26, Sept. 27, Sept. 29, Oct. 3, Oct. 4 (twice), Oct 7 (four times)		events in Appendix A

208   1967   October 7 $>20^h30^m$ $-2\,0\,0$

AIMP 2	El > 20 keV	$>20^h36^m$			30 El	(69)
IMP 4	Pr > 1 MeV	$\begin{bmatrix}8^d01^h30^m\\\pm30^m\Delta3\end{bmatrix}$	$13^h\Delta2$	$<15^h$	0.65 Pr	(181)
IMP 4	Pr > 0.8/>9 MeV	$\begin{bmatrix}8^d02^h45^m\\\pm05^m\Delta3/\pm15^m\end{bmatrix}$	$03^h\Delta1/04^h$	$13^h/25^h$	2.85/0.015 Pr	(107)

Source: $\odot$ flare 2044   S18   W62   2B   McM 9004

	1967	Oct. 8, Oct. 12, Oct. 19, Oct. 25 (four times), Oct. 26		events in Appendix A

209   1967   October 26 06^h40^m $-2\,0\,0$

AIMP 2	El > 45/>22 keV	06^h40^m			20/50 El	(117)
AIMP 2	El > 40 keV	$06^h40^m\pm01^m$			30 El	(69)
IMP 4	Pr > 9/>0.8 MeV	$\begin{bmatrix}06^h40^m\pm40^m/\\06^h45^m\pm15^m\Delta8\end{bmatrix}$	07^h30^m	$25^h/31^h$	0.011/2.4 Pr	(107)

Source: $\bullet$ flare 0608   N10   W38   1B   McM 9034

	1967	Oct. 27		event in Appendix A

210 1967 October 27 11^h35^m $-2\,0\,0$

AIMP 2	El $>$ 40 keV	$11^h35^m \pm 10^m$				50 El	(69)
AIMP 2	El $>$ 22 keV	$12^h\Delta3$	$14^h\Delta2$			170 El	(117)
IMP 4	Pr $>9/>0.8$ MeV	$15^h\Delta4$	$16^h\Delta4$		$26^h/40^h$	0.011/16 Pr	(107)

Source: ⊙ flare 1107 N09 W46 SN McM 9034

211 1967 October 29 04^h $-1\,0\,0$

AIMP 2	El $>$ 22 keV	$04^h\Delta2$	$18^h\Delta2$	21^h		150 El	(117)
IMP 4	Pr $>$ 1 MeV	$06^h \pm 60^m\Delta3$	$18^h\Delta1$			26.8 Pr	(181)
IMP 4	Pr $>0.8/>29$ MeV	⌈$07^h \pm 60^m\Delta3/$ ⌊$07^h05^m \pm 05^m\Delta2$	$19^h\Delta1/13^h30^m$	$20^h/17^h$		71/0.02 Pr	(107)
IMP 4	Pr 19/4 MeV	$07^h30^m \pm 60^m$	$14^h\Delta2$			⌈0.00485*/ ⌊1.46* Pr	(182)
IMP 4	Pr $>$ 9 MeV	$08^h \pm 90^m$	$16^h\Delta1$	21^h		0.2 Pr	(107)

Source: ⊙ flare 0258 N10 W80 1N McM 9034

212 1967 October 30 $>00^h$ $-1\,0\,0$

AIMP 2	El $>$ 40 keV	$00^h \pm 24^m$	$01^h\Delta1$	3^h		250 El	(69)
IMP 4	El $>$ 70 keV	00^h17^m	$02^h\Delta1$	50^h		100 El	(176)
AIMP 2	El $>45/>22$ keV	00^h20^m	$01^h\Delta1$	$48^h/3^h$		220/800 El	(117)
IMP 4	El $>$ 0.3 MeV	00^h37^m	05^h30^m	24^h		0.09 El	(166)
IMP 4	Pr 19/>29 MeV	⌈$01^h30^m \pm 60^m/$ ⌊$02^h\Delta2$	$09^h/10^h\Delta2$	$/30^h$		0.000275*/ 0.01 Pr	(182, 107)
IMP 4	Pr $>10/>2$ MeV	$02^h\Delta2/03^h\Delta1$	⌈$06^h\Delta2/$ ⌊$31^d02^h\Delta1$	$20^h/40^h$		0.43/16 Pr	(173, 176)
IMP 4	Pr $>0.8/4$ MeV	⌈$03^h\Delta3/$ ⌊$03^h30^m \pm 60^m$	$31^d05^h45^m/$ $30^d05^h\Delta2$	77^h		202/1.14* Pr	(107, 182)

Source: ● flare 29^d2347 N10 W90 2B McM 9034

213 1967 November 2 09^h $0\,1\,0$

f-min	PCA	$09^h\Delta1$	40^h				(203)
IMP 4	El $>$ 2.7 MeV	$09^h\Delta2$				0.18 El	(166)
IMP 4	Pr 19/>29 MeV	⌈$09^h15^m \pm 15^m/$ ⌊$09^h25^m \pm 05^m$	$18^h\Delta2/17^h45^m$	$/22^h$		0.000533*/ 0.85 Pr	(182, 107)
IMP 4	Pr $>9/>0.8$ MeV	$09^h30^m \pm 30^m/\Delta8$	$17^h45^m/18^h30^m$	22^h		9.7/162.5 Pr	(107)
IMP 4	⌈El $>$ 70 keV/ ⌊>0.3 MeV	$10^h/10^h15^m$	$/17^h30^m$			210/29 El	(176, 175)
IMP 4	Pr 4 MeV	$10^h15^m \pm 15^m$	$18^h\Delta2$			2.16* Pr	(182)
AIMP 2	El $>$ 40 keV	$10^h17^m \pm 15^m$				250 El	(69)
OGO 4	Pr $>$ 1.2 MeV	$11^h\Delta2$	$20^h\Delta2$			100 Pr	(180)
RIOM	PCA		18^h			0.9 dB	(169)

Source: ● flare 0852 S18 W02 2B McM 9047

(continues on next page)

213 continued

IMP 4	Pr 19/4 MeV	$\begin{bmatrix}3^d05^h/ \\ 3^d06^h15^m \pm 30^m\end{bmatrix}$	$\begin{bmatrix}3^d09^h20^m/ \\ 3^d10^h30^m\end{bmatrix}$		0.0305*/ 3.2* Pr	(182)
IMP 4	Pr $>30/>10$ MeV	$3^d07^h\Delta3$	$3^d10^h\Delta1/\Delta2$	$15^h/31^h$	0.37/5.26 Pr	(173)
IMP 4	Pr $>0.8/1$ MeV	$3^d07^h \pm 50^m\Delta3$	$\begin{bmatrix}3^d10^h30^m/ \\ 3^d09^h18^m\end{bmatrix}$	$33^h/$	203/11* Pr	(107, 188)

Source: ▲GMS begins with SC $3^d09\,14$ this storm is associated with the above flare

214	1967 November 4	12^h15^m				$-1\,0\,0$
IMP 4	$\begin{bmatrix}\text{El} >70\text{ keV}/ \\ >0.3\text{ MeV}\end{bmatrix}$	$12^h13^m/12^h26^m$	$12^h40^m/13^h$	$/12^h$	100/0.9 El	(176, 166)
AIMP 2	El >40 keV	$12^h15^m \pm 01^m$			200 El	(69)
IMP 4	Pr 19/>29 MeV	$\begin{bmatrix}12^h45^m \pm 15^m/ \\ 13^h05^m \pm 05^m\end{bmatrix}$	$15^h\Delta2/14^h30^m$	$/12^h$	0.00473*/ 0.07 Pr	(182, 107)
IMP 4	Pr $>10/4$ MeV	$\begin{bmatrix}14^h\Delta1/ \\ 13^h15^m \pm 15^m\end{bmatrix}$	$16^h/16^h30^m$	$15^h/$	0.6/0.363* Pr	(173, 182)
IMP 4	Pr $>1/0.5$ MeV	$15^h\Delta3/17^h30^m$	$21^h\Delta2/5^d07^h30^m$		12.8/28* Pr	(181, 175)

Source: ●flare 1151 S18 W33 1B McM 9047

	1967 Nov. 7	event in Appendix A				
215	1967 November 7	$\sim03^h$				$-1\,1\,0$
IMP 4	Pr $>0.8/>29$ MeV	$\begin{bmatrix}02^h \pm 30^m\Delta8/ \\ 03^h \pm 50^m\end{bmatrix}$	$07^h\Delta2/04^h50^m$	$\lvert24^h/13^h$	11.3/0.033 Pr	(107)
IMP 4	El >70 keV	02^h26^m			12 El	(176)
IMP 4	Pr $>19/>9$ MeV	$\begin{bmatrix}03^h \pm 30^m/ \\ 03^h30^m \pm 30^m\Delta3\end{bmatrix}$	$04^h30^m/06^h30^m$	$/35^h$	0.00189*/ 0.43 Pr	(182, 107)
IMP 4	Pr 4/>1 MeV	$03^h15^m \pm 15^m/\Delta3$	$07^h\Delta2/07^h15^m$	$/>50^h$	$\begin{bmatrix}0.0816*/ \\ 1.85\text{ Pr}\end{bmatrix}$	(182, 181)
RIOM	PCA		$06^h\Delta2$		0.5 dB	(169)

Source: ○flare 0157 S15 W51 SN McM 9047

	1967 Nov. 7, Nov. 8	events in Appendix A				
216	1967 November 10	19^h30^m				$-1\,0\,0$
IMP 4	El >0.3 MeV	19^h30^m	$20^h\Delta1$		12 El	(175)
IMP 4	Pr $>9/>1$ MeV	$19^h30^m \pm 30^m\Delta3$	20^h30^m	8^h	0.32/2.6 Pr	(107, 181)
IMP 4	Pr >0.5 MeV	19^h30^m	21^h30^m		4* Pr	(175)
IMP 4	El >70 keV	$20^h\Delta2$	$13^d12^h\Delta4$		230 El	(176)

(this max. includes the three following events)

Source: ▢ bright surge 1825–1850 at W limb where region 9047 is going over;
type II 1833–1848; type IV 1834–1912

217	1967	November 11	$02^h 30^m$					$-1\,0\,0$
AIMP 2	El >40 keV		$02^h 30^m \pm 06^m$				60 El	(69)
IMP 4	Pr $>9/>0.8$ MeV		$03^h \pm 30^m/03^h \Delta 3$	$13^h \Delta 2$	19^h		0.51/76 Pr	(107)
IMP 4	Pr >1 MeV		$03^h 30^m \pm 40^m \Delta 3$	$13^h 30^m$	18^h		1.96 Pr	(181)

Source: ○ flare 0211   S26   W90   ?   McM 9047
    □ region 9047 is just beyond W limb and gives frequent limb flares on the 11th

218	1967	November 11	21^h					$0\,0\,0$
IMP 4	Pr 19/4 MeV		$\lceil 18^h 45^m \pm 60^m/$ $\lfloor 20^h 45^m \pm 90^m$	$12^d 04^h/$ $12^d 21^h \Delta 2$			$0.00181*/$ $0.142*$ Pr	(182)
IMP 4	Pr $>29/>9$ MeV		$\lceil 22^h \Delta 2/$ $\lfloor 22^h \pm 30^m \Delta 3$	$12^d 19^h/$ $12^d 22^h \Delta 2$	29^h		0.05/1.4 Pr	(107)
IMP 4	Pr >0.8 MeV		$22^h \pm 10^m \Delta 3$	$12^d 17^h \Delta 2$	30^h		105 Pr	(107)

Source: ◇ moderate geomagnetic disturbance begins $11^d 19^h$
    ◇ return of active region 9034 near E limb (events of October 25–26)

219	1967	November 13	$03^h 30^m$					$0\,1\,0$
IMP 4	El >0.3 MeV		$03^h 30^m$	$12^h 30^m$			320 El	(175)
IMP 4	Pr 19/$>$29 MeV		$\lceil 03^h \pm 90^m/$ $\lfloor 03^h 30^m \pm 60^m \Delta 3$	$13^h/11^h \Delta 2$	$/61^h$		$0.0124*/$ 0.28 Pr	(182, 107)
IMP 4	Pr $>9/>1$ MeV		$03^h 30^m \pm 30^m \Delta 3$	$11^h \Delta 2/16^h \Delta 4$	$59^h/>59^h$		4.68/21.6 Pr	(107, 181)
IMP 4	Pr 0.5/4 MeV		$\lceil 03^h 30^m/$ $\lfloor 04^h 30^m \pm 30^m$	$11^h \Delta 1/13^h \Delta 4$			$46*/0.665*$ Pr	(175, 182)
RIOM	PCA		$15^d 03^h \Delta 2$				0.5 dB	(169)

Source: □ flare in region 9073 at E limb
    □ active region 9047 is 2 days beyond W limb

220	1967	November 15	$>14^h 30^m$					$0\,0\,0$
IMP 4	Pr $>9/19$ MeV		$\lceil 14^h 30^m \pm 30^m \Delta 3/$ $\lfloor 14^h 31^m \pm 60^m$	$16^d 00^h/$ $15^d 21^h \Delta 2$	$>8^d/$		$5.83/$ $0.0125*$ Pr	(107, 182)
IMP 4	Pr $>1/4$ MeV		$\lceil 15^h \pm 30^m \Delta 3/$ $\lfloor 15^h 15^m \pm 30^m$	$16^d 01^h \Delta 2/$ $15^d 23^h 30^m$	$>8^d/$		92.1/2.24* Pr	(181, 182)
IMP 4	Pr $>29/>0.8$ MeV		$\lceil 16^h \pm 60^m/$ $\lfloor 16^h 15^m \pm 15^m \Delta 3$	$18^h 40^m/$ $23^h \Delta 2$	$75^h/>8^d$		0.21/218 Pr	(107)

Source: ◇ GMS begins $15^d 15^h$, the event is sequential with event of October 19 (in Appendix A)

	1967	Nov. 23		event in Appendix A				

221	1967	November 27	$>05^h$				$-2\,0\,0$
IMP 4	Pr$>9/>1$ MeV	$\begin{bmatrix}05^h\Delta2/ \\ 07^h20^m\pm60^m\Delta3\end{bmatrix}$	$28^d00^h/13^h\Delta2$	$35^h/>6^d$		0.025/ 35.9 Pr	(107, 181)
IMP 4	Pr 0.5 MeV	08^h30^m	$28^d14^h\Delta1$			400* Pr	(175)

Source: origin unknown

	1967	Nov. 30 (twice), Dec. 1 (three times), Dec. 2		events in Appendix A			
222	1967	December 3	09^h10^m				$1\,2\,0$
IMP 4	El>70 keV	09^h10^m					(176)
AIMP 2	El>40 keV	$09^h18^m\pm02^m$				850 El	(69)
IMP 4	El$>2.7/>0.3$ MeV	09^h20^m	$/13^h\Delta1$			0.75/18.5 El	(166)
IMP 4	Pr$>29/>9$ MeV	$\begin{bmatrix}09^h25^m\pm30^m/ \\ \pm05^m\Delta8\end{bmatrix}$	$12^h50^m/16^h40^m$	$120^h/125^h$		9.5/30.9 Pr	(107)
IMP 4	Pr$>0.8/>94$ MeV	$\begin{bmatrix}09^h25^m\pm15^m\Delta8/ \\ 09^h35^m\pm10^m\end{bmatrix}$	$4^d03^h/3^d11^h\Delta2$	$125^h/40^h$		348/1.48 Pr	(107)
IMP 4	Pr>60 MeV	$09^h35^m\pm10^m$	12^h10^m	40^h		3.88 Pr	(173)
IMP 4	Pr 19/4 MeV	$\begin{bmatrix}10^h20^m\pm10^m/ \\ 10^h37^m\pm15^m\end{bmatrix}$	$12^h30^m/ 15^h30^m$			0.265*/ 6.12* Pr	(182)
PION 7	Pr$>165/>12$ MeV	$<11^h/11^h30^m\Delta5$	$15^h/4^d02^h\Delta2$			0.03/0.78 Pr	(119)
PION 7	Pr>73 MeV	$12^h\Delta5$	$19^h\Delta5$			0.1 Pr	(119)
RIOM	PCA	14^h				1.8 dB	(169)
(PION 7 at 1.08 AU, ES $-$ 39°)							

Source: □ active region 9091 is one day beyond W limb; type II 0854–0904

	1967	Dec. 9	event in Appendix A				
223	1967	December 11	$>08^h$				$-2\,0\,0$
IMP 4	Pr$>1/>9$ MeV	$08^h\Delta2/18^h\pm90^m$	$12^d05^h\Delta1$	$>22^h/>11^h$	0.2/0.04 Pr	(181, 107)
IMP 4	Pr>29 MeV	$23^h\Delta2$	$12^d05^h\Delta1$	06^h	0.007 Pr	(107)

Source: ◇ possibly particle stream related to development of region 9108 near CM

224	1967	December 12	$>02^h30^m$				$-1\,0\,0$
PION 7	Pr$>12/>0.6$ MeV	$02^h30^m/03^h\Delta3$	$12^h\Delta2/?\Delta5$		1.3 Pr/	(119)
IMP 4	Pr$>10/>29$ MeV	$04^h\Delta2/06^h\Delta3$	$17^h/10^h\Delta2$	$>2^d/8^h$	0.16/0.02 Pr	(173, 107)
IMP 4	Pr>1 MeV	$06^h30^m\pm60^m$	$13^d01^h\Delta2$	$>4^d$	5.1 Pr	(181)
(PION 7 at 1.09 AU, ES $-$ 41°)						

Source: ● flare 11^d2347 S22 W17 2B McM 9108

225	1967 December 16	04^h30^m				0 1 0
AIMP 2	El >40 keV	$>04^h20^m$			100 El	(69)
PION 8	Pr >14 MeV	$04^h30^m \pm 30^m \Delta 8$	21^h30^m	32^h	9.1 Pr	(183)
IMP 4	Pr >0.8 MeV	$04^h30^m \pm 30^m \Delta 8$	$15^h40^m \Delta 8$		132 Pr	(107)
IMP 4	El $>2.7/>0.3$ MeV	04^h37^m	$/16^h \Delta 1$		0.6/5 El	(166)
IMP 4	Pr $>29/19$ MeV	$\lceil 04^h40^m \pm 30^m/$	$06^h30^m/$		0.83/	(107, 182)
		$\lfloor 04^h45^m \pm 15^m$	23^h30^m		0.0216* Pr	
IMP 4	Pr >94 MeV	$05^h \Delta 1$	07^h (2nd max. 17^h)		0.16 Pr	(107)
IMP 4	Pr $>60/>10$ MeV	$05^h \pm 20^m$	$22^h \Delta 2/21^h \Delta 1$	$25^h/>39^h$	0.33/4.64 Pr	(173)
IMP 4	Pr 4/>1 MeV	$05^h30^m \pm 30^m$	$\lceil 17^d07^h \Delta 2/$	$/>39^h$	0.75*/	(182, 181)
			$\lfloor 16^d15^h30^m$		41.9 Pr	
f-min	PCA	$07^h \Delta 1$		88^h		(203)
RIOM	PCA		$17^h \Delta 1$		0.8 dB	(169)

(PION 8 at 0.99 AU, ES + 0.5°)

Source: ● flare 0247 N23 E66 3N McM 9118

226	1967 December 17	12^h				0 0 0
IMP 4	Pr $>9/>1$ MeV	$\lceil 12^h \pm 30^m/$	$12^h50^m/$	6^h	3/30 Pr	(107, 181)
		$\lfloor 12^h \pm 10^m \Delta 3$	$14^h \Delta 2$			
PION 8	Pr >14 MeV	$12^h \pm 20^m \Delta 3$	12^h30^m	7^h	5.3 Pr	(183)
IMP 4	Pr >29 MeV	$12^h05^m \pm 05^m \Delta 3$	12^h45^m	6^h	0.5 Pr	(107)

(PION 8 at 0.98 AU, ES + 1°)

Source: ○ flare 0440 N19 E03 1F McM 9115
□ active region 9108 one day beyond W limb

227	1967 December 17	19^h				0 1 0
IMP 4	Pr >94 MeV	$19^h \pm 15^m$	20^h50^m	18^h	0.3 Pr	(107)
IMP 4	El >2.7 MeV	19^h02^m			0.07 El	(166)
IMP 4	Pr >29 MeV	$19^h05^m \pm 20^m \Delta 3$	$21^h \Delta 1$	23^h	1.84 Pr	(107)
AIMP 2	El >40 keV	$19^h11^m \pm 03^m$			200 El	(69)
IMP 4	Pr 19 MeV	$19^h15^m \pm 15^m$	20^h30^m		0.0406* Pr	(182)
PION 8	Pr >14 MeV	$19^h20^m \pm 20^m \Delta 3$	20^h30^m	23^h	11.1 Pr	(183)
IMP 4	Pr $>60/>30$ MeV	$19^h30^m \pm 20^m$	$21^h \Delta 2$	$17^h/35^h$	0.5/1.64 Pr	(173)
IMP 4	Pr $>10/>0.8$ MeV	$\lceil 19^h30^m \pm 20^m/$	$22^h/21^h40^m$	$>23^h/24^h$	5.14/2880 Pr	(173, 107)
		$\lfloor \pm 70^m \Delta 3$				
IMP 4	Pr 4/>1 MeV	$\lceil 20^h45^m \pm 15^m/$	$21^h30^m/$	$/25^h$	0.805*/	(182, 181)
		$\lfloor 20^h \Delta 3$	$18^d17^h \Delta 2$		31.2 Pr	
RIOM	PCA		21^h		0.7 dB	(169)

Source: □ type II 1838–1841; 1843–1854; type IV 1900–1908, cont. (Dkm) 1841–1900; SWF, imp 1, 1834–2010
○ flare 1816 N19 W09 SB McM 9115

228	1967 December 18	$>18^h$				0 0 0
IMP 4	Pr $>0.8/29$ MeV	$\begin{bmatrix}18^h\Delta3/\\18^h20^m\pm50^m\Delta3\end{bmatrix}$	$\begin{matrix}19^d02^h/\\18^d22^h\Delta2\end{matrix}$	$160^h/60^h$	440/0.5 Pr	(107)
IMP 4	Pr $>10/>1$ MeV	$19^h/21^h\Delta3$	$\begin{bmatrix}23^h\Delta2/\\19^d01^h30^m\end{bmatrix}$	$50^h/>7^d$	3.3/188 Pr	(173, 181)

Source: ○ flare 1434   N20   E32   1N   McM 9118
          ○ flare $<$1434   N18   W15   SN   McM 9115
          △ GMS begins with sc 18^d0538 (associated with flare of event 225)

	1967	Dec. 23, Dec. 26, Dec. 28		events in Appendix A		
229	1967	December 29	$\sim01^h$			$-1\,0\,0$
IMP 4	Pr $>9/>0.8$ MeV	$00^h50^m\pm10^m\Delta8$	$\begin{bmatrix}01^h30^m/\\01^h10^m\Delta8\end{bmatrix}$	$32^h/33^h$	0.12/53 Pr	(107)
AIMP 2	El >40 keV	$01^h10^m\pm20^m$			60 El	(69)
PION 8	Pr >11 MeV	$01^h40^m\pm20^m\Delta3$	02^h40^m	35^h	0.1 Pr	(183)

(PION 8 at 0.94 AU, ES + 2°)

Source: ⊙ flare 0047   S27   W78   1N   McM 9120
          ⊘ flare 0106   S15   W22   1B   McM 9128

	1967	Dec. 29	event in Appendix A			
230	1967	December 30	$\sim09^h$			$-2\,0\,0$
IMP 4	Pr $>9/>1$ MeV	$\begin{bmatrix}09^h\pm100^m/\\09^h30^m\pm30^m\end{bmatrix}$	$17^h\Delta2/11^h\Delta1$	$50^h/16^h$	0.04/1.21 Pr	(107, 181)
PION 7	Pr >0.6 MeV	$28^d11^h\Delta5$	$\begin{bmatrix}28^d14^h\Delta2\\\text{2nd max. }30^d15^h\end{bmatrix}$	$>2^d$	3.1 Pr	(119)
IMP 4	Pr >29 MeV	$16^h\Delta2$	$31^d02^h\Delta4$	33^h	0.02 Pr	(107)

(PION 7 at 1.07 AU, ES − 43°)

Source: ◇ corotating structure sequential with geomagnetic disturbance, gradual geomagnetic storm begins 31^d03^h at
          Earth

231	1967	December 30	$<13^h$			(0) 0 0
PION 6	Pr $>175/>0.6$ MeV	$13^h\Delta5$	$14^h\Delta1/17^h\Delta5$		1.17/58 Pr	(104)
PION 6	Pr >13 MeV	$14^h\Delta5$	$18^h\Delta1$		9.5 Pr	(104)

(PION 6 at 0.85 AU, ES + 120°)

Source: □ flare invisible hemisphere? Region McM 9115 went over W limb December 23 for Earth

	1967	Dec. 31	event in Appendix A	

232	1968	January 5	14^h					$-1\,0\,0$

IMP 4	Pr $>9/>0.8$ MeV	$14^h\Delta2$	$22^h\Delta2/18^h30^m$	$24^h/48^h$	0.1/14.2 Pr	(107)
PION 8	Pr >11 MeV	$20^h\Delta2$	$6^d02^h\Delta2$	36^h	0.014 Pr	(183)
(PION 8 at 1 AU, ES $= 0°$)						

Source: ● flare 0458 N12 E72 2B McM 9146

	1968	Jan. 8, Jan. 9 (twice), Jan 11		events in Appendix A	

233	1968	January 11	$\sim18^h$			$-1\,(1)\,0$

AIMP 2	El >40 keV	17^h18^m			30 El	(226)
IMP 4	Pr $>29/>9$ MeV	$\begin{bmatrix}17^h30^m\pm90^m/\\18^h\pm30^m\end{bmatrix}$	$/20^h\Delta1$	$7^h/>7^h$	0.055/ 0.29 Pr	(107)
PION 8	Pr >14 MeV	$18^h30^m\pm60^m$	$12^d00^h\Delta2$	8^h	0.18 Pr	(183)
RIOM	PCA	18^h40^m				(184)
IMP 4	Pr >0.8 MeV	$19^h30^m\pm30^m\Delta3$	$21^h\Delta1$	$>6^h$	35 Pr	(107)
IMP 4	El >70 keV	$20^h\Delta2$			75 El	(176)

Source: ● flare 1659 S25 W38 1B McM 9145

234	1968	January 12	01^h			$0\,(1)\,0$

IMP 4	Pr $>29/>9$ MeV	$/01^h15^m\pm15^m\Delta3$	$\begin{bmatrix}02^h40^m/\\03^h30^m\end{bmatrix}$	$28^h/>17^h$	0.15/1.48 Pr	(107)
IMP 4	Pr >1 MeV	$01^h15^m\pm15^m\Delta3$	$04^h\Delta1$		14.9 Pr	(181)
IMP 4	Pr >0.8 MeV	$01^h20^m\pm20^m\Delta8$	03^h30^m	$>17^h$	49 Pr	(107)
AIMP 2	El >40 keV	01^h21^m			110 El	(226)
PION 8	Pr >14 MeV	$02^h\pm60^m\Delta3$	04^h30^m	16^h	1 Pr	(183)
RIOM	PCA		$03^h\Delta1$		0.7 dB	(184)

Source: ☉ flare 0019 N11 W24 SN McM 9146

235	1968	January 12	18^h30^m			$0\,0\,0$

IMP 4	Pr >9 MeV	$18^h30^m\pm25^m\Delta3$	20^h30^m	8^h	3.55 Pr	(107)
IMP 4	Pr >0.8 MeV	$18^h30^m\pm100^m\Delta8$	$19^h\Delta2$	12^h	3770 Pr	(107)
AIMP 2	El >40 keV	$19^h40^m\pm70^m$		$-\cdot\cdot$	50 El	(226)

Source: ● flare 1807 S25 W53 2B McM 9145

	1968	Jan. 12		event in Appendix A	

236 1968 January 14 13^h30^m -200

IMP 4	Pr >9 MeV	$13^h30^m\pm30^m$	$15^d09^h\Delta2$	48^h	0.05 Pr	(107)
IMP 4	Pr >1 MeV	$22^h\Delta3$	$15^d17^h\Delta2$	$>61^h$	0.92 Pr	(181)
PION 8	Pr >11 MeV	$22^h20^m\pm40^m\Delta3$	$15^d12^h\Delta2$	60^h	0.023 Pr	(183)

(PION 8 at 1 AU, ES $-0.2°$)

Source: ⊙ flare 0725 N24 W09 1B McM 9153

1968 Jan. 15 event in Appendix A

237 1968 January 17 07^h -100

AIMP 2	El >40 keV	$07^h10^m\pm40^m$			170 El	(226)
IMP 4	Pr >9 MeV	$09^h\Delta4$	$16^h\Delta4$	18^h	0.1 Pr	(107)
PION 8	Pr >14 MeV	$09^h\pm40^m$	$10^h\Delta1$	15^h	0.125 Pr	(183)
IMP 4	Pr >1 MeV	$11^h20^m\pm20^m\Delta4$	$15^h\Delta4$	53^h	3.1 Pr	(181)

Source: ● flare 0534 N08 W90 1N McM 9146

1968 Jan. 24, Jan. 30 events in Appendix A

238 1968 February 1 09^h -100

IMP 4	Pr >0.8 MeV	$09^h\pm60^m\Delta3$	17^h30^m	$>9^h$	6 Pr	(107)
IMP 4	Pr >9 MeV	$10^h20^m\pm30^m$	17^h30^m	$>8^h$	0.17 Pr	(107)

Source: ◇ increase in structure of permanent particle flux associated with active region 9184 at CM on Jan. 31
⦿ flare 31^d2129 N15 W18 2N McM 9184

---

239 1968 February 1 18^h30^m -100

AIMP 2	El >40 keV	18^h15^m			45 El	(226)
IMP 4	El >70 keV	18^h20^m			40 El	(176)
IMP 4	Pr $>9/>0.8$ MeV	$\begin{bmatrix}18^h25^m/\\18^h30^m\pm20^m\Delta8\end{bmatrix}$	19^h30^m	$>12^h$	0.15/30 Pr	(107)
IMP 4	El >0.3 MeV	$18^h34^m\pm3^m$			0.3 El	(166)
IMP 4	Pr >1 MeV	$19^h\pm60^m\Delta3$	$2^d00^h\Delta1$	$>12^h$	1.7 Pr	(181)
PION 8	Pr >14 MeV	$19^h\Delta1$	20^h30^m	4^h	0.09 Pr	(183)

(PION 8 at 1.01 AU, ES $-0.3°$)

Source: ⊙ flare 1756 S15 E48 SN McM 9193
⊙ flare 1802 $\begin{bmatrix}$ N09 W35 SN McM 9184 \\ N17 W20 $\end{bmatrix}$

1968 Feb. 1 event in Appendix A

240	1968 February 2	06^h					$-1\,0\,0$
AIMP 2	El $>40\,keV$	05^h59^m			110 El	(226)
IMP 4	Pr $>9/>0.8\,MeV$	$06^h\pm10^m\Delta3/\Delta8$	$07^h\Delta1/20^h\Delta2$	$30^h/>30^h$	0.16/54 Pr	(107)
PION 8	Pr $>14\,MeV$	$06^h\Delta2$	$07^h\Delta2$	5^h	0.094 Pr	(183)
IMP 4	Pr $>1\,MeV$	$07^h\Delta3$	$20^h\Delta2$	$>30^h$	1.9 Pr	(181)

Source: ● flare 0541   N15   W24   1B   McM 9184

	1968 Feb. 5	event in Appendix A					
241	1968 February 6	05^h					$-2\,0\,0$
IMP 4	Pr $>9/>1\,MeV$	$05^h/07^h\Delta2$	$19^h/14^h\Delta2$	$18^h/10^h$	0.06/0.29 Pr	(107, 181)
IMP 4	Pr $>0.8\,MeV$	$07^h\Delta2$	$09^h\Delta2$	18^h	7 Pr	(107)

Source: □ region 9184 is just beyond W limb
　　　　○ flare 2340   N17   W79   1F   McM 9184

	1968 Feb. 6, Feb. 8	events in Appendix A					
242	1968 February 8	$>14^h$					$0\,1\,0$
IMP 4	El $>70\,keV$	14^h10^m			140 El	(176)
AIMP 2	El $>40\,keV$	14^h14^m			130 El	(226)
IMP 4	Pr $>0.8\,MeV$	$14^h15^m\pm05^m\Delta8$	$20^h\Delta1$	3^d	55 Pr	(107)
IMP 4	El $>2.7/>0.3\,MeV$	14^h18^m	$/15^h\Delta1$		0.07/1.6 El	(166)
IMP 4	Pr $>94/>10\,MeV$	$14^h20^m\pm20^m$	$15^h/17^h\Delta1$	$5^h/13^h$	0.05/1.61 Pr	(107, 173)
PION 8	Pr $>14\,MeV$	$14^h20^m\pm20^m$	15^h30^m	$<10^h$	3.2 Pr	(183)
IMP 4	Pr $>29\,MeV$	$14^h35^m\pm20^m$	16^h20^m	20^h	0.48 Pr	(107)
RIOM	PCA	$15^h\Delta1$	$17^h\Delta1$		0.6 dB	(184)
IMP 4	Pr $>1\,MeV$	$16^h30^m\pm30^m\Delta3$	20^h30^m	3^d	23 Pr	(181)
(PION 8 at 1.02 AU, ES $-1°$)						

Source: □ type II 1402–1405; type III 1359–1410; region 9184 is two days beyond W limb
　　　　▲ geomagnetic storm modulation, sc begins 10^d1621

243	1968 February 12	09^h					$-1\,0\,0$
IMP 4	Pr $>0.8/>9\,MeV$	$09^h/12^h\Delta4$	$13^d17^h/16^h\Delta2$	$>68^h$	196/0.38 Pr	(107)
IMP 4	Pr $>1\,MeV$	$12^h\Delta4$	$13^d16^h\Delta2$	$>68^h$	92 Pr	(181)
PION 8	Pr $>14\,MeV$	$16^h30^m\pm30^m$		48^h	0.11 Pr	(183)

Source: ◇ particle stream related to CM transit of active region 9204

244	1968	February 15	08^h					−1 0 0
PION 8	Pr > 11 MeV	08^{h}Δ3	17^{h}Δ2	>43^h	0.09 Pr	(183)
IMP 4	Pr > 9/>0.8 MeV	08^{h}Δ3	13^h/20^{h}Δ2	>43^h	0.12/164 Pr	(107)

Source: ◇ geomagnetic storm modulation, storm begins 15^{d}07^h
  ◇ new region 9211 forms on the disk on 14 and is moderately active
  ○ flare 0301   S25   E22   SB   McM 9211
  ○ flare 0034   N20   W10   1N   McM 9204

245	1968	February 17	03^{h}10^m					0 1 0
IMP 4	El > 70 keV	03^{h}10^m			190 El	(176)
AIMP 2	El > 40 keV	03^{h}15^m ± 7^m			250 El	(226)
IMP 4	El > 0.3 MeV	03^{h}17^m ± 3^m	09^{h}Δ1	12^h	0.7 El	(166)
PION 8	Pr > 14 MeV	03^{h}20^m ± 20^m	09^{h}Δ5	26^h	1.4 Pr	(183)
IMP 4	Pr > 29/>9 MeV	⌈03^{h}25^m ± 5^m Δ3/ ⌊Δ8	05^{h}30^m/ 08^{h}30^m	38^h/45^h	0.47/3 Pr	(107)
IMP 4	Pr > 0.8 MeV	03^{h}25^m ± 5^m Δ8	09^{h}Δ1	65^h	373 Pr	(107)
IMP 4	Pr > 1 MeV	05^{h}Δ3	11^{h}Δ1	75^h	33 Pr	(181)
RIOM	PCA		09^{h}Δ1		0.4 dB	(172)

Source: ● flare 0252   N17   W47   1B   McM 9204

	1968	Feb. 18	event in Appendix A					
246	1968	February 21	20^{h}45^m					−1 0 0
PION 8	Pr > 14 MeV	20^h ± 60^m	23^{h}Δ5	12^h	0.09 Pr	(183)
IMP 4	Pr > 9/>1 MeV	⌈20^{h}45^m ± 15^m/ ⌊22^h ± 30^m	22^{d}01^{h}20^m/ 06^{h}Δ1	20^h/64^h	0.14/8.2 Pr	(107, 181)
PION 7	Pr > 0.6 MeV	22^{d}00^{h}Δ2	22^{d}06^{h}Δ2	>48^h	2.5 Pr	(107)

(PION 7 at 1.12 AU, ES − 54°; PION 8 at 1.03 AU, ES − 1.5°)

Source: □ active region McM 9204 is on invisible hemisphere, one day beyond W limb, type II 2025
  ○ flare 2010   N17   E73   SF   McM 9222

247	1968	February 26	~05^h					−1 0 0
IMP 4	Pr > 9/>29 MeV	04^{h}30^m/05^{h}Δ2	13^{h}20^m/14^{h}Δ2	40^h	0.14/0.015 Pr	(107)
IMP 4	Pr > 0.8 MeV	06^{h}Δ2	19^{h}30^m	45^h	12.5 Pr	(107)
PION 8	Pr > 11 MeV	09^{h}Δ5	27^{d}00^{h}Δ2	53^h	0.099 Pr	(183)

(PION 8 at 1.03 AU, ES − 1.5°)

Source: ◇ sequential particle stream related to return of active region 9184 to CM; this event is sequential with event of
  Feb. 1
  ○ flare   0441   S25   E24   1N   McM 9224
  ○ flare   0246   S29   E25   SN   McM 9224

248	1968 March 6	$>12^h$					$-1\,0\,0$
IMP 4	Pr >0.8 MeV	$12^h\Delta2$	$7^d22^h\Delta2$	$>46^h$	18 Pr	(107)	
PION 8	Pr >11 MeV	21^h	$7^d11^h\Delta2$	47^h	0.0034 Pr	(183)	
IMP 4	Pr >9 MeV	$7^d01^h\Delta2$	$7^a13^n\Delta5$	24^h	0.13 Pr	(107)	
(PION 8 at 1.04 AU, ES $-2°$)							

Source: $\Diamond$ sequential particle stream related to return of active region 9204 at E limb on Mar. 6; this event is sequential
with event of Feb. 8

	1968 March 8	event in Appendix A					
249	1968 March 10	00^h30^m					$-1\,0\,0$
IMP 4	Pr $>1/>0.8$ MeV	$00^h30^m \pm 10^m\Delta3$	$02^h\Delta1/00^h45^m$	$6^h/5^h$	2.1/27.4 Pr	(181, 107)	
IMP 4	Pr >9 MeV	$00^h35^m \pm 05^m$	00^h45^m	5^h	0.24 Pr	(107)	
PION 8	Pr >11 MeV	$02^h40^m \pm 20^m\Delta3$	$05^h\Delta2$	4^h	0.007 Pr	(183)	
(PION 8 at 1.04 AU, ES $-2.5°$)							

Source: $\blacktriangle$ geomagnetic storm modulation, sc begins 9^d2340

250	1968 March 10	05^h					$-1\,0\,0$
IMP 4	Pr >9 MeV	$05^h \pm 20^m\Delta3$	$08^h\Delta1$	$>9^h$	0.7 Pr	(107)	
AIMP 2	El >40 keV	05^h22^m			40 El	(226)	
IMP 4	Pr >0.8 MeV	$05^h25^m \pm 05^m\Delta8$	13^h30^m	$>9^h$	28.5 Pr	(107)	
IMP 4	Pr >1 MeV	$06^h30^m \pm 30^m\Delta3$	$13^h\Delta1$	$>9^h$	11.3 Pr	(181)	
PION 8	Pr >11 MeV	$06^h40^m \pm 20^m\Delta3$	09^h30^m	$>7^h$	0.051 Pr	(183)	

Source: origin unknown;
          type III + V, 0505–0508, no flare reported

251	1968 March 10	14^h					$-1\,0\,0$
PION 8	Pr >11 MeV	$14^h\Delta3$	$15^h\Delta1$	$>2^d$	0.051 Pr	(183)	
IMP 4	Pr >9 MeV	$14^h30^m \pm 20^m\Delta3$	15^h10^m	27^h	0.7 Pr	(107)	
IMP 4	Pr >0.8 MeV	$14^h30^m \pm 30^m\Delta3$	$11^d08^h30^m$	$>4^d$	143 Pr	(107)	
IMP 4	Pr >1 MeV	$15^h \pm 30^m\Delta3$	$11^d05^h\Delta2$	$>4^d$	35.9 Pr	(181)	

Source: origin unknown

252	1968	March 14	21^h					-200
PION 8	Pr > 11 MeV		$20^h \pm 20^m$	$16^d08^h\Delta2$	$>3^d$		0.0085 Pr	(183)
IMP 4	Pr > 1/>0.8 MeV		$21^h\Delta3$	$16^d02^h\Delta2$	$>3^d$		3.1/10 Pr	(181, 107)
IMP 4	Pr > 9 MeV		$22^h\Delta2$	$16^d00^h\Delta2$	40^h		0.02 Pr	(107)

(PION 8 at 1.04 AU, ES $- 3°$)

Source: ◊ sequential geomagnetic disturbance begins gradually 14^d00^h
  ○ flare 1000   N14   E35   SN,   McM 9262
  ○ flare 1008   S20   W53   1F   McM 9268

253	1968	March 21	15^h30^m					-200
IMP 4	Pr > 9 MeV		$15^h30^m \pm 10^m$	$22^d00^h10^m$	27^h		0.07 Pr	(107)
IMP 4	Pr > 0.8 MeV		$15^h30^m \pm 30^m$	$22^d07^h\Delta2$	35^h		51 Pr	(107)
IMP 4	Pr > 29 MeV		$16^h \pm 40^m$	$23^h\Delta2$	22^h		0.016 Pr	(107)
PION 8	Pr > 11 MeV		$16^h\Delta1$	$22^d01^h\Delta2$	$>44^h$		0.062 Pr	(183)
IMP 4	Pr > 1 MeV		$17^h\Delta2$	$22^d04^h\Delta2$	$>43^h$		1 Pr	(181)

Source: ⊙ flare 1421   N17   W54   1B   McM 9266

	1968	March 23	event in Appendix A					
254	1968	March 24	22^h					-200
IMP 4	Pr > 1/>0.8 MeV		$22^h\Delta2$	$\begin{bmatrix}27^d07^h15^m/ \\ 05^h\Delta2\end{bmatrix}$	$>5^d$		12.7/220 Pr	(181, 107)
PION 7	Pr > 12 MeV		$24^d00^h\Delta2$				0.25 Pr	(119)
PION 8	Pr > 11 MeV		$22^h\Delta2$	$26^d13^h\Delta5$	$>3^d$		0.0057 Pr	(183)

(PION 7 at 1.13 AU, ES $- 59°$; PION 8 at 1.04 AU, ES $- 3.5°$)

Source: ◊ CMP of region 9273 which is a return of 9221 + 9185
  ○ flare 1635   S12   W02   1B   McM 9273
  ○ flare 0737   S12   E03   1B   McM 9273

255	1968	March 27	12^h					-200
IMP 4	Pr > 1/>9 MeV		$12^h\Delta3/14^h\Delta2$	$15^h50^m/22^h\Delta2$	$6^h/>15^h$		12.8/0.04 Pr	(181, 107)
IMP 4	Pr > 0.8 MeV		$15^h30^m \cdot \pm 30^m\Delta3$	$20^h\Delta2$	$>15^h$		18 Pr	(107)

Source: origin uncertain, numerous subflares in various regions
  ⊘ flare < 1757   S12   W42   2B   McM 9273

	1968	March 28	event in Appendix A					

256	1968	March 30	01^h					$-1\,0\,0$
AIMP 2	El >40 keV	00^{h}52^m				130 El	(226)
IMP 4	El $>2.7/>0.3$ MeV	01^{h}01^m	/04^{h}30^m	/12^h	0.02/0.4 El	(166)
PION 8	Pr >14 MeV	01^{h}20^m ± 20^m Δ8	05^{h}Δ8	10^h	0.2 Pr	(183)

(PION 8 at 1.04 AU, ES $-4.5°$)

Source: ⊙ flare 0025   S15   W65   SF   McM 9273

257	1968	March 31	19^h					$-1\,0\,0$
IMP 4	Pr $>9/>0.8$ MeV	19^{h}Δ2	1^{d}00^{h}Δ2	30^h/55^h	0.1/100 Pr	(107)
IMP 4	Pr >1 MeV	23^{h}Δ2	1^{d}17^{h}Δ2	$>48^h$	0.02 Pr	(181)

Source: □ active region 9273 is just beyond W limb
◇ sequential geomagnetic disturbance begins 31^{d}20^h

258	1968	April 4	14^h					$(-1)\,0\,0$
IMP 4	Pr $>9/>1$ MeV	14^{h}Δ2	5^{d}03^h/02^{h}Δ2	14^h/34^h	0.15/0.14 Pr	(107, 181)
IMP 4	Pr >0.8 MeV	14^{h}Δ2	5^{d}03^{h}Δ2	40^h	32 Pr	(107)

Source: ◇ growth on disk of bright region 9302 at CM on April 3
◇ geomagnetic disturbance begins with sc 5^{d}1328

259	1968	April 6	7^h					$-2\,0\,0$
IMP 4	Pr $>0.8/>1$ MeV	07^{h}Δ2	08^h/7^{d}20^{h}Δ2	50^h/$>48^h$	20/0.8 Pr	(107, 181)
IMP 4	Pr >9 MeV	08^{h}Δ2	15^{h}Δ2	8^h	0.02 Pr	(107)

Source: ◇ geomagnetic storm modulation; sc begins 5^{d}1328
○ flare 0706   S22   W65   1N   McM 9289

	1968 ⎮ April 9	event in Appendix A						
260	1968 April 12	20^h						$-2\,0\,0$
IMP 4	Pr >1 MeV	20^{h}Δ3	14^{d}22^{h}Δ2	$>7^d$	5.5 Pr	(181)
IMP 4	Pr >0.8 MeV	20^{h}Δ3	15^{d}00^{h}Δ2	$>5^d$	12 Pr	(107)
PION 8	Pr >11 MeV	24^{h}Δ5	14^{d}15^{h}Δ2	$>5^d$	0.04 Pr	(183)
IMP 4	Pr >9 MeV	13^{d}09^{h}Δ4	14^{d}20^{h}Δ2	4^d	0.06 Pr	(107)

(PION 8 at 1.05 AU, ES $-5°$)

Source: ◇ sequential geomagnetic disturbance begins 12^{d}19^h
◇ particle stream related to CMP of new active regions (9306–9313–9315–9322)

261	1968 April 17	18^h				$-2\,0\,0$
IMP 4	Pr >0.8 MeV	$18^h\Delta4$	$18^d02^h\Delta2$	1^d	9.5 Pr	(107)
IMP 4	Pr >9 MeV	$19^h40^m\pm40^m$	$18^d03^h\Delta2$	1^d	0.02 Pr	(107)
PION 8	Pr >11 MeV	$18^d02^h\Delta3$	$19^d16^h\Delta2$	$>40^h$	0.0062 Pr	(183)

(PION 8 at 1.06 AU, ES $-\,6°$)

Source: □ bright region McM 9301 is 2 days beyond W limb
　　　　○ flare <0610　S12　E56　1N　McM 9327

262	1968 April 19	18^h				$-2\,0\,0$
IMP 4	Pr >9 MeV	$18^h\Delta2$	$20^h\Delta1$	20^h	0.015 Pr	(107)
IMP 4	Pr >0.8 MeV	$18^h\Delta2$	$20^d00^h\Delta2$	48^h	3.2 Pr	(107)
PION 8	Pr >11 MeV	$18^h40^m\pm20^m\Delta3$	$20^d21^h\Delta2$	$>75^h$	0.018 Pr	(183)
IMP 4	Pr >1 MeV	$19^h\Delta3$	$20^d01^h\Delta2$	$>3^d$	1.1 Pr	(181)

Source: ⊙ flare <1640　N20　W62　SN　McM 9313

263	1968 April 22	21^h				$-2\,0\,0$
PION 8	Pr >11 MeV	$21^h\Delta3$	$23^d03^h\Delta2$		0.0043 Pr	(183)
IMP 4	Pr >0.8 MeV	$23^d03^h\Delta1$	$23^d07^h\Delta1$	6^h	4 Pr	(107)
IMP 4	Pr >9 MeV	$23^d05^h30^m$	$23^d06^h30^m$	6^h	0.015 Pr	(107)

Source: □ type IV (Dkm) 1752–1853, imp. 1; SWF, imp. 3, 1800–2100; flare invisible hemisphere?

264	1968 April 23	12^h40^m				$-1\,0\,0$
IMP 4	El >0.3 MeV	$12^h40^m\pm15^m$	$15^h\Delta1$		0.4 El	(166)
IMP 4	Pr >0.8 MeV	$12^h40^m\pm10^m\Delta8$	15^h45^m	70^h	19 Pr	(107)
IMP 4	El >40 keV	12^h50^m			75 El	(226)
PION 8	Pr >14 MeV	$13^h\Delta2$	$16^h\Delta2$	24^h	0.13 Pr	(183)
IMP 4	Pr >9 MeV	$13^h25^m\pm10^m$	$16^h\Delta1$	65^h	0.16 Pr	(107)
IMP 4	Pr >1 MeV	$16^h\Delta2$	$25^d13^h\Delta2$	69^h	2.5 Pr	(181)

Source: □ active regions 9313 and 9315 are 2 days beyond W limb
　　　　○ flare 1129　N07　E34　SN　McM 9345

265	1968 April 26	15^h30^m				0 1 0
IMP 4	El >40 keV	15^h48^m			470 El	(226)
IMP 4	El $>2.7/>0.3$ MeV	15^h50^m	$/18^h05^m$	$/48^h$	0.6/14 El	(166)
IMP 4	Pr >0.8 MeV	$15^h55^m \pm 05^m \Delta 8$	18^h45^m	$>25^h$	298 Pr	(107)
IMP 4	Pr >10 MeV	$16^h20^m \pm 20^m$	$20^h\Delta 1$	35^h	6.26 Pr	(173)
PION 8	Pr >14 MeV	$16^h20^m \pm 20^m$	$21^h\Delta 1$	32^h	6.8 Pr	(183)
IMP 4	Pr >29 MeV	$16^h25^m \pm 05^m$	19^h30^m	35^h	0.52 Pr	(107)
RIOM	PCA	16^h30^m	18^h15^m		0.5 dB	(184)
IMP 4	Pr >1 MeV	$17^h30^m \pm 30^m$	$27^d08^h\Delta 2$	$>23^h$	4.7 Pr	(181)

(PION 8 at 1.06 AU, ES $- 6.5°$)

Source: ⊡ type III 1515–1525, type II (Dkm) 1529–1534; event on invisible hemisphere?

	1968 Apr. 27 (three times)	events in Appendix A				
266	1968 April 29	03^h				$-1\,0\,0$
IMP 4	Pr $>1/>0.8$ MeV	$03^h\Delta 3$	$19^h/17^h\Delta 2$	$>42^h/>40^h$	5.5/15 Pr	(181, 173)
IMP 4	Pr >9 MeV	$06^h\Delta 3$	$17^h\Delta 2$	$>36^h$	0.8 Pr	(107)

Source: ◇ growth in newly formed region McM 9358
◯ flare 2038   N09   W85   SN   McM 9337

267	1968 April 30	18^h45^m				$-1\,0\,0$
IMP 4	Pr >9 MeV	$18^h45^m \pm 15^m\Delta 3$	19^h20^m	10^h	0.3 Pr	(107)
IMP 4	Pr >0.8 MeV	$18^h50^m \pm 10^m\Delta 3$	20^h30^m	10^h	60 Pr	(107)
IMP 4	Pr >1 MeV	$21^h\Delta 3$	$22^h\Delta 1$	$>5^h$	3.7 Pr	(181)

Source: ◯ flare 0516   N19   E73   SN   McM 9364
▫ region 9337 is one day beyond W limb

268	1968 May 4	00^h				$-2\,0\,0$
IMP 4	Pr $>9/>0.8$ MeV	$00^h/02^h\Delta 2$	$06^h/12^h\Delta 2$	12^h	0.013/2 Pr	(107)
IMP 4	Pr >1 MeV	$03^h30^m \pm 30^m$	$07^h\Delta 1$	$>12^h$	0.4 Pr	(181)
PION 8	Pr >11 MeV	$<06^h\Delta 5$				(183)

(PION 8 at 1.07 AU, ES $- 7°$)

Source: ● flare 3^d2123 N18   E49   1B   McM 9372

269   1968   May 4 13^h45^m $(-2)\,0\,0$

IMP 4	Pr $>$ 1 MeV	$13^h45^m \pm 30^m$	$5^d19^h\Delta2$	72^h	1.5 Pr	(181)
IMP 4	Pr $>$ 0.8 MeV	$14^h\Delta3$	$6^d16^h\Delta2$	65^h	7.8 Pr	(107)
PION 7	Pr $>$ 12 MeV	$19^h\Delta5$	$5^d00^h\Delta2$	15^h	0.02 Pr	(104)
PION 8	Pr $>$ 11 MeV		$6^d10^h\Delta2$		0.0057 Pr	(183)

(PION 7 at 1.11 AU, ES $-$ 67°; PION 8 at 1.07 AU, ES $-$ 7°)

Source: ⊙ flare 0427   N20   E18   1B   McM 9364
⬦ region 9364 is transiting the CM on May 5–6

       1968   May 8, May 10   events in Appendix A

270   1968   May 11 06^h $-1\,0\,0$

IMP 4	Pr $>$ 0.8 MeV	$06^h\Delta2$	$13^d18^h\Delta2$	$>7^d$	44 Pr	(107)
IMP 4	Pr $>$ 1 MeV	$08^h30^m \pm 30^m$	$12^d19^h\Delta2$		7.9 Pr	(181)
IMP 4	Pr $>$ 9 MeV	$09^h\Delta2$	$13^d18^h\Delta2$	95^h	0.4 Pr	(107)
PION 8	Pr $>$ 11 MeV	$09^h40^m \pm 40^m$	$\Delta7$	5^d		(183)

(PION 8 at 1.08 AU, ES $-$ 8°)

Source: ⬦ sequential particle stream related to geomagnetic storm modulation, sequential geomagnetic storm begins gradually 10^d16^h
⊙ flare 10^d1937 N19   W76   SB   McM 9364

---

271   1968   May 22 $>09^h$ $-2\,0\,0$

PION 7	Pr $>$ 12 MeV	$02^h\Delta2$	$25^d05^h\Delta5$	7^d	0.7 Pr	(119)
IMP 4	Pr $>$ 0.8/$>$9 MeV	$09^h/10^h\Delta4$	$\begin{bmatrix} 28^d14^h/ \\ 26^d04^h\Delta2 \end{bmatrix}$	$12^d/9^d$	16.2/0.08 Pr	(107)
IMP 4	Pr 15 MeV	$12^h\Delta2$	$25^d20^h\Delta2$		0.001* Pr	(185)
PION 8	Pr $>$ 11 MeV	$16^h\Delta2$	$\Delta7$	12^d		(183)

(PION 7 at 1.11 AU, ES $-$ 71°; PION 8 at 1.08 AU, ES $-$ 9°)

Source: ⬦ corotating structure related to zone of active regions (McM 9399, 9400, 9401) transiting near CM
⊘ flare 21^d1952 N19   E90   SN   McM 9410

---

272   1968   June 3 00^h $-2\,0\,0$

IMP 4	Pr $>$ 9 MeV	$00^h \pm 30^m$	$4^d10^h\Delta2$	4^d	0.03 Pr	(107)
PION 7	Pr $>$ 12 MeV		$4^d12^h\Delta5$	$>4^d$	0.004 Pr	(119)
PION 8	Pr $>$ 11 MeV	$12^h\Delta2$	$4^d12^h\Delta2$	$>4^d$	0.0057 Pr	(183)
IMP 4	Pr $>$ 1 MeV	$4^d03^h\Delta4$	$5^d14^h\Delta2$		0.2 Pr	(181)
IMP 4	Pr $>$ 0.8 MeV	$\Delta4$	$4^d13^h\Delta2$	$>3^d$	2 Pr	(107)

(PION 7 at 1.08 AU, ES $-$ 72°; PION 8 at 1.08 AU, ES $-$ 10°)

Source: ⊙ flare 3^d2330 S30   W15   1N   McM 9420
⊙ flare 3^d0253 N09   W15   1N   McM 9419

273	1968	June 7	14^h					−1 0 0
IMP 4		Pr > 9 MeV	14^{h}Δ2	8^{d}14^{h}Δ4	35^h	0.08 Pr	(107)	
IMP 4		Pr > 0.8 MeV	14^{h}Δ2	8^{d}03^{h}Δ2	>36^h	5 Pr	(107)	
IMP 4		Pr > 1 MeV	16^h ± 30^m	8^{d}14^{h}Δ4	26^h	1.4 Pr	(181)	
AIMP 2		El > 22 keV	16^{h}19^m	18^{h}30^m			(189)	

Source: ⊙ flare 1226   S13   E16   1N   McM 9429
⊘ flare < 1610   N21   W35   1N   McM 9423

	1968	June 9	event in Appendix A				

274	1968	June 9	09^{h}30^m					2 3 0
IMP 4		El > 2.7/>0.3 MeV	09^{h}22^m/09^{h}20^m ± 3^m			1.7/22 El	(166)	
AIMP 2		El > 40 keV	09^{h}30^m			1000 El	(226)	
IMP 4		Pr > 94 MeV	09^{h}30^m	13^{h}Δ1	45^h	2.6 Pr	(107)	
IMP 4		Pr > 60 MeV	09^{h}35^m ± 05^m	13^{h}Δ1	42^h	5.38 Pr	(173)	
IMP 4		Pr > 30/>10 MeV	09^{h}45^m ± 15^m	14^h/10^{d}06^{h}Δ1	45^h/80^h	12.37/354 Pr	(173)	
Balloon		Pr > 130 MeV		10^{d}09^{h}Δ5		3 Pr	(43)	
OGO 5		Pr > 90 MeV	10^{h}Δ1	17^{h}Δ1	55^h	0.35 Pr	(186)	
RIOM		PCA	10^{h}Δ1	10^{d}08^{h}Δ1	63^h	6.5 dB	(146)	
PION 7		Pr > 73/15 MeV	10^{h}Δ2/Δ5	⎡ 20^{h}Δ5/ ⎣ 10^{d}06^{h}Δ5		1.3/2* Pr	(119, 185)	
OGO 5		El > 12 MeV	11^{h}Δ1	15^{h}Δ1	40^h	0.04 El	(186)	
PION 8		Pr > 14 MeV	11^{h}Δ5	14^{h}Δ1	70^h	43.2 Pr	(183)	

(PION 7 at 1.08 AU, ES − 73°; PION 8 at 1.08 AU, ES − 12°)

Source: ● flare 0830   S14   W09   3B   McM 9429
▲ major geomagnetic storm; sc begins 10^{d}2154

	1968	June 17, June 26	events in Appendix A				

275	1968	June 26	20^{h}30^m					−1 0 0
IMP 4		Pr > 9 MeV	20^{h}30^m ± 30^m	28^{d}05^{h}30^m	35^h	0.09 Pr	(107)	
PION 8		Pr > 11 MeV	27^{d}02^{h}Δ2	Δ7			(183)	

(PION 8 at 1.08 AU, ES − 13°)

Source: ○ flare 1320   N14   W72   SN   McM 9462
○ flare < 2250   S38   E65   2N   McM 9485
○ flare 2330   N14   W78   1N   McM 9462

	1968	June 29, July 5	events in Appendix A				

276   1968   July 6 10^h 0 1 0

PION 7	$Pr > 12\,MeV$	$10^h\Delta2$	$7^d02^h\Delta7$		32 Pr	(119)
PION 7	$Pr > 165/{>}73\,MeV$	$11^h\Delta2$	$20^h\Delta5$	$47^h/36^h$	5/4 Pr	(119)
IMP 4	$El > 40\,keV$				50 El	(226)
OGO 5	$El > 12\,MeV$	$12^h\Delta2$	$7^d12^h\Delta4$	6^d	0.0025 El	(186)
IMP 4	$Pr > 94\,MeV$	$12^h\Delta2$	$7^d06^h\Delta2$	120^h	0.25 Pr	(107)
PION 8	$Pr > 14\,MeV$	$14^h\Delta2$	$7^d13^h\Delta2$	$>61^h$	2.38 Pr	(183)
IMP 4	$Pr > 60/{>}10\,MeV$	$15^h\Delta2$	$7^d12^h\Delta1$	$100^h/{>}65^h$	0.4/1 Pr	(173)
IMP 4	$Pr > 0.8\,MeV$	$15^h \pm 60^m$	$8^d08^h\Delta2$	$>67^h$	35.5 Pr	(107)
IMP 4	$El > 0.3\,MeV$	$15^h15^m \pm 15^m$	$7^d12^h\Delta1$		1 El	(166)
IMP 4	$Pr > 30\,MeV$	$16^h\Delta2$	$7^d12^h\Delta1$	120^h	0.6 Pr	(173)
RIOM	PCA		$8^d18^h\Delta2$		0.4 dB	(172)

(PION 7 at 1.07 AU, ES $-\,77°$; PION 8 at 1.07 AU, ES $-\,14°$)

Source: ■ flare 0943   N13   E89   1N   McM 9503

---

277   1968   July 8 $\sim18^h$ (superposed on event 276 continuing)   1 1 0

AIMP 2	$El > 40\,keV$				250 El	(226)
PION 7	$Pr > 73/{>}165\,MeV$	$8^d18^h/{<}20^h\Delta2$	$\begin{bmatrix}9^d01^h\Delta2/\\00^h\Delta5\end{bmatrix}$	$25^h/40^h$	0.5/1.6 Pr	(119)
PION 8	$Pr > 14\,MeV$	$9^d03^h \pm 20^m\Delta3$	$10^d12^h\Delta2$		14.4 Pr	(183)
IMP 4	$Pr > 9\,MeV$	$9^d05^h\Delta3$	$11^d02^h\Delta2$	$>73^h$	12 Pr	(107)
RIOM	PCA	$9^d09^h\Delta2$	$11^d03^h\Delta1$		1.1 dB	(184)
IMP 4	$Pr > 29\,MeV$	$9^d10^h\Delta3$	$10^d14^h\Delta2$	$>80^h$	0.37 Pr	(107)
IMP 4	$El > 0.3\,MeV$	$9^d22^h\Delta3$	$10^d15^h\Delta2$		1.6 El	(190)

(PION 7 at 1.07 AU, ES $-\,77°$; PION 8 at 1.07 AU, ES $-\,14.5°$)

Source: ● flare $<$1708   N13   E58   3B   McM 9503
        ▲ geomagnetic storm begins with sc $9^d21^h54^m$
        ⊘ flare 9^d1809   N13   E40   2B   McM 9503

---

278   1968   July 12 06^h 0 1 0

IMP 4	$Pr > 10\,MeV$	$05^h30^m \pm 30^m\Delta3$	$09^h\Delta1$	14^h	2.46 Pr	(173)
IMP 4	$Pr > 0.8\,MeV$	$06^h \pm 40^m\Delta3$	$08^h\Delta2$	13^h	650 Pr	(107)
RIOM	PCA	$06^h\Delta3$	$11^h\Delta1$		0.9 dB	(172)

Source: ⊙ flare $<$0000   N12   E10   2N   McM 9503

279	1968 July 12	>14^h				1 2 0
AIMP 2	El >40 keV	14^{h}Δ1			800 El	(226)
IMP 4	El >0.3 MeV	14^{h}30^m	16^{h}Δ2	>4^h	1.2 El	(191)
PION 7	Pr 15 MeV	17^{h}Δ5	13^{d}03^{h}Δ2		0.047* Pr	(185)
IMP 4	El >0.3 MeV	18^{h}Δ3	13^{d}04^{h}Δ2	>22^h	7 El	(190)
IMP 4	Pr >29/>9 MeV	18^{h}Δ3	⌈13^{d}07^{h}Δ1/ ⌊13^{d}10^{h}Δ2	>25^h/>23^h	0.36/19 Pr	(107)
IMP 4	Pr >0.8 MeV	19^{h}Δ3	13^{d}11^{h}Δ2	>19^h	2120 Pr	(107)
PION 8	Pr >14 MeV	22^{h}Δ5	13^{d}04^{h}Δ1	4^d	62.5 Pr	(183)
RIOM	PCA	Δ3	13^{d}19^{h}Δ1		3 dB	(184)

(PION 7 at 1.04 AU, ES − 77°; PION 8 at 1.07 AU, ES − 15°)

Source: ☉ flare <1348   N11   W20   2N   McM 9499 (198)
        or problem (191, 185)

IMP 4	Pr >0.8 MeV	13^{d}14^{h}Δ8	13^{d}18^{h}Δ8	>7^d	1350 Pr	(107)
IMP 4	El >3/>0.3 MeV	13^{d}16^{h}Δ1/Δ3	13^{d}22^{h}Δ2/Δ1	22^h/48^h	3/120 El	(118)
IMP 4	Pr >9 MeV	⌈13^{d}16^{h}20^m ± ⌊20^{m}Δ8	13^{d}21^{h}Δ8	>10^d	95 Pr	(107)
IMP 4	Pr >30 MeV	13^{d}17^{h}Δ8	13^{d}22^{h}30^{m}Δ8	7^d	1.01 Pr	(173)
PION 6	Pr >13 MeV		13^{d}16^{h}Δ5		14 Pr	(104)
OGO 5	El >12 MeV	13^{d}18^{h}Δ2	13^{d}22^{h}Δ2	18^h	0.02 El	(186)

(PION 6 at 0.99 AU, ES + 160°)

Source: ▲ geomagnetic storm begins with sc 13^{d}1612

	1968 July 21	event in Appendix A				
280	1968 July 26	12^h				**−1 0 0**
IMP 4	Pr >0.8 MeV	12^h ± 60^m	17^{h}Δ2	>5^d	3.4 Pr	(107)
IMP 4	Pr >94 MeV	12^{h}50^m ± 10^m	19^{h}Δ2	20^h	0.15 Pr	(107)
IMP 4	El >0.3 MeV	13^{h}05^m	27^{d}00^{h}Δ1	35^h	0.2 El	(166)
IMP 4	Pr >60/>30 MeV	⌈13^{h}30^m ± 30^m/ ⌊±40^m	17^h/19^{h}Δ1	20^h/25^h	0.31/0.4 Pr	(173)
IMP 4	Pr >10 MeV	14^h ± 30^m	19^{h}Δ1	40^h	0.48 Pr	(173)
PION 8	Pr >14 MeV	14^{h}Δ2	27^{d}01^{h}Δ2	48^h	0.23 Pr	(183)
PION 7	Pr >12 MeV	23^{h}Δ5	28^{d}10^{h}Δ2	>4^d	0.14 Pr	(119)

(PION 8 at 1.06 AU, ES − 16°; PION 7 at 1.05 AU, ES − 75°)

Source: origin uncertain
        □ region 9503 is on invisible hemisphere near CM (185, 166)

281	1968 August 3	>06^h				−2 0 0
PION 7	Pr >165/>73 MeV	06^h/08^{h}Δ2	20^h/23^{h}Δ2	50^h/>40^h	0.35/0.4 Pr	(119)
PION 6	Pr >13 MeV	12^{h}Δ5	4^{d}18^{h}Δ5	>4^d	0.28 Pr	(104)
PION 8	Pr >14 MeV	19^{h}Δ2	4^{d}14^{h}Δ2	22^h	0.125 Pr	(183)
IMP 4	Pr >10 MeV	23^{h}Δ2	4^{d}02^{h}Δ2	25^h	0.04 Pr	(173)
IMP 4	Pr >29/>0.8 MeV	23^{h}30^{m}Δ4	4^{d}04^h/09^{h}Δ2	>36^h	0.02/0.7 Pr	(107)

(PION 6 at 0.98 AU, ES + 161°; PION 7 at 1.03 AU, ES − 77°; PION 8 at 1.06 AU, ES − 17°)

Source: ● flare 0313   N10   E75   2N   McM 9567

282	1968	August 5	>03^h				−1 0 0
PION 8	Pr > 14 MeV	04^{d}17^{h}Δ3	6^{d}02^{h}Δ2	3^d	0.36 Pr	(183)
IMP 4	Pr > 9 MeV	05^{h}Δ3	6^{d}14^{h}Δ2	>37^h	0.12 Pr	(107)
IMP 4	Pr > 29 MeV	10^{h}45^m ± 15^{m}Δ3	6^{d}09^{h}Δ2	>28^h	0.07 Pr	(107)
IMP 4	Pr > 0.8 MeV	11^{h}Δ3	16^{h}Δ2	>29^h	1.5 Pr	(107)

Source: ◇ geomagnetic storm modulation; geomagnetic disturbance begins gradually on Aug. 5

283	1968	August 6	15^h				−1 0 0
IMP 4	Pr > 29 MeV	15^{h}Δ3	20^{h}Δ2	>3^d	0.08 Pr	(107)
IMP 4	Pr > 9 MeV	16^{h}Δ3	7^{d}05^{h}Δ2	4^d	0.2 Pr	(107)
IMP 4	Pr > 0.8 MeV	16^{h}Δ3	7^{d}11^{h}Δ2	5^d	8.6 Pr	(107)

Source: ⊙ flare 1318   N15   W88   SN   McM 9545
⊙ flare 1315   N07   E31   SB   McM 9567
⊙ flare <1344   N13   E32   SN   McM 9567

284	1968	August 14	14^h				−1 0 0
AIMP 2	El > 40 keV	13^{h}43^m			130 El	(226)
IMP 4	Pr > 0.8 MeV	13^{h}55^m ± 05^{m}Δ8	16^{h}Δ2	81^h	29.9 Pr	(107)
IMP 4	Pr > 9 MeV	14^{h}05^m ± 05^{m}Δ8	19^{h}Δ2	74^h	0.74 Pr	(107)
IMP 4	El > 0.3 MeV	14^{h}10^m ± 08^m	16^{h}Δ1		0.7 El	(166)
IMP 4	Pr > 29 MeV	15^{h}15^m ± 05^m	19^{h}Δ2	30^h	0.13 Pr	(107)
PION 8	Pr > 11 MeV	17^{h}Δ3	15^{d}05^{h}Δ2	2^d	0.0113 Pr	(183)
(PION 8 at 1.05 AU, ES − 18°)						

Source: ● flare <1327   N13   W80   1B   McM 9567
● flare 1400   N13   W82   1N   McM 9567

	1968	Aug. 17	event in Appendix A				
285	1968	August 19	17^h				−1 0 0
IMP 4	Pr > 0.8 MeV	17^{h}Δ3	20^{d}15^{h}Δ2	>44^h	4.5 Pr	(107)
PION 8	Pr > 11 MeV	22^h ± 20^m	21^{d}00^{h}Δ7		0.025 Pr	(183)
IMP 4	Pr > 9 MeV	22^{h}Δ2	20^{d}16^{h}Δ2	4^d	0.2 Pr	(107)
IMP 4	Pr > 29 MeV	20^{d}01^h ± 60^m	20^{d}20^{h}Δ2	>2^d	0.03 Pr	(107)

Source: ◇ bright regions McM 9593 and 9597 are transiting CM
○ flare <1533   S11   W22   SN   McM 9593

	1968	Aug. 21	event in Appendix A				

286	1968 August 28	19^h				$-1\,0\,0$
IMP 4	Pr > 10 MeV	$19^h\Delta1$	$31^d04^h\Delta2$	70^h	0.16 Pr	(173)
IMP 4	Pr > 0.8 MeV	$19^h\pm60^m$	$30^d18^h\Delta2$	70^h	9.3 Pr	(107)
PION 8	Pr > 11 MeV	$23^h\Delta2$	$\Delta7$			(183)

(PION 8 at 1.05 AU, ES − 19°)

Source: origin unknown
◇ bright regions McM 9630 and 9634 are coming around E limb Aug. 28 and 29

	1968 Sept. 1, Sept. 2, Sept. 3, Sept. 4, Sept. 10, Sept. 23		events in Appendix A			
287	1968 September 26	$>01^h30^m$				$1\,1\,0$
VLF	PCA	01^h20^m				(192)
AIMP 2	El > 40 keV	01^h30^m			10 El	(226)
IMP 4	Pr > 9/>0.8 MeV	$01^h45^m\Delta8/\pm5^m$	$15^h/14^h30^m\Delta8$ $>56^h/54^h$		13/550 Pr	(107)
IMP 4	El > 2.7 MeV	06^h38^m			0.5 El	(166)
RIOM	PCA	$<08^h\Delta2$	$23^h\Delta1$		0.8 dB	(184)
AIMP 2	El > 40 keV	08^h15^m			1200 El	(226)
IMP 4	Pr > 1.8 MeV	$10^h\Delta2$	$23^h\Delta2$		400 c	(166)
IMP 4	Pr > 29 MeV	$10^h30^m\pm30^m$	$15^h\Delta1$	40^h	0.13 Pr	(107)
AIMP 2	Pr > 0.32 MeV	$11^h\Delta2$		325^h		(193)

Source: ● flare 0026 N14 E34 2B McM 9687

	1968 ⋮ Sept. 26, Sept. 28		events in Appendix A			
288	1968 September 28	10^h				$1\,1\,0$
AIMP 2	El > 40 keV	09^h30^m			600 El	(226)
IMP 4	Pr > 30/>9 MeV	$10^h\Delta2/\Delta3$	$\begin{bmatrix}17^h\Delta2/\\29^d04^h\Delta4\end{bmatrix}$ $>31^h$		0.1/11.2 Pr	(173, 107)
PION 8	Pr > 14 MeV	$10^h\pm20^m\Delta3$	$18^h\Delta2$	$>23^h$	7.1 Pr	(183)
IMP 4	Pr > 0.8 MeV	$11^h\Delta3$	$29^d19^h\Delta2$	$>47^h$	996 Pr	(107)
RIOM	PCA	$11^h\Delta2$	$21^h\Delta1$		1.2 dB	(184)
IMP 4	El > 2.7/>0.3 MeV	$11^h32^m\Delta2/\Delta3$	$/14^h\Delta1$	$/>23^h$	0.05/4.8 El	(166)

(PION 8 at 1.04 AU, ES − 21°)

Source: ● flare 0721 S18 E39 2B McM 9692

1968 Sept. 29   event in Appendix A

289   1968   September 29 17^h 1 2 1

$Pr > 1.5$ GeV	GLE (141)	$16^h45^m \pm 03^m$				Churchill
			$17^h30^m \pm 30^m$		$1.4 \pm 0.5\%$	Thule
				$4^h \pm 2^h$		South Pole
Balloon	$Pr > 100$ MeV	16^h50^m	$19^h\Delta2$	24^h	2 Pr	(194)
VLF	PCA	16^h50^m				(192)
IMP 4	$Pr > 94$ MeV	$16^h55^m \pm 5^m$	19^h30^m	40^h	5 Pr	(107)
AIMP 2	$El > 40$ keV	$17^h\Delta1$			2200 El	(226)
RIOM	PCA	$17^h\Delta1$	$23^h\Delta1$		1.7 dB	(184)
IMP 4	$El > 2.7/>0.3$ MeV	$17^h03^m/\pm02^m\Delta3$	$/20^h\Delta2$	$/80^h$	1.8/23 El	(166)
IMP 4	$Pr > 29/>9$ MeV	$[17^h15^m \pm 5^m/$	$21^h\Delta1/$	$75^h/104^h$	14.5/27 Pr	(107)
		$[\pm 15^m\Delta3$	$30^d01^h30^m$			
IMP 4	$Pr > 60$ MeV	$<17^h20^m$	$20^h\Delta1$	50^h	10.32 Pr	(173)
PION 8	$Pr > 14$ MeV	$18^h\Delta5$	$30^d00^h30^m$	30^h	21.4 Pr	(183)
IMP 4	$Pr > 0.8$ MeV	$30^d12^h\Delta3$	$[1^d02^h\Delta2$	94^h	1470 Pr	(107)
			$[\text{spike on } 2^d00^h$			
AIMP 2	$Pr > 0.32$ MeV				7600 Pr	(193)

(PION 8 at 1.04 AU, ES $- 21°$)

Source: ●flare 1617   N17   W51   2B   McM 9678
  ▲geomagnetic storm begins with sc $02^d00^h18^m$

	1968	Oct. 1, Oct. 3	events in Appendix A

290   1968   October 4 01^h 1 2 0

PION 8	$Pr > 14$ MeV	$00^h40^m \pm 20^m$	$\Delta5$			(183)
IMP 4	$El > 0.3/>2.7$ MeV	$[00^h44^m \pm 05^m/$	04^h30^m	$3^d/$	22/0.6 El	(166)
		$[00^h45^m$				
AIMP 2	$El > 40$ keV	$00^h58^m \pm 48^m$			470 El	(226)
IMP 4	$Pr > 0.8$ MeV	$01^h\Delta8$	$6^d07^h45^m$	$>6^d$	2870 Pr	(107)
VLF	PCA	$01^h\Delta1$				(192)
IMP 4	$Pr > 94$ MeV	$01^h \pm 30^m$	$03^h\Delta1$	20^h	0.24 Pr	(107)
IMP 4	$Pr > 29/>9$ MeV	$01^h \pm 10^m$	$05^h\Delta1/07^h\Delta2$	$80^h/160^h$	8.35/43.3 Pr	(107)
IMP 4	$Pr > 60$ MeV	$01^h20^m \pm 10^m$	$04^h\Delta1$	25^h	1.22 Pr	(173)
RIOM	PCA	01^h20^m	$15^h\Delta1$		1.6 dB	(184)
IMP 4	$El > 2.7/>0.3$ MeV	$16^h\Delta1$			0.2/7 El	(166)

(PION 8 at 1.04 AU, ES $- 21.5°$)

Source: ●flare 3^d2343 S17   W36   2B   McM 9692
  ▲geomagnetic storm begins with sc $6^d06^h28^m$

	1968	Oct. 17	event in Appendix A

291   1968   October 24 02^h 0 0 0

IMP 4	$Pr > 0.8$ MeV	$02^h\Delta3$	$26^d02^h\Delta2$	$>55^h$	80 Pr	(107)
IMP 4	$Pr > 9$ MeV	$10^h\Delta2$	$26^d06^h\Delta2$	$>2^d$	1.25 Pr	(107)

Source: ●flare 23^d2352 S13   E59   2N   McM 9740

292	1968 October 26	08^h				−1 0 0
IMP 4	Pr > 29/>10 MeV	08^{h}Δ2	16^{h}Δ2/Δ1	16^h/10^h	0.006/0.26 Pr	(107, 173)
IMP 4	Pr > 0.8 MeV	09^{h}Δ3	20^{h}Δ2	>3^d	359 Pr	(107)
AIMP 2	Pr > 0.32 Pr	09^{h}Δ1				(193)

Source: ⊙ flare 0046   S20   E32   1N   McM 9740

293	1968 October 27	14^h				−1 0 0
IMP 4	Pr > 9 MeV	14^{h}Δ3	28^{d}02^{h}Δ2	>49^h	0.25 Pr	(107)
IMP 4	Pr > 29 MeV	16^{h}20^m ± 20^m	21^{h}Δ1	47^h	0.04 Pr	(107)

Source: ● flare   1232   S17   E17   1B   McM 9740
              <1318   S17   E17   2N   McM 9740

294	1968 October 29	15^{h}30^m				0 0 0
AIMP 2	El > 40 keV	15^{h}15^m ± 45^m			1800 El	(226)
IMP 4	El > 2.7/>0.3 MeV	15^{h}30^m/±08^m	/18^{h}Δ1	/30^h	0.03/9 El	(166)
PION 8	Pr > 14 MeV	15^{h}40^m ± 20^m	19^{h}Δ1	>1^d	9.38 Pr	(183)
IMP 4	Pr > 29 MeV	16^h ± 15^m	19^{h}Δ1	15^h	0.03 Pr	(107)
IMP 4	Pr > 9 MeV	16^h ± 30^{m}Δ3	18^{h}30^m	>21^h	3.6 Pr	(107)
IMP 4	Pr > 0.8 MeV	16^{h}Δ3	30^{d}04^{h}Δ2	>32^h	602 Pr	(107)
(PION 8 at 1 AU, ES − 23°)						

Source: ⊙ flare <1222   S16   W12   2B   McM 9740
           ⊙ flare 1515   N15   W82   ·SN   McM 9735
           ⊙ flare 1515   S14   W19   1N   McM 9740

295	1968 October 30	12^{h}30^m				0 0 0
IMP 4	Pr > 29 MeV	12^{h}30^m ± 30^m	14^{h}Δ1	>13^h	0.15 Pr	(107)
IMP 4	Pr > 9 MeV	12^{h}50^m ± 10^{m}Δ3	15^{h}Δ2	>12^h	1.46 Pr	(107)

Source: ● flare <1235   S18   W26   2B   McM 9740

296	1968 October 31	00^{h}30^m				2 3 0
IMP 4	Pr > 0.8 MeV	00^{h}Δ8	09^{h}Δ1	>50^h	5936 Pr	(107)
IMP 4	El > 2.7/>0.3 MeV	⌈00^{h}29^m/± 03^m	/09^{h}Δ1	/37^h	0.6/21 El	(166)
VLF	PCA	⌊01^{h}Δ1				(192)
IMP 4	Pr > 9/>29 MeV	⌈01^h ± 30^m/	14^{h}Δ1	>37^h	169/14.2 Pr	(107)
		⌊01^{h}50^m ± 05^{m}Δ3				
IMP 4	Pr > 60 MeV	02^{h}Δ1	15^{h}Δ1	>36^h	1.6 Pr	(173)
AIMP 2	El > 40 keV	04^{h}50^m			5000 El	(226)
RIOM	PCA	05^{h}Δ2	17^{h}Δ1		5.5 dB	(184)
IMP 4	Pr > 94 MeV	05^{h}Δ2	14^{h}Δ1	20^h	0.4 Pr	(107)
IMP 4	Pr > 1.8 MeV	05^{h}Δ3	1^{d}02^{h}Δ2		1000 Pr	(179)

(continues on next page)

296 continued

PION 8	Pr > 14 MeV	$\Delta 5$	$16^h \Delta 2$		3.75 Pr	(183)

Source: ● flare $30^d 2340$ S14 W37 3B McM 9740
    ▲ major geomagnetic storm begins with sc $1^d 09^h 16^m$

297	1968 November 1	$13^h 30^m$				2 3 0
RIOM	PCA	$12^h \Delta 3$	$2^d 18^h \Delta 1$		4 dB	(184)
IMP 4	El > 2.7/>0.3 MeV	$13^h 30^m /\Delta 3$	$/21^h \Delta 1$	$/65^h$	1.0/46 El	(166)
AIMP 2	El > 40 keV				2500 El	(226)
IMP 4	Pr > 30 MeV	$13^h 30^m \pm 10^m$	$21^h \Delta 1$	38^h	11.21 Pr	(173)
IMP 4	Pr > 10 MeV	$13^h 30^m \pm 30^m \Delta 3$	$2^d 06^h \Delta 1$	65^h	151.67 Pr	(173)
IMP 4	Pr > 60 MeV	$14^h \pm 30^m \Delta 3$	$20^h \Delta 1$	20^h	1.12 Pr	(173)
IMP 4	Pr > 94 MeV	$14^h 50^m \pm 10^m$	$22^h \Delta 2$	21^h	0.32 Pr	(107)
IMP 4	Pr > 0.8 MeV	$02^d 02^h \Delta 3$	$2^d 19^h \Delta 4$	77^h	5470 Pr	(107)
IMP 4	Pr > 1.8 MeV	$02^d 03^h \Delta 3$	$2^d 09^h \Delta 2$	52^h	550 Pr	(179)
RIOM	PCA		$2^d 17^h \Delta 1$		5.9 dB	(195)

Source: ● flare 0814 S16 W47 2N McM 9740

298	1968 November 4	$05^h 50^m$				1 2 0
IMP 4	El > 0.3 MeV	$05^h 39^m \pm 03^m$	$07^h \Delta 1$	45^h	11.2 El	(166)
IMP 4	El > 2.7 MeV	$05^h 41^m$			0.4 El	(166)
IMP 4	Pr > 60/>94 MeV	$05^h 50^m \pm 20^m /$ $55^m \pm 5^m$	$08^h /07^h \Delta 1$	$30^h /28^h$	1.6/0.54 Pr	(173, 107)
IMP 4	Pr > 10 MeV	$05^h 55^m \pm 5^m$	$09^h \Delta 1$	45^h	19.27 Pr	(173)
AIMP 2	El > 40 keV	$06^h \Delta 4$				(226)
IMP 4	Pr > 30 MeV	$06^h \pm 10^m$	$08^h \Delta 1$	40^h	4.9 Pr	(173)
RIOM	PCA	$06^h 15^m$	$09^h \Delta 1$		1.6 dB	(184)
IMP 4	Pr > 1.8 MeV	$07^h \Delta 3$	$18^h \Delta 2$	$>2^d$	220 Pr	(179)
IMP 4	Pr > 0.8 MeV	$07^h \pm 70^m$	$18^h \Delta 1$	$>5^d$	1290 Pr	(107)
PION 8	Pr > 14 MeV	$08^h \Delta 2$	$17^h \Delta 2$		5.62 Pr	(183)

Source: ■ flare <0524 S15 W90 1B McM 9740

299	1968 November 9	12^h				−1 0 0
IMP 4	Pr > 0.8 MeV	$12^h \Delta 2$	$13^d 04^h \Delta 2$	$>90^h$	10 Pr	(107)
PION 9	Pr > 10 MeV	$16^h \Delta 3$	$13^d 03^h \Delta 2$	$>85^h$	0.1 Pr	(217)
PION 8	Pr > 14 MeV	$11^d 07^h \Delta 2$	$12^d 08^h \Delta 2$	60^h	0.56 Pr	(183)
IMP 4	Pr > 9 MeV	$12^d 16^h \Delta 2$	$13^d 04^h \Delta 2$	$>14^h$	0.13 Pr	(107)

(PION 8 at 1 AU, ES − 22°; PION 9 at 1 AU, ES = 0°)

Source: origin unknown, possibly disk transit of some active regions

300   1968   November 13 $>01^h$ −1 (1) 0

AIMP 2	Pr > 0.32 MeV	$01^h\triangle1$		72^h	370 Pr	(193)
PION 9	Pr > 14 MeV	$05^h20^m \pm 20^m$	$10^h\triangle1$	48^h	0.41 Pr	(217)
IMP 4	El > 0.3 MeV	$05^h30^m \pm 20^m$	$13^h\triangle2$		0.4 El	(166)
VLF	PCA	$06^h\triangle1$				(192)
AIMP 2	El > 40 keV	06^h10^m			80 El	(226)
IMP 4	Pr > 9 MeV	$06^h30^m \pm 30^m\triangle3$	$11^h\triangle2$	90^h	0.5 Pr	(107)
IMP 4	Pr > 0.8 MeV	$06^h30^m \pm 40^m\triangle3$	$15^h\triangle2$		101 Pr	(107)
IMP 4	Pr > 29 MeV	$08^h \pm 10^m$	$12^h\triangle1$	9^h	0.04 Pr	(107)

Source: origin unknown; no flare patrol 0550–0603

---

301   1968   November 18 10^h30^m 2 3 3

AIMP 2	Pr > 0.32 MeV	$10^h\triangle2$	$20^d10^h\triangle2$	230^h	31200 Pr	(193)
IMP 4	El > 1 MeV	10^h30^m	11^h15^m	110^h	90 c	(196)
IMP 4	El > 500/>300 keV	10^h30^m	11^h15^m		300/70 c	(196)
OGO 5	El > 0.3 MeV				150 El	(166)
Pr > 3.6 GeV	GLE (141)	$10^h35^m \pm 05^m$				⌜Dallas, ⌞Kerguelen
			$10^h55^m \pm 05^m$		14%	Goose Bay
				$9^h30^m \pm 1^h$		⌜Calgary, ⌞Sulphur Mt.
AIMP 2	El > 40 keV	10^h42^m	11^h30^m		18000 El	(226)
IMP 4	Pr > 0.8 MeV	$10^h45^m \pm 05^m\triangle8$	$20^d09^h\triangle1$	163^h	11840 Pr	(107)
RIOM	PCA	10^h45^m	$14^h\triangle1$		12.5 dB	(184)
VLF	PCA	10^h45^m				(192)
IMP 4	Pr > 29/>94 MeV	$10^h50^m \pm 05^m$	$\left[\begin{array}{l}13^h30^m/ \\ 11^h30^m\end{array}\right.$	$200^h/50^h$	433/37 Pr	(107)
AIMP 2	El > 50 keV	10^h54^m		$>4^d$	6300 El	(197)
IMP 4	Pr 5 MeV	$11^h\triangle2$	$19^d17^h\triangle2$		520* Pr	(196)
PION 9	Pr > 14 MeV	$11^h \pm 10^m$	$22^h\triangle1$	6^d	320 Pr	(217)
PION 8	Pr > 7 MeV	$11^h\triangle2$	$19^h\triangle2$	8^d	1100 Pr	(199)
IMP 4	Pr > 60 MeV	$12^h\triangle5$	$14^h\triangle1$	55^h	96.15 Pr	(173)
IMP 4	Pr > 0.7 MeV	$14^h\triangle5$	$19^d20^h\triangle2$		5400 Pr	(174)
PION 8	Pr > 11 MeV		$21^h\triangle5$		129.5 Pr	(183)
Balloon	Pr > 100 MeV		$13^h\triangle5$		6 Pr	(43)

(PION 9 at 0.98 AU, ES − 1°)

Source: ■ flare < 1026   N21   W87   1B   McM 9760
▲ geomagnetic storm begins with sc $20^d09^h04^m$

302	1968 November 20	21^h				1 0 0
IMP 4	Pr > 30 MeV	$21^h\Delta 3$	$22^h\Delta 2$	4^h	1.3 Pr	(181)
IMP 4	Pr > 10 MeV	$21^h\Delta 3$	$23^h\Delta 2$	3^h	14 Pr	(181)
PION 9	Pr > 14 MeV	$21^h\Delta 3$	$23^h\Delta 2$	3^h	23.4 Pr	(217)
IMP 4	Pr > 0.8 MeV	$21^h30^m\Delta 3$	$23^h\Delta 1$	4^h	1375 Pr	(107)

Source: origin uncertain
    □ active region McM 9760 is on invisible hemisphere, 2 days beyond W limb

303	1968 November 21	03^h				1 0 0
PION 9	Pr > 14 MeV	$02^h20^m \pm 20^m\Delta 3$	$05^h\Delta 2$	6^h	28.8 Pr	(217)
IMP 4	Pr > 29 MeV	$02^h50^m \pm 10^m\Delta 3$	$05^h\Delta 2$	7^h	2.5 Pr	(107)
IMP 4	Pr > 10 MeV	$03^h\Delta 3$	04^h30^m	5^h	18.5 Pr	(181)
IMP 4	Pr > 0.8 MeV	$03^h\Delta 3$	$06^h\Delta 2$	8^h	1812 Pr	(107)

Source: □ active region McM 9760 is 3 days beyond W limb
    ○ flare 0033   N14   W17   1N   McM 9772

	1968 Nov. 25	event in Appendix A				

304	1968 December 3	04^h				1 2 0
PION 9	Pr > 14 MeV	$03^h\Delta 2$	$05^d02^h\Delta 2$	$>52^h$	31 Pr	(217)
IMP 4	Pr > 1 MeV	$04^h\Delta 2$				(174)
OGO 5	Pr > 90 MeV	$06^h\Delta 2$	$4^d20^h\Delta 4$	40^h	0.09 Pr	(186)
IMP 4	Pr > 9/>0.8 MeV	$06^h\Delta 2/\Delta 3$	$5^d07^h/06^h\Delta 2$	$>50^h$	15/614 Pr	(107)
PION 9	Pr > 7 MeV	$06^h\Delta 2$	$5^d20^h\Delta 2$		170 Pr	(199)
IMP 4	Pr > 30 MeV	$07^h\Delta 2$	$5^d08^h\Delta 1$	$>50^h$	1.66 Pr	(173)
IMP 4	El > 0.3 MeV	$10^h\Delta 2$	$5^d21^h\Delta 2$	10^d	50 c	(190)
IMP 4	Pr > 60 MeV	$10^h\Delta 2$	$5^d10^h\Delta 2$	$>49^h$	0.12 Pr	(173)
OGO 5	El > 12 MeV	$12^h\Delta 2$	$4^d16^h\Delta 2$	40^h	0.005 El	(186)
IMP 4	El > 3 MeV	$12^h\Delta 2$	$5^d18^h\Delta 2$	5^d	5 El	(190)
RIOM	PCA	$19^h\Delta 2$	$5^d09^h42^m$		4.3 dB	(195)

(PION 9 at 0.97 AU, ES − 2°)

Source: ● flare 2^d2115 N20   E89   1N   McM 9802
             $2^d<2202$ N19   E80   3N   McM 9802

305	1968 December 5	$>06^h$				2 3 0
PION 8	Pr > 14 MeV	$4^d21^h40^m \pm 20^m$	$5^d11^h\Delta 2$	$>6^d$	172 Pr	(183)
PION 9	Pr > 10 MeV	$5^d06^h\Delta 3$	$6^d04^h\Delta 2$	$>13^d$	253 Pr	(217)
IMP 4	Pr > 0.8 MeV	$08^h\Delta 3$	$23^h\Delta 2$	17^d	3820 Pr	(107)
AIMP 2	Pr > 0.32 MeV				1960 Pr	(193)
IMP 4	Pr > 30 MeV	$09^h\Delta 3$	$6^d01^h\Delta 1$	105^h	31 Pr	(173)
IMP 4	Pr > 10 MeV	$09^h\Delta 3$	$6^d04^h\Delta 1$	160^h	152 Pr	(173)
OGO 5	El > 12 MeV	$10^h\Delta 4$	$21^h\Delta 2$	50^h	0.015 El	(186)
IMP 4	Pr > 60 MeV	$11^h\Delta 2$	$6^d01^h\Delta 1$	60^h	5.1 Pr	(173)
IMP 4	Pr > 1 MeV		$6^d06^h\Delta 2$		1250 Pr	(174)

(*continues on next page*)

305 continued

IMP 4	Pr > 94 MeV	$13^h\Delta2$	$5^d23^h\Delta2$	50^h	0.8 Pr	(107)
RIOM	PCA		$6^d05^h\Delta2$		4.7 dB	(184)
(PION 9 at 0.96 AU, ES −2°)						

Source: ♦ geomagnetic storm modulation; sc begins $5^d06^h33^m$

	1968 Dec. 16, Dec. 23	events in Appendix A	
306	1968 December 24	22^h30^m	−1 0 0

IMP 4	Pr > 9 MeV	$22^h30^m \pm 30^m$	$25^d02^h45^m$	12^h	0.3 Pr	(107)
IMP 4	Pr > 0.8 MeV	$23^h \pm 10^m\Delta3$	$25^d03^h\Delta2$	$>36^h$	40 Pr	(107)
PION 9	Pr > 10 MeV	$\Delta5$	$25^d04^h\Delta2$	$>2^d$	0.0113 Pr	(217)
(PION 9 at 0.95 AU, ES − 2.5°)						

Source: ☉ flare 2227   N19   E29   1B   McM 9842

	1968 Dec. 26	event in Appendix A	
307	1968 December 27	12^h	−1 0 0

PION 8	Pr > 14 MeV	$12^h30^m \pm 30^m$	$18^h\Delta2$	$>1^d$	0.3 Pr	(183)
IMP 4	Pr > 29 MeV	$13^h\Delta1$	$28^d04^h\Delta2$	55^h	0.1 Pr	(107)
IMP 4	Pr > 9 MeV	$14^h\Delta2$	$28^d13^h\Delta4$	80^h	0.14 Pr	(107)
IMP 4	Pr > 0.8 MeV	$14^h \pm 60^m\Delta3$	$29^d03^h\Delta4$	$>62^h$	4 Pr	(107)

Source: ● flare <1050   N16   E03   2B   McM 9842

	1968 Dec. 28 (twice), Dec. 30 (twice)	events in Appendix A	
	1969 Jan. 2, Jan. 3 (twice), Jan. 9, Jan. 14	events in Appendix A	
308	1969 January 17	14^h	0 0 0

IMP 4	Pr > 9/>29 MeV	$14^h/16^h\Delta2$	$18^d20^h\Delta2$	$34^h/38^h$	1.7/0.022 Pr	(107)
IMP 4	Pr > 0.8 MeV	$14^h\Delta2$	$18^d19^h\Delta2$	$>6^d$	120 Pr	(107)
PION 8	Pr > 14 MeV	$18^h\Delta2$	$18^d04^h\Delta2$	$>1^d$	0.18 Pr	(183)
PION 9	Pr > 10 MeV	$18^d01^h\Delta2$	$19^d07^h\Delta2$	4^d	0.011 Pr	(217)
		other max. 18^d11^h				
(PION 8 at 1 AU, ES − 22°; PION 9 at 0.9 AU, ES − 0.5°)						

Source: ☉ flare <1242   N16   E48   SB   McM 9873
         ⊘ flare 1703   N16   E45   SB   McM 9873

	1969 Jan. 24	event in Appendix A

309   1969   January 24 07^h50^m 0 1 0

PION 9	Pr > 14 MeV	$07^h40^m \pm 20^m$	$09^h\Delta5$	40^h	11.3 Pr	(217)
IMP 4	El > 40 keV	07^h50^m			230 El	(226)
VLF	PCA	07^h52^m				(192)
IMP 4	El > 0.3/>2.7 MeV	$07^h55^m \pm 3^m$/ $08^h\Delta1$	$09^h\Delta1$/		2.8/0.04 El	(166)
IMP 4	Pr > 10 MeV	$08^h \pm 10^m$	12^h30^m	40^h	3.17 Pr	(200)
VEN 6	Pr > 1 MeV	$08^h\Delta2$	$20^h\Delta2$	228^h	1300 Pr	(43)
IMP 4	Pr > 29 MeV	$08^h10^m \pm 5^m$	12^h30^m	50^h	0.62 Pr	(107)
IMP 4	Pr > 0.8 MeV	$08^h15^m \pm 15^m\Delta3$	$25^d01^h\Delta2$	$>8^d$	**1980 Pr**	(107)
HEOS	Pr > 6 MeV	$08^h54^m \pm 15^m$	$17^h\Delta2$	90^h	23.8 Pr	(201)
AIMP 2	Pr > 0.32 MeV	$10^h\Delta1$	$25^d06^h\Delta2$	119^h	2950 Pr	(193)
RIOM	PCA	$14^h\Delta2$	$17^h\Delta1$	24^h	1.2 dB	(146)

(PION 9 at 0.9 AU, ES = 0°; VEN 6 at 0.98 AU, ES − 1.5°)

Source: ● flare <0706   N20   W08   3B   McM 9879

1969   Feb. 2, Feb. 9, Feb. 11, Feb. 12   events in Appendix A

310   1969   February 23 05^h −1 0 0

PION 9	Pr > 14 MeV	$05^h \pm 20^m$	$06^h\Delta1$	12^h	0.25 Pr	(217)
VLF	PCA	$05^h\Delta1$				(192)
AIMP 2	El > 40 keV	05^h10^m			35 El	(226)
HEOS	Pr > 6 MeV	$06^h\Delta2$	$09^h\Delta2$	15^h	0.07 Pr	(202)
VEN 6	Pr > 1 MeV	$06^h\Delta2$	$10^h\Delta2$	32^h	4 Pr	(43)
IMP 4	Pr >9/>0.8 MeV	$06^h\Delta4$	$06^h20^m/09^h\Delta2$	$8^h/42^h$	0.4/3.4 Pr	(107)
AIMP 2	Pr > 0.32 MeV	10^h30^m		38^h	8.6 Pr	(193)

(PION 9 at 0.83 AU, ES + 5°; VEN 6 at 0.95 AU, ES − 3°)

Source: ● flare 0442   N12   W09   1N   McM 9946

/1969   Feb. 24   event in Appendix A

311   1969   February 24 23^h30^m −1 0 0

VEN 6	Pr > 30 MeV	$22^h\Delta2$		$>64^h$		(43)
				(duration includes two next events)		
VLF	PCA	23^h30^m				(192)
IMP 4	El > 0.3 MeV	$23^h35^m \pm 3^m$	$25^d01^h\Delta1$		1.2 El	(166)
AIMP 2	El > 40 keV	23^h35^m			500 El	(226)
PION 9	Pr > 14 MeV	$23^h40^m \pm 20^m$	$25^d01^h\Delta1$	10^h	1.5 Pr	(217)
IMP 4	Pr > 9 MeV	$23^h45^m \pm 10^m$	$25^d01^h30^m$	$>10^h$	0.46 Pr	(107)
IMP 4	Pr > 0.8 MeV	$23^h45^m \pm 10^m\Delta8$	$25^d00^h30^m$	$>10^h$	71 Pr	(107)
IMP 4	Pr > 29 MeV	$23^h50^m \pm 10^m$	$25^d01^h30^m$	10^h	0.12 Pr	(107)
HEOS	Pr > 6 MeV	$25^d00^h30^m \pm 30^m$	$25^d04^h\Delta2$	9^h	0.22 Pr	(201)
VEN 6	Pr > 1 MeV	$25^d02^h\Delta2$		216^h		(43)
				(duration includes the next 4 events)		
AIMP 2	Pr > 0.32 MeV	25^d05^h		167^h		(193)
				(duration includes the next 4 events)		

(PION 9 at 0.83 AU, ES + 5°; VEN 6 at 0.94 AU, ES − 3°)

Source: ● flare 2305   N12   W31   2B   McM 9946

312	1969 February 25	$\sim09^h30^m$				1 2 3
Pr > 3.6 GeV	GLE	09^h15^m				Climax
(141)			$09^h45^m \pm 10^m$		16%	Goose Bay
				> 5^h		several
PION 9	Pr > 14 MeV	$09^h20^m \pm 20^m \Delta3$	$\Delta5$	20^h		(217)
HEOS	El > 7.5 MeV	$09^h26^m \pm 6^m$	10^h04^m	24^h	112 c	(204)
IMP 4	El > 2.7 MeV	09^h26^m			6.5 El	(166)
IMP 4	El > 0.3 MeV	$09^h26^m \pm 5^m \Delta3$	$11^h\Delta1$	20^h	105 El	(166)
VLF	PCA	09^h28^m				(192)
OGO 5	El > 12 MeV	$09^h30^m \Delta2$	$10^h\Delta1$	20^h	0.07 El	(186)
IMP 4	Pr > 94 MeV	$09^h30^m \pm 5^m$	$11^h\Delta1$	> 21^h	5.4 Pr	(107)
IMP 4	Pr > 30/>10 MeV	$09^h30^m \pm 10^m$	$12^h/13^h\Delta1$	$22^h/>21^h$	41.4/88.4 Pr	(200)
HEOS	Pr > 6 MeV	$09^h30^m \pm 10^m \Delta3$	$17^h\Delta2$	21^h	26.6 Pr	(201)
IMP 4	Pr > 0.8 MeV	$09^h30^m \pm 30^m \Delta3$	11^h30^m	23^h	2000 Pr	(107)
HEOS	Pr > 360 MeV	$09^h33^m \pm 2^m$	$11^h\Delta1$	22^h	1.5 Pr	(205)
AIMP 1	El > 50 keV	09^h34^m	11^h16^m		1970 El	(197)
IMP 4	Pr > 60 MeV	$09^h35^m \pm 5^m$	11^h30^m	> 20^h	24.2 Pr	(200)
HEOS	Pr > 41 MeV	09^h40^m	$12^h\Delta1$	19^h	13.7 Pr	(206)
AIMP 2	El > 40 keV	$09^h45^m \pm 15^m$			5000 El	(226)
RIOM	PCA	$12^h\Delta2$	$15^h\Delta1$		2.1 dB	(146)
Balloon	Pr > 100 MeV		$14^h\Delta5$		4 Pr	(43)
VEN 6	Pr > 30 MeV		$10^h\Delta2$		31 Pr	(43)

Source: ● flare 0900   N13   W37   2B   McM 9946

313	1969 February 26	05^h				1 1 0
IMP 4	El > 0.3 MeV	$04^h49^m \pm 15^m \Delta3$	$09^h\Delta2$	10^h	10 El	(179)
PION 9	Pr > 14 MeV	$05^h20^m \pm 20^m \Delta3$	07^h30^m	> 9^h	90.6 Pr	(217)
IMP 4	El > 1.1 MeV	$05^h30^m \Delta3$	08^h30^m	27^h	4 c	(187)
AIMP 2	El > 40 keV	$05^h30^m \Delta3$			1500 El	(226)
Balloon	Pr > 100 MeV		$07^h\Delta5$		1 Pr	(43)
IMP 4	Pr > 60 MeV	$06^h \pm 20^m \Delta3$	$09^h\Delta1$	30^h	2.44 Pr	(200)
IMP 4	Pr > 30 MeV	$06^h10^m \pm 10^m \Delta3$	$09^h\Delta1$	30^h	5.21 Pr	(200)
IMP 4	Pr > 10 MeV	$06^h50^m \pm 10^m \Delta3$	$12^h\Delta1$	> 33^h	14.4 Pr	(200)
IMP 4	Pr > 94 MeV	$06^h50^m \pm 10^m \Delta3$	08^h30^m	25^h	0.4 Pr	(107)
VEN 6	Pr > 1 MeV		$14^h\Delta2$		1000 Pr	(43)
IMP 4	Pr > 0.8 MeV	$09^h\Delta3$	$22^h\Delta2$	31^h	1140 Pr	(107)
AIMP 2	Pr > 0.3 MeV		$12^h\Delta2$		3000 Pr	(193)
RIOM	PCA		$17^h\Delta1$		0.9 dB	(146)

Source: ● flare 0418   N13   W46   2B   McM 9946

314	1969 February 27	14^h45^m				1 1 0
HEOS	El > 7.5 MeV	$14^h20^m \pm 6^m$	$18^h\Delta2$		3.4 c	(204)
IMP 4	El > 1.1 MeV	$14^h30^m\Delta3$	18^h45^m		12 c	(187)
IMP 4	El > 0.3 MeV	$14^h37^m \pm 11^m\Delta3$	$19^h\Delta1$	85^h	34 El	(166)
AIMP 2	El > 40 keV	$14^h45^m\Delta3$			2000 El	(226)
IMP 4	El > 2.7 MeV	14^h50^m			1.9 El	(166)
IMP 4	Pr > 94/>30 MeV	$15^h \pm 20^m$	$22^h/21^h\Delta2$	$22^h/50^h$	0.54/9.31 Pr	(107, 200)
IMP 4	Pr > 60/>10 MeV	$15^h \pm 10^m/\Delta3$	$19^h/22^h\Delta1$	$25^h/>23^h$	3.64/28.1 Pr	(200)
HEOS	Pr > 6 MeV	$15^h\Delta2$	$28^d00^h\Delta2$	$>23^h$	26.6 Pr	(201)
RIOM	PCA	$15^h\Delta3$	$17^h\Delta1$		1.1 dB	(146)
HEOS	Pr > 41 MeV	$15^h15^m \pm 15^m$	21^h30^m	30^h	5.5 Pr	(206)
IMP 4	Pr > 0.8 MeV	$16^h \pm 60^m\Delta3$	$28^d04^h\Delta1$	$>23^h$	1490 Pr	(107)
PION 9	Pr > 14 MeV	$\Delta5$	$22^h\Delta1$	$>3^d$	440 Pr	(217)

Source: ● flare 1348   N13   W65   2B   McM 9946
          △ major magnetic storm begins 27^d03^h, max. 27^d18^h

315	1969 February 28	13^h				1 1 0
IMP 4	Pr > 9 MeV	$13^h\Delta3$	$16^h\Delta2$	140^h	10.4 Pr	(107)
IMP 4	Pr > 29 MeV	$13^h20^m \pm 10^m\Delta3$	$17^h\Delta1$	140^h	4.55 Pr	(107)
HEOS	Pr > 6 MeV	$14^h\Delta3$	$19^h\Delta2$	125^h	21 Pr	(202)
RIOM	PCA	$14^h\Delta3$	$18^h\Delta2$		1 dB	(146)
IMP 4	Pr > 0.8 MeV	$15^h\Delta3$	$16^h\Delta2$	170^h	718 Pr	(107)

Source: ○ flare 1231   S21   E09   SN   McM 9957
          ○ flare 0544   N16   E74   1N   McM 9966
          □ active region 9946 is going over W limb

	1969 March 8	event in Appendix A				
316	1969 March 12	18^h30^m				0 1 0
AIMP 2	El > 40 keV	$18^h14^m \pm 12^m$			200 El	(226)
PION 8	Pr > 42/>7 MeV	18^h30^m	$19^h/22^h\Delta1$	$/90^h$	0.028/15 Pr	(207, 199)
HEOS	El > 7.5 MeV				0.63 c	(204)
IMP 4	El > 2.7/>0.3 MeV	18^h45^m			0.02/0.6 El	(166)
HEOS	Pr > 6 MeV	$19^h \pm 6^m$	$22^h\Delta2$	$>85^h$	4.2 Pr	(201)
PION 9	Pr > 32 MeV	$19^h\Delta1$	23^h30^m		0.0015 Pr	(207)
PION 9	Pr > 14 MeV	$19^h \pm 20^m$	$13^d00^h\Delta2$	58^h	3.1 Pr	(217)
IMP 4	Pr > 30/>10 MeV	$19^h30^m \pm 30^m$	$21^h\Delta1$	$50^h/65^h$	0.83/2.18 Pr	(200)
IMP 4	Pr > 60 MeV	$20^h\Delta2$	$21^h\Delta1$	20^h	0.1 Pr	(200)
PION 6	Pr > 7.5 MeV	$20^h\Delta2$			2 Pr	(207)
AIMP 2	Pr > 0.32 MeV	20^h30^m			120 Pr	(193)
RIOM	PCA		$22^h\Delta1$		0.7 dB	(184)

(PION 6 at 0.93 AU, ES − 151°; PION 8 at 1.04 AU, ES − 24°; PION 9 at 0.78 AU, ES + 14°)

Source: ● flare 1739   N12   W80   2B   McM 9966

317	1969	March 16	06^h30^m				$-1\,0\,0$
PION 9	$Pr > 14\,MeV$	$06^h20^m \pm 20^m$	$09^h\Delta2$	19^h	3.7 Pr	(217)	
IMP 4	$Pr > 0.8\,MeV$	$06^h45^m \pm 15^m\Delta3$	$10^h\Delta2$	$>18^h$	5 Pr	(107)	
IMP 4	$Pr > 29\,MeV$	$07^h\Delta3$	$13^h\Delta1$	25^h	0.1 Pr	(107)	
PION 9	$Pr > 7\,MeV$	$07^h\Delta1$	$09^h\Delta1$		2.2 Pr	(199)	
IMP 4	$Pr > 10\,MeV$	$07^h45^m \pm 30^m$	$10^h\Delta1$	20^h	0.19 Pr	(200)	

(PION 9 at 0.78 AU, ES + 15°)

Source: □ active region McM 9966 is on invisible hemisphere, 3 days beyond W limb
○ flare 0122   N17   E90   SN   McM 9944

318	1969	March 17	01^h				$0\,0\,0$
IMP 4	$Pr > 9\,MeV$	$00^h45^m \pm 15^m\Delta3$	$02^h\Delta1$	10^h	5.5 Pr	(107)	
HEOS	$Pr > 6\,MeV$	$01^h\Delta3$	$05^h\Delta2$	20^h	0.26 Pr	(202)	
IMP 4	$Pr > 0.8\,MeV$	$01^h \pm 20^m\Delta3$	$03^h\Delta2$	3^d	110 Pr	(107)	
AIMP 2	$Pr > 0.32\,MeV$	02^h30^m		59^h	780 Pr	(193)	

Source: ○ flare 16^d2334 N16   E30   1B   McM 9994
$\mathcal{O}$ flare 16^d2108 N23   W11   1B   McM 9995
○ flare 16^d2011 N16   E07   1F   McM 9996
◇ geomagnetic storm modulation; geomagnetic storm begins with sc 17^d0300

319	1969	March 21	$\sim05^h$				$0\,1\,0$
AIMP 2	$El > 40\,keV$	$04^h55^m \pm 25^m$			200 El	(226)	
VLF	PCA	05^h27^m				(192)	
IMP 4	$El > 2.7/>0.3\,MeV$	05^h33^m	$/12^h\Delta2$		0.02/1.4 El	(166)	
HEOS	$Pr > 6\,MeV$	$05^h36^m \pm 30^m$	$17^h\Delta2$	130^h	10.5 Pr	(201)	
IMP 4	$Pr > 30\,MeV$	$05^h45^m \pm 30^m$	$17^h\Delta2$	30^h	0.4 Pr	(200)	
IMP 4	$Pr > 10\,MeV$	$06^h \pm 30^m$	$18^h\Delta2$	85^h	4.74 Pr	(200)	
PION 9	$Pr > 14\,MeV$	$06^h\Delta5$	$22^h\Delta5$	$>2^d$	6.2 Pr	(217)	
AIMP 2	$Pr > 0.32\,MeV$	$08^h\Delta1$	$23^d03^h\Delta2$	68^h	1780 Pr	(210)	
VEN 6	$Pr > 1\,MeV$	$09^h\Delta2$	$22^d06^h\Delta2$	204^h	1180 Pr	(209)	
IMP 4	$Pr > 0.8\,MeV$	$10^h\Delta2$	$22^d17^h\Delta2$	$>82^h$	575 Pr	(107)	
RIOM	PCA		$21^h\Delta1$		0.8 dB	(184)	

(PION 9 at 0.77 AU, ES + 20°; VEN 6 at 0.88 AU, ES − 1°)

Source: ● flare 0139   N20   E17   2B   McM 9994
⊘ flare 1312   N19   E09   2B   McM 9994

	1969	March 24 (twice)		events in Appendix A			
320	1969	March 27	14^h				$-1\,0\,0$
IMP 4	$Pr > 29\,MeV$	$14^h\Delta2$	$28^d04^h\Delta2$	40^h	0.07 Pr	(107)	
IMP 4	$El > 0.3\,MeV$	$14^h40^m \pm 15^m$			0.3 El	(166)	
IMP 4	$Pr > 9/>0.8\,MeV$	$15^h\Delta2/\Delta3$	$\begin{cases}28^d05^h\Delta1/ \\ 00^h\Delta2\end{cases}$	$40^h/45^h$	0.22/14 Pr	(107)	
HEOS	$Pr > 6\,MeV$	$17^h\Delta2$	$28^d03^h\Delta2$	40^h	0.27 Pr	(202)	

Source: ● flare 1315   N20   W69   2B   McM 9994

321	1969	March 30	03^h				112
PION 9	Pr > 14 MeV	$02^h40^m \pm 20^m$	$13^h\Delta5$	$>5^d$	12.5 Pr	(217)	
HEOS	Pr > 63/>24 MeV	02^h50^m	$17^h/31^d03^h\Delta1$	$190^h/$	4/5 Pr	(208)	
IMP 4	Pr > 94/>60 MeV	$03^h\Delta2$	$16^h/18^h\Delta2$	$150^h/190^h$	1.4/8.75 Pr	(107, 173)	
AIMP 2	El > 40 keV	03^h15^m			1200 El	(226)	
IMP 4	Pr > 29 MeV	$03^h \pm 45^m$	$31^d03^h\Delta1$	12^d	8.2 Pr	(107)	
IMP 4	Pr > 9 MeV	$03^h \pm 30^m$	$16^h\Delta1$	12^d	22 Pr	(107)	
PION 9	Pr > 7 MeV	$03^h\Delta2$	$08^h\Delta2$	11^d	120 Pr	(199)	
Pr > 2.3 GeV	GLE (141)	$04^h \pm 60^m$				several	
			$16^h \pm 60^m$		8.8%	Sanae	
				48^h		South Pole	
PION 8	Pr > 7 MeV	$04^h\Delta2$	$10^h\Delta5$		12 Pr	(199)	
VEN 6	Pr > 1 MeV	$04^h\Delta2$	$12^h\Delta2$	$>5^d$	630 Pr	(209)	
VLF	PCA	$04^h\Delta1$				(192)	
OGO 5	El > 12 MeV	$06^h\Delta2$		$>10^d$	0.24 El	(186)	
IMP 4	El > 3/>0.3 MeV	$06^h\Delta2$	$/20^h\Delta2$	$/>11^d$	4.3/26 El	(179)	
AIMP 2	El > 50 keV	$06^h\Delta1$	16^h48^m		210 El	(197)	
AIMP 2	Pr > 0.32 MeV	06^h30^m	$31^d01^h12^m$	90^h	820 Pr	(193)	
HEOS	El > 7.5 MeV	$06^h53^m \pm 6^m$	$16^h\Delta1$	$>2^d$	84 c	(204)	
IMP 4	Pr > 0.8 MeV	$07^h\Delta8$	$20^h\Delta2$	12^d	810 Pr	(107)	
Balloon	Pr > 160 MeV	$08^h\Delta2$	$18^h\Delta1$	120^h	2 Pr	(43)	
RIOM	PCA	$11^h\Delta2$	$19^h\Delta2$	$>5^d$	1.3 dB	(146)	

(PION 8 at 1.05 AU, ES $-24°$; PION 9 at 0.76 AU, ES $+24°$; VEN 6 at 0.86 AU, ES $+0.5°$)

Source: ▢ flare invisible hemisphere
  <0332   N19   W90   1N   McM 9994

322	1969	April 11	00^h				330
IMP 4	El > 0.3/>3 MeV	$00^h\Delta3/$			160/4 El	(179)	
HEOS	El > 7.5 MeV	$01^h\Delta2$	$13^d04^h\Delta2$		8.6 c	(204)	
IMP 4	Pr > 94/>29 MeV	$01^h\Delta3$	$13^d06^h/07^h\Delta2$	$120^h/>14^d$	2.2/500 Pr	(107)	
AIMP 2	El > 40 keV	$02^h\Delta1$			100000 El	(226)	
IMP 4	Pr > 10/>0.8 MeV	$02^h\Delta3$	$\left[\begin{array}{l}13^d03^h\Delta1/\\06^h\Delta2\end{array}\right.$	$320^h/13^d$	1350/ 37500 Pr	(200, 107)	
PION 9	Pr > 14 MeV	$02^h\Delta2$	$13^d\Delta5$	$>9^d$	312 Pr	(217)	
IMP 4	Pr > 60 MeV	$03^h\Delta2$	$13^d03^h\Delta1$	190^h	16.0 Pr	(173)	
PION 8	Pr > 7 MeV	$07^h\Delta2$	$12^d\Delta7$	12^d	9000 Pr	(199)	
VLF	PCA	$09^h\Delta1$				(192)	
AIMP 2	Pr > 0.32 MeV	$<12^h30^m$	$13^d06^h\Delta2$	312^h	21500 Pr	(193)	
Balloon	Pr > 100 MeV	$13^h\Delta5$	$13^d09^h\Delta2$	50^h	1 Pr	(43)	
RIOM	PCA	$13^h\Delta2$	$\Delta7$	$>10^d$	>12 dB	(146, 184)	
PION 9	Pr > 7 MeV	$\Delta5$	$13^d22^h\Delta2$	12^d	350 Pr	(199)	
VEN 6	Pr > 1 MeV	$\Delta5$	$13^d03^h\Delta2$	308^h	9400 Pr	(209)	

(VEN 6 at 0.83 AU, ES $+3.5°$; PION 9 at 0.76 AU, ES $+32°$; PION 8 at 1.05 AU, ES $-25.5°$)

Source: Active region McM 10035 (this region is a return of active region McM 9994) is beyond E limb
  ■ flare 10^d0410 N11   E90   1N   McM 10035
  ■ SWF, imp. 2+, 10^d1057, SPA 10^d1055–1100; 3 GHz burst 10^d1052–1059, 1.9 and 1106–1130, 1.2;
    200 MHz burst 10^d1059–1106, 4; DS type IV 10^d1100–1215, DS type II (Dkm) 10^d1111–1120;
    X-rays burst 10^d1056–1102, 10^d1104–11^d0429

323	1969 April 24	10^h				0 0 0
PION 9	Pr > 14 MeV	$10^h\Delta2$	$25^d02^h\Delta2$	$>3^d$	3.7 Pr	(217)
AIMP 2	Pr > 0.32 MeV	$18^h\Delta3$	$25^d00^h\Delta2$		35 Pr	(193)
HEOS	Pr > 6 MeV	$20^h\Delta2$	$25^d08^h\Delta2$	$>31^h$	0.67 Pr	(202)
IMP 4	Pr > 9 MeV	$20^h\Delta4$	$25^d01^h\Delta2$	$>33^h$	1.7 Pr	(107)
IMP 4	Pr > 0.8 MeV	$<20^h\Delta4$	$25^d03^h\Delta2$	$>33^h$	49 Pr	(107)
(PION 9 at 0.77 AU, ES + 35°)						

Source: ◇ transit across the disk of active region McM 10057
      ○ flare 0308   N23   W64   2N   McM 10035
      ○ flare 1245   N07   E72   SN   McM 10057

324	1969 April 26	01^h30^m				−1 0 0
AIMP 2	Pr > 0.32 MeV	01^h30^m	$28^d03^h\Delta2$	87^h	15200 Pr	(193)
HEOS	Pr > 6 MeV	$03^h\Delta3$	$10^h\Delta2$		0.5 Pr	(202)
IMP 4	Pr > 9/>0.8 MeV	$\begin{bmatrix}05^h\pm40^m/\\\pm30^m\,\Delta3\end{bmatrix}$	$07^h/28^d03^h\Delta3$	$>75^h/>74^h$	0.6/900 Pr	(107)
VEN 6	Pr > 1 MeV		$27^d18^h\Delta2$		316 Pr	(43)
(VEN 6 at 0.78 AU, ES + 9°)						

Source: origin unknown, perhaps particle emission related to region McM 10057 transit, or geomagnetic storm
      modulation

325	1969 April 29	07^h				−1 0 0
HEOS	Pr > 6 MeV ·	$07^h\Delta2$	$30^d03^h\Delta2$	85^h	0.13 Pr	(202)
IMP 4	Pr > 9/>0.8 MeV	$08^h/12^h\Delta3$	$30^d03^h/05^h\Delta2$	$>3^d$	0.3/28 Pr	(107)

Source: ◇ CM transit of active region McM 10057
      ○ flare <0223   N09   W48   1F   McM 10055

326	1969 May 2	19^h				−2 0 0
PION 8	Pr > 14 MeV	$19^h\Delta5$	$3^d00^h\Delta2$	16^h	0.14 Pr	(183)
HEOS	Pr > 6 MeV	$19^h15^m\pm4^m$	$3^d00^h\Delta2$	20^h	0.13 Pr	(201)
(PION 8 at 1.07 AU, ES −27.5°)						

Source: ● flare 1745   N09   W40   1B   McM 10057

327	1969 May 5	12^h50^m				−2 0 0
AIMP 2	El > 40 keV	12^h50^m			110 El	(226)
VEN 6	Pr > 1 MeV	$14^h\Delta2$	$6^d10^h\Delta2$	56^h	8 Pr	(43)
PION 8	Pr > 14 MeV	$13^h20^m\pm20^m$	$16^h\Delta5$	$>1^d$	0.21 Pr	(183)
HEOS	Pr > 6 MeV	$\Delta5$	$22^h\Delta5$		0.13 Pr	(202)
(PION 8 at 1.07 AU, ES −27.5°; VEN 6 at 0.75 AU, ES + 15°)						

Source: ● flare 1237   N08   W77   SN   McM 10057

	1969	May 6, May 8	events in Appendix A				
328	1969	May 13	05^h				1 1 0
PION 8	Pr > 14 MeV	05^{h}Δ2				(183)
HEOS	Pr > 6 MeV	10^{h}Δ2	14^{d}24^{h}Δ5	80^h	26 Pr	(202)
AIMP 2	Pr > 0.3 MeV	20^{h}Δ2	15^{d}03^{h}Δ2	6^d	16500 Pr	(210)
RIOM	PCA		15^{d}00^{h}Δ2		1.2 dB	(184)

Source: □ Activity on invisible hemisphere from regions 10065 and 10057 (210)
      ○ flare 1450   S02   W80   SN   McM 10071
      ◇ geomagnetic storm begins 12^{d}18^h, sc 14^{d}1930

	1969	May 23, May 28 (three times)		events in Appendix A			
329	1969	May 28	>13^h				−2 0 0
AIMP 2	El > 40 keV	13^{h}10^m			650 El	(226)
HEOS	Pr > 6 MeV	14^{h}15^m ± 10^m	17^{h}Δ1	>11^h	0.21 Pr	(201)
VELA 4	Pr 0.6 MeV	16^{h}Δ2	29^{d}00^{h}Δ2	>12^h	20* Pr	(168)
HEOS	El > 0.5 MeV		14^{h}Δ1	>11^h	15 El	(202)

Source: ● flare 1241   N10   W59   1B   McM 10109

	1969	May 28	event in Appendix A				
330	1969	May 29	01^h				−2 0 0
AIMP 2	El > 40 keV	00^{h}37^m			800 El	(226)
HEOS	El > 0.5 MeV	01^{h}Δ3	02^{h}Δ1	>21^h	20 El	(202)
PION 8	Pr > 14 MeV	01^{h}Δ5	06^{h}Δ2		0.56 Pr	(183)
HEOS	Pr > 6 MeV	01^{h}46^m ± 6^{m}Δ3	04^{h}Δ1	>19^h	0.27 Pr	(201)
VELA 4	Pr 0.6 MeV	08^{h}Δ3	16^{h}Δ2		30* Pr	(168)
(PION 8 at 1.07 AU, ES −29°)						

Source: ○ flare 0020   N11   W64   1B   McM 10109
      ○ flare 0022   S12   W90   SN   McM 10099

	1969	May 29 (four times)		events in Appendix A			
331	1969	May 29	20^h				−2 0 0
AIMP 2	El > 40 keV	19^{h}55^m			750 El	(226)
PION 8	Pr > 14 MeV	20^{h}40^{m}Δ3	23^{h}Δ1	10^h	0.16 Pr	(183)
HEOS	Pr > 6 MeV	21^{h}14^m ± 6^{m}Δ3	22^{h}Δ2	25^h	0.33 Pr	(201)
HEOS	El > 0.5 MeV	22^{h}Δ3	23^{h}Δ1	>15^h	40 El	(202)
VELA 4	Pr 0.6 MeV	30^{d}00^{h}Δ3	30^{d}08^{h}Δ2	36^h	60* Pr	(168)

Source: ○ flare 1939   N10   W76   1B   McM 10109
      ○ flare 1931   S16   W82   SN   McM 10108

	1969	May 30 (four times), May 31 (twice), June 1, June 2, June 5, June 7			events in Appendix A	
332	1969	June 7	$\sim 15^h$			1 1 0
PION 8	Pr > 14 MeV	$15^h\Delta5$	$20^h\Delta2$	$>2^d$	3.75 Pr	(183)
OGO 6	Pr 1.1 MeV	$15^h\Delta2$	$8^d23^h\Delta1$		2500* Pr	(211)
OGO 6	Pr 2/5 MeV	$16^h/17^h\Delta2$	$8^d16^h/21^h\Delta1$		1000*/80* Pr	(211)
VELA 4	Pr 0.6 MeV	$18^h\Delta3$	$8^d14^h\Delta2$	$>7^d$	5000* Pr	(168)
HEOS	Pr > 6 MeV	$18^h45^m \pm 10^m\Delta3$	$8^d18^h\Delta2$	170^h	60 Pr	(201)
HEOS	Pr > 24 MeV	$20^h\Delta2$	$8^d14^h\Delta1$	100^h	0.6 Pr	(202)
RIOM	PCA	21^h45^m	$8^d18^h\Delta2$	$>5^d$	1.4 dB	(184)
OGO 6	Pr > 5 MeV	$20^h\Delta2$	$8^d15^h40^m$	$>5^d$	130 Pr	(184)
AIMP 2	Pr > 0.32 MeV			182^h	10000 Pr	(193)
PION 9	Pr > 14 MeV	$8^d12^h\Delta2$	$8^d19^h\Delta5$	51^h	0.5 Pr	(217)

(PION 8 at 1.07 AU, ES − 30°; PION 9 at 0.85 AU, ES + 60°)

Source: ● flare <0945   N11   E34   1N   McM 10134

	1969	June 11, June 16, June 28, July 2, July 13, July 20, July 24, Aug. 4, Aug. 7, Aug. 10, Aug. 16, Aug. 17, Aug. 19, Sept. 3, Sept. 7, Sept. 14, Sept. 17 events in Appendix A				
333	1969	September 25	$\sim 07^h50^m$			1 1 0
AIMP 2	El > 40 keV	07^h33^m			330 El	(226)
IMP 5	Pr > 30 MeV	$07^h50^m \pm 10^m$	$10^h\Delta1$	15^h	1.05 Pr	(200)
IMP 5	Pr > 60/>10 MeV	$08^h \pm 10^m/\pm 20^m$	$10^h/12^h\Delta1$	$10^h/37^h$	0.16/14.92 Pr	(200)
HEOS 1	Pr > 6 MeV	$08^h09^m \pm 7^m$	$16^h\Delta2$	48^h	66 Pr	(201)
IMP 5	Pr > 1 MeV	$08^h40^m \pm 28^m$	$26^d09^h\Delta1$	48^h	286 Pr	(212)
AIMP 2	Pr > 0.32 MeV	10^h		230^h		(193)
				(duration includes next four events)		
RIOM	PCA	$10^h\Delta2$	$13^h\Delta2$	$>12^h$	0.7 dB	(214)
PION 8	Pr > 14 MeV	$13^h\Delta5$	$17^h\Delta5$	$>24^h$	8.6 Pr	(183)

(PION 8 at 1.05 AU, ES − 41°)

Source: ● flare <0658   N13   W15   3N   McM 10326

	1969	Sept. 25	event in Appendix A			
334	1969	September 27	08^h			−1 1 0
IMP 5	Pr > 1 MeV	$08^h\Delta3$	$21^h\Delta1$	$>16^h$	640 Pr	(212)
HEOS 1	El > 0.5 MeV	$09^h\Delta3$	$15^h\Delta1$	15^h	200 El	(202)
HEOS 1	Pr > 6 MeV	$09^h \pm 100^m\Delta3$				(201)
IMP 5	Pr > 10 MeV	$10^h\Delta1$	$19^h\Delta2$	$>12^h$	0.9 Pr	(200)
AIMP 2	Pr > 0.32 MeV	continuing				(193)
PION 9	Pr > 14 MeV	$11^h\Delta5$	$14^h\Delta2$	40^h	0.62 Pr	(217)
RIOM	PCA	continuing				

(PION 9 at 0.99 AU, ES + 62°)

Source: ● flare <0350   N09   E02   3B   McM 10333

335 1969 September 27 $\sim 19^h$ 1 2 0

PION 8	Pr > 14 MeV	$18^h40^m \pm 20^m$	20^h30^m	$>24^h$	35 Pr	(183)
IMP 5	Pr $> 30/> 10$ MeV	$22^h\Delta 1/\Delta 3$	$28^d03^h\Delta 1$	$20^h/45^h$	0.06/11.01 Pr	(200)
HEOS 1	Pr > 5 MeV	$22^h\Delta 3$	$28^d03^h\Delta 1$	48^h	800 Pr	(202)
RIOM	PCA	$22^h\Delta 3$	$28^d03^h\Delta 2$	47^h	1.7 dB	(214)
IMP 5	Pr > 1 MeV	$28^d00^h \pm 10^m\Delta 3$	$29^d04^h\Delta 1$	$>60^h$	9240 Pr	(212)
HEOS 1	Pr > 24 MeV	$28^d02^h\Delta 2$	$28^d06^h\Delta 2$	25^h	0.6 Pr	(202)
AIMP 2	Pr > 0.32 MeV	29^d			2160 Pr	(193)

(PION 8 at 1.05 AU, ES $-41.5°$)

Source: $\triangle$ geomagnetic storm modulation; a severe magnetic disturbance follows the sc which begin 27^d2125 and 29^d0453 (events 333 and 334)
  ○ flare 2042   N07   W72   1N   McM 10325

1969	Sept. 30, Oct. 7, Oct. 10, Oct. 11	events in Appendix A

336 1969 October 14 01^h $-2\,0\,0$

PION 9	Pr > 14 MeV	$01^h \pm 40^m$		$>2^d$		(217)
HEOS 1	Pr > 6 MeV	$03^h\Delta 2$	$05^h\Delta 1$	$>3^h$	0.1 Pr	(202)
IMP 5	Pr > 1 MeV	$03^h \pm 20^m$	05^h30^m	$>3^h$	1 Pr	(212)

(PION 9 at 0.98 AU, ES $+60°$)

Source: origin uncertain
  □ type II, 13^d2336–2350; type III, 13^d2333–2335; 3 GHz burst 2333; region McM 10351 is two days beyond W limb, on invisible hemisphere

---

337 1969 October 14 06^h $-1\,0\,0$

HEOS 1	Pr > 6 MeV	$06^h\Delta 3$	$10^h\Delta 5$		0.4 Pr	(202)
AIMP 2	El > 40 keV	06^h06^m			60 El	(226)
IMP 5	Pr > 1 MeV	$06^h20^m \pm 20^m\Delta 3$	$10^h\Delta 2$	$>6^h$	6.8 Pr	(212)
IMP 5	Pr $> 10/> 30$ MeV	$\lceil 07^h + 60^m/$ $\lfloor 08^h \pm 90^m$	$09^h/10^h\Delta 2$	$>4^h/>3^h$	0.3/0.1 Pr	(200)

Source: ● flare 0539   N25   W71   2N   McM 10352

1969	Oct. 14	event in Appendix A

338 1969 October 14 11^h $0\,1\,0$

AIMP 2	El > 40 keV	$10^h53^m\Delta 3$			2000 El	(226)
IMP 5	Pr > 60 MeV	$11^h \pm 10^m$	11^h30^m	7^h	0.17 Pr	(200)
IMP 5	Pr $> 30/> 10$ MeV	$11^h \pm 10^m\Delta 3$	11^h30^m	$7^h/>9^h$	0.6/3.5 Pr	(200)
HEOS 1	Pr > 6 MeV	$\Delta 5$	$14^h\Delta 5$		6 Pr	(202)
IMP 5	Pr > 1 MeV	$12^h\Delta 3$	$13^h\Delta 2$	$>99^h$	10.2 Pr	(212)
PION 8	Pr > 14 MeV	$\Delta 3$	$11^h\Delta 1$		0.15 Pr	(183)
HEOS 1	El > 7.5 MeV	$14^h\Delta 5$	$18^h\Delta 5$	9^h	2.94 c	(204)
RIOM	PCA				0.4 dB	(228)

(PION 8 at 1.03 AU, ES $-43°$)

Source: ○ flare <1034 S06   W13   1F   McM 10361
  □ region 10351 is on invisible hemisphere, two days beyond W limb

339	1969 October 14	19^h15^m					0 0 0
AIMP 2	El >40 keV	19^h12^m			450 El	(226)	
IMP 5	Pr >10 MeV	$19^h45^m \pm 15^m \Delta3$	$21^h\Delta1$	20^h	1.6 Pr	(200)	
IMP 5	Pr >30 MeV	$20^h\Delta1$	$15^d00^h\Delta1$	10^h	0.12 Pr	(200)	
HEOS 1	Pr >6 MeV	$20^h\Delta3$	$15^d01^h\Delta2$	45^h	5 Pr	(202)	
PION 9	Pr >14 MeV	$\Delta3$	$15^d02^h\Delta2$		3.1 Pr	(217)	

(PION 9 at 0.98 AU, ES + 60°)

Source: origin uncertain
□ DS type II 1853–1857, type III 1846–1848, 1855–1856; region McM 10351 is two days beyond W limb on invisible hemisphere

	1969 Oct. 18, Oct. 20	events in Appendix A					
340	1969 November 2	10^h30^m					3 3 0
IMP 5	El >0.35 MeV	$10^h34^m \pm 1^m$	12^h30^m		10000 El	(215)	
OV 5–6	El >45 keV	$10^h34^m \pm 1^m$				(216)	
AIMP 2	El >40 keV	10^h35^m			30000 El	(226)	
IMP 5	El $>1.1/>0.6$ MeV	$10^h37^m \pm 3^m$	11^h45^m		2000/6000 El	(215)	
VLF	PCA	10^h37^m				(192)	
IMP 5	Pr $>60/>30$ MeV	$10^h45^m \pm 10^m$	$12^h30^m/13^h\Delta1$	$80^h/100^h$	201/740 Pr	(200)	
IMP 5	Pr >10 MeV	$10^h45^m \pm 10^m$	$14^h\Delta1$	$>119^h$	1320 Pr	(200)	
IMP 5	Pr 17 MeV	$10^h59^m \pm 3^m$	$23^h\Delta2$	100^h	60* Pr	(215)	
IMP 5	Pr >1 MeV	$11^h \pm 10^m$	$3^d07^h\Delta2$	$>119^h$	1610 Pr	(212)	
IMP 5	Pr 2.5 MeV	11^h15^m	$3^d04^h\Delta2$		400* Pr	(215)	
RIOM	PCA	$11^h\Delta2$	$17^h\Delta2$	4^d	14.5 dB	(227)	
AIMP 2	Pr >0.32 MeV	11^h30^m		$>116^h$	3200 Pr	(193)	
HEOS 1	El >7.5 MeV	$11^h59^m\Delta5$	17^h52^m		17.4 c	(204)	
HEOS 1	Pr >200 MeV	$11^h59^m\Delta5$	$13^h\Delta2$	80^h	20 Pr	(208)	
PION 9	Pr >14 MeV	$14^h\Delta5$	$3^d14^h20^m$		645 Pr	(217)	

(PION 9 at 0.92 AU, ES + 68°)

Source: ■ flare <1102 N14   W90   1N   McM 10385

341	1969 November 7	08^h					0 1 0
AIMP 2	El >40 keV	$08^h\Delta2$			30 El	(226)	
IMP 5	Pr >10 MeV	$08^h \pm 30^m \Delta3$	$8^d4^h\Delta2$	45^h	7.9 Pr	(200)	
HEOS 1	Pr >6 MeV	$08^h30^m \pm 30^m \Delta3$	$8^d4^h\Delta1$	55^h	33 Pr	(201)	
AIMP 2	Pr >0.32 MeV	$08^h30^m\Delta3$		$>135^h$	11400 Pr	(193)	
HEOS 1	Pr >1 MeV	$09^h\Delta3$	$8^d17^h\Delta2$	$>65^h$	3000 Pr	(218)	
PION 9	Pr >14 MeV	$09^h40^m \pm 20^m$	$14^h\Delta2$	2^d	2.1 Pr	(217)	
RIOM	PCA	$12^h\Delta3$	$8^d07^h\Delta2$	40^h	1.4 dB	(227)	

Source: ☉ flare 0322   N13   E11   2N   McM 10406

342	1969	November 10	03^h				−1 0 0
IMP 5	Pr > 1 MeV		03^{h}Δ3	12^{h}Δ2	>82^h	5.1 Pr	(212)
IMP 5	Pr > 10 MeV		07^{h}Δ5	12^{h}Δ1	22^h	0.25 Pr	(200)

Source: ◇ geomagnetic storm modulation, storm begins with sc 8^{d}1837, increases in intensity 10^{d}03^h
◇ bright region McM 10418 forms on the disk in west 10^d; also transit of bright region McM 10412

343	1969	November 13	05^h				−1 0 0
IMP 5	Pr > 10 MeV		05^{h}Δ4	13^{h}Δ2	25^h	0.15 Pr	(200)
HEOS 1	Pr > 6 MeV		05^{h}Δ2	00^{h}Δ2	>40^h	0.13 Pr	(202)
IMP 5	Pr > 1 MeV		13^{h}Δ3	14^{d}03^{h}Δ2	70^h	3.3 Pr	(212)

Source: origin unknown

	1969	Nov. 16	event in Appendix A

344	1969	November 18	03^{h}38^m				−1 0 0
AIMP 2	El > 40 keV		03^{h}38^m			65 El	(226)
IMP 5	Pr > 10 MeV		05^{h}Δ2	10^{h}Δ2	>12^h	0.13 Pr	(200)
IMP 5	Pr > 1 MeV		07^{h}30^m ± 30^{m}Δ3	12^{h}Δ1	>11^h	11.2 Pr	(212)
PION 9	Pr > 14 MeV		13^{h}Δ5	18^{h}Δ2	30^h	0.9 Pr	(217)
HEOS 1	Pr > 6 MeV		Δ5	11^{h}Δ5		0.16 Pr	(202)
(PION 9 at 0.91 AU, ES + 68°)							

Source: ○ flare 0045   N10   E40   SB   McM 10432
□ SWF, imp. 1, 0337–0430; DS type III 0223, 0229; active region McM 10412 is on invisible hemisphere, one day beyond W limb

345	1969	November 18	16^h				−1 0 0
HEOS 1	Pr > 6 MeV		16^{h}Δ3	20^{h}Δ5	56^h	0.2 Pr	(202)
IMP 5	Pr > 10 MeV		17^{h}Δ3	19^{d}04^{h}Δ2	2^d	0.15 Pr	(200)
IMP 5	Pr > 1 MeV		18^{h}Δ3	19^{d}08^{h}Δ2	53^h	3.6 Pr	(212)
PION 8	Pr > 14 MeV		21^{h}Δ2	19^{d}16^{h}Δ2		1.2 Pr	(183)
(PION 8 at 1.01 AU, ES − 44.5°)							

Source: ● flare 1633   N14   E40   2B   McM 10432

346	1969	November 20	23^{h}30^m				−2 0 0
HEOS 1	Pr > 5 MeV		23^{h}30^m	21^{d}02^{h}Δ2	>18^h	0.6 Pr	(202)
IMP 5	Pr > 1 MeV		23^{h}30^m ± 30^{m}Δ3	21^{d}07^{h}Δ2	>20^h	9 Pr	(212)
AIMP 2	Pr > 0.32 MeV		23^{h}50^m		192^h		(193)
					(duration includes next four events)		

Source: ⊙ flare 1619   N07   E07   2B   McM 10432
○ flare 2309   N09   E12   SN   McM 10432

347	1969	November 21	$\sim 15^h 30^m$				$-1\,0\,0$
PION 9	Pr > 14 MeV	$15^h 30^m \Delta 5$	$18^h \Delta 2$				(183)
HEOS 1	Pr > 6 MeV	$17^h \Delta 3$	$22^d 12^h \Delta 2$	53^h	0.2 Pr		(202)
IMP 5	Pr > 1 MeV	$19^h 30^m \pm 30^m \Delta 3$	$22^d 10^h \Delta 2$	$>62^h$	100 Pr		(212)
AIMP 2	Pr > 0.32 MeV				950 Pr		(193)

(PION 9 at 0.87 AU, ES + 69°)

Source: origin unknown

	1969	Nov. 23	event in Appendix A				
348	1969	November 24	$\sim 10^h$				$0\,1\,0$
AIMP 2	El > 40 keV	$09^h 48^m \pm 18^m$			60 El		(226)
HEOS 1	Pr > 63 MeV	$10^h \pm 30^m$	$14^h \Delta 2$	20^h	0.5 Pr		(208)
IMP 5	Pr > 10 MeV	$10^h \pm 10^m$	$17^h \Delta 1$	50^h	3.5 Pr		(200)
HEOS 1	Pr > 6 MeV	$10^h \pm 30^m$	$18^h \Delta 2$	110^h	4 Pr		(201)
IMP 5	Pr > 30 MeV	$10^h 20^m \pm 20^m$	$15^h \Delta 5$	35^h	0.9 Pr		(200)
IMP 5	Pr > 60 MeV	$10^h 30^m \pm 30^m$	$15^h \Delta 5$	20^h	0.4 Pr		(200)
VLF	PCA	$<11^h \Delta 2$					(192)
RIOM	PCA		24^d		0.7 dB		(228)
PION 9	Pr > 14 MeV	$14^h \Delta 2$	$19^h \Delta 5$		0.3 Pr		(217)

(PION 9 at 0.89 AU, ES + 69°)

Source: ● flare 0914   N15   W31   2B   McM 10432

	1969	Nov. 25	event in Appendix A				
349	1969	November 30	18^h				$-1\,0\,0$
PION 9	Pr > 14 MeV	$17^h 40^m \pm 20^m$	$1^d 01^h \Delta 5$	2^d	4.6 Pr		(217)
HEOS 1	Pr > 6 MeV	$18^h \Delta 2$	$1^d 18^h \Delta 2$	2^d	0.3 Pr		(202)
PION 9	Pr > 10 MeV	$18^h 40^m \pm 20^m$		3^d			(217)
IMP 5	Pr > 1 MeV	$19^h \pm 60^m \Delta 3$	$4^d 05^h \Delta 2$	6^d	1.4 Pr		(212)

Source: □ DS type II 1713–1716, 1724–1729; DS type IV 1717–1748; continuum 1748–2155; active region
         McM 10432 is two days beyond W limb on invisible hemisphere
       ○ flare 1700   N23   E02   SN   McM 10447

	1969	Dec. 7, Dec. 9, Dec. 13	events in Appendix A

350 1969 December 18 15^h50^m 0 1 0

AIMP 2	El >40 keV	15^h42^m			70 El	(226)
IMP 5	Pr $>60/>30$ MeV	$15^h50^m/16^h\pm10^m$	$20^h\Delta1$	$15^h/20^h$	0.55/1.19 Pr	(200)
IMP 5	Pr >10 MeV	$16^h20^m\pm40^m$	$20^h\Delta1$	25^h	1.52 Pr	(200)
HEOS 1	Pr >6 MeV	$16^h22^m\pm6^m$	$20^h\Delta2$	$>32^h$	0.8 Pr	(201)
IMP 5	Pr >1 MeV	$17^h\pm30^m$	$22^h\Delta1$	$>33^h$	43.2 Pr	(212)
AIMP 2	Pr >0.32 MeV	18^h30^m		90^h	300 Pr	(193)
RIOM	PCA				0.6 dB	(228)

Source: □ flare behind W limb? DS type IV 1450–1504, 3 GHz burst 1445–1545; extensive prominence activity observed in Hα at NW limb at 1445

351 1969 December 20 00^h 0 1 0

AIMP 2	El >40 keV	$00^h05^m\pm35^m$			250 El	(226)
IMP 5	Pr $>60/>30$ MeV	$00^h40^m\pm40^m/\pm60^m\Delta4$	$05^h\Delta4$	$10^h/20^h$	0.25/1.2 Pr	(200)
IMP 5	Pr >10 MeV	$01^h\pm20^m\Delta4$	$05^h\Delta4$	55^h	8 Pr	(200)
HEOS 1	Pr >6 MeV	$01^h\Delta3$	$05^h\Delta2$	2^d	17.5 Pr	(202)
IMP 5	Pr >1 MeV	$02^h30^m\pm20^m\Delta3$	$08^h\Delta2$	65^h	103 Pr	(212)
RIOM	PCA				1.3 dB	(228)

Source: origin unknown
no flare patrol $19^d\,1920$–2230; DS type II $19^d\,2330$–2358; no known active region on invisible hemisphere 1–3 days beyond W limb

1969	Dec. 23	event in Appendix A

352 1969 December 30 $>20^h$ 0 1 0

AIMP 2	El >40 keV	20^h14^m			150 El	(226)
IMP 5	Pr >30 MeV	$21^h\Delta1$	$31^d03^h\Delta4$	15^h	0.19 Pr	(200)
IMP 5	Pr $>10/>1$ MeV	$21^h30^m\pm40^m/22^h\pm30^m$	$31^d08^h/12^h\Delta1$	$35^h/80^h$	1.09/203 Pr	(200, 212)
HEOS 1	Pr >6 MeV	$23^h\Delta5$	$31^d10^h\Delta2$	70^h	2.1 Pr	(202)
AIMP 2	Pr >0.32 MeV	$31^d01^h30^m$		84^h	1140 Pr	(193)
RIOM	PCA				0.4 dB	(228)

Source: ● flare 1927 S14 W85 1N McM 10491

APPENDIX A

LIST OF UNCONFIRMED AND LOW-ENERGY PROTON EVENTS AND OF PURE ELECTRON EVENTS, OMITTED IN PART 1

APPENDIX A

Chronological List of Unconfirmed (UN) and Low-Energy (L) Proton Events and of 'Pure' (PE) Electron Events Omitted in Part 1

1955

Feb. 1	UN	(3)	f-min PCA, onset 09^h, dur. 12^h, weak; origin unknown
Nov. 19	UN	(3)	f-min PCA, onset 12^h, dur. 12^h, weak; $\diamond$ GMS resurges with an sc 1320
Dec. 6	UN	(3)	f-min PCA, onset 04^h, dur. 24^h, weak; $\triangle$ sc 5^d2216 after flare imp. 3, 3^d1058, E08, McM 3342

1956

Apr. 15	UN	(3)	f-min PCA, onset 01^h, dur. 24^h, weak; may be particle stream remnants after Events 2 and 3
May 14	UN	(3)	f-min PCA, onset 05^h, dur. 24^h, weak; $\circ$ flare imp. 2, <0425, W20, McM 3488; $\diamond$ GMS resurges $\sim 13^d22$
Aug. 28	UN	(3)	f-min PCA, onset 23^h, dur. 24^h, weak; $\circ$ ambiguous flares imp. 2, McM 3643, <2220, E51, and 1520, E60
Nov. 8	UN	(3)	f-min PCA, onset unknown, dur. 72^h, weak; $\odot$ flare imp. 3, 7^d1109, E32, McM 3751
Dec. 25	UN	(3)	f-min PCA, onset unknown, dur. 72^h, weak; $\diamond$ GMS starts with an sc 0754 if onset late on 25^d, then $\circ$ flare imp. 2, 2150, W02, McM 3800

1957

Mar. 28	UN	(3)	f-min PCA, onset 09^h, dur. 48^h, weak; origin uncertain
Apr. 2	UN	(1)	f-min PCA, onset 23^h, dur. 168^h, including Event 9, about 11 h prior to Event 9; origin unknown
Apr. 6	UN	(2)	VHF PCA, onset 08^h, 3.2 dB; part of Event 9, possibly $\square$: McM 3907 one day beyond W limb
May 5	UN	(3)	f-min PCA, onset 02^h, dur. 48^h, weak; origin uncertain
May 19	UN	(12)	RIOM PCA, onset 02^h, dur. $> 10^h$, 1 dB; $\square$ type II, 0008–0016, no flare known; $\circ$ flare imp. 1+, 18^d0810, W15, McM 3979
May 30	UN	(3)	f-min PCA, onset unknown, dur. 72^h, weak; unless $\triangle$ sc 0822 associated, origin unknown
June 28	UN	(1)	f-min PCA, onset unknown, dur. 120^h; $\circ$ ambiguous flares; in McM 4024: 27^d2322, imp. 1, W62; in McM 4039: 27^d2330, imp. 1, E32; 28^d0658, imp. 2, E28
July 19	UN	(3)	f-min PCA, onset unknown, dur. 48^h, weak; $\diamond$ GMS starts gradually 19^d13^h, probably caused by $\circ$ flare imp. 1+, 16^d1742, W28, McM 4061 (type IV)
Sept. 3	UN	(14)	f-min PCA, onset 15^h, dur. $> 24^h$; $\odot$ flare imp. 3, 1412, W30, McM 4124 part of Event 25
Sept. 10	UN	(14)	f-min PCA, onset 06^h, dur. $< 12^h$; $\circ$ flare imp. 3, 0223, E16, McM 4134
Sept. 20	UN	(14)	f-min PCA, onset 03^h, dur. $< 24^h$; $\odot$ flare imp. 3, $19^d < 0350$, E02, McM 4151; part of Event 27
Oct. 5	UN	(14)	f-min PCA, onset 03^h, dur. $< 12^h$; origin unknown
Oct. 20	UN	(18)	RIOM PCA, onset 13^h, about 6 h prior to Event 30 and 3.5^h before the source flare; origin unknown

1957 (cont.)

| Nov. 24 | UN | (23) | f-min PCA, onset 02^h, dur. 60^h; ◇ gradual GMS starts 23^d22^h–24^d10^h; 7 h after the onset ⊘ flare imp. 3, 0848, E37, McM 4263 |
| Nov. 26 | UN | (14) | f-min PCA, onset 12^h, dur. $< 12^h$; ◇ GMS resurgence ~ 14^h, possibly related to the flare 24^d0848 mentioned above, or ○ flare imp. 1, 24^d1611; E23, McM 4263 (type IV) |

1958

Jan. 25	UN	(10)	f-min PCA, onset 16^h, dur. 24^h; ambiguous flares: ⊙ imp. 2, 1205, E11, McM 4382; ⊙ imp. 3, < 0915, W70, McM 4372; ○ imp. 2, 0956, W52, McM 4376
Mar. 3	UN	(101)	Balloon, protons, no onset given, 2500 counts (a very high flux; but this is the first event ever observed in the U.S.S.R.); ⊙ flare imp. 3, < 1005, E60, McM 4445; △ sc 0930 starts a GMS
Mar. 21	UN	(10)	f-min PCA, onset unknown, dur. 60^h; part of Event 37 in (1, 3); ◇ increase in GMS intensity; several flares, most likely ○ flare imp. 1, < 0656, E58, McM 4469
Sept. 14	UN	(3)	f-min PCA, onset 10^h45^m, dur. 24^h, weak; ● flare imp. 2+, 0822, W80, McM 4741
Sept. 22	UN	(23)	f-min PCA, onset 05^h30^m, about 9 h prior to Event 49 and 2 h before the source flare; origin unknown
Oct. 3	UN	(43, 97)	Balloon Pr > 100 MeV, onset $< 09^h$, 0.3 Pr; ○ ambiguous flares: imp. 2, 2^d1806, E52, McM 4794; imp. 1, 2^d2143, W38, McM 4781 (type II); imp. 1+, 3^d0552, W47, McM 4796.

1959

July 9	UN	(43, 97)	Balloon Pr > 100 MeV, onset $< 09^h$, 0.2 Pr, a doubtful increase; about 20 h prior to Event 55; origin unknown
July 9	UN	(1)	f-min PCA, onset 20^h, dur. 312^h; about 8 h prior to Event 55; ⊙ flare imp. 2, 1930, E67, McM 5265 (but PCA onset too early)
Aug. 2	UN	(3)	f-min PCA, onset unknown, dur. 96^h, weak; origin uncertain
Sept. 12	UN	(34)	LUNIK 2, heavy particles, onset 11^h27^m, unless of instrumental origin, the source is unknown
Nov. 9	UN	(42)	EXPL 7, Pr > 30 MeV, onset $11^h\Delta2$, 0.77 Pr; origin unknown
Nov. 30	UN	(42)	EXPL 7, Pr > 30 MeV, onset early on 30^d, dur. 60^h, 0.02 Pr; ⊙ flare imp. 2+, 0247, E16, McM 5476; ⊘ flare imp. 3, 1720, E06, McM 5476 (type II+IV)
Dec. 2	UN	(56)	EXPL 7, Pr > 30 MeV, onset unknown, 0.07 Pr; may be part of the preceding event; ○ ambiguous flares in McM 5476
Dec. 21	UN	(3)	f-min PCA, onset 06^h, dur. 24^h, very weak; ● flare imp. 2, 0043, W53, McM 5494

1960

Jan. 16	UN	(35)	f-fix PCA, onset 03^h, dur. 24^h; ⊙ flare imp. 2, 15^d1336, W68, McM 5525
Feb. 7	UN	(35)	f-fix PCA, onset 07^h, dur. 96^h; ◇ recurrent particle stream after the Events 61 and/or 62 (60)
Feb. 15	UN	(35)	f-fix PCA, onset 10^h, dur. 96^h; ◇ coincident with recurrent gradual GMS
Feb. 29	UN	(35)	f-fix PCA, onset 16^h, dur. 190^h; ◇ coincident with recurrent weak GMS
Mar. 10	UN	(35)	f-fix PCA, onset 18^h, dur. 60^h; ⊙ flare imp. 1, 1716, E07, McM 5592; ◇ coincident with recurrent weak GMS
Mar. 28	UN	(37)	PION 5, Pr > 25 MeV, onset $00^h\Delta2$, max. $03^h\Delta2$, dur. 6^h, 0.016 Pr; 30 h prior to Event 64, origin unknown

1960 (cont.)

Mar. 28	UN	(37)	PION 5, Pr > 25 MeV, onset $22^h\Delta2$, max. $29^d01^h\Delta2$, dur. 4^h, 0.016 Pr; 9 h prior to Event 64; ● flare imp. 2, 2042, E37, McM 5615
Apr. 3	UN	(37)	PION 5, Pr > 25 MeV, onset $15^h\Delta2$, max. $18^h\Delta2$, 0.08 Pr; ○ flare imp. 2, < 0317, W33, McM 5615; ○ flare imp. 2, 1140, W38, McM 5615
Apr. 4	UN	(37)	PION 5, Pr > 25 MeV, onset $09^h\Delta2$, max. $23^h\Delta2$, 0.0012c, about 20 h prior to Event 68; ○ flare imp. 1+, 0846, W51, McM 5615
Apr. 15	UN	(35)	f-fix PCA, onset 10^h, dur. 96^h; ◇ coincident with recurrent GMS
May 12	UN	(43)	Balloon Pr > 100 MeV, onset < 09^h, 0.2 Pr; about one day prior to Event 74; remnants of Event 72, or changing geomagnetic cutoff
May 12	UN	(3)	f-min PCA, onset unknown, dur. 12^h, weak; about half-a-day prior to Event 74; ○ flare imp. 1+, < 1342, W59, McM 5654
May 28	UN	(15)	f-min PCA, onset 01^h, dur. 26^h, weak; part of Event 76, possibly due to ○ flare imp. 1, 27^d1414, W26, McM 5669
June 15	UN	(35)	f-fix PCA, onset 10^h, dur. 48^h; ◇ in recurrent sequence with events 75 (May 17) – Apr. 15 – 63 (March 17) – Feb. 15
June 28	UN	(35)	f-fix PCA, onset 19^h, dur. 36^h; part of Event 79, possibly due to ○ flare imp. 1+, 1214, W37, McM 5713 (type II); a possible contributor is also ⊘ flare imp. 1, 29^d0125, W50, McM 5713
July 14	UN	(15)	f-min PCA, onset unknown, dur. 24^h, very weak; ◇ great GMS onset 1702, preliminary sc 0447; ○ flare imp. 2, 1057, W52, McM 5740
Aug. 29	UN	(15)	f-min PCA, onset 09^h, dur. 14^h; RIOM absorption of 0.3 dB recorded after the sc (128); ◇ major GMS begins with an sc 0022
Sept. 1	UN	(3)	f-min PCA, onset unknown, dur. 48^h, weak; origin uncertain
Sept. 25	UN	(35)	f-fix PCA, onset 21^h, dur. 120^h, about 11 h prior to Event 84 and 8 h before the source flare; origin unknown
Oct. 11	UN	(35)	f-fix PCA, onset ~06^h; ● flare imp. 2, 0517, W36, McM 5880

1961

Feb. 13	UN	(3)	f-min PCA, onset unknown, dur. 24^h; ◇ GMS begins with an sc 0253
Feb. 18	UN	(3)	f-min PCA, onset unknown, dur. 96^h; ◇ a gradual GMS starts 17^d06^h and ends on 21^d, in a series of recurrent GMS's
Mar. 17	UN	(3)	f-min PCA, onset unknown, dur. 72^h; ◇ GMS in the same recurrent series as Feb. 17^d–21^d
July 17	UN	(15)	f-min PCA, onset 10^h, dur. 17^h; ⊙ flare imp. 2+, 0710, 45W, McM 6171; △ sc at 18^h26^m starts a major GMS
July 28	UN	(58)	Balloon Pr, onset $03^h\Delta1$, max. $04^h\Delta1$; f-min PCA in (3) continues; ● flare imp. 2+, < 0244, W38, McM 6178
Aug. 10	UN	(60)	f-min PCA, onset 10^h, dur. ~50^h; ◇ recurrent particle stream, remnants of Events 96–100 (60)
Sept. 18	L	(85, 142)	D 31, Pr 1.5–4.4 MeV, onset $00^h\Delta2$, max. $09^h\Delta2$, 3.2 Pr; EXPL 12, Pr > 3 MeV, 18^d, no time given; ⊙ flare imp. 2+, 16^d1057, E77, McM 6227
Dec. 1	UN	(142)	EXPL 12, Pr > 3 MeV, onset < 03^h, 16 Pr; also f-min PCA according to (60), but no data given; ◇ major GMS starts ~03^h; recurrent particle stream, remnants of Event 106

1962

Oct. 6	UN	(3)	f-min PCA, onset unknown, dur. 216^h; ◇ moderate GMS begins with an sc 7^d2026, recurrent since June 23

1963

Feb. 15/16	L	(44)	EXPL 14, Pr > 2.9 MeV, onset > 15^d18^h, max. $18^d00^h\Delta2$, dur. 120^h, 0.7 Pr; Pr > 5.9 MeV, onset $16^d00^h\Delta2$, max. $17^d18^h\Delta2$, dur. 96^h, 0.1 Pr; origin uncertain, type II bursts 15^d2019 and 2048, ○ flare imp. 1−, 15^d2020, E72, McM 6701
Mar. 8	L	(44)	EXPL 14, Pr > 10/ > 5.4 MeV, onset $12^h\Delta2$, max. $10^d18^h\Delta2$, dur. 96^h, 0.03/0.2 Pr; Pr > 3.1 MeV, onset $12^h\Delta2$, max. $9^d03^h\Delta2$, dur. 96^h, 1.5 Pr; ◇ a recurrent GMS of 5^d duration begins ~7^d18^h, in the sequence S1
Apr. 4	L	(44)	EXPL 14, Pr > 6.0 MeV, onset $00^h\Delta2$, max. $5^d00^h\Delta2$, dur. 48^h, 0.09 Pr; Pr > 9 MeV, 0.01 Pr; Pr > 3.4 MeV, onset $00^h\Delta2$, max. $5^d12^h\Delta2$, dur. 48^h, 0.8 Pr; ◇ a recurrent GMS begins 4^d, in the sequence S1
Apr. 24	L	(44)	EXPL 14, Pr > 3.4 MeV, onset $12^h\Delta2$, max. $25^d12^h\Delta2$, dur. 48^h, 2.0 Pr; Pr > 6.0 MeV, onset $18^h\Delta2$, max. $25^d06^h\Delta2$, dur. 24^h, 0.01 Pr, □ McM 6766, ~ 3 days beyond W limb
May 27	L	(44)	EXPL 14, Pr > 10/ > 3.5 MeV, onset $00^h\Delta2$, max. $06^h\Delta4$, dur. 24^h, 0.02/0.15 Pr; ◇ a minor GMS begins at 20^h in sequence S1; □ McM 6805 goes over the limb on 27^d (in this region ○ flare imp. 1, 25^d1622, W85, with type II+IV)
May 29	UN	(3)	ƒ-min PCA, onset unknown, dur. 96^h, very weak (not confirmed by EXPL 14); error in date?
June 25	L	(44)	EXPL 14, Pr > 10/ > 5.7/ > 3.6 MeV, onset $00^h\Delta2$, max. $26^d00^h\Delta2$, dur. $24^h/24^h/48^h$, 0.07/0.25/0.4 Pr; ◇ a recurrent GMS begins at 01^h in sequence S1
Sept. 14	UN	(151)	RIOM PCA, onset ~20^h (very slight absorption at Mc Murdo) about 14 h prior to Event 117; ◇ major GMS begins 13^d19^h
Sept. 19	UN	(3)	ƒ-min PCA, onset 05^h43^m, dur. 48^h, weak; no RIOM absorption > 0.3 dB at McMurdo and Shepherd Bay (151); ALOU 1, Pr > 1.3 MeV, increased flux on 20^d (157); △ sc 0543 during a long-lived geomagnetic disturbance
Oct. 12	UN	(3)	ƒ-min PCA, onset unknown, dur. 24^h, weak; ◇ recurrent gradual GMS begins 11^d06^h in sequence S1
Dec. 3	UN	(50, 68)	IMP 1, Pr > 0.9 MeV, onset $08^h\Delta2$, max. $5^d12^h\Delta2/\Delta4$, dur. 168^h, 3.4 Pr; ◇ recurrent GMS of long duration in sequence S1 begins with an sc 2^d2116

1964

Jan. 2	UN	(50, 68)	IMP 1, Pr > 0.9 MeV, onset $08^h\Delta2$, max. $3^d11^h\Delta2$, dur. 120^h, 0.62 Pr; ◇ gradual recurrent GMS 1^d–5^d in sequence S1
Jan. 23	UN	(50, 68)	IMP 1, Pr > 0.9 MeV, onset < 23^d07^h, max. $27^d10^h\Delta2$, dur. 120^h, 0.31 Pr, very small gradual event; origin uncertain, ◇ co-rotating stream (50)?
Jan. 28	UN	(50, 68)	IMP 1, Pr > 0.9 MeV, onset $13^h\Delta4$, max. $29^d01^h\Delta5$, dur. 96^h, 3.8 Pr; ◇ recurrent geomagnetic storminess 28^d–1^d, in sequence S1; ○ flare imp. 2, 26^d1238, W08, McM 7108
Jan. 28	UN	(68)	IMP 1, Pr > 0.9 MeV, onset $18^h\Delta3$, max. $23^h\Delta1$, dur. 8^h, 33.2 Pr, superposed on the preceding event; △ 1922 starts a GMS in sequence S1
Feb. 18	UN	(50, 68)	IMP 1, Pr > 0.9 MeV, onset $12^h\Delta2$, max. $25^d06^h\Delta2$, dur. 10 days, 0.62 Pr; co-rotating stream (50)
Mar. 3	UN	(50, 68)	IMP 1, Pr > 0.9 MeV, onset $22^h\Delta2$, max. $5^d02^h\Delta2$, dur. 100^h; 4.0 Pr; ◇ recurrent gradual GMS 3^d–8^d in sequence S2 (cf. Event 112)
Mar. 16	UN	(50)	IMP 1, Pr > 0.9 MeV, onset $02^h\Delta2$, max. $09^h\Delta2$, dur. 7^h, 0.25 Pr; 15 h prior to Event 122; ◇ co-rotating stream (50), sequential with events Feb. 18, Jan. 23
Mar. 16	UN	(68)	IMP 1, Pr > 0.9 MeV, onset $09^h\Delta2$, max. $12^h\Delta2$, dur. 6^h, 6.0 Pr; about 8 h prior to Event 122 and 7 h before the source flare; part of the preceding event, or ○ flare imp. 1−, 0431, W68, McM 7182 (SWF imp. 2)

1964 (cont.)

Mar. 19	UN	(68)	IMP 1, Pr > 0.9 MeV, onset 20^h, max. 24^h, dur. 10^h, 5.6 Pr; □McM 7182 about 2 days beyond W limb
Mar. 22	L	(50, 68)	IMP 1, Pr > 0.9 MeV, onset $19^h\Delta2$, max. $23^d15^h\Delta2$, dur. 48^h, 0.62 Pr; ◇ recurrent gradual GMS 22^d-26^d in sequence S1
Mar. 27	L	(50, 68)	IMP 1, Pr > 0.9 MeV, onset $09^h\Delta2$, max. $28^d07^h\Delta2$, dur. 48^h, 0.50 Pr; □McM 7189 just beyond W limb;
Mar. 31	L	(50, 68)	IMP 1, Pr > 0.9 MeV, onset $20^h\Delta2$, max. $1^d15^h\Delta2$, dur. 48^h, 0.31 Pr; ◇ co-rotating stream (50) in sequence S2 (but no magnetic disturbance)
Apr. 15	L	(50, 68)	IMP 1, Pr > 0.9 MeV, onset $13^h\Delta2$, max. $19^d04^h\Delta2$, dur. 7^d, 0.81 Pr; ◇ recurrent GMS 15^d-21^d begins with an sc 17^d00^h in sequence S1
May 10	UN	(68)	IMP 1, Pr > 0.9 MeV, onset $22^h\Delta2$, max. $11^d04^h\Delta2$, dur. 20^h, 12 Pr; △ sc 0035 starts ◇ GMS
Oct. 4	UN	(76)	IMP 2, Pr > 0.9 MeV, onset $< 10^h$, max. $19^h\Delta2$, dur. $> 48^h$, 2.6 Pr; ◇ interval of geomagnetic storminess 3^d-9^d beginning with an sc 3^d1243, in sequence S2; of interest: McM 7512 is born on 4^d

1965

Jan. 8 (MAR 4 at 1.1 AU, ES + 2°)	L	(76, 77, 87)	IMP 2, Pr > 0.9 MeV, onset $10^h\Delta2$, max. $9^d15^h\Delta4$, dur. 100^h, 2.6 Pr; MAR 4, Pr > 0.5 MeV, onset $15^h\Delta2$, max. $10^d18^h\Delta2$, dur. 120^h, 4.3 Pr; Pr > 1.0 MeV, onset $20^h\Delta1$, dur. 120^h; origin uncertain
Jan. 10	UN	(3)	f-min PCA, onset 09^h, dur. 24^h; it coincides with the max. of the preceding event, but the particle flux seems to be too low to produce a PCA
Feb. 2	PE	(66)	IMP 2, El > 40 keV, onset $< 21^h35^m\Delta1$, 300 El (Pr > 0.9 MeV no event); ⊙ flare imp. 1−, 2043, E09, McM 7661 (type III)
Apr. 17	UN	(86)	PCA RIOM, onset unknown, dur. 96^h, 'small sporadic absorption' (not confirmed by MAR 4); ◇ strong GMS begins with an sc 1312
Apr. 20 (MAR 4 at 1.4 AU, ES − 24°)	UN	(79)	MAR 4, Pr > 0.5 MeV, onset $00^h\Delta2$, max. $06^h\Delta2$, dur. 24^h, 0.4 Pr; part of the preceding event; origin unknown, unless: McM 7779 newly formed 18^d; it starts a recurrent particle sequence
May 7 (MAR 4 at 1.5 AU, ES − 32°)	UN	(79)	MAR 4, Pr > 0.5 MeV, onset $00^h\Delta2$, max. $8^d03^h\Delta2$, dur. 72^h, 0.8 Pr; ◇ co-rotating structure, giving rise to weak gradual GMS 8^d-10^d
May 16 (MAR 4 at 1.5 AU, ES − 36°)	UN	(79)	MAR 4, Pr > 0.5 MeV, onset $18^h\Delta2$, max. $21^h\Delta2$, dur. 48^h, 0.4 Pr; ◇ weak GMS 16^d-17^d with an sc 0036, sequential with event on Apr. 20; ⊙ flare imp. 1+, < 0755, E90 (E54 at MAR 4), McM 7812 (SWF imp. 3)
May 20 (MAR 4 at 1.5 AU, ES − 38°)	UN	(79)	MAR 4, Pr > 0.5 MeV, onset $06^h\Delta2$, max. $21^d03^h\Delta2$, dur. 48^h, 0.17 Pr; origin unknown
May 25 (MAR 4 at 1.5 AU, ES − 41°)	L	(77, 79, 81)	MAR 4, El > 150/ > 40 keV, onset ?/$23^h20^m \pm 05^m$, max. $26^d01^h30^m\Delta1$, dur. ?/67^h, 2/80 El; Pr > 1.0/ > 0.5 MeV, onset $23^h50^m\Delta1$/$26^d07^h30^m\Delta1$, max. ?/$27^d09^h\Delta2$, dur. 28^h/72^h, ?/1.1 Pr (the early onset at 2350 may be due to El > 150 keV contamination); ● flare imp. 1, 2239, W73, McM 7809
May 31 (MAR 4 at 1.5 AU, ES − 43°)	UN	(79)	MAR 4, Pr > 0.5 MeV, onset $00^h\Delta2$, max. $12^h\Delta2$, dur. 24^h, 0.15 Pr (not confirmed by IMP 3); origin unknown
June 1 (MAR 4 at 1.5 AU, ES − 44°)	UN	(77, 79)	MAR 4, Pr > 1.0/> 0.5 MeV, onset $02^h\Delta1$/$06^h\Delta2$, max. ?/$2^d00^h\Delta2$, dur. 72^h/24^h, ?/3.0 Pr (not confirmed by IMP 3); ◇ co-rotating structure, sequential with event on May 7, giving rise to GMS 3^d12^h at Earth
June 5 (MAR 4 at 1.5 AU, ES − 45°)	L	(77, 80, 81)	MAR 4, El > 40 keV, onset $19^h10^m \pm 08^m$, max. $20^h45^m\Delta1$, dur. 28^h, 58 El; Pr > 1.0 MeV, onset $20^h\Delta1$, dur. 96^h; IMP 3, Pr > 0.9 MeV, onset $< 19^h30^m$, max. $19^h30^m\Delta5$, dur. $> 20^h$, 3.6 Pr; ● flare imp. 1−, 1807, W50, McM 7842
June 12 (MAR 4 at 1.5 AU, ES − 48°)	UN	(79)	MAR 4, Pr > 0.5 MeV, onset $10^h\Delta2$, max. $15^d08^h\Delta2$, dur. 92^h, 1.0 Pr; ◇ recurrent particle stream, sequential with Apr. 20 and May 16

1965 (cont.)

June 13	L	(66, 77, 81)	MAR 4, El > 40 keV, onset > 02^h58^m < 05^h20^m, max. $14^h\Delta1$, 10 El; Pr > 1.0 MeV, onset $04^h\Delta1$; IMP 3, Pr > 0.9 MeV, onset $08^h\Delta2$, max. $12^h\Delta2$, dur. 40^h, 0.5 Pr; $\Diamond$ same as for June 12; $\odot$ flare imp. 1, 0257, W03, McM 7847
June 15 (MAR 4 at 1.55 AU, ES − 49°)	L	(77, 79)	MAR 4, Pr > 1.0/> 0.5 MeV, onset $10^h\Delta1/20^h\Delta3$, max. ?/$16^d06^h\Delta2$, dur. $96^h/144^h$, ?/17 Pr; IMP 3, Pr > 0.9 MeV, onset $12^h\Delta2$, max. $15^h\Delta2$, dur. 24^h, 1.0 Pr; $\triangle$ sc 1058, possibly associated with the source flare of the preceding event
June 17	UN	(77, 80)	IMP 3, Pr > 0.9 MeV, onset $21^h\Delta2$, max. $18^d02^h\Delta1$, dur. 6^h, 1.1 Pr (not observed on MAR 4); origin unknown
June 28 (MAR 4 at 1.55 AU, ES − 56°)	UN	(79)	MAR 4, Pr > 0.5 MeV, onset $12^h\Delta2$, max. $29^d00^h\Delta2$, dur. 36^h (not observed on IMP 3); $\Diamond$ co-rotating structure, sequential with events on June 1 and May 7
June 29 (MAR 4 at 1.55 AU, ES − 57°)	L	(77, 79)	MAR 4, Pr > 1.0/> 0.5 MeV, onset $12^h\Delta1/\Delta3$, max. ?/$30^d03^h\Delta2$, dur. 72^h; IMP 3, Pr > 0.9 MeV, onset $30^d06^h\Delta2$, max. $1^d00^h\Delta2$, dur. 48^h, 0.25 Pr; $\Diamond$ continuation of the preceding event?
July 2	UN	(79)	MAR 4, Pr > 0.5 MeV, onset $21^h\Delta2$, max. $3^d00^h\Delta2$, dur. 20^h, 2.1 Pr (not observed on IMP 3); origin uncertain
July 4/6 (MAR 4 at 1.55 AU, ES − 59°)	L	(77, 79)	MAR 4, Pr > 0.5 MeV, onset $15^h\Delta2$, max. $6^d06^h\Delta2$, dur. 72^h, 2.1 Pr; Pr > 1.0 MeV, onset $5^d22^h\Delta1$, dur. 24^h; IMP 3, Pr > 1.0 MeV, onset $6^d04^h\Delta1$, max. $6^d07^h\Delta2$, dur. 48^h; $\Diamond$ co-rotating stream and $\triangle$ sc 6^d0450? A complex situation, difficult to understand
July 10 (MAR 4 at 1.58 AU, ES − 62°)	UN	(79)	MAR 4, Pr > 0.5 MeV, onset $00^h\Delta2$, max. $15^h\Delta2$, dur. 24^h, 0.4 Pr; origin unknown
July 13 (MAR 4 at 1.6 AU, ES − 64°)	L	(77, 79)	MAR 4, Pr > 0.5 MeV, onset $15^h\Delta2$, max. $14^d03^h\Delta2$, dur. 72^h, 6.4 Pr; Pr > 1.0 MeV, onset $14^d15^h\Delta1$ (not observed on IMP 3); $\Diamond$ sequential with the events on June 15 and May 20

(No 'pure' MAR 4 events considered after this date)

July 28 (ZOND 3 at 1.00 AU, ES + 1°)	UN	(43, 103)	ZOND 3, Pr > 1.0 MeV, onset $13^h\Delta2$, max. $29^d12^h\Delta7$, dur. 36^h, 5.5 Pr; origin unknown
Aug. 16 (ZOND 3 at 1.03 AU, ES + 1°) (MAR 4 at 1.6 AU, ES − 81°)	L	(43, 77, 80, 103)	IMP 3, Pr > 0.9 MeV, onset $13^h\Delta1$, max. $17^h\Delta1$, dur. 30^h, 1.3 Pr; ZOND 3, Pr > 1.0 MeV, onset $15^h\Delta2$, max. $23^h\Delta2$, dur. 52^h, 3.0 Pr; also seen as doubtful event on MAR 4 (79), Pr > 0.5 MeV, onset $18^h\Delta2$, max. $17^h00^h\Delta2$, dur. 24^h, 0.4 Pr (no Pr > 1.0 MeV on MAR 4); origin uncertain
Sept. 1 (ZOND 3 at 1.05 AU, ES + 1°) (MAR 4 at 1.6 AU, ES − 87°)	L	(43, 77, 80, 103)	IMP 3, Pr > 0.9 MeV, onset $00^h\Delta2$, max. $3^d16^h\Delta2$, dur. 96^h, 0.15 Pr; ZOND 3, Pr > 1.0 MeV, onset $10^h\Delta2$, max. $4^d14^h\Delta2$, dur. 86^h, 1.8 Pr; MAR 4 (77, 79) recorded an increase five days earlier: Pr > 0.5 MeV on Aug. $26^d12^h\Delta2$, max. $30^d00^h\Delta2$, dur. 132^h, 6.0 Pr; and Pr > 1.0 MeV on Aug. $27^d04^h\Delta1$, max. $30^d00^h\Delta2$; this indicates $\Diamond$ co-rotating stream (newly formed) which comes again on Sept. 21 to MAR 4 and Sept. 29 to the Earth
Sept. 29 (ZOND 3 at 1.12 AU, ES − 2°) (MAR 4 at 1.6 AU, ES − 100°)	L	(43, 77, 80, 103)	IMP 3, Pr > 0.9 MeV, onset < $04^h\Delta4$, max. $30^d04^h\Delta2$, dur. 45^h, 0.6 Pr; ZOND 3, Pr > 1.0 MeV, onset $08^h\Delta2$, max. $01^d00^h\Delta2$, dur. 116^h, 3.0 Pr; MAR 4 (77, 79) recorded an increase 7 days earlier: Pr > 0.5 MeV on $22^d05^h\Delta2$, max. $23^d03^h\Delta2$, dur. 15 days, 0.5 Pr; and Pr > 1.0 MeV on $22^d10^h\Delta1$, dur. 96^h; $\Diamond$ co-rotating stream, sequential with the event on Sept. 1
Oct. 8	L	(66, 80)	IMP 3, El > 40 keV, onset $16^h22^m \pm 01^m$, max. $17^h00^m\Delta1$, 230 El; Pr > 0.9 MeV, onset $16^h30^m\Delta1$, max. $17^h\Delta1$, dur. 18^h, 2.0 Pr; $\odot$ flare imp. 1−, 1603, W84, McM 8005 (proton arrival too early; probably electron contamination)
Nov. 23 (VEN 2 at 0.99 AU, ES − 1°)	UN	(43, 103)	VEN 2, Pr > 1.0 MeV, onset $16^h\Delta2$, max. $24^d12^h\Delta5$, dur. 24^h, 1.3 Pr; no Pr > 0.9 MeV on IMP 3; origin unknown

1965 (cont.)

Nov. 25	L	(43, 80, 103)	VEN 2, Pr > 1.0 MeV, onset $22^h\Delta2$, max. $26^d16^h\Delta2$, dur. 124^h, 3.6 Pr;
(VEN 2 at 0.99 AU, ES − 1°)			ZOND 3, Pr > 1.0 MeV, onset $22^h\Delta2$, max. $27^d13^h\Delta2$, dur. 70^h, 4.8 Pr;
(ZOND 3 at 1.24 AU, ES − 16°)			IMP 3, Pr > 0.9 MeV, onset $26^d00^h\Delta4$, max. $27^d15^h\Delta2$, dur. 96^h, 1.25 Pr;
			◊ co-rotating stream, sequential with events on Sept. 1 and Sept. 29
Dec. 7	L	(43)	VEN 2, Pr > 1.0 MeV, onset $12^h\Delta2$, max. $19^h\Delta2$, dur. 12^h, 2.0 Pr; IMP 3,
(VEN 2 at 0.98 AU, ES − 2°)			Pr > 0.9 MeV, onset $15^h\Delta1$, max. $17^h\Delta1$, dur. 10^h, 12 Pr (no Pr > 6.5 MeV); origin unknown
Dec. 17	UN	(104)	PION 6, Pr > 0.6 MeV, onset $20^h\Delta5$, max. $18^d08^h\Delta2$, dur. 110^h, 0.17 Pr
(PION 6 at 1.0 AU, ES = 0°)			(no Pr > 0.9 MeV on IMP 3); origin unknown
Dec. 24	UN	(70)	PION 6, Pr > 0.6 MeV, onset $22^h\Delta2$, max. $25^d21^h\Delta2$, dur. 55^h, 0.05 Pr
(PION 6 at 0.99 AU, ES − 0.5°)			(no Pr > 0.9 MeV on IMP 3); ◊ co-rotating stream, sequential with events on Nov. 26, Sept. 1, and Sept. 29
Dec. 27	L	(66, 80, 84, 104, 154)	IMP 3, El > 40 keV, onset $06^h46^m \pm 01^m$, max. 07^h15^m, 250 El; Pr > 0.9 MeV, onset $07^h\Delta1$; max. $08^h\Delta1$, dur. 35^h, 9.4 Pr; PION 6, Pr > 13/> 7.5
(PION 6 at 0.99 AU, ES − 1°)			MeV, onset $09^h/08^h\Delta1$, max. $15^h\Delta1$, dur. $27^h/20^h$, 0.009/0.02 Pr;
			Pr > 0.6 MeV, onset $07^h\Delta1$, max. $14^h\Delta2$, dur. 55^h, 1.9 Pr; ⊙ flare imp. 1−,
			0620, W29, McM 8105 (low-energy proton arrival too early; electron contamination?)
Dec. 29	L	(66, 80, 84, 103, 154)	IMP 3, El > 40 keV, onset $12^h15^m\Delta1$, max. $13^h00^m\Delta1$, 25 El; Pr > 0.9
(VEN 2 at 0.91 AU, ES − 2°)			MeV, onset $13^h\Delta2$, max. $17^h\Delta2$, dur. 7^h, 1.6 Pr; PION 6, Pr > 7.5/> 0.6
			MeV, onset $14^h10^m/12^h\Delta3$, max. $17^h/18^h\Delta1$, dur. $10^h/13^h$, 0.01/2.1 Pr;
			VEN 2, Pr > 1.0 MeV, onset $18^h\Delta2$, dur. 100^h (including the following events); ⊙ flare imp. 1+, 1123, W60, McM 8105

1966

Jan. 2	L	(70, 80)	PION 6, Pr > 0.6 MeV, onset $14^h\Delta3$, max. $15^h\Delta1$, dur. 50^h, 24 Pr (Pr
(PION 6 at 0.96 AU, ES − 1°)			> 7.4 MeV no event); IMP 3, Pr > 0.9 MeV, onset $16^h\Delta3$, max. $18^h\Delta2$, dur. 45^h, 4.6 Pr (Pr > 6.5 MeV no event); ▣ McM 8105 is 1~2 days beyond W limb
Feb. 3	L	(84, 104)	PION 6, Pr > 7.5 MeV, onset $02^h\Delta1$, max. $12^h\Delta1$, 0.005 Pr; Pr > 0.6 MeV,
(PION 6 at 0.94 AU, ES − 1.5°)			onset $00^h\Delta2$, max. $21^h\Delta1$, dur. 115^h, 0.6 Pr (Pr > 0.9 MeV no event on IMP 3); ⊙ flare imp. 1N, 2^d1510, E27, McM 8154
Feb. 7	UN	(104)	PION 6, Pr > 0.6 MeV, onset $22^h\Delta1$, max. $8^d04^h\Delta2$, dur. 36^h, 0.27 Pr
(PION 6 at 0.93 AU, ES − 1°)			(Pr > 0.9 MeV no event on IMP 3); ⊙ flare imp. SN, 1558, W38, McM 8154
Feb. 9	UN	(104)	PION 6, Pr > 0.6 MeV, onset $10^h\Delta3$, max. $10^d00^h\Delta1$, dur. 50^h, 0.62 Pr;
(PION 6 at 0.93 AU, ES − 1°)			continuation of the preceding event, origin unknown
Feb. 12	UN	(104)	PION 6, Pr > 0.6 MeV, onset $04^h\Delta2$, max. $08^h\Delta2$, dur. 30^h, 0.034 Pr
			(Pr > 0.9 MeV no event on IMP 3); ⊙ flare imp. 1B, 11^d2335, W60, McM 8158
Feb. 19	UN	(104)	PION 6, Pr > 0.6 MeV, onset $10^h\Delta1$, max. $20^d00^h\Delta2$, dur. 62^h, 0.55 Pr
(PION 6 at 0.93 AU, ES − 0.5°)			(Pr > 0.9 MeV no event on IMP 3); ◊ a weak GMS on 19^d−21^d
Feb. 22	UN	(104)	PION 6, Pr > 0.6 MeV, onset $00^h\Delta2$, max. $15^h\Delta2$, dur. 60^h, 0.55 Pr (Pr
(PION 6 at 0.92 AU, ES = 0°)			> 0.9 MeV no event on IMP 3); ◊ a gradual GMS on 22^d−25^d
Feb. 27	L	(80, 104)	PION 6, Pr > 0.6 MeV, onset $19^h\Delta2$, max. $28^d07^h\Delta5$, dur. 80^h, 20.7 Pr
(PION 6 at 0.92 AU, ES + 0.5°)			(during the past 44 h the flux has gradually increased one order of magnitude); IMP 3, Pr > 0.9 MeV, onset $20^h\Delta1$, max. $28^d11^h\Delta5$, dur. 25^h, 1.0 Pr; ⊙ flare imp. SB, 0015, W50, McM 8174
Feb. 28	PE	(66)	IMP 3, El > 40 keV, onset $08^h50^m \pm 10^m$, 40 El; may be related to the
			maximum > 0.9 MeV Pr flux of the preceding event; ⊙ flare imp. SN, < 0750, W70, McM 8174
Mar. 10	UN	(104)	PION 6, Pr > 0.6 MeV, onset $00^h\Delta2$, max. $11^d03^h\Delta2$, dur. 45^h, 0.034 Pr
(PION 6 at 0.91 AU, ES + 1)			(no unique Pr > 0.9 MeV event on IMP 3); origin unknown, unless ◊ very weak GMS on 10^d

1966 (cont.)

Mar. 17 (PION 6 at 0.89 AU, ES + 3°)	UN	(70)	PION 6, Pr > 0.6 MeV, onset $13^h\Delta2$, dur. 17 days (!) (Pr > 0.9 MeV no event on IMP 3 until 19^d); ⌀ ambiguous flares in McM 8207; imp. SB, 16^d2252, E58 (DS type IV); imp. 1B, 16^d1918, E58 (DS type IV); imp. 1B, 16^d1603, E60 (DS type IV); imp. 2B, 16^d0912, E65; ◇ onset of the 'permanent particle flux' associated with the transit of McM 8207
Mar. 18	UN	(104)	PION 6, Pr > 0.6 MeV, onset $20^h\Delta2$, max. $19^d12^h\Delta2$, dur. 28^h, 1.0 Pr – rise to a fairly constant value, holding until the onset of Event 130 (Pr > 0.9 MeV no event on IMP 3); ⊙ flare imp. 1N, 0420, E40, McM 8207 (hard X-rays)
Mar. 19	L	(66, 80, 125)	IMP 3, El >`40 keV, onset $01^h04^m \pm04^m$, max. $01^h30^m\Delta1$, 12 El; Pr > 0.9 MeV, onset $06^h\Delta2$, max. $15^h\Delta2$, 1.2 Pr; PION 6, Pr > 7.5 MeV, onset $03^h\Delta2$, max. $14^h\Delta5$, 0.005 Pr; ● flare imp. 1N, 18^d2345, E27, McM 8207 and ⌀ flare imp. 3N, 19^d0338, E33, McM 8207; alternately, the first flare might have produced the electron, and the second the proton flux; in this case, however, the proton arrival seems to be too early
Mar. 19	PE	(66)	IMP 3, El > 40 keV, onset $14^h51^m \pm02^m$, max. $15^h20^m\Delta1$, 100 El (the preceding proton event in progress); ● flare imp. 1N, 1357, E27, McM 8207
Mar. 21	UN	(80)	IMP 3, Pr > 0.9 MeV, onset $12^h\Delta3$, max. $16^h\Delta1$, dur. 7^h, 4.0 Pr (no data on PION 6); ⊙ flare imp. 2B, 0925, W02, McM 8207
Mar. 24	PE	(66)	IMP 3, El > 40 keV, onset $19^h30^m \pm02^m$, max. 19^h35^m, 640 El (proton enhancement may be masked by the flux remaining after Event 135); ⊙ flare imp. SB, 1858, W51, McM 8207
Mar. 26	PE	(66)	IMP 3, El > 40 keV, onset $19^h16^m \pm01^m$, max. 19^h30^m, 280 El (proton enhancement may be masked by the decaying flux after Event 135); ● flare imp. 1B, < 1843, W70, McM 8207
Mar. 31 (PION 6 at 0.85 AU, ES + 6°)	UN	(70)	PION 6, Pr > 0.6 MeV, onset $14^h\Delta5$, max. $3^d10^h\Delta2$, dur. 11 days (!), 5.9 Pr; ⊙ flare imp. 2N, 30^d1241, E50, McM 8223; ◇ onset of the 'permanent particle flux' associated with the transit of McM 8223
Apr. 1	UN	(80)	IMP 3, Pr > 0.9 MeV, onset $16^h\Delta2$, max. $2^d01^h\Delta2$, dur. 80^h, 8.3 Pr; it overlaps with the preceding event, starting later and ending earlier; △ sc 12^h37^m starts a GMS; ⊙ flare imp. 2N, 31^d1808, E35, McM 8223
Apr. 12 (PION 6 at 0.84 AU, ES + 9°)	UN	(104)	PION 6, Pr > 0.6 MeV, onset $12^h\Delta5$, max. $14^h\Delta2$, dur. 7^h, 1.0 Pr (no Pr > 0.9 MeV event on IMP 3); unusual short impulsive bursts; ▢ McM 8223?
Apr. 12	L	(80, 104)	PION 6, Pr > 0.6 MeV, onset $19^h\Delta3$, max. $13^d04^h\Delta2$, dur. 89^h, 6.0 Pr; IMP 3, Pr > 0.9 MeV, onset $13^d02^h\Delta1$, max. $13^d12^h\Delta6$, dur. 17^h, 10 Pr; ambiguous flares in McM 8240: ⊙ imp. SN, 1526, W80; ⌀ imp. 1N, 1621, W83; ⌀ imp. 1N, 1717, W80
Apr. 25 (PION 6 at 0.83 AU, ES + 14°)	UN	(104)	PION 6,.Pr > 0.6 MeV, onset $00^h\Delta2$, max. $27^d22^h\Delta2$, dur. 108^h, 5.2 Pr; onset of the 'permanent particle flux' associated with the transit of McM 8279
Apri 26	UN	(84)	PION 6, Pr > 7.5 MeV, onset $13^h\Delta1$, max. $27^d06^h\Delta1$, 0.001 Pr; it overlaps with the preceding event; origin unknown
May 7	PE	(66)	IMP 3, El > 40 keV, onset $02^h35^m \pm05^m$, max. 02^h50^m, 70 El; ○ flare imp. SF, 0125, W41, McM 8284
May 26	UN	(80)	IMP 3, Pr > 0.9 MeV, onset $19^h\Delta1$, max. $20^h\Delta1$, dur. < 6^h, 8.0 Pr (also see next event on 27^d); ○ flare imp. 1B, 25^d1530, E04, McM 8310 (type II and IV); ○ flare imp. SB, 26^d0953, W06, McM 8310 (type III);
May 27	PE or L	(66, 80)	IMP 3, El > 40 keV, onset $01^h\Delta2$, 150 El; according to (66) also Pr > 0.5 MeV present; for Pr > 0.9 MeV an increase might have been superimposed on the previous event, max. $02^h\Delta1$, 10 Pr (80) (not uniquely discernible as a separate event); ○ flare imp. SF 26^d2347, W17, McM 8310, or delayed event (66) related to the 25^d flare of the previous event

1966 (cont.)

June 2	PE	(66)	IMP 3, El > 40 keV, onset $14^h45^m \pm 10^m$, max. 15^h45^m, 10 El; (Pr > 0.9 MeV no event); ⊙ flare imp. SB, 1422, W47, McM 8318
June 28	PE	(66)	IMP 3, El > 40 keV, onset > 0000 < 0230, 25 El; (Pr > 0.9 MeV no event); ⊙ flare imp. 2N, 27^d2358, W58, McM 8344
July 3	PE	(66)	IMP 3, El > 40 keV, onset $22^h04^m \pm 01^m$, max. 22^h40^m, 90 El (also see next event); ▫ McM 8348 is 2 days beyond W limb
July 3	UN	(80)	IMP 3, Pr > 0.9 MeV, onset $22^h\Delta2$, max. $4^d11^h\Delta2$, dur. 20^h, 0.25 Pr; in spite of time coincidence with the preceding event, this enhancement is supposed to be independent: ◇ onset of a 'permanent particle stream' associated with McM 8362 (also see Event 147)
July 25	UN	(80)	IMP 3, Pr > 0.9 MeV, onset $20^h\Delta2$, max. $27^d13^h\Delta2$, dur. 78^h, 0.38 Pr; ◇ onset of 'permanent particle stream' as McM 8362 (now 8413) appears at the E limb (strengthened in Event 152); ○ flare imp. 1B, 0457, E15, McM 8408
Aug. 6	UN	(80)	IMP 3, Pr > 0.9 MeV, onset $06^h\Delta2$, max. $12^h\Delta2$, dur. 42^h, 2.9 Pr; origin unknown (McM 8427, three days old, produces frequent small flares near W limb)
Aug. 18	UN	(80)	IMP 3, Pr > 0.9 MeV, onset $00^h\Delta2$, a low enhancement lasting for 20 days; ◇ 'permanent flux' from McM 8454, 8459 + 8461 which are approaching E limb
Aug. 23 (PION 7 at 1.03 AU, ES = 0°)	UN	(119)	PION 7, Pr > 0.6 MeV, onset $< 00^h\Delta1$, max. $02^h\Delta1$, dur. 10^h, 0.034 Pr; after that a gradual continuous enhancement lasting ~ 19 days; ◇ see the preceding event; of interest: a 'stable spot' in McM 8454, living for 6 solar rotations, is born on 22^d, with CMP on 23^d
Aug. 27 (PION 7 at 1.04 AU, ES = 0°)	UN	(119)	PION 7, Pr > 0.6 MeV, onset $\leq 20^h$, 20^h prior to Event 160; ○ flare imp. 1N, 1855, W52, McM 8454; 1820–2035 increase in intensity of radio continuum
Aug. 31	PE	(66)	IMP 3, El > 40 keV, onset $03^h40^m \pm 15^m$, max. 04^h30^m, 250 El; ● flare imp. 1N, 0250, W30, McM 8461
Sept. 11	PE	(69)	IMP 3, El > 40 keV, onset $07^h50^m \pm 10^m$, max. 04^h30^m, 250 El (Pr > 0.9 MeV no event); ⊙ flare imp. SB, 0715, W60, McM 8484
Oct. 4	UN	(118)	IMP 3, El > 4 MeV, onset unknown, dur. 3 days (Pr > 0.9 MeV no event); PION 7, Pr > 0.6 MeV, complex structure from 3^d22^h to 12^d; origin unknown
Oct. 19	UN	(118)	IMP 3, El > 4 MeV, onset unknown, dur. 6 days (Pr > 0.9 MeV no event); origin unknown
Oct. 20	PE	(69)	IMP 3, El > 40 keV, onset $22^h28^m \pm 03^m$, 25 El; ● flare imp. SB, 2152, W50, McM 8546
Oct. 20	UN	(80)	IMP 3, Pr > 0.9 MeV, onset $22^h30^m\Delta1$, max. $23^h\Delta1$, dur. 22^h, 1.25 Pr; Pr > 6.5 MeV no event, hence the flare at 2152 (mentioned above) cannot be the source of this rise; electron contamination (doubtful, due to the small flux of 25 El) or origin unknown
Oct. 24	PE	(69)	IMP 3, El > 40 keV, onset $10^h56^m \pm 04^m$, 30 El (Pr > 0.9 MeV no event); ⊙ flare imp. SB, 0942, W90, McM 8546
Oct. 30 (PION 7 at 1.09 AU, ES − 5°)	L	(80, 119)	PION 7, Pr > 0.6 MeV, onset 12^h, max. 3^d04^h, dur. 12 days, 2.24 Pr; IMP 3, Pr > 0.9 MeV, onset $31^d00^h\Delta4$, max. $2^d04^h\Delta2$, dur. 66^h, 1.0 Pr; gradual enhancement; ◇ 'permanent flux' related to McM 8566 + 8567 newly formed on the disk; ◇ a sequential GMS begins gradually at 30^d12^h and lasts until Nov. 2^d
Nov. 1	PE	(69)	IMP 3, El > 40 keV, onset $13^h00^m \pm 30^m$, 10 El (Pr > 0.9 MeV event of Oct. 30 continues); ⊙ flare imp. 2N, < 1053, W15, McM 8567
Nov. 2	PE	(69)	IMP 3, El > 40 keV, onset $18^h06^m \pm 20^m$, 15 El (max. flux of the Pr > 0.9 MeV event of Oct. 30); ● flare imp. SN, 1707, W35, McM 8567

1966 (cont.)

Nov. 3	PE	(69)	IMP 3, El > 40 keV, onset > 18^h15^m < 21^h40^m, 10 El (Pr > 0.9 MeV no event); ● flare imp. SN, 1854, W49, McM 8567
Nov. 4	PE	(69)	IMP 3, El > 40 keV, onset 13^h45^m ±05^m, 20 El (Pr > 0.9 MeV no event); ☉ flare imp. SN, 1306, W60, McM 8567
Nov. 6	PE	(69)	IMP 3, El > 40 keV, onset 06^h25^m ±05^m, 15 El (Pr > 0.9 MeV no event); ▫ McM 8567 on W limb

Nov. 16 PE or L (69, 119) IMP 3, El > 40 keV, onset 18^h18^m ±02^m, 100 El (Pr > 0.9 MeV no event);
(PION 7 at 1.10 AU, ES − 7°) PION 7, Pr > 0.6 MeV, onset $18^h\Delta1$, max. $17^d03^h\Delta2$, dur. 30^h, 0.34 Pr
(most probably electron contamination); ☉ flare imp. SN, 1753, W54,
McM 8573

Nov. 20 UN (119) PION 7, Pr > 0.6 MeV, onset < 12^h, max. $22^d12^h\Delta2$, dur. 94^h, 0.05 Pr
(PION 7 at 1.10 AU, ES − 8°) (IMP 3: Pr > 0.9 MeV no event); ▫ McM 8573?

Nov. 30 UN (104) PION 6, Pr > 0.6 MeV, onset $17^h\Delta2$, max. $01^d12^h\Delta5$, 2.6 Pr (IMP 3; Pr
(PION 6 at 0.98 AU, ES + 36°) > 0.9 MeV no event); origin unknown

Dec. 12 UN (119) PION 7, Pr > 0.6 MeV, onset $15^h\Delta3$, max. $23^h\Delta5$, 50 Pr; ◊ co-rotating
(PION 7 at 1.11 AU, ES − 11°) stream which gave rise to GMS at the Earth at 13^d01^h (cf. Event 165)

Dec. 19 UN (119) PION 7, Pr > 0.6 MeV, onset $05^h\Delta3$, max. $07^h\Delta2$, dur. 20^h, 2.1 Pr; 12 h
(PION 7 at 1.12 AU, ES − 12°) prior to > 0.9 MeV rise at IMP 3 (next event); origin unknown

Dec. 19 L (69, 80) IMP 3, El > 40 keV, onset 14^h52^m ±10^m, 35 El; Pr > 0.9 MeV, onset
$17^h\Delta2$, max. $20^d05^h\Delta1$, dur. 15^h, 1.33 Pr; ☉ flare imp. 1N, < 1419, W67,
McM 8613; ▫ bright regions mentioned in Event 165 just beyond W limb

Dec. 22 UN (106) PION 7, Pr > 0.6 MeV, onset $00^h\Delta2$, six-fold rise to 23^d00^h, then no more
(PION 6 at 0.84 AU, ES + 37°) data; PION 6, Pr > 0.6 MeV, $11^h−20^h$ apparently saturated; ▫ 21^d1450
radio continuum, type II and type III; △ sc 22^d0440 during a weak GMS

Dec. 29 L (80, 119) PION 7, Pr > 0.6 MeV, onset $12^h\Delta2$, max. $16^h\Delta2$, dur. 39^h, 3.44 Pr;
PION 6, Pr > 0.6 MeV, onset 30^d11^h, rise until 30^d19^h, then no more data
IMP 3, Pr > 0.9 MeV, onset $30^d23^h\Delta2$, max. $31^d20^h\Delta2$, dur. 37^h, 0.25 Pr;
origin unknown

Dec. 31	PE	(69)	IMP 3, El > 40 keV, onset 00^h00^m ±20^m, 10 El (Pr > 0.9 MeV: preceding event in progress); ● flare imp. 1B, 30^d2231, E32, McM 8629
Dec. 31	PE	(69)	IMP 3, El > 40 keV, onset 09^h00^m ±30^m, 10 El (Pr > 0.9 MeV: event in progress); ● flare imp. 1B, 0842, E27, McM 8629
Dec. 31	PE	(69)	IMP 3, El > 40 keV, onset 15^h30^m ±20^m, 15 El (Pr > 0.9 MeV: event in progress); ☉ flare imp. SF, 1516, E22, McM 8629
Dec. 31	PE	(69)	IMP 3, El > 40 keV, onset 18^h35^m ±20^m, 20 El (Pr > 0.9 MeV: event in progress); ● flare imp. 1B, 1836, E22, McM 8629

1967

Jan. 3	L	(69, 80)	IMP 3, El > 40 keV, onset 11^h30^m ±15^m, 10 El; Pr > 0.9 MeV onset $15^h\Delta2$, max. $18^h\Delta2$, dur. 6^h, 1 Pr; ☉ flare imp. SN, 1044, W22, McM 8629
Jan. 8	PE	(69)	IMP 3, El > 40 keV, onset 06^h48^m±03^m, 100 El; no data on PION 6, nor on PION 7; origin unknown; ▫ McM region 8629 at W limb
Jan. 14	PE	(69)	IMP 3, El > 40 keV, onset 02^h10^m ±15^m, 75 El; ☉ flare imp. SN, 0118, W89, McM 8632
Jan. 17	L	(80)	IMP 3, Pr > 6.5/> 0.9 MeV, onset $05^h/03^h\Delta3$, max. $06^h\Delta2$, dur. ?/40^h, 0.04/12.5 Pr; ▫ McM regions 8631 and 8632 behind W limb 4−3 days
Jan. 31	UN	(80)	IMP 3, Pr > 0.9 MeV, onset $06^h\Delta3$, max. $09^h\Delta7$, dur. 62^h, > 748.9 Pr; ☉ flare imp. 1N, < 0037, W60, McM 8659; ☉ flare imp. 3F?, 0204, E90, McM 8680; ☉ flare imp. SN, 0512, W90, McM 8659.
Feb. 7	UN	(152)	RIOM PCA, onset $15^h\Delta2$, max. $17^h\Delta1$, 1.6 dB; ☉ flare imp. 1B, 1025, E30, McM 8684; ☉ flare imp. 2N, 6^d1825, E85, McM 8687, △ sc 6^d1636 followed by a GMS

1967 (cont.)

Mar. 9	L	(80)	IMP 3, Pr > 6.5/> 0.9 MeV, onset $12^h\Delta3$, max. $14^h\Delta2/\Delta1$, dur. $15^h/34^h$, 0.022/10 Pr; ○ flare imp. 1N, 1015, W04, McM 8715
Mar. 9	PE	(69)	IMP 3, El > 40 keV, onset $16^h31^m \pm 02^m$, 70 El; ○ flare imp. SN, 1600, W58, McM 8714
Mar. 13	PE	(69)	IMP 3, El > 40 keV, onset $09^h29^m \pm 04^m$, 400 El; ○ flare imp. SN, 0821, W57, McM 8715
Mar. 13	L	(80)	IMP 3, Pr > 6.5/> 0.9 MeV, onset $19^h/18^h\Delta3$, max. $20^h/19^h\Delta1$, dur. $90^h/> 160^h$, 0.8/150 Pr; origin uncertain
Mar. 14	PE	(69)	IMP 3, El > 40 keV, onset $07^h\Delta2$, 250 El; origin unknown
Mar. 24	L	(117, 80, 169)	IMP 3, El > 45 keV, onset $15^h\Delta2$, max. $25^d12^h\Delta5$, dur. 72^h, 80 El; Pr > 0.9 MeV, onset $14^h\Delta2$, max. $25^d08^h\Delta2$, dur. 65^h, 4 Pr; RIOM P.CA, max. $25^d01^h\Delta2$, 0.9 dB; ◇ co-rotating structure associated with region 8740
Mar. 27	UN	(80)	IMP 3, Pr > 0.9 MeV, onset $15^h\Delta1$, max. 19^h, dur. $> 6^h$, 1.12 Pr; no data on PION 6, nor PION 7; ◇ co-rotating structure; ○ flare imp. 1N, 1444, W05, McM 8740
Apr. 1	PE	(69)	IMP 3, El > 40 keV, onset $09^h19^m \pm 03^m$, 70 El; ○ flare imp. SB, 0835, W74, McM 8740
Apr. 1	L	(67, 69, 80)	IMP 3, El > 3 MeV/> 40 keV, onset $< 16^h/14^h\Delta2$, max. $18^h\Delta2/?$, dur. $12^h/?$, 0.07/150 El; AIMP 1, Pr > 15 MeV, onset 15^h, 0.008 c; IMP 3, Pr > 0.9 MeV, onset $20^h\Delta3$, max. $23^h\Delta2$, dur. 20^h, 4 Pr; ● flare imp. 1B, 1410, W80, McM 8740
Apr. 1	PE	(69)	IMP 3, El > 40 keV, onset 23^h35^m, 400 El; ○ flare imp. SF, 2325, W79, McM 8740
Apr. 3	L	(80)	IMP 3, Pr > 6.5/> 0.9 MeV, onset $01^h30^m\Delta5$, max. $3^h/4^h\Delta2$, dur. $6^h/20^h$, 0.18/15 Pr; no data on PION 6, nor PION 7; ○ flare imp. SF, 2^d1935, W79, McM 8739; □ region 8740 on invisible hemisphere
Apr. 4	UN	(80)	IMP 3, Pr > 0.9 MeV, onset $03^h\Delta2$, max. $14^h\Delta2$, dur. 45^h, 2.5 Pr; no data on PION 7, △ geomagnetic storm begins with sc 0304
Apr. 8	UN	(80)	IMP 3, Pr > 0.9 MeV, onset $05^h\Delta2$, max. $9^d18^h\Delta2$, dur. 60^h, 1.79 Pr; no data on PION 7 and PION 6; ◇ co-rotating stream sequential with event of Mar. 11; ○ flare imp. SN, 07^d1845, E28, McM 8760
May 7	L	(69, 172)	IMP 3, El > 40 keV, onset 08^h21^m, 1000 El; AIMP 1, Pr > 15 MeV, onset $9^h\Delta2$, 0.003 c; OGO 3, Pr > 6 MeV, onset $09^h\Delta2$; origin unknown; △ GMS with sc 0105
May 19 (PION 6 at 0.87 AU, ES + 96°)	UN	(104)	PION 6, Pr > 0.6 MeV, onset $01^h\Delta2$; ○ flare imp. SN, 18^d1855, W18, McM 8809
June 1	L	(181, 107)	IMP 4, Pr > 1 MeV, onset $15^h \pm 40^m\Delta3$, max. $19^h\Delta2$, dur. 42^h, 14.3 Pr; Pr > 0.8 MeV, onset $14^h\Delta3$, max. $19^h\Delta2$, dur. 40^h, 33.6 Pr; ○ flare imp. SB, 1116, E30, McM 8831; ○ flare, imp. SN, 1148, E29, McM 8831; □ region 8818 at W limb
June 4	L	(181, 107)	IMP 4, Pr > 1 MeV, onset $12^h\Delta3$, max. $5^d11^h\Delta2$, dur. 24^h, 22.1 Pr; Pr > 0.8 MeV, onset $12^h\Delta3$, max. $5^d11^h\Delta2$, dur. 24^h, 94 Pr; ◇ particle stream related to CMP of active region McM 8831; minor geomagnetic disturbance 4^d12^h
July 9	L	(69, 117)	IMP 4, El > 20 keV, onset $03^h07^m \pm 02^m$, 25 El; Pr > 0.5 MeV, onset $04^h\Delta3$, max. $10^h\Delta2$, dur. 24^h, 20 Pr; ○ flare, imp. SN, 0110, W64, McM 8875
July 11	L	(181, 175)	IMP 4, Pr > 1/0.5 MeV, onset $16^h/17^h\Delta1$, max. $21^h30^m/22^h$, dur. $58^h/?$, 4.8/290* Pr; □ region McM 8875 is on invisible hemisphere, one day beyond W limb
July 16	PE	(69)	IMP 4, El > 20 keV, onset $23^h37^m \pm 0.3^m$, 10 El; ○ flare, imp. SF, 2205, E90, McM 8901; ○ flare, imp. SF, 2340, E90, McM 8901; □ region 8880 beyond W limb

1967 (cont.)

July 23	L	(181, 175)	IMP 4, Pr > 1/0.5 MeV, onset $19^h\Delta2/24^d00^h\Delta1$, max. $25^d06^h\Delta2/21^h\Delta1$, dur. $> 120^h/?$, 0.36/24* Pr; origin uncertain, $\Diamond$ possibly a particle stream related to transit of active regions 8905 and 8907, which are near E limb on July 23
July 28	PE	(69)	AIMP 2, El > 40 keV, onset $02^h10^m \pm05^m$, 10 El; $\odot$ flare, imp. SF, 0053, E08, McM 8905
July 28	PE	(69)	AIMP 2, El > 40 keV, onset $17^h50^m \pm15^m$, 10 El; $\circ$ numerous subflares in regions 8905 and 8907
July 28	PE	(69)	IMP 4, El > 40 keV, onset $21^h08^m \pm07^m$, 15 El; $\circ$ flare, imp. SN, 2038, W05, McM 8905
July 29	PE	(69)	AIMP 2, El > 40 keV, onset $08^h29^m \pm02^m$, 85 El; $\circ$ flare, imp. SN, 0820, W14, McM 8905
July 29	PE	(69)	AIMP 2, El > 40 keV, onset $19^h46^m \pm02^m$, 50 El; $\circ$ numerous subflares in regions 8905 and 8911
July 29	PE	(69)	AIMP 2, El > 40 keV, onset $20^h02^m \pm02^m$, max. 20^h15^m, 350 El; $\circ$ flare, imp. 1B, 1944, W22, McM 8905
July 30	PE	(69)	AIMP 2, El > 40 keV, onset $02^h18^m \pm01^m$, 200 El; $\circ$ flare, imp. SN, 0151 W18, McM 8905
July 30	PE	(178, 69)	IMP 4, El > 0.5/> 0.17 MeV, onset 16^h33^m, dur. 10^m, 30/150 c; AIMP 2, El > 40 keV, onset $16^h37^m \pm01^m$, max. 16^h48^m, 800 El; IMP 4, El > 22 keV, onset $16^h42^m \pm02^m$, max. 16^h55^m, 120 El; $\odot$ flare, imp. 1N, 1612, W36, McM 8905
July 30	PE	(178, 69)	IMP 4, El > 0.5/> 0.17 MeV, onset 20^h, dur. 50^m, 60/270c; AIMP 2, El > 40 keV, onset $19^h53^m \pm01^m$, max. 20^h01^m, 1200 El; $\odot$ flare, imp. SF, 1945, W10, McM 8907; $\odot$ flare, imp. SN, 1946, W33, McM 8905; $\odot$ flare, imp. SB, 1555, W07, McM 8907
July 31	PE	(69)	AIMP 2, El > 40 keV, onset $21^h36^m \pm01^m$, 150 El; $\odot$ flare, imp. 1B, 2047, W50, McM 8905
Aug. 2	PE	(176, 69)	IMP 4, El > 70 keV, onset 17^h43^m, max. 17^h50^m, 90 El; AIMP 2, El > 40 keV, onset $17^h46^m \pm01^m$, 550 El; $\odot$ flare, imp. SN, 1726, W76, McM 8905
Aug. 4	L	(69, 107)	IMP 4, El > 40 keV, onset $15^h42^m \pm01^m$, 30 El; Pr > 0.8 MeV, onset $16^h20^m \pm60^m\Delta3$, max. $17^h\Delta2$, dur. $> 7^h$, 16.5 Pr; $\odot$ flare, imp. 1N, 1511, W87, McM 8905
Aug. 12	PE	(69)	IMP 4, El > 20 keV, onset $14^h09^m \pm03^m$, 30 El; $\circ$ flare, imp. SF, 1328, W72, McM 8916; $\circ$ flare, imp. SN, 1210, W40, McM 8921
Aug. 13	PE	(69)	IMP 4/AIMP 2, El > 20 keV, onset $19^h30^m \pm04^m/19^h38^m \pm02^m$, 30 El; $\circ$ flare, imp. SF, 1809, W53, McM 8921; $\square$ active region McM 8916 just beyond W limb
Aug. 13	PE	(69)	IMP 4, El > 20 keV, onset $20^h57^m \pm06^m$, 30 El; $\circ$ flare, imp. SN, 2056, W89, McM 8916
Aug. 15 (VEN 4 at 0.89 AU, ES + 0.5°)	L	(181, 175, 43)	IMP 4, Pr > 1 MeV, onset $12^h \pm20^m$, max. $16^d09^h\Delta2$, dur. 35^h, 1 Pr; Pr 0.5 MeV, onset 20^h30^m, max. 16^d10^h, 65* Pr; VEN 4, Pr > 1 MeV, onset $22^h\Delta3$, max. $22^h\Delta2$, dur. 24^h, 2 Pr; $\circ$ flare, imp. 1N, < 0653, W41, McM 8926; $\circ$ flare, imp. SF, 0212, W63, McM 8921; $\circ$ flare, imp. SN, 0210, W39, McM 8926; $\square$ region 8916 is one day beyond W limb
Aug. 17 (VEN 4 at 0.89 AU, ES + 1°)	UN	(43)	VEN 4, Pr > 1 MeV, onset $06^h\Delta2$, max. $18^d18^h\Delta2$, dur. 72^h, 24 Pr; $\Diamond$ geomagnetic storm modulation, gradual beginning 1704
Aug. 18	L	(181, 107)	IMP 4, Pr > 1 MeV, onset $11^h \pm60^m$, max. $19^h\Delta2$, dur. 48^h, 6.7 Pr; Pr > 0.8 MeV, onset $11^h\Delta2$, max. $20^h\Delta4$, dur. 90^h, 14 Pr; $\odot$ flare, imp. SB, 0239, E89, McM 8942

1967 (cont.)

Aug. 24	L	(181, 107)	IMP 4, Pr > 1 MeV, onset 01^{h}Δ2, max. 25^{d}01^{h}Δ2, dur. > 24^h, 0.9 Pr; Pr > 0.8 MeV, onset 01^{h}30^m±50^m, max. 25^{d}01^{h}45^m, dur. 30^h, 4.1 Pr; ○ flare, imp. 1N?, 0016, E20, McM 8942
Aug. 26	L	(69, 107)	AIMP 2, El > 20 keV, onset 02^{h}Δ2, 10 El; IMP 4, Pr > 0.8 MeV, onset 05^{h}Δ8, max. 15^{h}Δ2, dur. 60^h, 0.72 Pr; ● flare, imp. 1B, 0014, E00, McM 8949
Sept. 3	PE	(69)	IMP 4/AIMP 2, El > 20 keV, onset 05^{h}31^m ±03^m/05^{h}45^m ±01^m, 35 El; □ active region McM 8942 on invisible hemisphere, 2 days beyond W limb
Sept. 9	PE	(69)	AIMP 2, El > 20 keV, onset 16^{h}24^m±02^m, 20 El; ○ flare, imp. SF, 1556, W62, McM 8961; ○ flare, imp. SN, 1514, W37, McM 8962
Sept. 10	PE	(69)	AIMP 2, El > 20 keV, onset 06^{h}30^m, 8 El; ○ flare, imp. SF, 0601, W34, McM 8963
Sept. 10	L	(69, 107)	AIMP 2, El > 20 keV, onset 09^{h}15^m, 5 El; IMP 4, Pr > 0.8 MeV, onset 09^{h}15^m ±05^{m}Δ8, max. 09^{h}25^m, dur. 17^h, 0.3 Pr; ☉ flare, imp. SB, 0852, W46, McM 8962
Sept. 10	PE	(69)	AIMP 2, El > 20 keV, onset 15^{h}10^m, 7 El; ○ flare, imp. SN, 1401, W35, McM 8963
Sept. 11	L	(69, 107)	AIMP 2, El > 40 keV, onset 13^{h}47^m ±02^m, 10 El; IMP 4, Pr > 0.8 MeV, onset 13^{h}50^m ±10^{m}Δ8, max. 14^h, dur. 9^h, 0.17 Pr; ○ flare, imp. SF, 1323, W87, McM 8961
Sept. 12	L	(181, 107)	IMP 4, Pr > 1 MeV, onset 00^{h}Δ2, max. 13^{d}17^{h}Δ2, dur. > 48^h, 0.18 Pr; Pr > 0.8 MeV, onset 00^{h}Δ3, max. 13^{d}09^{h}Δ2, dur. 90^h, 0.63 Pr; ○ numerous subflares in regions 8963, 8961 and 8973; ◊ particle stream related to region 8963
Sept. 12	PE	(69)	AIMP 2, El > 20 keV, onset 03^{h}07^m ±08^m, 35 El; origin unknown; □ region 8961 is just beyond W limb
Sept. 18	PE	(69)	AIMP 2, El > 40 keV, onset 05^{h}Δ2, 150 El; ○ flare, imp. SF, 0407, E4₁ McM 8985; ○ flare, imp. SB, 0139, W51, McM 8973
Sept. 26	L	(107, 181)	IMP 4, Pr > 0.8 MeV, onset 07^{h}45^m ±15^m, max. 27^{d}11^{h}Δ2, dur. 60^h, 2.9 Pr; Pr > 1 MeV, onset 08^{h}20^m±30^m, max. 27^{d}11^{h}Δ2, dur. 47^h, 0.81 Pr; ◊ particle structure related to increase of activity in region 8985; ○ flare, imp. SB, 0711, W62, McM 8985
Sept. 27	PE	(69)	AIMP 2, El > 20 keV, onset 07^{h}08^m ±02^m, 7 El; ○ flare, imp. SN, < 0640, W75, McM 8985
Sept. 29	UN	(107, 203)	IMP 4, Pr > 0.8 MeV, onset 19^h ±60^m, max. 30^{d}15^{h}Δ2, dur. 35^h, 0.5 Pr; f-min PCA, onset 30^{d}03^{h}Δ1, dur. 11^h; ◊ particle structure related to growth of new regions on the disk; ○ flare, imp. SN, 1542, E80, McM 9005
Oct. 3	L	(181, 175)	IMP 4, Pr > 1 MeV, onset 06^{h}Δ2, max. 4^{d}03^{h}Δ5, dur. 72^h, 0.17 Pr; Pr 0.5 MeV, onset 03^{h}30^m, max. 4^{d}07^h, 1.4* Pr; ◊ structure related to formation of new plage on the disk; ○ flare, imp. 1N, 2^d 2151, E46, McM 9006; ○ flare, imp. SN, 0019, W10, McM 9002
Oct. 4	PE	(69)	AIMP 2, El > 40 keV, onset 08^{h}47^m ±01^m, 15 El; ○ flare, imp. SN, 0826, W68, McM 8998
Oct. 4	PE	(69)	AIMP 2, El > 20 keV, onset 17^{h}10^m ±05^m, 10 El; ○ flare, imp. SN, 1651, W70, McM 8998
Oct. 7 (VEN 4 at 0.73 AU, ES + 22.5°)	L	(69, 107, 177)	AIMP 2, El > 40 keV, onset 02^{h}28^m ±13^m, 15 El; IMP 4, Pr > 0.8 MeV, onset 05^{h}15^m±05^m, max. 05^{h}45^m, dur. 8^h, 0.2 Pr; VEN 4, Pr > 1 MeV, onset 06^{h}Δ2, dur. 112^h, duration includes next four proton events; ○ flare, imp. 1B?, 0231, W47, McM 9004
Oct. 7	L	(181, 107)	IMP 4, Pr > 1/> 0.8 MeV, onset 14^{h}Δ2/13^h ±100^{m}Δ3, dur. 3^d; ◊ growth and development on the disk of regions 9004 and 9021
Oct. 7	PE	(69)	AIMP 2, El > 20 keV, onset 13^{h}28^m ±09^m, 10 El; origin uncertain

1967 (cont.)

Oct. 7	L	(69, 107, 181)	AIMP 2, El > 40 keV, onset $17^h02^m \pm 01^m$, 10 El; IMP 4, Pr > 9 MeV, $20^h30^m \pm 30^m$, max. 20^h50^m, 0.004 Pr; Pr > 1 MeV, onset $21^h\Delta3$, max. $22^h\Delta2$, dur. < 5^h, 0.02 Pr; ○ flare, imp. SN, 1625, W59, McM 9004; ○ flare, imp. SN, 1653, W60, McM 9004
Oct. 8	L	(69, 181)	AIMP 2, El > 20 keV, onset $15^h52^m \pm 02^m$, 25 El; IMP 4, Pr > 1 MeV, onset $15^h55^m \pm 05^m\Delta3$, max. $17^h\Delta1$, dur. 13^h, 3.2 Pr; ○ flare, imp. SF, 1534, W68, McM 9004
Oct. 12	L	(181, 107, 203)	IMP 4, Pr > 1 MeV, onset $18^h\Delta2$, max. $14^d15^h\Delta2$, dur. > 72^h, 0.1 Pr; > 0.8 MeV, onset $12^h\pm60^m$, max. $14^d17^h\Delta4$, dur. 125^h, 0.95 Pr; f-min PCA, onset $13^d00^h\Delta1$, dur. 9^h; origin uncertain, ◊ possibly particle stream related to transit of region 9018
Oct. 19	L	(181, 107)	IMP 4, Pr > 1 MeV, onset $23^h\Delta2$, max. $20^d16^h\Delta2$, dur. > 96^h, 0.1 Pr; Pr > 0.8 MeV, onset $23^h\Delta2$, max. $20^d13^h\Delta2$, dur. 100^h, 0.51 Pr; origin unknown, perhaps a structure related to disk transit of several plages
Oct. 25	PE	(69)	AIMP 2, El > 20 keV, onset 06^h45^m, 7 El; origin unknown, no flare patrol 0344−0711
Oct. 25	L	(181, 107)	IMP 4, Pr > 1 MeV, onset $14^h\Delta2$, max. $23^h\Delta2$, 0.04 Pr; Pr > 0.8 MeV, onset $14^h\Delta2$, max. $26^d02^h\Delta2$, 0.2 Pr; ◊ onset of a particle stream related to central meridian transit of region 9034
Oct. 25	PE	(69)	AIMP 2/IMP 4, El > 20 keV, onset $14^h27^m \pm16^m/14^h49^m\pm1^m$, 20 El; ⊙ flare, imp. 1N, < 1327, W24, McM 9034
Oct. 25	PE	(69)	AIMP 2, El > 20 keV, onset 21^h50^m, 5 El; ⊙ flare, imp. SB, 2129, W34, McM 9034
Oct. 26	L	(69, 107)	AIMP 2, El > 40 keV, onset $00^h09^m \pm5^m$, 10 El; IMP 4, Pr > 0.8 MeV, onset $02^h\Delta2$, max. $06^h\Delta1$, dur. 4^h; ⊙ flare imp. 1B, 25^d < 2312, W28, McM 9034
Oct. 27	L	(181, 117, 175)	IMP 4, Pr > 1 MeV, onset $06^h55^m \pm20^m\Delta3$, max. $12^h\Delta2$, 8.5 Pr; AIMP 2, El > 22 keV, onset $08^h\Delta2$; IMP 4, Pr 0.5 MeV, onset 07^h, max. 13^h, 73* Pr; ◊ geomagnetic storm modulation, gradual beginning 27^d08^h
Nov. 7	PE	(69)	AIMP 2, El > 20 keV, onset $01^h\Delta1$, 20 El; ○ flare, imp. SN, 0005, W22, McM 9047; ○ flare, imp. SF, 0041, W45, McM 9047
Nov. 7	PE	(69)	AIMP 2, El > 40 keV, onset $22^h47^m \pm01^m$, 20 El; ⊙ flare, imp. SN, 2142, S22 W60, McM 9047; ⊙ flare, imp. 1F, 2218, W63, McM 9048
Nov. 8	L	(107, 176, 181)	IMP 4, Pr > 0.8 MeV, onset $14^h\Delta2$, dur. 53^h; El > 70 keV, onset $9^d03^h\Delta2$, dur. 12^d; Pr > 1 MeV, onset $9^d09^h\Delta2$, max. $10^d04^h\Delta2$, dur. 44^h, 0.7 Pr; ◊ onset of a particle stream related to CM transit of region 9054; ◊ co-rotating structure sequential with event of Oct. 12
Nov. 23	L	(175, 181)	IMP 4, Pr 0.5 MeV, onset 23^h15^m, max. $24^d12^h\Delta1$, 11* Pr; Pr > 1 MeV, onset $24^d00^h\Delta3$, max. $24^d12^h\Delta2$, dur. 3^d, 1.1 Pr; origin unknown
Nov. 30	PE	(69)	AIMP 2, El > 40 keV, onset $05^h20^m \pm20^m$, 20 El; ○ flare, imp. SN, 0445, W78, McM 9093
Nov. 30	PE	(69)	AIMP 2, El > 40 keV, onset $08^h44^m \pm14^m$, 10 El; ○ flare, imp. SN, 0839, W60, McM 9091
Dec. 1	PE	(69)	AIMP 2, El > 20 keV, onset $04^h\Delta2$, 50 El; ⊙ flare, imp. 1N, < 0335, W70, McM 9091
Dec. 1	PE	(69)	AIMP 2, El > 40 keV, onset $10^h41^m \pm01^m$, 10 El; ⊙ flare, imp. SN, 0942, W69, McM 9091
Dec. 1	PE	(69)	AIMP 2, El > 20 keV, onset $22^h58^m \pm6^m$, 25 El; ○ flare, imp. SF, 2230, W83, McM 9091
Dec. 2	PE	(69)	AIMP 2, El > 40 keV, onset $06^h26^m \pm02^m$, 40 El; ⊙ flare, imp, SN, 0535, W87, McM 9091; ⊙ flare, imp. SN, 0601, W90, McM 9091

1967 (cont.)

Dec. 9	UN	(104)	PION 6, Pr > 13 MeV, event in decay $9^d12^h\Delta5$, 10 Pr; Pr > 175 MeV,
(PION 6 at 0.87 AU, ES + 117°)			event in decay $9^d12^h\Delta5$, 0.86 Pr; no event on IMP 4; □ region 9091
			is 6 days beyond W limb for Earth but near CM for PION 6; ◇ co-rotating
			structure?
Dec. 23	PE	(69)	AIMP 2, El > 40 keV, onset $01^h35^m \pm 10^m$, 30 El; ○ flare, imp. SB, 0120,
			W85, McM 9115; ○ flare, imp. SN, 0119, W18, McM 9118
Dec. 26	L	(107)	IMP 4, Pr > 9 MeV, onset $04^h35^m \pm 05^m$, max. $06^h\Delta1$, dur. 8^h, 0.2 Pr;
			Pr > 0.8 MeV, onset $04^h35^m\pm05^m$, max. 06^h30^m, dur. 8^h, 30 Pr; not
			discernible on 1 MeV channel on IMP 4; □ region 9115 is 2 days beyond
			W limb
Dec. 28	PE	(69)	AIMP 2, El > 20 keV, onset $14^h26^m \pm 01^m$, 35 El; ● flare, imp. 1N, 1335,
			W19, McM 9128
Dec. 29	PE	(69)	AIMP 2, El > 40 keV, onset $19^h18^m \pm 01^m$, 10 El; ⊙ flare, imp. SN, 1859,
			W36, McM 9128
Dec. 31	L	(69, 107, 183)	AIMP 2, El > 40 keV, onset $15^h19^m \pm 01^m$, 65 El; IMP 4, Pr > 0.8 MeV,
(PION 8 at 0.95 AU, ES + 3°)			onset $15^h30^m \pm 30^m\Delta8$, max. $16^h30^m\Delta8$, dur. 10^h, 9 Pr; PION 8, Pr > 11
			MeV, onset $17^h40^m \pm 20^m$, max. $23^h\Delta2$, dur. 50^h, 0.04 Pr; ○ flare, imp.
			SF, 1431, W86, McM 9124; □ regions McM 9118+9120, 2 and 3 days
			beyond W limb

1968

Jan. 8	UN	(107)	IMP 4, Pr > 0.8 MeV, onset $04^h \pm 70^m$, max. 07^h30^m, dur. 24^h, 468 Pr;
			Pr > 9 MeV, onset $04^h30^m\pm30^m$, max. 07^h30^m, dur. 12^h, 1.5 Pr;
			no data on PION 7, no event on PION 8, nor on IMP 4 1 MeV channel;
			origin uncertain; three flares (7^d2153, 7^d2242, 8^d0216) in regions 9145 +
			9146
Jan. 9	L	(119, 107)	PION 7, Pr > 0.6 MeV, onset $06^d20^h\Delta2$; IMP 4, Pr > 0.8 MeV, onset
(PION 7 at 1.1 AU, ES − 45°)			$9^d03^h\Delta2$, max. $10^d11^h\Delta2$, dur. > 4^h, 63.2 Pr; beginning of a long interval
			of enhanced flux; ◇ particle stream related to CMP of active regions 9145
			and 9146; ○ flare, imp. 1B, 0012, E22, McM 9146; ○ flare, imp. 1N, 0109,
			W05, McM 9145
Jan. 9	L	(107, 174)	IMP 4, Pr > 0.8 MeV, onset $07^h05^m \pm 15^m\Delta3$, max. $09^h\Delta2$, 2.5 Pr;
			Pr > 1 MeV, onset $09^h\Delta2$, 16 Pr; this event is surimposed on preceding
			event; ⊙ flare, imp. 2N, 0507, E26, McM 9146
Jan. 11	L	(107, 188)	IMP 4, Pr > 0.8 MeV, onset $11^h45^m \pm 15^m\Delta3$, max. 12^h50^m, dur. 6^h,
			447 Pr; Pr 1 MeV, max. 12^h51^m, 130* Pr; ▲GMS begins with sc
			11^d1251; ⊘ flare, imp. 1B, 10^d2145, W26, McM 9145
Jan. 12	PE	(226)	AIMP 2, El > 40 keV, onset $22^h38^m \pm 28^m$, 75 El; ○ flare, imp. SF, 2150,
			W33, McM 9146; □ type III, radioburst 2211
Jan. 15	L	(119, 107)	PION 7, Pr > 0.6 MeV, onset $10^h\Delta5$, max. $17^h\Delta2$, 2 Pr; IMP 4, Pr > 0.8
(PION 7 at 1.1 AU, ES − 47°)			MeV, onset $16^d08^h \pm 30^m$, max. $13^h\Delta2$, dur. 20^h, 54 Pr; □ region 9146 at
			W limb; ◇ geomagnetic storm begins gradually 16^d15^h; ○ flare, imp. 2B,
			W40, McM 9146; ○ flare, imp. 2B, W45, McM 9146
Jan. 24	L	(119, 107)	PION 7, Pr > 0.6 MeV, onset $22^d12^h\Delta2$, max. $25^d00^h\Delta2$, dur. 7^d, 5 Pr;
(PION 7 at 1.1 AU, ES − 47°)			IMP 4, Pr > 0.8 MeV, onset $03^h\Delta2$, max. $17^h\Delta2$, dur. > 6^d, 40 Pr; ◇ par-
			ticle stream coincident with region McM 9184 at E limb
Jan. 30	L	(107, 183)	IMP 4, Pr > 0.8 MeV, onset $08^h\Delta4$, max. $31^d01^h\Delta2$, dur. 38^h, 13.7 Pr;
			PION 8, Pr > 11 MeV, onset $10^h\Delta2$, max. $31^d04^h\Delta2$, dur. 44^h, 0.006 Pr;
(PION 8 at 1.01 AU, ES − 0.3°)			● flare, imp. 1B, 29^d1537, E28, McM 9184
Feb. 1	PE	(226)	AIMP 2, El > 40 keV, onset 19^h35^m, 80 El; ● flare, imp. 1N 1915, W16,
			McM 9184
Feb. 5	L	(183, 181)	PION 8, Pr > 11 MeV, onset $19^h \pm 20^m$, max. $21^h\Delta2$, dur. > 15^h, 0.0042
(PION 8 at 1.01 AU, ES − 0.4°)			Pr; IMP 4, Pr > 1 MeV, onset $19^h45^m \pm 15^m$, max. 20^h30^m, dur. 5^h, 0.2 Pr;
			□ region 9184 on W limb; ○ flare, imp. SN, 1833, W79, McM 9184

1968 (cont.)

Feb. 6	L	(226, 183)	AIMP 2, El > 40 keV, onset 09^h02^m, 45 El; PION 8, Pr > 11 MeV, onset $10^h\Delta3$, max. $13^h\Delta1$, dur. 6^h, 0.002 Pr; ○ flare, imp. SN, 0848, W79, McM 9185; □ region 9184 is just beyond W limb
Feb. 8	L	(107, 183)	IMP 4, Pr > 0.8 MeV, onset $00^h\Delta4$, max. $13^h\Delta1$, dur. $> 14^h$, 1.5 Pr; PION 8, Pr > 11 MeV, onset $01^h\Delta1$, max. $02^h\Delta1$, 0.0034 Pr; origin uncertain; ◊ geomagnetic storm begins 7^d17^h, sequential with storm on Jan. 11
Feb. 18	PE	(226)	AIMP 2, El > 40 keV, onset 09^h14^m, 170 El; ● flare, imp. 1N, 0851, W47, McM 9206
Mar. 8 (PION 8 at 1.04 AU, ES $- 2°$)	L	(181, 183)	IMP 4, Pr > 1 MeV, onset $10^h\Delta3$, max. 9^d21^h, dur. $> 38^h$, 0.82 Pr; PION 8, Pr > 11 MeV, onset $20^h\Delta3$, max. $9^d08^h\Delta2$, dur. $> 30^h$, 0.0051 Pr; ◊ additional structure in permanent particle flux, part of preceding Event 248
Mar. 23 (PION 8 at 1.04 AU, ES $- 3°$)	L	(183, 107)	PION 8, Pr > 11 MeV, onset $12^h\Delta3$, max. $18^h\Delta2$, dur. 24^h, 0.0057 Pr; IMP 4, Pr > 0.8 MeV, onset $12^h30^m \pm 15^m\Delta3$, max. $20^h\Delta2$, dur. 27^h, 6 Pr; ⊙ flare imp. 1B, 0935, E33, McM 9279
Mar. 28 (PION 8 at 1.04 AU, ES $- 4°$)	L	(107, 183)	IMP 4, Pr > 9 MeV, onset $05^h \pm 30^m\Delta3$, max. 08^h30^m, dur. 15^h, 0.05 Pr; PION 8, Pr > 11 MeV, onset $05^h40^m \pm 20^m$, max. 08^h30^m, dur. 28^h, 0.017 Pr; IMP 4, Pr > 0.8 MeV, onset $06^h\Delta3$, max. $08^h30^m\Delta2$, 12 Pr; ⊙ flare, imp. 1N, 0321, W51, McM 9273
Apr. 9 (PION 8 at 1.05 AU, ES $- 5°$)	L	(107, 183, 181)	IMP 4, Pr > 0.8 MeV, onset $08^h\Delta2$, max. $12^d03^h\Delta2$, dur. $> 84^h$, 0.8 Pr; PION 8, Pr > 11 MeV, onset $12^h\Delta2$, max. $11^d23^h\Delta2$, dur. $> 72^h$, 0.0018 Pr; IMP 4, Pr > 1 MeV, onset $15^h\Delta2$, max. 12^d02^h, dur. $> 77^h$, 0.5 Pr; ◊ particle stream related to regions 9301 (CMP, Apr. 9) and 9313 (CMP, Apr. 15)
Apr. 27	PE	(226)	AIMP 2, El > 40 keV, onset 03^h10^m, 130 El; origin uncertain; type III 0216 (m+Dkm), 0224, 0301−0303
Apr. 27	L	(107)	IMP 4, Pr > 0.8 MeV, onset $17^h\Delta3$, max. $28^d05^h\Delta1$, dur. $> 34^h$, 41.5 Pr; origin unknown
Apr. 27	PE	(226)	AIMP 2, El > 40 keV, onset $17^h40^m \pm 15^m$, 80 El; ⊙ flare, imp. SN, 1729, W65, McM 9337
May 8 (PION 8 at 1.07 AU, ES $- 7.5°$)	L	(183, 107)	PION 8, Pr > 11 MeV, onset $16^h40^m \pm 40^m$, max. $20^h\Delta2$, dur. 50^h, 0.0062 Pr; IMP 4, Pr > 0.8 MeV, onset $9^d06^h\Delta4$, max. $10^d00^h\Delta2$, dur. 40^h, 3 Pr; ⊙ flare, imp. SF, < 1415, E58, McM 9382,
May 10	PE	(189)	AIMP 2, El > 22 keV, onset 08^h53^m, max. 09^h15^m; ⊙ flare, imp. 1N, 0840, W70, McM 9364
June 9 (PION 7 at 1.08 AU, ES $- 73°$)	L	(107, 119)	IMP 4, Pr > 0.8 MeV, onset $02^h\Delta3$, max. $06^h\Delta2$, dur. $> 8^h$, 4.5 Pr; PION 7, Pr > 13 MeV, onset $< 05^h\Delta5$, max. $07^h\Delta5$, 0.019 Pr; ⊙ flare, imp. 2N, 0025, E38, McM 9443
June 17	UN	(107)	IMP 4, Pr > 0.8 MeV, onset $< 09^h\Delta4$, max. 17^h30^m, dur. 100^h, 81 Pr; ⊙ flare, imp. 1N, < 0255, E41, McM 9459; □ active regions are 2 or 3 days beyond limb
June 26 (PION 8 at 1.08 AU, ES $- 13°$)	L	(107, 183)	IMP 4, Pr > 0.8 MeV, onset $05^h45^m \pm 15^m$, max. $08^h\Delta1$, dur. 18^h, 5.1 Pr; PION 8, Pr > 11 MeV, onset $06^h\Delta2$, max. $10^h\Delta2$, dur. 15^h, 0.0020 Pr; ⊙ flare, imp. SB, < 0509, W68, McM 9462
June 29	UN	(107)	IMP 4, Pr > 0.8 MeV, onset $06^h\Delta3$, max. $30^d12^h\Delta2$, dur. 110^h, 6.4 Pr; origin unknown; □ active region McM 9462 is 2 days beyond W limb
July 5	PE	(189)	AIMP 2, El > 22 keV, onset 10^h27^m, max. 11^h10^m; □ SWF 1021−1032, imp. 1−, minor 10 cm burst; no DS data
July 21	UN	(107)	IMP 4, Pr > 0.8 MeV, onset $03^h \pm 15^m\Delta3$, max. $22^d06^h\Delta2$, dur. 75^h, 5.7 Pr; ○ flare, imp. 2N, 08^h10^m, E54, McM 9530; □ region McM 9503 is two days beyond W limb

1968 (cont.)

Aug. 17	L	(107)	IMP 4, Pr > 9 MeV, onset $16^h\Delta2$, max. $18^d21^h\Delta2$, dur. 52^h, 0.018 Pr; Pr > 0.8 MeV, onset $22^h\Delta2$, max. $19^d12^h\Delta2$, dur. > 43^h, 0.18 Pr; origin unknown
Aug. 21	UN	(107)	IMP 4, Pr > 0.8 MeV, onset $13^h \pm 40^m\Delta3$, max. 17^h20^m, dur. 15^h, 19.6 Pr; spike on this channel 23^d1715; ● flare, imp. 1N, 0146, W43, McM 9593; ▲ geomagnetic storm begins with sc $23^d17^h15^m$
Sept. 1	PE	(226, 189)	AIMP 2, El > 40 keV, onset 16^h36^m, max. $18^h\Delta2$, dur. 5^h, 70 El; AIMP 2, El > 22 keV, onset $16^h36^m \pm 02^m$, max. 17^h50^m, dur. 6^h; ● flare, imp. SN, 1622, W90, McM 9611
Sept. 2	L	(107)	IMP 4, Pr > 9 MeV, onset $12^h\Delta2$, max. $14^h\Delta2$, dur. 25^h, 0.02 Pr; Pr > 0.8 MeV, onset $12^h\Delta2$, max. $03^d07^h\Delta2$, dur. > 36^h, 11.5 Pr; ◇ CMP of active regions 9630 and 9634; ○ flare, imp. 1N, 0658, E04, McM 9630; ○ flare, imp. 1N, 1039, E42, McM 9640
Sept. 3	UN	(107)	IMP 4, Pr > 0.8 MeV, onset $18^h30^m \pm 30^m\Delta3$, max. $20^h\Delta1$, dur. 22^h, 9 Pr; ⊙ flare, imp. 1N, 1553 and 1715, E07, McM 9634
Sept. 4 (PION 8 at 1.04 AU, ES − 20°)	UN	(183)	PION 8, Pr > 14 MeV, onset $02^h20^m \pm 20^m$, max. 03^h30^m, dur. 20^h, 0.18 Pr; ● flare, imp. 1N, < 0031, W14, McM 9630; ▲ geomagnetic storm modulation, sc begins $6^d14^h38^m$ (detectable on IMP 4 Pr > 0.8 MeV)
Sept. 10	UN	(107)	IMP 4, Pr > 0.8 MeV, onset $19^h15^m \pm 15^m$, max. $11^d05^h\Delta1$, dur. 50^h, 1.6 Pr; ○ flare, imp. SN, 1135, E66, McM 9658; □ active region McM 9634 is going around W limb; ◇ bright region McM 9655 is at CM on Sept. 9
Sept. 23	UN	(107)	IMP 4, Pr > 0.8 MeV, onset $12^h\Delta2$, max. $25^d23^h\Delta2$, dur. > 61^h, 0.5 Pr; origin unknown; ◇ co-rotating structure, active region McM 9687 (return of active region 9630) is on E limb at this time
Sept. 26 (PION 8 at 1.04 AU, ES − 20.5°)	UN	(183)	PION 8, Pr > 14 MeV, onset $16^h40^m \pm 20^m\Delta3$, max. $19^h\Delta2$, dur. > 39^h, 12.5 Pr; ○ flare, imp. 1B, 1121, W80, McM 9680; ○ flare, imp. 1N, < 1035, W60, McM 9688
Sept. 28 (PION 8 at 1.04 AU, ES − 21°)	UN	(183)	PION 8, Pr > 14 MeV, onset $06^h40^m \pm 20^m\Delta3$, dur. > 4^h; ● flare, imp. 1B, < 0155, E06, McM 9687
Sept. 29	PE	(166)	IMP 4, El > 2.7 MeV, onset 10^h42^m, 0.02 El; IMP 4, El > 0.3 MeV, onset 10^h42^m, max. 13^h30^m, dur. > 7^h, 0.5 El; ● flare, imp. 2B, < 0920, W13, McM 9687
Oct. 1	UN	(183)	PION 8, Pr > 14 MeV, onset $14^h \pm 20^m\Delta3$, max. 17^h30^m, dur. 40^h, 1.8 Pr; no additional event discernible on IMP 4; origin uncertain; ○ flare, imp. 1N, 0033, W34, McM 9687
Oct. 3	PE	(226, 189)	AIMP 2, El > 40 keV, onset 13^h02^m, 80 El; AIMP 2, El > 22 keV, onset $13^h02^m \pm 02^m$; ⊙ flare, imp. SN, 1239, W68, McM 9687
Oct. 17	L	(107)	IMP 4, Pr > 9 MeV, onset $02^h\Delta2$, max. $20^h\Delta2$, dur. > 3^d, 0.03 Pr; Pr > 0.8 MeV, onset $03^h\Delta2$, max. $19^d01^h\Delta2$, dur. > 7^d, 1.7 Pr; ◇ sequential particle stream related to disk transit of region 9735
Nov. 25	UN	(107)	IMP 4, Pr > 0.8 MeV, onset $05^h \pm 30^m\Delta3$, max. $11^h\Delta2$, dur. > 4^d, 43 Pr; ◇ CM transit of active region 9780 (return of active region 9740, events 291 to 297); ◇ Forbush decrease 25^d03^h sequential with Event 294; ○ flare, imp. SN, 0335, E00, McM 9780
Dec. 16	PE	(226)	AIMP 2, El > 40 keV, onset 07^h25^m, 60 El; origin uncertain; □ region 9802 is just beyond W limb; ○ flare imp. 1F, < 0510, W35, McM 9811
Dec. 23	UN	(107)	IMP 4, Pr > 0.8 MeV, onset $16^h\Delta2$, max. $24^d01^h\Delta2$, dur. > 31^h, 600 Pr; Pr > 10 MeV, onset $20^h\Delta2$, max. $24^d01^h\Delta2$, dur. 26^h, 4 Pr; origin uncertain; ○ flare, imp. 1N, 1112, E50, McM 9842; ◇ particle stream sequential with event of Nov. 25 (region 9826 is at CM on Dec. 22)
Dec. 26	UN	(107)	IMP 4, Pr > 0.8 MeV, onset $11^h \pm 30^m\Delta3$, max. 12^h40^m, dur. > 27^h, 2.5 Pr; ○ flare, imp. 1N, < 1000, W54, McM 9826; ○ flare, imp. SF, 1012, E12, McM 9842

1968 (cont.)

Dec. 28	PE	(189)	AIMP 2, El > 22 keV, onset 04^h01^m, max. 04^h20^m; $\circ$ flare, imp. SF, 0341, W13, McM 9842
Dec. 28	PE	(226)	AIMP 2, El > 40 keV, onset 10^h32^m, 80 El; $\bullet$ flare, imp. SN, 0957, W38, McM 9838
Dec. 30	UN	(107)	IMP 4, Pr > 0.8 MeV, onset $04^h \pm 30^m \Delta 3$, max. $08^h \Delta 1$, dur. $> 15^h$, 5.5 Pr; origin uncertain; $\circ$ flare, imp. 1B, $29^d 1914$, W33, McM 9842; $\circ$ flare, imp. 1N, 0217, W24, McM 9842
Dec. 30	UN	(107)	IMP 4, Pr > 0.8 MeV, onset $19^h \Delta 3$, max. $20^h \Delta 2$, dur. $> 2^d$, 25 Pr; $\odot$ flare, imp. SN, 1424, E26, McM 9847

1969

Jan. 2	PE	(189)	AIMP 2, El > 22 keV, onset $12^h \pm 40^m$, max. $15^h \Delta 2$; $\odot$ flare, imp. 1N, 1140, W83, McM 9841
Jan. 3 (PION 9 at 0.93 AU, ES $-2°$)	L	(217, 107)	PION 9, Pr > 10 MeV, onset $01^h \pm 20^m$, max. $04^h \Delta 2$, dur. 23^h, 0.017 Pr; IMP 4, Pr > 0.8 MeV, onset $02^h \Delta 3$, max. $11^h \Delta 1$, dur. 17^h, 18 Pr; $\circ$ flare, imp. 1N, $2^d 2115$, W75, McM 9842; $\circ$ flare, imp. SN, $2^d 2248$, E72, McM 9855; $\square$ region McM 9842 is beyond W limb and region McM 9855 near E limb; $\oslash$ flare, imp. 1N, 0439, E72, McM 9855
Jan. 3	UN	(107)	IMP 4, Pr > 0.8 MeV, onset $18^h30^m \pm 10^m \Delta 3$, max. $20^h \Delta 1$, dur. 26^h, 8 Pr; origin unknown; $\square$ region McM 9842 is on invisible hemisphere beyond W limb
Jan. 9 (PION 9 at 0.9 AU, ES $-1°$)	L	(107, 217)	IMP 4, Pr > 0.8 MeV, onset $12^h \Delta 2$, max. $11^d 11^h \Delta 2$, dur. 3^d, 4 Pr; Pr > 9 MeV, onset $10^d 02^h 30^m \pm 30^m$, max. $05^h \Delta 1$, dur. 5^h, 0.06 Pr; PION 9, Pr > 10 MeV, onset $10^d 13^h \Delta 2$, max. $15^h \Delta 2$, dur. 12^h, 0.006 Pr; origin uncertain, numerous flares in region McM 9861, 9^d at 0041, 0306, 0508, 1323, 1831
Jan. 14 (PION 9 at 0.9 AU, ES $-0.5°$; PION 8 at 1AU, ES $-22°$; VEN 6 at 0.98 AU, ES $-0.5°$)	L	(107, 43, 217, 183)	IMP 4, Pr > 0.8 MeV, onset $02^h \Delta 2$, max. $15^d 10^h \Delta 2$, dur. 65^h, 21 Pr; VEN 6, Pr > 1 MeV, onset $09^h \Delta 2$, max. $15^d 01^h \Delta 2$, dur. 56^h, 15 Pr; PION 9, Pr > 10 MeV, onset $10^h \Delta 2$, max. $14^h \Delta 2$, dur. 2^d, 0.007 Pr; PION 8, Pr > 11 MeV, onset $\Delta 5$, max. $14^h \Delta 2$, dur. 36^h, 0.004 Pr; origin uncertain; $\square$ type IV Dkm $13^d 2105-2347$; type II $13^d 2229-2245$; SWF, imp. 1, $0120-0226$; $\diamond$ region McM 9871 forms on disk 13^d-14^d
Jan. 24	L	(107)	IMP 4, Pr > 0.8 MeV, onset $04^h30^m \pm 15^m$, max. $07^h \Delta 1$, dur. $> 3^h$, 36 Pr; Pr > 9 MeV, onset $04^h35^m \pm 5^m$, max. 04^h45^m, dur. $> 3^h$, 0.2 Pr; precursor to next major event (Event 309); $\diamond$ probably related to particles arriving prior to onset of a geomagnetic storm which begins gradually 24^d12^h
Feb. 2	L	(107)	IMP 4, Pr > 0.8 MeV, onset $10^h \Delta 2$, max. 15^h15^m, 80 Pr; Pr > 9 MeV, onset $14^h \Delta 2$, max. $17^h \Delta 2$, dur. 8^h, 0.1 Pr; $\circ$ flare, imp. 2N, 0508, E73, McM 9911; $\circ$ flare, imp. 1B, $1^d 2125$, E77, McM 9911; $\triangle$ major geomagnetic storm begins with sc $2^d 1502$
Feb. 9	UN	(107)	IMP 4, Pr > 0.8 MeV, onset $23^h \Delta 2$, max. $11^d 02^h \Delta 2$, dur. 50^h, 1 Pr; $\circ$ flare, imp. SB, 1723, E00, McM 9918; $\diamond$ bright region McM 9926 forms on disk 9^d
Feb. 11	L	(107)	IMP 4, Pr > 0.8 MeV, onset $05^h15^m \pm 15^m \Delta 3$, max. $12^h \Delta 2$, dur. 12^h, 200 Pr; IMP 4, Pr > 9 MeV, onset $10^h \Delta 2$, max. $13^h \Delta 2$, dur. 10^h, 0.2 Pr; $\diamond$ geomagnetic storm modulation, geomagnetic storm begins with sc $10^d 2024$
Feb. 12	L	(107)	IMP 4, Pr > 9 MeV, onset $18^h \Delta 2$, max. $13^d 00^h \Delta 2$, dur. 11^h, 0.06 Pr; Pr > 0.8 MeV, onset $18^h30^m \pm 10^m$, max. $20^h \Delta 2$, dur. 11^h, 63 Pr; origin unknown
Feb. 24	PE	(226)	AIMP 2, El > 40 keV, onset 14^h50^m, 25 El; $\bullet$ flare, imp. SN, 1431, W27, McM 9946

1969 (cont.)

Mar. 8	L	(107)	IMP 4, Pr > 0.8 MeV, onset $18^h\Delta2$, max. $9^d16^h\Delta2$, dur. $> 4^d$, 3 Pr; Pr > 9 MeV, onset $19^h\pm90^m$, max. $9^d12^h\Delta2$, dur. $> 4^d$, 0.15 Pr; ○ flare, imp. SB, 1806, W22, McM 9966; ◊ increase of activity in region McM 9966 on 7^d-13^d
Mar. 24 (PION 9 at 0.77 AU, ES + 20°)	UN	(217)	PION 9, Pr > 14 MeV, onset $01^h\Delta1$, max. 02^h30^m, 0.31 Pr; △ geomagnetic storm begins with sc 23^d1826 (this storm is associated with flare of 21^d, Event 319)
Mar. 24	L	(210, 107)	AIMP 2, Pr > 0.3 MeV, onset $18^h\Delta3$, max. $25^d06^h\Delta1$, dur. 95^h, 700 Pr; IMP 4, Pr > 0.8 MeV, onset $20^h\Delta3$, max. $25^d02^h\Delta2$, dur. $> 61^h$, 41 Pr; ○ flare, imp. 1N, 1449, W42, McM 9994
May 6	PE	(226)	AIMP 2, El > 40 keV, onset $07^h\Delta1$, 60 El; ○ flare, imp. SF, < 0637, W90, McM 10057
May 8	L	(202)	HEOS, Pr > 6 MeV, onset $10^h\Delta2$, max. $9^d04^h\Delta2$, dur. 55^h, 0.1 Pr; Pr > 5 MeV, onset $10^h\Delta2$, max. $9^d04^h\Delta2$, dur. 55^h, 0.5 Pr; ○ flare, imp. 1N, 0419, E39, McM 10078; ◊ particle stream related to CM transit of new and bright regions
May 23 (PION 9 at 0.82 AU, ES + 55°)	UN	(217)	PION 9, Pr > 10 MeV, onset $\Delta5$, max. $23^d<07^h\Delta5$; no event on PION 8, nor HEOS; origin unknown; ○ flares, imp. 2B, 22^d1900, E40, 22^d1933, E21, McM 10109
May 28	PE	(226)	AIMP 2, El > 40 keV, onset 04^h40^m, 30 El; ○ flare, imp. SF, 28^d0416, W80, McM 10099; ○ flare, imp. SF, < 0430, W54, McM 10109
May 28	PE	(226)	AIMP 2, El > 40 keV, onset 07^h50^m, 80 El; ⊙ flare, imp. SN, 0737, W56, McM 10109
May 28	PE	(226)	AIMP 2, El > 40 keV, onset 10^h57^m, 100 El; ○ flare, imp. 1N, 1038, W58, McM 10109
May 28	PE	(226)	AIMP 2, El > 40 keV, onset 23^h30^m, 80 El; ○ flare, imp. SN, 2313, W64, McM 10109; ○ flare, imp. SN, 2256, W34, McM 10109
May 29	PE	(226)	AIMP 2, El > 40 keV, onset 04^h20^m, 170 El; ● flare, imp. 1N, 0406, W67, McM 10109
May 29	PE	(226)	AIMP 2, El > 40 keV, onset $13^h43^m\pm25^m$, 100 El; ○ flare, imp. SN, 1330, W77, McM 10108; ○ flare, imp. SB, < 1405, W73, McM 10109
May 29	PE	(226)	AIMP 2, El > 40 keV, onset $14^h54^m\pm19^m$, 100 El; ⊙ flare, imp. 1B, < 1442, W73, McM 10109; ○ flare, imp. 1N, 1421, W78, McM 10108
May 29	PE	(226)	AIMP 2, El > 40 keV, onset 17^h45^m, 60 El; ○ flare, imp. 1N, 1728, W78, McM 10108; ○ flare, imp. SN, 1718, W73, McM 10109; ○ flare, imp. 1N, 1704, W78, McM 10108
May 30	PE	(226)	AIMP 2, El > 40 keV, onset 04^h30^m, 80 El; ⊙ flare, imp. 1N, 0420, W90, McM 10108
May 30	PE	(226)	AIMP 2, El > 40 keV, onset 06^h10^m, 500 El; ○ flare, imp. 1N, 0610, W83, McM 10109; ○ flare, imp.?, < 0534, W84, McM 10108
May 30	PE	(226)	AIMP 2, El > 40 keV, onset 11^h18^m, 200 El; ○ flare, imp. SB, 1025, W90, McM 10108; ○ flare, imp. SB, < 1046, W82, McM 10109
May 30	PE	(226)	AIMP 2, El > 40 keV, onset 20^h24^m, 40 El; ○ flare, imp. SB, 1958, E90, McM 10128; □ region McM 10108 and 10109 are just beyond W limb on invisible hemisphere
May 31	PE	(226)	AIMP 2, El > 40 keV, onset 00^h38^m, 80 El; ○ flare, imp. SF, < 0001, W90, McM 10109; most of active region McM 10109 is beyond W limb
May 31	PE	(226)	AIMP 2, El > 40 keV, onset 21^h20^m, 125 El; origin uncertain; □ active region McM 10109 is beyond W limb; SWF 2057–2305, imp. 3; type III 2101–2105, cont. 2117–2209
June 1	PE	(226)	AIMP 2, El > 40 keV, onset 15^h35^m, 80 El; origin uncertain; ○ flare, imp. 1N, 1507, E90, McM 10130; □ regions McM 10108 and 10109 are on invisible hemisphere beyond W limb

1969 (cont.)

June 2	L	(226, 202, 168)	AIMP 2, El > 40 keV, onset $09^h43^m \pm 5^m$, 350 El; HEOS , El > 0.5 MeV, onset $10^h\Delta1$, max. $11^h\Delta1$, dur. 30^h, 6 El; VELA 4, Pr 0.6 MeV, onset $18^h\Delta2$, max. $3^d00^h\Delta2$, dur. 34^h, 20* Pr; ○ flare, imp. **SF**, 0848, E79, McM 10130; □ active regions McM 10108 and 10109 are on invisible hemisphere, 3 days beyond W limb
June 5 (PION 8 at 1.07 AU, ES $- 30°$)	L	(183, 202)	PION 8, Pr > 14 MeV, onset $20^h\Delta2$, max. $7^d06^h\Delta2$, dur. $> 35^h$, 0.18 Pr; HEOS, Pr > 6 MeV, onset $6^d04^h\Delta2$, max. $6^d07^h\Delta2$, dur. $> 38^h$, 0.2 Pr; ⊙ flare, imp. 2B, 0952, E64, McM 10134; ⦰ flare, imp. 3B, < 1442, E53, McM 10134
June 7	L	(202, 193, 168)	HEOS, Pr > 5 MeV, onset $03^h\Delta2$, max. $12^h\Delta2$, dur. $> 15^h$, 0.8 Pr; AIMP 2, Pr > 0.32 MeV, onset $< 05^h\Delta2$; VELA 4, Pr 0.6 MeV, onset $06^h\Delta2$, max. $12^h\Delta2$, 40* Pr; ○ flare, imp. 1B, 0018, E45, McM 10135; ○ flare, imp. 1N, 6^d2303, E45, McM 10135; ○ flare, imp. 2B, 6^d1604, E19, McM 10130
June 11 (PION 8 at 1.07 AU, ES $- 30°$)	L	(226, 183)	AIMP 2, El > 40 keV, onset 16^h48^m, 200 El; PION 8, Pr > 14 MeV, onset $17^h\Delta2$, max. $22^h\Delta2$, dur. $> 1^d$, 0.375 Pr; ● flare, imp. 2B, 1615, W20, McM 10134
June 16	L	(193, 168)	AIMP 2, Pr > 0.32 MeV, onset 05^h30^m, dur. 114^h, 110 Pr; VELA 4, Pr 0.6 MeV, onset $06^h\Delta5$, max. $22^h\Delta2$, dur. $> 5^d$, 50* Pr; origin uncertain; flare invisible hemisphere?; active regions 10134 and 10135 are crossing W limb; ◇ geomagnetic storm modulation, geomagnetic disturbance begins 16^d06^h
June 28	UN	(212)	IMP 5, Pr > 1 MeV, onset $15^h\Delta2$, max. $1^d04^h\Delta2$, dur. $> 11^d$, 0.4 Pr; origin unknown; ◇ particle stream related to formation of new region McM 10185 on the disk
July 2 (PION 9 at 0.79 AU, ES $+ 50°$)	UN	(217)	PION 9, Pr > 14 MeV, onset $\Delta5$, max. $13^h\Delta5$, 0.62 Pr; □ large regions McM 10146 and 10148 are on invisible hemisphere, 2 ad 3 days beyond W limb for PION 9; ◇ new region 10185 is near CM for PION 9 on 2^d
July 13	PE	(189)	IMP 5, El > 22 keV, onset $07^h06^m \pm 3^m$, max. 07^h20^m; ● flare, imp. SN, 0647, W90, McM 10181
July 20	L	(193, 212)	AIMP 2, Pr > 0.32 MeV, onset $15^h\Delta1$, dur. 46^h, 125 Pr; IMP 5, Pr > 1 MeV, onset $15^h\Delta2$, max. $21^d05^h\Delta2$, dur. 45^h, 5 Pr; □ DS type IV, 19^d1814–1842, 1845–2041, 2041–2221, active regions McM 10191 and 10192 are on invisible hemisphere, 1 or 2 days beyond W limb; ◇ growth of new region 10209 on disk after CMP on July 18
July 24	L	(212, 193)	IMP 5, Pr > 1 MeV, onset $06^h\Delta1$, max. $25^d16^h\Delta1$, dur. 43^h, 16.2 Pr; AIMP 2, Pr > 0.32 MeV, onset $25^d13^h\Delta1$, dur. 62^h, 300 Pr; origin uncertain; ◇ region McM 10223 develops on the disk on 23^d–24^d; bright region 10223 is near CM on 24^d
Aug. 4	L	(202, 212)	HEOS 1, El > 0.5 MeV, onset $18^h\Delta2$, max. $06^d03^h\Delta2$, dur. 20^h, 0.6 El; IMP 5, Pr > 1 MeV, onset $5^d14^h\Delta2$, max. $6^d14^h\Delta1$, dur. $> 44^h$, 0.44 Pr; ○ flare, imp. SN, 5^d0939, W58, McM 10232; ○ flare, imp. SN, 1849, W50, McM 10232; ◇ particle stream related to zone of plages near CM 4^d–7^d
Aug. 7	L	(202, 212)	HEOS 1, El > 0.5 MeV, onset $12^h\Delta2$, max. $13^h\Delta2$, dur. 20^h, 0.6 El; IMP 5, Pr > 1 MeV, onset $20^h\Delta3$, max. $22^h\Delta2$, dur. 30^h, 0.3 Pr; ⊙ flare, imp. 2N, 0901, E28, McM 10253; □ flare, imp. SN, 0926, W89, McM 10232
Aug. 10	L	(212, 193)	IMP 5, Pr > 1 MeV, onset $09^h30^m \pm 30^m$, max. $11^d06^h\Delta1$, dur. 50^h, 6 Pr; AIMP 2, Pr > 0.32 MeV, onset 13^h30^m, dur. 36^h, 76 Pr; origin unknown
Aug. 16	L	(202, 212, 193)	HEOS 1, Pr > 5 MeV, onset $09^h\Delta2$, max. $10^h\Delta2$, dur. 15^h, 0.25 Pr; IMP 5, Pr > 1 MeV, onset $09^h30^m \pm 30^m$, max. 13^h30^m, dur. $> 31^h$, 4.4 Pr; AIMP 2, Pr > 0.32 MeV, onset $13^h\Delta1$; origin unknown
Aug. 17	L	(212, 193)	IMP 5, Pr > 1 MeV, onset $16^h\Delta3$, max. $18^d07^h\Delta1$, dur. $> 34^h$, 15.4 Pr; AIMP 2, Pr > 0.32 MeV, 300 Pr; origin unknown

1969 (cont.)

Aug. 19	UN	(212)	IMP 5, Pr > 1 MeV, onset $02^h\Delta3$, max. $07^h\Delta2$, dur. 30^h, 0.65 Pr; ⊙ flare, imp. 1B, 18^d2201, W53, McM 10262
Sept. 3	UN	(212)	IMP 5, Pr > 1 MeV, onset $12^h\Delta4$, max. $5^d10^h\Delta2$, dur. 4^d, 1.36 Pr; ○ flare, imp. 1N, 2144, W16, McM 10289; ○ flare, imp. SN, 0925, W02, McM 10292
Sept. 7	L	(202, 212)	HEOS 1, El > 0.5 MeV, onset $08^h\Delta2$, max. $15^h\Delta2$, dur. 35^h, 0.7 El; IMP 5, Pr > 1 MeV, onset $21^h\Delta2$, max. $8^d10^h\Delta1$, dur. 28^h, 0.25 Pr; ○ flare, imp. 2N, 0331, W75, McM 10289; ○ flare, imp. SN, 1749, W58, McM 10292; other flares 1645, 1020 in regions 10298, and 10304
Sept. 14	UN	(212)	IMP 5, Pr > 1 MeV, onset $16^h\Delta2$, max. $15^d03^h\Delta1$, dur. > 70^h, 1.2 Pr; ◇ geomagnetic storm begins with sc on 14^d1518, geomagnetic storm modulation
Sept. 17	UN	(212)	IMP 5, Pr > 1 MeV, onset $14^h \pm 60^m\Delta3$, max. $22^h\Delta2$, dur. 48^h, 1.5 Pr; ⊙ flare, imp. SB, 1207, W06, McM 10317; ⊙ flare, imp. SN, 1012, W06, McM 10317; ⊙ flare, imp. 1N, 0935, W27, McM 10309; ○ flare, imp. SN, 0836, W06, McM 10317
Sept. 25	PE	(226, 189)	AIMP 2, El > 40 keV, onset 14^h47^m, 70 El; IMP 5, El > 22 keV, onset $14^h47^m \pm 3^m$, max. 14^h55^m; ● flare, imp. 1N, 1414, W39, McM 10325
Sept. 30	UN	(212)	IMP 5, Pr > 1 MeV, onset $12^h\Delta3$, max. $15^h\Delta1$, dur. 150^h, 89 Pr; ○ flare, imp. 1N, 0619, E42, McM 10344; □ active region McM 10325 is on invisible hemisphere, 2 days beyond W limb; ◇ recovery from major magnetic storm (Event 335)
Oct. 7	UN	(212)	IMP 5, Pr > 1 MeV, onset $20^h \pm 40^m$, max. $8^d08^h\Delta2$, dur. 50^h, 0.56 Pr; □ numerous type III (1821, 1747, 1624), regions McM 10335 and 10337 are on invisible hemisphere, one or two days beyond W limb
Oct. 10 (PION 9 at 0.98 AU, ES + 60°)	L	(217, 212)	PION 9, Pr > 14 MeV, onset < $10^h\Delta5$, max. $12^h\Delta2$, dur. 16^h, 0.18 Pr; IMP 5, Pr > 1 MeV, onset $10^h\Delta4$, max. $13^h\Delta2$, dur. 20^h, 0.37 Pr; ○ flare, imp. SN, 9^d2056, W52, McM 10351; ◇ geomagnetic storm modulation, geomagnetic storm begins with sc 9^d1648
Oct. 11	UN	(217)	PION 9, Pr > 14 MeV, onset $10^h30^m\Delta5$, max. $12^d08^h\Delta5$, no event on IMP 5; origin uncertain; ○ flare, imp. SF, < 0935, W71, McM 10351 (near CM for PION 9)
Oct. 14	PE	(226)	AIMP 2, El > 40 keV, onset 09^h47^m, 220 El; origin unknown, active region McM 10351 is two days beyond W limb
Oct. 18	UN	(212)	IMP 5, Pr > 1 MeV, onset $15^h\Delta3$, max. $22^h\Delta2$, dur. 25^h, 1.1 Pr; origin unknown; numerous regions are on the invisible hemisphere beyond W limb
Oct. 20	UN	(212)	IMP 5, Pr > 1 MeV, onset $13^h\Delta4$, max. $21^d06^h\Delta1$, dur. > 6^d, 1.51 Pr; ⊙ flare, imp. 1F, < 0326, E80, McM 10385; ○ flare, imp. SN, 1048, E78, McM 10385
Nov. 16	UN	(212)	IMP 5, Pr > 1 MeV, onset $17^h\Delta1$, max. $17^d04^h\Delta2$, dur. 38^h, 1.74 Pr; ○ flare, imp. 1N, 0953, W88, McM 10412; □ numerous type III bursts between 1037–1513, region McM 10412 is going over W limb on 16^d
Nov. 23 (PION 9 at 0.89 AU, ES + 69°)	UN	(217)	PION 9, Pr > 14 MeV, onset $14^h\Delta5$, max. $18^h\Delta1$, dur. 8^h, 1.2 Pr; no additional event on IMP 5, nor HEOS 1; ● flare, imp. 1B, 0958, W19, McM 10432
Nov. 25	UN	(212)	IMP 5, Pr > 1 MeV, onset $16^h\Delta3$, max. $26^d06^h\Delta2$, dur. > 123^h, 26.1 Pr; origin uncertain; ◇ geomagnetic storm modulation, storm begins with sc 26^d1507; numerous flares on 25^d in region McM 10432
Dec. 7	UN	(212)	IMP 5, Pr > 1 MeV, onset $03^h\Delta2$, max. $8^d05^h\Delta1$, dur. 60^h, 2.8 Pr; □ active region McM 10447 is just beyond W limb on 7^d; ◇ region McM 10459 is crossing CM on 7^d

1969 (cont.)

Dec. 9	L	(193, 212)	AIMP 2, Pr > 0.32 MeV, onset $16^h50^m \pm 30^m$, dur. 72^h, 300 Pr; IMP 5, Pr > 1 MeV, onset $17^h20^m \pm 30^m$, max. $10^d18^h\Delta2$, dur. 65^h, 32 Pr; origin unknown, numerous active regions on the invisible hemisphere one or two days beyond W limb
Dec. 13	UN	(212)	IMP 5, Pr > 1 MeV, onset $15^h \pm 30^m$, max. $23^h\Delta1$, dur. 30^h, 2.9 Pr; ○ flare, imp. 1B, 1340, E58, McM 10478; ○ flare, imp. 1N, 0339, E88, McM 10477; ○ flare, imp. SF, 1752, E90, McM 10477
Dec. 23	L	(212, 193)	IMP 5, Pr > 1 MeV, onset $12^h\Delta2$, max. $24^d07^h\Delta2$, dur. 45^h, 14 Pr; AIMP 2, Pr > 0.32 MeV, onset 16^h30^m, dur. 32^h, 204 Pr; ○ flare, imp. 1N, 0613, W48, McM 10477; ○ flare, imp. 1N, 0355, E41, McM 10499

PART 2

LIST OF FLARES WHICH HAVE BEEN IDENTIFIED AS SOURCES OF PARTICLE EVENTS

1955	January 16	⊙	McM 3065		Event 1	(X 2 0)	CFI ≥ 9
Hα	<2130	≤2130	2220	N33 W41	3		$\Delta t \sim 1^h$
SWF	2110		2210		weak	(5)	
2.8 GHz	2105		2135				
DS continuum	2150		2215				

1956	February 23	●	McM 3400		Event 2	(X 3 4)	CFI ≥ 12
Hα	≤0334	0342	0510	N23 W80	3	(4, 51) $\Delta t \sim 10^m$
SWF	0330		0610		3+	(4, 100)
3.7 GHz	0334		0349		>4.2	] type IV (99),
200 MHz	0335		>0600		>4.3	] no DS data
White light	≤0345		0350			(7)
Spray ejection	0335					(41)

1956	March 10	●	McM 3432		Event 3	(X 2 0)	CFI ≥ 7
Hα	<0515		0640	N16 E88	2		$\Delta t \sim 4^h$
SWF	0448		0635		3−	(4, 100)
3.0 GHz	0443	0518	0603		2.9	(4, 51)
No measurements at metric wavelengths, no DS data						

1956	April 27	⊙	McM 3474		Event 4	(X (1) 0)	CFI ≥ 7
Hα	2050	2100	>2150	N16 W27	1+		$\Delta t \sim 1^h$
SWF	2053		2117		1+	(100)
2.8 GHz	2051		2101		>2.5	
167 MHz	burst; no DS data					

1956	August 31	●	McM 3643		Event 5	(X 2 1)	CFI ≥ 12
Hα	1226	1243	1626	N15 E15	3		$\Delta t \sim 20^m$
SWF	1239		1400		3	(100)
2.8 GHz	1231		1310			] type IV (99),
200 MHz	1231		1333		4.3	] no DS data
White light	seen at Arosa					

1956	November 7	⊙	McM 3751		Appendix A		CFI ≥ 10
Hα	1109	1135	1354	S17 E32	3+		$\Delta t \sim 13^h$
SWF	1127		1400		3−	(100)
9.4 GHz	1103		1325		>2.8	] probably
						] type IV (99),
200 MHz	1115		1200		3.3	] no DS data

1956	November 13	⊙	McM 3753			Event 6	(X 3 0)	CFI ≧ 7
Hα		< 1430	1501	1555		N16 W10	2	Δt ∼ 0–6ʰ
SWF		1430		1630			2+	(100)
9.4 GHz		1433		> 1445			> 2.5	⎤ probably
								type IV (99),
545 MHz		1425		1445			3.0	⎦ no DS data

1956	November 14	⦰	McM 3751			Event 6	(X 3 0)	CFI ≧ 10
Hα		1037	1055	1400		S20 W55	3	
SWF		1037		1155			2+	(100)
9.4 GHz		1035		1230			> 3.0	⎤ type IV (99),
200 MHz		1037		1110			3.6	⎦ no DS data
Loops		1126		1408				(33)

1957	January 20	●	McM 3820			Event 7	(X 2 0)	CFI ≧ 6
Hα		< 1100	1119	1417		S30 W18	3	Δt ∼ 4ʰ
SWF		1113		1126			1+	(100)
3.0 GHz		1100		1124			2.3	⎤ possibly
								type IV (99),
536 MHz		1057		1230			> 2.4	⎦ no DS data
Loops		1250		1320				(33)

1957	February 21	⊙	McM 3856			Event 8	(X (1) 0)	CFI = 7
Hα		1605	1930	> 2205		N20 W33	3+	(4, 11) Δt ∼ 2–6ʰ
SWF		small if any						(5)
2.8 GHz		1750		2150			> 1.3	
167 MHz		1827		2245			2.7	
DS type II		2008		2011			2	(25)

1957	April 3	●	McM 3907			Event 9	(X 2 0)	CFI ≧ 8
Hα		0825	0835	> 1026		S14 W60	3	Δt ∼ 2–5ʰ
SWF		0833		0908			2	(100)
9.4 GHz		0828		1245			> 2.8	⎤ probably
								type IV (99),
545 MHz		0830		0930			∼ 3.2	⎦ no DS data

1957	April 11	⦰	McM 3923			Event 10	(X (1) 0)	CFI = 9
Hα		1722	1738	> 1850		S23 E05	2+	
SWF		1731		1835			3	(100)
2.8 GHz		1725		1840			> 2.1	
167 MHz		1726		1739			2.4	
DS type III		1733		1744				

1957	April 12	⊘	McM 3916		Event 10	(X (1) 0) CFI = 10

Hα	1850		2010	S25 W73	2	
SWF	1856		2025		3+	(100)
2.8 GHz	1856		1919		> 2.7	
167 MHz	1858		1913		3.2	
DS continuum	1858		1902 plus type III; close-to-limb type IV possible			
DS type II	1905		1916		3	(25)

1957	April 17	⊙	McM 3941		Event 11	(X (1) 0) CFI = 14

Hα	2000	2116	2300	N20 E69	3+	(11)	$\Delta t \sim 30^{\mathrm{h}}$
SWF	1937		2220		3+	(100)	
2.8 GHz	2006		2125		> 2.7		
200 MHz	2021		2059		> 2.2		
DS type IV	2011		2055		2	(25)	
DS type II	2032		2039		3	(25)	
DS type III	2027		2034				

1957	April 18	⊙	McM 3944		Event 11	(X (1) 0) CFI = 9

Hα	< 1310	1323	> 1353	S16 E64	2		$\Delta t \sim 13^{\mathrm{h}}$
SWF	1304		1340		2+	(5)	
3.0 GHz	1304		1310		2.7		
200 MHz	1305		1311		2.9		
DS continuum	1302		1304				
DS type II	1304		1312		3	(25)	
DS type III	1302		1306				

1957	June 19	⊙	McM 4024		Event 13	(X (1) 0) CFI = 11

Hα	1609	1613	1649	N20 E45	2		$\Delta t \sim 6^{\mathrm{h}}$
SWF	1608		1652		3	(11)	
2.8 GHz	1609		1619		> 2.3		
545 MHz	1610		1655		2.6		
DS type II	1615		1620		3	(25)	
DS type III	1609		1613				
DS	1622 onset of strong noise storm						

1957	June 30	⊙	McM 4039		Event 15	(X (1) 0) CFI ≧ 5

Hα	0924		1316	N10 W02	2+		$\Delta t \sim 15^{\mathrm{h}}$?
SWF	0927				very weak (99)		
3.0 GHz	0953		1002		2.1	possibly type IV (99),	
169 MHz	1006		1237		2.3	no DS data	

1957	July 3	●	McM 4039		Event 16	(X 3 0)	CFI ≧ 11
Hα	0712 0830	0840	>0745 1145	N14 W40 N10 W42	3+	(4, 17)	$\Delta t \sim 1^h$
SWF	0729 0830		0830 0914		2+ 3	(11)	
9.4 GHz	0729 0831	0742 0841	0815 0851		>3.4	type IV (99), no DS data	
600 MHz	0722 0824		0800 0945		>2.5		
Loops		0830	0925			(33)	

1957	July 24	●	McM 4070		Event 17	(X (1) 0)	CFI = 14
Hα	1712 1801	1737 1828	2025	S24 W27	3	(4, 17)	$\Delta t \sim 3^h$
SWF	1727 1759		1750 1920		1 3−	(11)	
9.4 GHz	1730	1736	1748		>2.5		
3.0 GHz	1733 1801		1742 >1925		3.1		
167 MHz	1810	1832	1849		3.0		
DS type IV	1802		1915		3	(25)	

1957	August 9	☉	McM 4082		Event 19	(X 2 0)	CFI = 5
Hα	1330	1355	1442	S33 W77	1		$\Delta t \sim 1.5^h$
SWF	1340		1700		3	(11)	
2.8 GHz	1454	1458	1507		>1.2		
536 MHz	1433	1437	1440		1.8		
81 MHz	1330		>1400		1.5		
DS type III	1312, 1315, 1517–1519, 1547						
Loops	1437		2357			(33)	

1957	August 28	☉	McM 4125		Events 20 and 21	(X (1) 0)	CFI = 14
Hα	<0913	0925 0955	1404	S31 E33	3	(4, 17)	$\Delta t \sim 5^h$?
SWF	0917		1135		3	(11)	$\Delta t \sim 13^h$
9.4 GHz	0915	0950	1245		>2.8	type IV (99),	
210 MHz	0917	0945	1045		2.0	no DS data	

1957	August 28	⊙	McM 4125		Event 21	(X 2 0)	CFI = 10	
Hα	2010	2024	2048		S28 E30	2+		Δt ~ 2^h
SWF	2020		2038			2+	(11)	
2.8 GHz	2018	2019.5	2023			> 2.8		
167 MHz	2021	2023.6	2025			> 3.7		
DS type II	2022		2026			3	(25)	
DS type III	2018		2026					
DS noise storm	2019		2028					

1957	August 31	⊘	McM 4125		Event 23	(X 3 0)	CFI ≥ 10	
Hα	0521	0552	0645		S32 W04	2+		Δt ~ 8^h

Associated events concurrent with the next flare

1957	August 31	⊘	McM 4124		Event 23	(X 3 0)	CFI ≥ 10	
Hα	0544	0551	0626		N13 E02	2		Δt ~ 8^h
SWF	0544		0710			3	(11)	
9.4 GHz	0548	0548	0600			> 2.4	type IV (99),	
200 MHz	0540	0615	0640			2.9	no DS data	

1957	August 31	●	McM 4124		Event 23	(X 3 0)	CFI = 15	
Hα	⌈1257	1312	> 1455		N25 W02	3	(4, 17)	Δt ~ 1^h
	⌊1338	1352	> 1420		N12 W02			
SWF	1303		1607			3+	(11)	
9.4 GHz	1302		1552			> 2.9		
200 MHz	1300	1310	1420			> 3.1		
DS type IV	< 1301		> 1600			3	(25)	
DS type III	many groups							

1957	September 2	⊙	McM 4124		Event 24	(X 3 0)	CFI = 6	
Hα	1257	1303	1346		N10 W26	2		Δt ~ 2^h
SWF	1259		1407			2−	(11)	
9.4 GHz	1255	1302	1336			> 2.5	possibly	
231 MHz	1300	1324	1402			2.8	type IV (99)	
DS type III	1259–1301, 1313–1315, 1346–1348, no type IV (28)							

1957	September 3	⊘ and ⊙	McM 4134		Event 24 and Appendix A	(X 3 0)	CFI = 11
Hα	1412	⌈1426 ⌊1431	1727	N24 W30	3		
SWF	1420		1603		3		(11)
9.4 GHz	1415	1423	1537		> 2.7		⌉type IV characteristics only on microwaves
450 MHz	1424	1428	~ 1434		2.6		
DS continuum	1424		1428⌉ type V ?				
	1455		1456⌋				
DS type III	1424–1425, 1455						
White light	1424		1430				(139)

1957	September 11	⊙	McM 4134		Event 25	(X 1 0)	CFI = 15
Hα	< 0236	0300	0722	N13 W02	3		$\Delta t \sim 24$–30^{h}
SWF	0244		0424		3		(11)
9.5 GHz	0247	0305	0457		2.8		⌉
200 MHz	0300	0308	0526		3.6		⌋type IV (99)
DS continuum	0330		0715				
DS	0300		0310 unclassified activity				

1957	September 12	⊘	McM 4134		Event 25	(X 1 0)	CFI = 11
Hα	1510	1516	1638	N11 W18	2		
SWF	1513		1552		2+		(11)
9.4 GHz	1514	1516	1525		> 3.0		
200 MHz	1515	1528	1645		3.0		
DS type IV	1515		2025		3		(25)
DS type II	1516		1528		3+		(25)
DS type III	1515		1521				

1957	September 18	⊙	McM 4151		Event 26	(X (1) 0)	CFI = 11
Hα	⌈< 1303 ⌊< 1425	1325 1530	> 1600	N23 E10 N20 E04	3	(4, 17)	$\Delta t \sim 7^{\text{h}}$
SWF	1245		1420		3−	(11)	
2.8 GHz	1258	1330	1650		> 1.5		
167 MHz	< 1245	1600	1820		> 3.5		
DS continuum	1315		1521				

1957	September 18	●	McM 4151			Event 26		(X (1) 0) CFI = 13	
Hα		⌈ < 1722 ⌊ < 1815	1740 1840	 2110		N23 E08 N20 E03	3+	(4, 17)	Δt ~ 3ʰ
SWF		⌈ 1730 ⌊ 1823		1813 1920			3+	(11)	
2.8 GHz		1805	1825	2115			> 2.4		
167 MHz		1820	2100	> 2450			3.3		
DS type IV		1810		2440			3	(25)	
DS type III		1741 and 1835							

1957	September 19	⊘ ⊙	McM 4151			Event 26 and Appendix A		(X (1) 0) CFI = 15	
Hα		< 0350	0410	0555		N23 E02	3		
SWF		0359		0453			3	(11)	
9.4 GHz		0359	0406	0419			> 3.1		
200 MHz		< 0408	0510	0640			2.8		
DS		0427		0730 storm with continuum, possibly type IV (47)					

1957	September 21	⊙	McM 4152			Event 27		(X 3 0) CFI > 11	
Hα		1330	1335	1510		N10 W06	3		Δt ~ 2ʰ
SWF		1330		1430			3−	(11)	
9.4 GHz		1330	1336	1437			> 3.0		
200 MHz		1330	1331	1347			> 3.1		
DS type IV		1330		1345			3	(25)	
DS type III		1330		1339 and several groups later on					

1957	September 26	●	McM 4159			Event 28		(X 2 0) CFI = 11	
Hα		1907	1952	2202		N22 E15	3		Δt ~ 2ʰ
SWF		1925		2105			2+	(11)	
2.8 GHz		1928	1939	2028			> 1.8		
167 MHz		1927	2200	> 2435			> 3.6		
DS type IV		1927		27ᵈ0350			3	(25)	
DS type III		1927							

1957	October 20	●	McM 4189			Event 29		(X 3 0) CFI = 14	
Hα		⌈ 1637 ⌊ 1644	1642 1647	 > 1804		S26 W45 S26 W35	3+	(17)	Δt ~ 2ʰ
SWF		1639		1915			3+	(11)	
2.8 GHz		1644	1651	1740			> 3.6		
167 MHz		1646	1700	1815			> 3.5		
DS type IV		1651		2013			3	(25)	
DS type II		1650		1658			3+	(25)	
DS type III		1647–1651, 1701–1702							

1957	November 5	⊘	McM 4207		Event 30	(X 2 0)	CFI ≧ 10
Hα	<1205	⌈1207 ⌊1237	1257	S24 W54	2	
SWF	1207		1221		2+	(11)
2.8 GHz	1205	1207	1213		>2.7	⌉ no DS data
200 MHz	1205		1217		~4.6	⌋

1957	November 24	⊘	McM 4263		Appendix A		CFI ≧ 13
Hα	⌈0848 ⌊1100	0911 1109	1202	S14 E37 S12 E35	3	(17)
SWF	⌈0901 ⌊1107		0933 1123		3− 1	(11)
9.4 GHz	⌈0857 ⌊1105	0903.5 1108	1003 1120		>2.7 >2.5	type IV (99) no DS data
2000 MHz	0850		0955		>4.7	

1957	December 16	⊙	McM 4314		Event 31	(X (1) 0)	CFI ≧ 9
Hα	1125	1140	1238	N17 E50	2	Δt ~ 16^h?
SWF	1129		1202		1+	(11)
9.4 GHz	1150	1154	1206		>2.5	possibly type IV (99),
3.0 GHz	1133	1153	1210		2.6	no DS data
200 MHz	1136		1145		>4.2 indicates strong type III	

1957	December 17	⊘	McM 4314		Event 31	(X (1) 0)	CFI ≧ 8
Hα	<0734	⌈0737 ⌊0907	1004	N20 E41	2	
SWF	0732		0830		2+	(11)
3.0 GHz	<0800		>1045		?	possibly type IV (99),
231 MHz	0745	0754	0759		3.8	no DS data

1957	December 28	●	McM 4321		Event 32	(X (1) 0)	CFI > 10
Hα	2229	2230	2331	N25 W50	2	Δt ~ 0.5^h
SWF	2230		2300		2+	(11)
9.5 GHz	2229	<2233	2249		>3.1	
200 MHz	2230		2238		>3.4	
DS type IV	2232		2255		3	(25)
DS type II	2230		2242		3+	(25)
DS type III	2230, 2233, 2300, 2301					
DS	2230		2231 unclassified activity			

1958	January 25	⊙	McM 4372		Appendix A		CFI ≧ 10
Hα	<0915	1005	1107	S23 W70	3		
SWF	0938		1052		3	(11)	
9.4 GHz	0935	⌈1001 ⎢1004 ⌊1015	1121		2.6	⎤ ⎢ no DS data ⌋	
208 MHz	⌈0934 ⌊1000	0937 1015	0939 1018		2.1 1.5		

1958	January 25	⊙	McM 4382		Appendix A		CFI ≧ 8
Hα	1205		1333	S20 E11	2		
SWF	1208		1236		2	(11)	
9.4 GHz	1204	1206	1306		2.5	⎤ possibly type ⎢ IV (99),	
231 MHz	1210	1212	1214		>4	⌋ no DS data ⌊ indicates a strong type III	

1958	February 9	●	McM 4400		Event 33	(X 2 0) CFI = 10	
Hα	2108	2141	2302	S12 W14	2+		Δt ∼ 8ʰ
SWF	2124		2144		1	(11)	
1.4 GHz	2113	⌈2136 ⌊2152	2210		2.9		
200 MHz	2109		2245		>3.5		
DS type IV	2116		2302		3	(25)	
DS type III	2119, 2120, 2159						

1958	March 3	⊙	McM 4445		Appendix A		CFI = 16
Hα	<1005	1020	1250	S16 E60	3		
SWF	1010		1145		3+	⌐(11)	
9.4 GHz	1010	1016	1131		3.1	⎤ type IV (99), no ⎢ DS data prior to	
231 MHz	1012	1016	1052		3.7	⌋ 1350	
DS type IV	<1350		1645		3	(25)	

1958	March 11	⊙	McM 4449		Event 34	(X (1) 0) CFI = 7	
Hα	<0030	0034	>0042	N11 E12	1		Δt ∼ 3ʰ
SWF	0048		0320		3	(11)	
3.7 GHz	0020	0025	0052		>1.9	⎤ no records	
1.0 GHz	0022	0027	0059		>2.4	⌋ below 1.0 GHz	
DS type II	0032		0046				
DS type III	0029–0036, 0038, 0048						
DS	0028, 0031, 0042–0046			unclassified activity			

1958	March 14	⊙	McM 4445			Event 35	(X (1) 0)	CFI = 8
Hα		1454	1507	> 1541		S21 W85	2+	$\Delta t \sim 0^{\mathrm{h}}$
SWF		1455		1705			3	(11)
SEA		1457	1521	1640			3	(11)
9.4 GHz		1457	1501	> 1501			2.5	⎤ close–to–limb
231 MHz		1458	1502	1513			2.0	⎦ type IV possible
DS		1458			1512	unclassified activity		

1958	March 23	●	McM 4476			Event 37	(X 2 0)	CFI = 14
Hα		0947	1005	1445		S14 E78	3+	$\Delta t \sim 5^{\mathrm{h}}$
SWF		0953		1309			3	(11)
3.0 GHz		0953		1140			> 3.5	⎤ type IV (99),
209 MHz		0957	1006	1200			3.3	⎦ no DS data
White light		1001	1004	1007				(139)
Loops		1030		2000				(33)

1958	June 4	●	McM 4578			Event 41	(X (1) 0)	CFI = 13
Hα		< 2147	2152	2356		N14 W58	2	$\Delta t \sim 1^{\mathrm{h}}$
SWF		2142		2240			2+	(11)
9.5 GHz		2141	2150	2217			> 3.0	
200 MHz		2140	2148	2216			3.1	
DS type IV		2148		2209			3	(25)
DS type III		2144, 2147–2152						(4)
DS		2149				unclassified activity		

1958	June 6	⊙	McM 4578			Event 42	(X (1) 0)	CFI = 11
Hα		0436	⎡ 0448 ⎣ 0505	0614		N16 W78	2	(17)
SWF		0436		0526			2	(11)
9.4 GHz		0436	⎡ 0438.4 ⎣ 0450.6	0500			> 2.9	⎤ probably
200 MHz		0434	0458	0514			2.8	type IV
DS continuum		0434		0521				⎦ (20, 98, 99)
DS type III		0433–0438, 0445, 0456						
DS		0445		0456		unclassified activity		

1958	July 7	●	McM 4634			Event 43	(X 4 0)	CFI = 15
Hα		0020	0110	0414		N25 W08	3+	(17) $\Delta t \sim 1^{\mathrm{h}}$
SWF		0025		0200			3	(11)
9.5 GHz		⎡ 0026.7 ⎣ 0055	0028 0112	0039 0205			3.1 3.3	
200 MHz		0027	0028	0227			3.2	
DS type IV		0052		0221			3	(25)
DS type II		⎡ 0032.5 ⎣ 0101		0048 0117				
DS type III+V		0027		0029.5		many type III groups afterwards		

1958	July 29	●	McM 4659		Event 44	(X 2 0)	CFI = 14
Hα	<0259	0304	0408	S14 W44	3		$\Delta t \sim 1^h$
SWF	0240		0449		3+	(11)	
9.5 GHz	0243.8	0304.2	0344		3.8		
200 MHz	0304	0305	0640		>4.4		
DS type IV	0321		0500				
DS type II	0304		0319				
DS type III	⌈0319 ⌊0334.5		0321 0336 (III+V)				

1958	August 16	●	McM 4686		Event 45	(X 3 0)	CFI = 16
Hα	0433	0440	0831	S14 W50	3+		$\Delta t \sim 1^h 30^m$
SWF	0432		0720		3+	(11)	
9.5 GHz	0434	0440	0544		3.9	⌉ type IV (99),	
200 MHz	0438	0440.5	0535		4.3	⌋ no DS data	

1958	August 19	☉	McM 4708		Event 46	(X 2 0)	CFI = 10
Hα	2118	2256	2411	N18 E26	2		$\Delta t \sim 42^h$
SWF	⌈2130 ⌊2200		2200 2305		1 2	(11) (11)	
9.5 GHz	⌈2159 2204 ⌊2305	2201 2210 2310	2202 2218 2314		2.7 2.9 2.7		
200 MHz	2150		2520		>2.3		
DS type IV	2207		0001		3	(25)	
DS type III	2200–2207, 2238						

1958	August 20	☉	McM 4708		Event 46	(X 2 0)	CFI = 10
Hα	0042	0045	0128	N16 E18	2+		$\Delta t \sim 48^h$
SWF	0042		0115		2+	(11)	
SEA	0045	0050	0145		2+	(11)	
9.5 GHz	0042	0044	0052		3.6		
200 MHz	0042		0048		2.2		
DS type II	0046		0105		3	(4, 25)	
DS type III	0040		0046				

1958	August 22	●	McM 4708		Event 47	(X 3 0)	CFI = 14
Hα	1428	1450	1717	N18 W10	3	(17)	$\Delta t \sim 1^h$
SWF	1425		1725		3+	(11)	
SCNA	1442	1450	1530		2+	(11)	
9.4 GHz	1423	⌈1439 1442 ⌊1451	1715		2.9	⌉ type IV (99), no DS data prior to	
231 MHz	1436	1503	1810		3.3	⌋ 1540	
DS type IV	<1540		>2517		2	(4, 25)	

1958	August 26	●	McM 4708		Event 48	(X 4 0)	CFI = 16
Hα		0005	0027	0124	N20 W54	3	$\Delta t \sim 1^h$
SWF		0010		0410		3+	(11)
SEA		0020	0039	> 0110		3	(11)
9.5 GHz		0018	[0026 0041]	0118		3.8	
200 MHz		0019		0230		4.9	
DS type IV		0020		0430		3	(4, 25)
DS type II		0021		0045			
DS type III		0020		0024 and several groups later on			

1958	September 14	●	McM 4741		Appendix A		CFI ≧ 10
Hα		0822	[0835 0900]	1030	S10 W80	2+	$\Delta t \sim 2^h30^m$
SWF		0851		0949		3	(11)
9.4 GHz		0832	[0835 0858 0904]	1015		3.1	probably type IV (99), no DS data
231 MHz		0852	0904	0927		2.5	

1958	September 22	⊙	McM 4765		Event 49	(X 3 0)	CFI ≧ 7
Hα		0738	0750	> 0910	S19 W42	2	$\Delta t \sim 7^h$
SWF		0755		0830		2	(11)
9.4 GHz		0741	0805	0809		2.5	possibly type IV (99), no DS data
208 MHz		0735	0738	1135 series of bursts		2.3	

1959	January 26	⊙	McM 4969		Event 50	(X (1) 0)	CFI ≧ 9
Hα		0842	0900	1000	N16 W61	3	$\Delta t \sim 5^h$
SEA		0847	0859	0924		2	(11)
SWF		0855		0915		2	(11)
9.4 GHz		0836	0856	next flare		2.8	no DS data
536 MHz		0856	0856	0907		1.9	

1959	January 26	⊙	McM 4969		Event 50	(X (1) 0)	CFI ≧ 7
Hα		1027	1050	1315	N16 W61	3	$\Delta t \sim 3.5^h$
SEA		1052		1112		1+	(11) (no SWF reported)
9.4 GHz		prec. flare	1036	1158		?	no DS data
3.0 GHz		1032		1106		3.0	
536 MHz		1030	1041	1105		2.4	

1959	February 12	☉	McM 5009			Event 51	(X 2 0)	CFI = 12	
Hα		< 2301	2325	2515		N13 E48	3		Δt ~ 9^h
SCNA		2305	2319	2350			1	(11)	
SWF		2308		2348			2	(11)	
9.4 GHz		2257	2313	2357			> 2.5		
200 MHz		2257	2319	2347			≧ 2.7		
DS type IV		2303		> 2409			3	(25)	
DS type III		2302		2400 series of many bursts					

1959	May 10	●	McM 5148			Event 52	(X 4 0)	CFI = 15	
Hα		2102	2140	2610		N18 E47	3+	(17)	Δt ~ 2^h
SCNA		2103	2123	2400			3	(11)	
SWF		2110		11^{d}0630			3+	(11)	
SEA		2115	2135	2330			3	(11)	
2.8 GHz		2100	2149	> 2340			> 3.4		
200 MHz		2120	2148	~ 2335			2.6		
DS type IV		2116		> 2530			3	(25)	
DS type II		2123		2141			3+	(25)	
DS type III		2104		2125 series of many bursts, some isolated bursts afterwards					
Loops		2120		2232				(33)	

1959	May 11	◑	McM 5148			Event 52	(X 4 0)	CFI = 14	
Hα		2006	2030	> 2150		N10 E41	3	(17)	
SCNA		2015	2029	2159			3	(11)	
SWF		2015		2122			3−	(11)	
2.8 GHz		2010	2022	> 2330			> 2.9		
200 MHz		2022	2022	~ 2044			3.2		
DS type IV		2028		2046			3	(25)	
DS type II		2020		2039			3+	(25)	
DS type III		2016		2131 series of many bursts					

1959	May 13	◑	McM 5148			Event 52	(X 4 0)	CFI = 12	
Hα		0509	0515	0553		N22 E26	2+	(17)	
SWF		0511		0547			2	(11)	
9.4 GHz		0510	0513	0515			> 3.0		
200 MHz		0512		0532			> 2.9		
DS type IV		0525		0630					
DS type II		0516		0525					
DS type III		0512–0515 (+ type V), 0520–0522							

1959	June 9	●	McM 5204		Event 53	(X (1) 0) CFI = 12	
Hα	1707		1900	N17 E90	2	(17)	Δt ~ 11ʰ
SWF	1635		1935		3+	(11)	
SCNA	1638	1655	?		2+	(11)	
9.4 GHz	1635	1652	1835		2.9		
2.8 GHz	1635	1652 / 1740	2330		> 3.3 / > 3.2		
200 MHz	1651 / 1735	1702	1722 / 1747		~ 1.8 / 1.5		
DS type IV	1714		1800		2	(25)	
DS type III	1647, 1651–1653, 1706, 1841						
Loops	1732		2346			(33)	

1959	June 12	☉	McM 5204		Event 54	(X 1 0) CFI ≧ 5	
Hα	0735 / 0904	0830 / 0910	1020	N21 E65 / N17 E67	2+	(17)	Δt ~ 23ʰ?
SWF	0755		0918		1+	(11)	
19.0 GHz	0905	0905	0908		2.9		
9.4 GHz	0745 / 0904	0749 / 0906	0805 / 0914		> 1.5 / 2.6	no DS data	
231 MHz	0737 / 0949		0810 / 1230		> 1.8 / 1.7		

1959	July 9	☉	McM 5265		Event 55 and Appendix A	(X 4 0) CFI = 11	
Hα	1930 / 2115 / 2155	1957 / 2130 / 2229	2320	N18 E67 / N19 E48 / N21 E55	2	(17) (Δt ~ 6–8ʰ)	
SEA	1947	1953	> 2115		2	(11)	
SWF	1943 / 2040		2012 / 2104		1+ / 1+	(11) / (11)	
9.4 GHz	2219	2224	2232		> 1.4		
2.8 GHz	2042 / 2112 / 2218	2046 / 2129 / 2227	2102 / 2208 / 2236		> 2.6 / > 2.7 / > 1.4		
200 MHz	2023		2213		3.4		
DS type IV	2044		> 2400		3	(25)	
DS type III	1934		2020 series of many bursts				

1959	July 10	●	McM 5265		Event 55	(X 4 0)	CFI = 12
Hα	0206	0230	> 1000	N20 E60	3+	(17) Δt ~ 2–4ʰ	
SWF	0200		0510		3+	(11)	
9.4 GHz	< 0209	0224	0245		> 4.4	⎤ type IV	
209 MHz	⎡ 0209	0221	0305		> 3.1	⎥ deduced	
	⎣ 0318		0500		1.6	⎦ (99, 102)	
DS type II	0222		0306				
DS type III+V	0210		0212				
Loops	0504		0900			(33)	

1959	July 14	●	McM 5265		Event 56	(X 4 0)	CFI = 16
Hα	< 0325	0349	0901	N17 E04	3+	(17) Δt ~ 1.5ʰ	
SWF	0328		0628		3+	(11)	
9.4 GHz	0330	0349	0435		> 3.8		
200 MHz	0337		1337		4.0		
DS type IV	0401		0610				
DS type II	0338		0412				
Loops	0530		0900			(33)	

1959	July 16	●	McM 5265		Event 57	(X 4 2)	CFI = 15
Hα	2114	2132	2430	N16 W31	3+	(17) Δt ~ 03ʰ	
SEA	2115	2140	2345		3+	(11)	
SWF	2118		2415		3+	(11)	
9.4 GHz	< 2207		2252		> 2.8		
2.8 GHz	2118	2154	> 2420		> 3.8		
200 MHz	2120		2530		3.0		
DS type IV	2121		> 2543		3	(25)	
DS type III	2120		2122				
Loops	2152		2310			(33)	

1959	August 18	●	McM 5323		Event 58	(X 2 0)	CFI = 14
Hα	1014	1030	1350	N12 W33	3	Δt ~ 0.5–1ʰ	
SWF	1025		1225		3	(11)	
19 GHz	1025		1045		> 3.3	⎤ type IV (99),	
9.4 GHz	1022	1030	1214		3.0	⎥ no DS data	
200 MHz	1025	1033	1225		3.7	⎦	

1959	August 18	☉	McM 5329		Event 59	(X (1) 0)	CFI = 8
Hα	1654	1725	1822	N05 E16	2+		Δt ~ 16ʰ
SWF	1620		1655		2	(11)	
2.8 MHz	⌈1619	1727	1919		> 1.3	⌉no other	
	⌊1655	1657	1705		> 1.2	⌋bursts	
D.S. type IV	⌈1353		2243		3	(25)	
	⌊1717		1810 a phase corresponding			(4)	
			to this flare				

DS type III all throughout the day

1959	September 1	☉	McM 5355		Event 60	(X (1) 0)	CFI ≧ 8
Hα	1923	1938	2216	N12 E60	2+		Δt ~ 8.5ʰ
SEA	1928	⌈1940 ⌊2022	2100		2+	(11)	
SWF	1945		2058		2	(11)	
2.8 GHz	1928	⌈2009 ⌊2023	2300		> 1.7		
545 MHz	1927.5		2000		> 2.5		
DS type IV	1914		1950		3	(25)	
DS type II	1939		1945		3	(25)	
DS type III	1923, 1935, 1946, 1952–2006						

1959	November 30	☉	McM 5476		Appendix A		CFI =14
Hα	0247		0356	N08 E16	2+		
SWF	0249		0320		3−	(11)	
9.4 GHz	0247	0250	0259		> 3.6		
200 MHz	0247	0254	0303		3.8		
DS type IV	0312		0350				
DS type II	0251		0328				
DS type III	0241, 0248, 0252						

1959	November 30	⊘	McM 5476		Appendix A		CFI = 11
Hα	1720	1744	1906	N08 E06	3		
SEA	1726	1745	1915		3−	(11)	
SWF	1735		1822		3−	(11)	
2.8 GHz	1738		1756		> 2.2		
167 MHz	1740	⌈1740 ⌊1833	1923		> 3.0		
DS type IV	1739		> 2330		3	(25)	
DS type II	1741		1810		3+	(25)	
DS type III	1738		1746				

1959	December 21	●	McM 5494			Appendix A		CFI = 12
Hα		0043	0052	> 0350		S04 W53	2	
SWF		not reported						
9.5 GHz		0049	0050	0055			2.9	
208 MHz		0043	0049	0138			2.1	
DS type IV		0120		0350				
DS type II		0055		0102				
DS type III(+V)		0043		0051				

1960	January 11	●	McM 5527			Event 61	((0) (2) 0)	CFI = 13
Hα		< 2040	2126	> 2355		N22 E02	3	Δt ~ 1ʰ?
SEA		2050	2115	?			1	(11)
SWF		2100		2124			2−	(11)
2.8 GHz		2056	2108	> 2131			2.3	
167 MHz		2056		> 2343			> 3.0	
DS type IV		2105		> 2355			3	(25)
DS type II		2103		2118			3	(25)

1960	January 15	☉	McM 5525			Appendix A		CFI ≧ 7
Hα		1336		> 1455		S20 W68	2	Δt ~ 13ʰ
SWF		1340		1425			1+	(11)
9.4 GHz		1335	1357	1502			2.3	⎤ type IV (99),
200 MHz		1347		1412			2.7	⎦ no DS data

1960	March 10	☉	McM 5592			Appendix A		CFI = 8
Hα		1716	1719	1810		N24 E07	1	Δt ~ 1ʰ
SWF		1719		1740			2−	(11)
SEA		1719	1725	1755			2	(11)
2.8 GHz		1717	1718.5	1724			2.5	
200 MHz		1718		1723			> 2.8	
DS type II		1720		1726			3	(25)
DS type III		1717		1719 (many type III throughout day)				

1960	March 28	● and ⊘	McM 5615			Event 64 and Appendix A	((−1) 2 0)	CFI = 11
Hα		2042	2056	> 2150		N14 E37	2	Δt ~ 10ʰ
SEA		2045	2050	2140			2	(11)
SWF		2050		2140			2+	(11)
SCNA		2048	2102	−			3	(11)
2.8 GHz		2048		2158			> 2.9	
167 MHz		2051	2130	> 2500			> 3.0	
DS type IV		2050		> 2447			3	(25)
DS type II		2057		2112			3	(25)

1960	March 29	●	McM 5615		Event 64	((−1) 2 0)	CFI = 15
Hα	0650	0710	1220	N12 E30	2+		$\Delta t \sim 0-1^h$
SWF	0652		0853		3+		(11)
9.5 GHz	0656.5	0733.5	0758		> 3.8		⎤ strong type
1.0 GHz	0656		0856		5.4		⎟ IV (99),
200 MHz	0700		0920		4.6		⎦ no DS data

1960	March 30	●	McM 5615		Event 65	(X (1) 0)	CFI ≧ 9
Hα	0216	0219	0240	N09 E15	1+		$\Delta t \sim 7^h$
SWF	0220		0249		1		(11)
9.4 GHz	0217	0219.5	0222		2.0		
545 MHz	0214		0223		2.1		
DS type IV	0325		> 0740				
Possible type II	0325		0337				
DS type III	0219–0220(+ type V), 0232–0233						

1960	March 30	●	McM 5615		Event 66	(X 3 0)	CFI = 14
Hα	⎡ 1455			N12 E11	2		(17) $\Delta t \sim 5^h$
	⎣ 1520	1540	2034				
SWF	1520		1800		3		(11)
SCNA	1522	1537	—		3		(11)
9.1 GHz	1521	1527	1533		3.3		
2.8 GHz	1518	1556	1900		3.2		
167 MHz	1529	1550	1930		> 3.0		
DS type IV	1526		> 2300		3		(25)
DS type II	1529		1540		3		(25)
DS type III	1553		1557				
Loops	1511		2040				(33)

1960	April 1	●	McM 5615		Event 67	(1 2 0)	CFI = 15
Hα	0843	0859	1320	N12 W11	3		(17) $\Delta t \sim 1^h$
SWF	0850	0947	0947		3		(11)
9.1 GHz	0846.5	0858	1007		3.9?		⎤ type IV (sp.
							⎟ diagram in
							⎟ (122)), no DS
200 MHz	0848		1048		3.7		⎦ data
Loops	0977						(33)

1960	April 5	●	McM 5615		Event 68	(1 2 0)	CFI = 13
Hα	≪ 0215	0245	> 0530		N12 W63	≧ 2	Δt ~ 2–4^h
SWF	0140		0417			3+	(11)
9.4 GHz	0143	⌈ 0201 / ⌊ 0226	0253			4.1	
200 MHz	0124		0224			> 2.1	
DS type IV	0207		0300				
Possible type II	0152		0207				
DS type III	0137–0221 series, 0246, 0247						
Loops	1304		1617				(33)

1960	April 28	●	McM 5645		Event 69	((2) 2 0)	CFI = 11
Hα	< 0130	0137	> 0145		S05 E34	3	Δt ~ 0.5–1.0^h
SWF	0120		0300			3+	(11)
9.5 GHz	0124.5	⌈ 0124.5 / ⌊ 0130	0140			2.8	
545 MHz	0135		0140			2.2	
DS type IV	0145		0230			weak	
DS type II	0120		0146			3	(4, 25)
DS type III	0117		0119				

1960	April 29	●	McM 5642		Event 70	((1) 3 0)	CFI = 11
Hα	< 0107	⌈ 0210 / ⌊ 0400	> 0908		N14 W21	2+	(17)
							Δt ~ 0–2^h
SWF	⌈ 0205 / ⌊ 0355		0355 / 0500			2+ / 2+	(11) / (11)
9.4 GHz	⌈ 0140 / 0203 / 0357 / ⌊ 0527	0140.8 / 0247 / 0414.7 / 0532	0144 / 0308 / 0450 / 0551			1.7 / 1.4 / 2.3 / 1.6	
545 MHz	⌈ 0140 / 0358 / ⌊ 0525		0256 / 0503 / 0631			4.5 / 2.4 / ≫ 2.6	
200 MHz	0346		0606			> 2.3	
DS type IV	⌈ 0200 / ⌊ 0350		0305 / > 0645				
DS type II	0214		0225				
Possible type II	0417		0425				
DS type III	0140–0148, 0357–0423 series						

1960	May 4	●	McM 5642			Event 71	((1) 2 4)	CFI $\geqslant$ 9
Hα		1000	1016	> 1200		N13W90	3	(17) $\Delta t \lesssim 0.5^{\text{h}}$
SWF		1015		1050			3	(11)
SEA		1016	1029	1125			1	(11)
9.4 GHz		1013.4	1033.0	1233			2.6	type IV, (sp. diagram in (122)), no DS
536 MHz		1010	1046	1101			2.1	data
Loops		1035		1314			0.8	(33)

1960	May 6	●	McM 5653			Event 72	((1) 3 0)	CFI = 15
Hα		1404	1440	2020		S08 E07	3+	$\Delta t \sim 2$–4^{h}
SWF		1427		1658			3	(11)
SEA		1430	1453	> 1630			2	(11)
2.8 GHz		1406.5	1434.5	1537			2.8	
545 MHz		1414		1830			2.4	
200 MHz		1414 onset storm						
DS type IV		1414		> 1610			3	(25)
DS type II		1438		1445			3	(25)
Loops		1540		1700				(33)

1960	May 9	⊙	McM 5657			Event 73	(X (1) 0)	CFI $\geqq$ 6
Hα		< 0704	0734	> 1021		S11 E52	3	$\Delta t \sim 1$–4^{h}
SWF		0700		0838			2	(11)
3.8 GHz		0710		> 0810			> 1.6	no DS data
536 MHz		0644	0657.5	0718			2.0	

1960	May 12	⊘ and ⊙	McM 5654			Event 74 and Appendix A	((1) 2 0)	CFI = 11
Hα		< 1342	1400	> 1546		N30 W59	1+	(17)
SEA		1345	1404	1531			2	(11)
SWF		1348		1722			3	(11)
9.1 GHz		1340		1440			2.5	
200 MHz		1345		1430			2.1	
DS		no event						

1960	May 13	●	McM 5654		Event 74	((1) 2 0)	CFI = 15
Hα	0519	0532	0733	N30 W67	3		$\Delta t \sim 1^h$
SWF	0512		0853		3+		(11)
9.4 GHz	0517	0532	0719		4.3		spectral dia-
							gram in (102)
200 MHz	0517	0553	0730		2.7		and (122)
DS type IV	0530		>0609				
DS type II	0523		0538				
DS type III	0517		0525				
White light	~0530						(4) Ondřejov
Loops	0546		2135				(33)

1960	May 17		?		Event 75	((0) (1) 0)	CFI = 4
Hα	1726		1743	S09 E33			(Sacramento
	(no flare, only surge activity in McM 5663;						Peak α Climax)
	no other events)						
SID	no event						(11)
600 MHz	1742.3		1745.2		1.6		
200 MHz	1743		1744		1.9		
DS type II	1743		1810		3		(25)
DS type IV	1755		1829		weak (no given in (25))		

1960	May 26	●	McM 5669		Event 76	((0) (1) 0)	CFI > 10
Hα	0850	0928	1050	N14 W15	2+		(17)
							$\Delta t \sim 1–3^h$
SEA	0908	0934	1024		?		(11)
SWF	0914		1000		2		(11)
9.1 GHz	0909		0934		>3.0		type IV (sp.
							diagram in
							(122)), no DS
200 MHz	0909		0949		3.0		data

1960	June 1	●	McM 5680		Event 77	((1) (1) 0)	CFI ≧ 12
Hα	0823	0900	1340	N29 E46	3+		(17) $\Delta t \sim 2^h$
SWF	0837		0952		3		(11)
9.1 GHz	0834		0937		>3.5		type IV (sp.
							diagram in
							(122)), no DS
200 MHz	0838.5		0943		3.5		data

1960	June 25	⊙	McM 5713			Event 78	(X (1) 0)	CFI = 13
Hα		1136	1215	1530	N21 E06	3	(17) $\Delta t \sim 5^h$
SEA		1200	1213	1300		1	(11)
SWF		1203		1310		2	(11)
9.1 GHz		1159		1229		2.5	
200 MHz		1200		1410		3.5	
DS type IV		< 1215		1500		3	(25)

1960	June 25	⊙	McM 5713			Event 78	(X (1) 0)	CFI = 8
Hα		1659	1707	1740	N19 W01	1	$\Delta t \sim 0^h$
SWF		1705		1730		1+	(11)
SEA		1705	1718	1735		2	(11)
2.8 GHz		1701	1705	1716		2.2	
545 MHz		1703		1843		2.7	
DS type IV		1717		1923		3	(25)
DS type III		1700		1713			
Spray associated							(41)

1960	June 25	⊘	McM 5713			Event 78	(X (1) 0)	CFI = 11
Hα		2039	2046	2140	N18 W04	2+	
SWF		2040		2110		2−	(11)
SEA		2047	2050	2124		2+	(11)
2.8 GHz		2037	2046	2117		2.8	
545 MHz		2040		2125		$\geqslant 2.4$	
DS type IV		2045		2153		3	(25)
DS type II		2048		2105		3	(25)
DS type III		2030		2046			
Spray associated							(41)

1960	June 26	⊙	McM 5719			Event 79	(X (1) 0)	CFI = 12
Hα		2358	2415	2600	S08 E34	3	$\Delta t \sim 23^h$
SWF		2403		2510		2−	(11)
9.4 GHz		2405	2412.4	2445		1.7	
200 MHz		2402		2452		> 2.4	
DS type IV		2413		2449		3	(25)
DS type II		2404	2409				
DS type III all the time (type IV with type III structure)							

1960	June 27	⊙	McM 5713		Event 79	(X (1) 0)	CFI = 9
Hα	0418	0430	0615	N20 W19	1+		$\Delta t \sim 19^{\text{h}}$
SWF	0417		0453		1+	(11)	
3.8 GHz	?	[0421.5 / 0424.5]	0430		>2.6		
200 MHz	0421.5	0502	0615		2.3		
DS type IV	0425		0539				
DS type II	0422–0443, 0453–0454.5, 0503–0505						
DS type III	0420–0421, 0443, 0450–0452, 0500–0503 (+ type V), 0508–0509						

1960	June 27	⊙	McM 5713		Event 79	(X (1) 0)	CFI = 12
Hα	2140	2156	2345	N22 W27	3		$\Delta t \sim 1^{\text{h}}$
SWF	2140		2358		2+	(11)	
SEA	2144	2158	2245		2+	(11)	
9.4 GHz	?	2154	>2239		2.0		
200 MHz	2144		2259		~2.4		
DS type IV	2150		2255		3	(25)	
Like type II	2159		2212				

1960	June 29	⊘	McM 5713		Appendix A		CFI = 12
Hα	0125	0148	0247	N20 W50	1		
SCNA	0135	0150	0300		2	(11)	
SWF	0138		0346		2	(11)	
3.8 GHz	0135	0148	0156		2.9		
200 MHz	0138		0155		>2.4		
DS type IV	0140		0230			(4, 25)	
DS type II	0149.5		0158				
DS type III	0122.5, 0133.5, 0135–0142						

1960	August 11	⊙	McM 5794		Event 80	((0) (1) 0)	CFI = 12
Hα	0223	0255	0356	N21 E35	2		$\Delta t \sim 21^{\text{h}}$
SWF	0225		0355		2	(11)	
9.4 GHz	0222	0253.2	0302		2.4		
200 MHz	0249.5	0251.2	0254		>3.2		
DS type IV	0307		>0615				
DS type II	0257.5		0314				
DS type III+V+U	0248		0257				

1960	August 11	☉	McM 5794		Event 80	((0) (1) 0)	CFI = 12	
Hα	1916	1929	2055	N22 E26	2+			$\Delta t \sim 5^{\mathrm{h}}$
SWF	1925		2030		2	(11)		
2.8 GHz	1923.5	1928	2000		3.0			
200 MHz	1926		1938		≥ 3.0			
DS type IV	1926		2019		2	(25)		
DS type II	1929		1938		3+	(25)		
DS type III	1926		1930					

1960	August 14	☉	McM 5794		Event 81	((0) (1) 0)	CFI > 11	
Hα	0511	0525	0655	N22 W06	2+			$\Delta t \sim 8^{\mathrm{h}}$
SEA	0506		0554		2	(11)		
SWF	0515		0600		3	(11)		
9.4 GHz	0515	0518.3	0535		3.2			possibly type IV (102), no
200 MHz	0517		0532		> 3.3			DS data

1960	September 3	●	McM 5837		Event 83	(2 2 1)	CFI = 14	
Hα	0037	0108	> 0154	N18 E88	2+			$\Delta t \sim 1^{\mathrm{h}}$
SWF	0045		0251		3+	(11)		
SCNA	0103	0109	> 0230		3	(11)		
9.4 GHz	0039	0108.0	0154		4.2			
200 MHz	0103		0136		> 3.0			
DS type IV	0038		> 0054		2	(25)		
DS type III	0103		0105					
Loops	0215		1200			(33)		

1960	September 26	●	McM 5858		Event 84	((1) (2) 0)	CFI = 12	
Hα	0525	0537	> 0616	S22 W64	1+	(17)		$\Delta t \sim 2^{\mathrm{h}}$
SWF	0520		0721		3+	(11)		
9.4 GHz	0530	0538.6	0600		3.3			
545 MHz	0531.5		0542		1.3			
DS type IV	< 0554		> 0611					
DS type II	0543		0604					
DS type III	0526, 0537–0541							

1960	October 11	●	McM 5880		Appendix A		CFI = 12	
Hα	0517	0535	0755	S17 W36	2			$\Delta t \sim 1^{\mathrm{h}}$
SEA	0524	0536	0651		2			
SWF	0525		0628		3	(11)		
9.4 GHz	0524	0529	0559		3.4			
200 MHz	0527	0528	0549		2.7			
DS type IV	0532		> 0613					
DS type II	0530		0547					
DS type III	0516, 0519–0530, 0538							

1960	October 29	●	McM 5909			Event 86	(X (1) 0)	CFI > 11
Hα	1026	1030	> 1252		N22 E27	3		(17) Δt ~ 2^h
SWF	1029		1149			3		
9.4 GHz	1025	~ 1041	> 1325			2.9		type IV (99),
200 MHz	1029		1125			3.3		no DS data

1960	November 10	●	McM 5925			Event 87	(X (1) 0)	CFI > 11
Hα	1009	1023	> 1635		N28 E28	3		(17) Δt ~ 8^h
SWF	1022		1152			2		(11)
9.1 GHz	1012	1019	1026			2.8		
	1119		1152			> 3.2		type IV (sp.
								diagrams in
								(102) and
200 MHz	1020		1116			2.7		(122)), no DS
	1116		1200			4.4		data

1960	November 11	●	McM 5925			Event 88	(X (1) 0)	CFI = 15
Hα	0305	0340	> 0428		N28 E12	2+		Δt ~ 1^h
SWF	0311		0616			3+		(11)
SEA	0315	0320	0414			1+		(11)
9.5 GHz	0315	0333	0500			3.8		
200 MHz	0321		~ 0730			4.0		
DS type IV	0330		> 0709 (Sp. diagram in (122) and (137).)					
Possible type II	0330		0345					
DS type II	0349		0357					
DS type III	0318–0339, 0357–0358							

1960	November 12	●	McM 5925			Event 89	(4 4 4)	CFI = 15
Hα	1315	1330	1922		N27 W04	3+		Δt ≤ 15^m
SEA	1325	1345	1530			2+		(11)
SWF	1326		1600			3+		(11)
9.1 GHz	1322	1332	1500			> 3.9		
200 MHz	1327.5		1800			> 3.3		
DS type IV	< 1345		> 1800			3		(25) (sp. dia-
								gram in
								(122))

DS type III structure in the type IV, no individual type III

| Loops | 1405 | | 1940 | | | | | (33) |

1960	November 14	●	McM 5925			Event 90	(X (1) 0)	CFI = 13
Hα		0246	0304	0520		N27 W20	2+	$\Delta t \sim 19^{\text{h}}$
SWF		0300		0500			3	(11)
SEA		0301	0327	> 0452			2	(11)
9.4 GHz		0258	0350.5	0520			3.9	
200 MHz		0319		0540			> 2.8	
DS type IV		0305		0500				
DS type II		0009		0011			weak	
DS type İII		0240–0241, 0302, 0335–0344						

1960	November 15	●	McM 5925			Event 91	(4 4 3)	CFI = 16
Hα		0207	0221	0427		N25 W35	3	$\Delta t \sim 15^{\text{m}}$
SWF		0217		0630			3+	(11)
SCNA		0218	0223	0315			3	(11)
SEA		0219	0245	0403			2	(11)
9.4 GHz		0218	0228.4	0343			4.4	
200 MHz		0221		0619			> 3.4	
DS type IV		0221		> 0608 (sp. diagram in (122) and (137))				
DS type II		0221		0248				
Loops		0223		0300				(33)

1960	November 19	⊘	McM 5925			Event 92	(X (1) 0)	CFI = 6
Hα		1522	1527			N27 W90	2	(17) $\Delta t \sim$?
		1543	1556	1649				
SID		no event						(11)
2.8 GHz		1554	1555	1556			0.6	
108 MHz		1557.5		1558.5			> 1.4	
DS type IV		1636		1723 with type III structure	3			(25)
DS type III		1559–1602, 1659–1701						

1960	November 20	●	McM 5925			Event 93	((3) 3 2)	CFI = 12
Hα		⌈ 2017	2020			N28 W90	2	(17) $\Delta t \sim 0.5^{\text{h}}$
		⌊ 2126	2135	2258				
SWF		2023		2145			3–	(11)
SEA		2023	2041	2140			2+	(11)
2.8 GHz		2023	2026.5	> 2110			2.6	
200 MHz		2028		2033			2.3	
DS type IV		2027		2046			2	(25)
DS type II		2028		2035			3	(25)
Unclassified		2041		2043				
Spray		2022						(41)
Loops		2114		2257				(33)

1960	December 5	●	McM 5929		Event 94	(X (1) 0)	CFI = 14
Hα	1825	1838	2350	N26 E74		3+	$\Delta t \sim 11^{\text{h}}$
SWF	1830		2010			3	(11)
SEA	1835	1850	1945			2+	(11)
2.8 GHz	1828	1837.5	1855			2.5	
200 MHz	1835		1843			>3.0	
DS type IV	1834		1858 with type III structure		3	(25)
DS type II	1834		1850			3	(25)

1961	July 11	●	McM 6171		Event 96	(X 1 0)	CFI = 14
Hα	1615	1700	2040	S07 E32		3	(17) $\Delta t \sim 4^{\text{h}}$
SWF	1648		2053			3+	(11)
SCNA	1650	1704	1750			3	(11)
SEA	1653	1711	1838			2	(11)
9.1 GHz	1652	1704.5	>1730			3.1	
200 MHz	1657		1847			>2.9	
DS type IV	1655		1845			3	(25)
DS type II	1702		1718			3+	
Loops	1615		2000				(33)

1961	July 12	●	McM 6171		Event 97	((2) 4 0)	CFI =16
Hα	1000	1025	>1300	S07 E23		3	(17) $\Delta t \sim 2^{\text{h}}$
SWF	1023		1200			3	(11)
SEA	1024	1038	1100			2	(11)
9.1 GHz	1018	1029	1145			3.8	type IV (sp. diagram in (122)), no DS
200 MHz	1022		1142			4.3	data
Loops	1000		1230				(33)

1961	July 15	⊘	McM 6172		Event 98	(X (2) 0)	CFI = 11
Hα	1433	1558	1929	N13 E15		3	$\Delta t \sim 1^{\text{h}}$
SEA	1435		1448			1	(11)
9.1 GHz	1430		1445			1.1	
200 MHz	1435.5		1440			2.4	
DS type IV	see next flare						
DS type III	1433		1441				

1961	July 15	●	McM 6171			Event 98	(X (2) 0)	CFI = 12
Hα		1508	1512	1549		S07 W20	2	(4) $\Delta t \sim 1^{h}$
SCNA		1512	1530	1517			1	(11)
SWF		1512		1705			3	(11)
9.4 GHz		1510	1512.4	1527			2.4	
108 MHz		1505		1845			> 2.5	
DS type IV		1533		> 1623			3	(25)

1961	July 17	⊙	McM 6171			Appendix A		CFI
Hα		0710		0926		S07 W45	2+	
SWF		0731		0800			1	(11)
SEA		0732	0742	0814			1	(11)
9.4 GHz		0710	0758	1055			1.8	⎤
3.8 MHz		0718	0759	> 0843			2.1	no DS data
600 MHz		0729		0805			1.7	⎦

1961	July 18	●	McM 6171			Event 99	(3 3 3)	CFI > 9
Hα		0920	1005	1250		S07 W59	3+	(17) $\Delta t \lesssim 0.5^{h}$
SEA		0943	1030	1051			2+	(11)
SWF		1000		1153			3	(11)
9.1 GHz		0939		1050			⩾ 3.4	type IV (sp. diagram in (122)), no DS
200 MHz		0944		1044			3.0	data
Loops		0921		1313				(33)

1961	July 20	●	McM 6171			Event 100	((1) 2 2)	CFI = 15
Hα		⎡1553 ⎣1823	1847	1942		S06 W90	3	(17) $\Delta t \sim 20^{m}$
SWF		1550		2200			3+	(11)
9.1 GHz		1552	1553.6	1637			~ 3.6	
200 MHz		1554		1613			3.6	
DS type IV		1552		1804			3	(25)
DS type II		1554–1556, 1557–1619					3	(25)
Loops		1605		2330				(33)
Spray		1553						(41)

1961	July 24	●	McM 6178		Event 101	((0) (1) 0)	CFI = 10
Hα	⎡ 0410			N15 E15	2+	(17) $\Delta t \sim 5^h$
	⎣ 0449	0504	0612			
SWF	0455		0620		2+	(11)
9.4 GHz	0430	0510	0610		1.5	
200 MHz	0437		0850		> 2.6	
DS type IV	0507		0632			
DS type II	0454.5		0514.5			
DS type III	0429.5–0433.5, 0436–0449 several groups, 0508–0511, 0524–>0635 several groups					

1961	July 28	●	McM 6178		Appendix A		CFI = 11
Hα	< 0244	0248	> 0431	N12 W38	2+	$\Delta t \sim 0.5^h$
SWF	0227		0331		2+	(11)
SEA	0232	0247	0316		1+	(11)
9.4 GHz	0230	0236	0300		2.3	
3.8 GHz	0226	0235	0321		2.6	
200 MHz	0231		0327		> 2.5	
DS type IV	0303		0344			
DS type II	0233		0258			
DS type III	0230.5					

1961	September 10	●	McM 6212		Event 103	((2) 2 0)	CFI = 9
Hα	⎡ 1958	2010		N08 W80	1	(17) $\Delta t \sim 0.5^h$
	⎣ < 2018	2030	2054	N15 W90		
SEA	1940	2003	> 2125		2	(11)
SWF	1942		2123		3	(11)
2.8 GHz	1930	2001	2031		2.9	
108 MHz	1934	1939	2014		> 1.5	
DS type IV	1937		2017		3	(25)
DS type II	1947		2014		3	(25)

1961	September 16	☉	McM 6227		Appendix A		CFI ≧ 9
Hα	1057	1110	1159	N18 E77	2+	$\Delta t \sim 37^h$
SWF	1102		1152		2	(11)
9.1 GHz	1102	1104	1127		2.8	⎤ type IV (102),
200 MHz	1103		1111		~ 3.7		no DS data
111 MHz	1103	1104	1118		4.0	⎦

1961	September 28	●	McM 6235		Event 104	(2 2 0)	CFI = 12
Hα	2202	2223	2530	N13 E29	3	$\Delta t = 23^m \pm 05^m$
Balloon X-rays > 20 keV ⎤	2216	2217	2221			(46)
SEA	2216	2224	2258		2	(11)
SWF	2218		2320		2	(11)
9.4 GHz	2213	2217.3	2253		3.2	
200 MHz	2213		~ 2340		> 2.9	
DS type IV	2212		2249		3	(25)
DS type II	2217		2231		3+	(25)
DS type III	2341		2343			(4)

1961	November 10	·●	McM 6264		Event 106	(1 1 0)	CFI ≧ 8
Hα	1434	1444	1450	N19 W90	1+	$\Delta t \sim 30^m$
SEA	1435	1456	?		2	(11)
SWF	1436		1534		2+	(11)
9.1 GHz	1432	1439.8	1452		2.1	
108 MHz	1432	⎡1435 ⎣1441	1501		> 2.5	
DS type II	⎡1433 ⎣1439		1437 1502		3 3	(25) (25)
DS type III	1432		1437		strong	
Loops	1450		1900			(33)

1962	October 23	☉	McM 6581		Event 110	(0 1 0)	CFI = 7
Hα	1642	1703	1745	N03 W70	2	$\Delta t \sim 0.5^h$
SID	no event					(11)
2.8 GHz	1642	1658	> 2112		1.2	
222 MHz	1645	1656	1833		1.8	
DS type II	1649		1650			
DS type III	1649–1653, 1734–1737					
DS	1656		1706 unclassified, similar to types II and IV			

1963	April 15	☉	McM 6766		Event 112	(1 1 0)	CFI > 6
Hα	1034	1125	1230	S11 W06	2	$\Delta t \sim 1^h$
SWF	1124		1140		2	(11)
SCNA	1124	1126	1144		2+	(11)
9.4 GHz	⎡1045.6 ⎣1122.5	1046.0 1124.7	1057 > 1222		0.6 2.4	⎤
600 MHz	1124		~ 1145		2.0	no DS data
23 MHz	1144.5	1145	1203		3.9	⎦

1963	May 1	⊘	McM 6790			Event 113	(−1 (1) 0)	CFI = 11
Hα		0525	0608	0835		N15 E46	2	
SWF		0530		0609			2+	(11)
9.4 GHz		0532	0537.6	0600			3.2	]type IV
200 MHz		0534	0535	0609			2.4	]deduced (102)
DS type II		0536		0558				
DS type III		0535–0537, 0548–0600						

1963	June 14	⊙	McM 6832			Event 114	((−1) (0) 0)	CFI = 5
Hα		[<0225	0228			N09 W34	1	Δt ~ 3.5^h
		[0247		0330				
SWF		0223		0248			1+	(11)
9.4 GHz		0221	0226.6	0231			1.7	
1000 MHz		0219	0227.6	0231			1.4	
DS continuum		0254		0607				
DS type II		0234.5		0252.5				
DS type III		0216–0224, 0245–0301, and other series later on						

1963	August 6	⊙	McM 6909			Event 115	((0) (1) 0)	CFI ≧ 6
Hα		0855	0910	1027		N13 W12	2	Δt ~ 2^h
SWF		0859		0936			1+	(11)
9.4 GHz		0859	0916	0950			1.0	]possibly type
200 MHz		[0858		0901			>1.8	]IV (102), no
		[0910	~0930	~1210		storm	2.1	]DS data
23 MHz		0903.5	0905	0918			4.3	

1963	August 9	●	McM 6908			Event 116	(0 (1) 0)	CFI = 8
Hα		2234	2245	2340		N07 W80	1	Δt ~ 0.5^h
SWF		2234		2340			1+	(11)
9.4 GHz		2239	2246.2	2309			2.1	
2.8 GHz		2234	2246.3	2304			2.5	
208 MHz		2233	2243	2248			1.7	
DS type IV		2237		2335			2	(11)
DS type III		2237.5		2238.3				(11)
DS continuum		2335		2443			1	(11)

1963	September 15	●	McM 6964		Event 117	(X 1 0)	CFI = 11
Hα	0015	0042	0219	N15 E75	2	$\Delta t \sim 9^{\mathrm{h}}$
SWF	0015		0315		3+	(11)
SEA	0017	0024	0235		3	(11)
9.4 GHz	0015	0047.5	0145		4.2	⎤type IV
200 MHz	0025		0145		> 2.9	⎦deduced (102)
DS type II	0027		0049		2	(11)
DS type III	0024		0025			(11)

1963	September 16	●	McM 6964		Event 118	(X 1 0)	CFI ≧ 9
Hα	1430	1505	1532	N12 E48	2	$\Delta t \sim 1^{\mathrm{h}}$
SWF	1440	1600	1645		3	(11)
9.1 GHz	1438	1510	~1628		2.9	
550 MHz	1435	1507	1605		2.5	
DS continuum	< 1352		1640 (weaker later on)		2	(11)
DS type III+V	1510		1512		very strong	

Type IV deduced (102), sp. diagram in (122)

1963	September 20	●	McM 6964		Event 119	(1 2 0)	CFI = 15
Hα	⎡2314			N10 W09	2	(17)
	⎣2351	2403	2601			$\Delta t \sim 0-3^{\mathrm{h}}$
SWF	2351		2725		3	(11)
SEA	2352	2404	2452		2+	(11)
9.4 GHz	⎡2347	2358.8	2404		3.6	
	⎢2429	2432.8	2435		2.4	
	⎣2502	2516	2547		3.5	
200 MHz	2350		2640		> 4.0	
DS type IV	2410		> 2455		3	(11)
DS type II	2400		2426		3	(4, 11)
DS type III	2321–2323, 2342–2344					
DS type III+V	2346		2357		very strong	

1963	September 26	●	McM 6964		Event 120	((1) 2 0)	CFI ≧ 11
Hα	< 0638	0717	0944	N13 W78	3	$\Delta t \sim 15^{\mathrm{m}}$
SWF	0709	0730	0851		3+	(11)
9.4 GHz	0705	0714	1000		3.6	⎤type IV (sp. diagram in (122)), no DS
200 MHz	0705	0710	0745		2.4	⎦data

1963	October 28	●	McM 7003			Event 121	((−1) (1) 0) CFI ≧ 8	
Hα		<0135	0158	0335		N12 W24	3	$\Delta t \sim 6^h$
SEA		0138	0150	0348			3+	(11)
SWF		0140	0150	0400			3	(11)
9.4 GHz		0140	0156.7	0220			2.2	
200 MHz		0142		0322			? (500 MHz > 3.3)	
Type IV (102), no DS data								

1964	March 16	●	McM 7182			Event 122	(0 1 0)	CFI = 11
Hα		1553	1608	1700		N05 W73	1+	$\Delta t \sim 0.5^h$
SWF		1556	1625	1730			3	(11)
2.8 GHz		1553	1614	1652			2.8	
200 MHz		1557		1558			> 2.6	
		1602	1607	1639			2.1	
DS type IV		1605		1608				
DS type II		1558		1608			very strong	
DS type III		1604		1616				

1965	February 2	☉	McM 7661			Appendix A		CFI
Hα		2043	2053	2109		N07 E09	1−	(11)
SID		no event						(11)
200 MHz		2051	2052.5	2053			1.9	
DS type III		2051		2053				

1965	February 5	●	McM 7661			Event 123	(1 2 0)	CFI = 9
Hα		1750	1810	2024		N08 W25	2	$\Delta t \cong 0.5^h$
SPA		1757	1812	1820		no other SID		(11)
2.8 GHz		1753	1826	1930			1.6	
200 MHz		1753		1943			≫ 2.4	
DS type IV		1800		1940				
DS type III		1800		1803				
DS		1800		1811 unclassified, with type II characteristics				

1965	May 25	●	McM 7809			Appendix A		CFI = 8
Hα		2239	2243	2253		N19 W73	1	
SWF		2242		2252			1−	(11)
SPA		2242	2247	2310				(11)
3.8 GHz		2242	2242.7	2243.5			1.0	
200 MHz		2241	2243	2245			2.1	
DS type IV		2241		2252			1+	(11)
DS type II		2246.5		2250			1	(11)
DS type III+V		2241		2245				

1965	**June 5**	●	McM 7842		Appendix A		CFI = 8
Hα		1807	1813	1835	S10 W50	1−	
SWF		1810	1815	1840		2	(11)
10.7 GHz		1807	1813	1821		2.2	(11)
2.8 GHz		1807	⌈1809.5 ∣1813 ⌊1821.3	1910		1.7	
200 MHz		⌈1812 ⌊1820	1823	1816 1831		> 2.3 2.3	
DS type IV		1821		1829			
DS type II		1817.7		1825			
DS type III+V		1813		1816			

1965	**June 13**	☉	McM 7847		Appendix A		CFI ≧ 5
Hα		0257		0430	N23 W03	1	
SPA		0300	0332	0431			(11)
SWF		0300		0435		1+	(11)
9.4 GHz		0300	0340	0420		1.1	⌉ no DS data
200 MHz		≦ 0300	0302	0358		2.1	⌋

1965	**October 4**	●	McM 8012		Event 124	(0 (1) 0)	CFI > 5
Hα		0937	0955	1026	S21 W30	2	
SID		no event					(11)
9.4 GHz		0940	1001	1140		1.1	type IV (sp. diagram in (102, 132)),
234 MHz		0947	0949	> 1049		~ 2.7	no DS data

1965	**October 8**	☉	McM 8005		Appendix A		CFI = 0
Hα		1603	1607	1612	N21 W84	1−	
SFD		1602	1605	1624	no other SID		(11)
2.8 GHz		1603	1605	1607		0.5	
DS type III		1603		1607			(11)

1965	**December 27**	☉	McM 8105		Appendix A		CFI ≧ 4
Hα		0620	0623	0632	N09 W29	1−	
SID		no event					(11)
17.0 GHz		0622.3	0622.4	0622.6		1.9	⌉
9.2 GHz		0622	0623	0629		2.4	no DS data
200 MHz		0621	0621	0623		3.0	⌋

1965	December 29	⊙	McM 8105			Appendix A		CFI ≧ 4
Hα		1123	1150	1235		N11 W60	1+	
SPA		1206	1208	1228		no other SID		(11)
9.4 GHz		1125	1157.5	1335			1.3	type IV (sp. diagram in (102)), no DS
200 MHz		1125	1200	1343			2.0	data

1965	December 30	●	McM 8105			Event 125	(−1 (0) 0)	CFI ≧ 5
Hα		0006	0040	0130		N09 W70	1?	
SID		no event						(11)
9.4 GHz		0019	0021 / 0032	0054			1.6	type IV (102),
200 MHz		0018	0022	0028			2.6	no DS data

1966	January 17	●	McM 8131			Event 127	(−2 0 0)	CFI ≧ 9
Hα		< 1029	1123	1250		N19 E27	2B	$\Delta t \sim 1^h$
SID		no event						(11)
Soft X-rays		1020	1035 / 1105	> 1430			SIG LG	(159) no HXR data
3.0 GHz		1032	1111	1327		3\	3.0	(124) type IV (102), no DS
200 MHz		1033	1144	1255			2.6	data

Spectral diagram in (122) and (123).

1966	January 18	●	McM 8131			Event 129	(−2 0 0)	CFI = 7
Hα		2253	2328	19^d0115		N20 E07	2B	$\Delta t \sim 1^h$
SWF		2258		19^d0001			2	(11)
Soft X-rays		2250	2335	$> 19^d0500$			VLG	(159) no HXR data
3.0 GHz		2256	2319	19^d0056		P2(1.9)	1.8	(124) DS no event
200 MHz		2300	2419	19^d0100			2.4	

1966	February 2	☉	McM 8154*		Appendix A		CFI = 4
Hα	1510	1525	1607		N26 E27	1N	$\Delta t \sim 9^{\rm h}$
SID	no event						(11)
Soft X-rays	no data						(159) no HXR data
2.8 GHz	1505		1755			0.8	
606 MHz	1512	1520	1549			0.7	
DS	1517		1523 unclassified activity				
DS type IV	1543		1559			1	(11)
DS type II	< 1518		1548			3	(11)
unclassified	1517		1523				
DS type III	1514		1515				

* a region almost without spots

1966	February 7	☉	McM 8154*		Appendix A		CFI = 1
Hα	1558	1628	1700		N27 W38	SN	
SID	no event						(11)
Soft X-rays	1600	1630	1800			SM	(159) no HXR data
2.8 MHz	1545		1650			0.5 no other burst	
DS type II	1645		1703			2	(11)

* a region almost without spots

1966	March 16	⊘	McM 8207		Appendix A		CFI ≥ 4
Hα	0912	0914	1007		N22 E65	2B	$\Delta t \sim 28^{\rm h}$
SID	no event						(11)
Soft X-rays	0912	0915	> 0925			SIG	(159) no HXR data
3.0 GHz	0912.5	0914.5	> 1530		U1/9.5(2.1)	1.7	⎤ no DS data
200 MHz	0914		0921			> 2.8	⎦

1966	March 16	⊘	McM 8207		Appendix A		CFI = 7
Hα	1603	1627	1705		N18 E60	1B	$\Delta t \sim 21^{\rm h}$
SWF	1625		1630			1	(11)
Soft X-rays	< 1628	1700	1700			SIG	(159)
Hard X-rays	1626	1627	1629			SM	(161)
2.8 GHz	⎡ 1616.8	1617.3	1619.5		P8.8(2.0)	0.9	
	⎣ 1624.7	1627	1633		P1.4(2.3)	1.9	
200 MHz	1626.8		1628			2.7	
DS type IV	1621.5		1800			3	(11)
DS type III	1615–1619.5, 1621–1624, 1625.5–1630						
DS type V		1627					

1966	March 16	⦶	McM 8207			Appendix A		CFI = 8
Hα	1918	[1923 1937	1950			N18 E58	1B	$\Delta t \sim 18^h$
SEA	1925		2002				1	(11)
SWF	1940	1942	1952				1	(11)
Soft X-rays	1922	1925	1945				SM	(159)
Hard X-rays	1922	1922.5	1924				SM	(161)
2.8 GHz	1921	1923	1928			U2.8/10.7(2.0)	2.0	
200 MHz	1920		1927				>2.7	
DS type IV	1921		2059				3	(11)
DS type III	1918–1919, 1921–1928, 1936–1937, 1941–1942							
DS type V		1922						

1966	March 16	⦶	McM 8207			Appendix A		CFI = 6
Hα	2252	2258	2352			N15 E58	SB	$\Delta t \sim 14^h$
SFD	2255	2256	2301					(11)
Soft X-rays	2250	2258	2320				SM	(159 no HXR data
2.8 GHz	2254	2256.5	2300			U2.8/9.4(2.0)	1.6	
200 MHz	2255.5		2259				2.9	
DS type IV	2255		2425				3	(11)
DS type III	2251–2252, 2255–2300, 2302–2309							

1966	March 18	⊙	McM 8207			Appendix A		CFI = 3
Hα	0420	0427	0510			N16 E40	1N	$\Delta t \sim 15^h$
SWF	0423		0515				1	(11)
Soft X-rays	0420	[0427 0450	0530				SIG	(159)
Hard X-rays	0421		0422				VSM	(161)
3.8 GHz	0420	0421.4	0530			P2.0(2.1)	2.0	]DS no event
1.0 GHz	0423	0423.3	0427				1.2	

1966	March 18	●	McM 8207			Appendix A		CFI = 5
Hα	2345	2355	19^d0010			N15 E27	1N	$\Delta t \sim 1^h$
SID	no event							(11)
Soft X-rays	2340	[2342 2350 2352	19^d0100				SM	(159) no HXR data
3.0 GHz	2347	2349	2359			0.5\(2.5)	1.4	(124)
208 MHz	2344	2350	19^d0014				2.6	
DS type III+V	2344		2355					
DS type II	19^d0006		19^d0010				3	(124)
DS cont.	19^d0000		19^d0050				2	(124)

1966	March 19	∅	McM 8207		Appendix A		CFI
Hα	0338	0348	0527	N21 E33	3N	$\Delta t \sim 0^{\text{h}}$?
SWF	0340	0350	0415		3	(11)
Soft X-rays	0333	0345	0800		LG	(159) no HXR data
3.0 GHz	0338	0343	0358	U1/9(3.6)	2.9	(124) ⎤ DS no
500 MHz	0340	0342	0410		3.0	⎦ event

1966	March 19	●	McM 8207		Appendix A		CFI = 4
Hα	⎡ 1357	1402	1418	N20 E27	SB	$\Delta t = 50^{\text{m}}$
	⎣ 1422	1431	1444	N14 E20	1N	
SWF	1400	1402	1420		1−	(11)
Soft X-rays	1355	⎡ 1402 ⎣ 1428	1445		SIG SM	(159)
Hard X-rays	1359.5	1400.8	1403		SM	(161)
3.0 GHz	⎡ 1356 ⎣ 1419	1401 1426	1406 1436	U2/9(2.0)	1.7 0.9	(124)
200 MHz	1419		1433		1.5	
DS type III	1402–1404, 1419–1432 (type III all throughout the day)					

1966	March 19	●	McM 8207		Event 130	(−2 0 0)	CFI = 3
Hα	2131	⎡ 2157 ⎣ 2209	2240	N15 E15	1B	$\Delta t \sim 0.5^{\text{h}}$
SPA	2142	2204	2310			(11)
Soft X-rays	2140	2157	2230		SIG	(159)
Hard X-rays	no event					(163)
3.0 GHz	2140		2230	\3	0.7	(124)
200 MHz	⎡ 2140 ⎢ 2156 ⎣ 2206	2143.5 2157.5 2206.5	2145 2159 2212		2.7 3.0 2.7	
DS type III	2127		2211 several groups (type III all throughout the day)			
DS U-burst	2132 and 2145					

1966	March 20	●	McM 8207		Event 131	(−2 0 0)	CFI ≧ 11
Hα	⎡ 0928 ⎣ 1011	0957 ⎡ 1015 ⎣ 1037	1255 1130	N21 E25 N15 E08	2B 2B	$\Delta t \sim 2^{\text{h}}$
SWF	0955	1002	1020		3	(11)
Soft X-rays	0930	⎡ 0950 ⎣ 0958	∼1230		SIG LG	(159)
Hard X-rays	0955	0957	1005		VLG	(161)
3.0 GHz	⎡ 0953 ⎣ 1028	0956 1035	1010 1052	P5(3.4)	3.3 2.0	(124) ⎤ type IV ⎥ (102),
204 MHz	0957	1035	1057		2.4	⎥ no DS
23 MHz	1009	1035	1047		4.2	⎦ data

1966	March 21	⊙	McM 8207		Appendix A		CFI ≧ 6
Hα	0925	0944	1125	N21 W02	2B	$\Delta t \sim 2.5^h$
SWF	0934		1000		1+	(11)
Soft X-rays	0925	0940	1000		SM	(159) no HXR data
3.0 GHz	0931	0945	1100	P7(1.9)	1.7	(124) no DS
200 MHz	0942		0947		2.3	data

1966	March 21	●	McM 8207		Event 133	(−2 0 0)	CFI = 8
Hα	2138	⌈ 2203 ⌊ 2230	2358	N19 W10	2B	$\Delta t \sim 2^h$
SWF	2227	2235	2300		1+	(11)
Soft X-rays	⌈ 2150 ⌊ 2218	2205 2233	> 2215 2345		SM SIG	(159) no HXR data
3.0 GHz	⌈ 2151 ⌊ 2220	2158 2258	2218 2411	U3/9(1.2) /9(1.4)	1.1 1.1	
200 MHz	⌈ 2150 | 2157 ⌊ 2242	2151 2158 2258	2156 2201 2309		3.3 3.1 2.8	
DS type III	2141		2206 several groups (type III all throughout the day)			
DS type II	⌈ 2209 ⌊ 2217.5		2214 2226.5		2 2'	(11) (11)

1966	March 23	⊙	McM 8207		Event 134	(−1 0 0)	CFI = 8
Hα	2248	⌈ 2259 ⌊ 2323	2335	N16 W33	1N	$\Delta t \sim 1^h$
SPA	2255	2300	2350			(11)
Soft X-rays	2245	2258	2330		VSM	(159)
Hard X-rays	no event					(163)
3.0 GHz	⌈ 2251 ⌊ 2318	2252 2319	2254 2320		1.3 0.3	
200 MHz	2326	2328	2330		3.3	
DS type IV	24^d0030		24^d0523			
DS type II	2326		2330			
DS type III	2243		2257 several groups (type III all throughout the day)			
DS type V	2326		2327			

1966	March 24	●	McM 8207			Event 135	(120)	CFI = 12
Hα		0225	0237	0414		N20 W42	2N	$\Delta t \sim 20^{m}$
SEA		0228	0249	0312			1+	(11)
Soft X-rays		0220	0240	>0400			LG	(159)
Hard X-rays		0230	0240	>0250			LG	(161)
3.0 GHz		0226	0231	0420		U1/9(3.0)	2.7	(124)
200 MHz		0233	0234.5	0238			3.3	
DS type IV			continues					
DS type II		0234		0253				
DS type III		0236–0236.5, 0304, 0308–0310, 0312						

1966	March 24	☉	McM 8207			Appendix A		CFI = 2
Hα		1858	1915	1930		N17 W51	SB	(11) $\Delta t \sim 15^{m}$
SPA		1915	1919	1940				(11)
X-rays		no event						(159, 160, 163)
3.0 GHz		1916	1916	1917		U3/10(1.8)	−0.1	(124)
200 MHz		1915		1916			2.0	
DS type III		1909		1918				
DS type V			1916					

1966	March 25	☉	McM 8207			Event 136	(−2 0 0)	CFI = 10
Hα		0145	0200	0340		N20 W54	2N	$\Delta t \sim 8^{h}$
SWF		0146	0154	0329			3	(11)
Soft X-rays		0150	0205	0400			LG	(159) no HXR data
3.8 GHz		0152	0156.5	0201		U1/9(2.3)	2.1	
200 MHz		0156	0158.5	0200			2.7	
DS type II		0157		0215				
DS type III		0137–0140, 0143, 0146–0146.5, 0155–0157						

1966	March 26	●	McM 8207			Appendix A		CFI ≥ 4
Hα		<1843	1852	1916		N15 W70	1B	$\Delta t \sim 25^{m}$
SWF		1848	1859	1920			1	(11)
Soft X-rays		<1847	1854	1920			SIG	(159) no HXR data
3.0 GHz		1850	1851	1855		U1.5P5(1.3)	1.0	(124)
108 MHz		1848	1850	1852			>2.4	
DS type III		1849.5	1852	1855.5				(4, 11)
DS type V			1852					

1966	March 30	☉	McM 8223		Appendix A		CFI = 14
Hα	1241	1252 / 1330	1423	N28 E50	2N		$\Delta t \sim 25^h$
SWF	1245	1253	1335		3		(11)
Soft X-rays	1243	1254	1630		LG		(159)
Hard X-rays	1246	1249.5	1315		LG		(161)
3.0 GHz	1244	1250	1436	U2/9(3.3)	2.7		(124)
200 MHz	1249		1323		4.4		
DS type IV	1307		1450		3		(11) (spectral diagram
DS type II	1253		1301		3		(11) in (102)).
DS type III	1249.5		1252				(11)

1966	April 12	☉	McM 8240		Appendix A		CFI = 3
Hα	1526	1530	1535	N22 W80	SN		(11) $\Delta t \sim 3.5^h$
SID	no event						(11)
Soft X-rays	no event						(159) no HXR data
18 MHz	1531		1533		?		the only burst
DS type III	1531		1535 or unclassified activity				(4, 11)
DS type II	1537		1552		3		(4, 11)
DS type IV	1552		1900		1−		(11)

1966	April 12	⊘	McM 8240		Appendix A		CFI ≥ 3
Hα	1621	1631	1643	N21 W83	1N		$\Delta t \sim 2.5^h$
SWF	1625	1630	1703		1		(11)
Soft X-rays	1625	1632	>1705		SM		(159) no HXR data
2.8 GHz	1630	1631	1636	/11(1.5)	1.3 no lower frequency		
DS type IV		continues					(11)

1966	April 12	⊘	McM 8240		Appendix A		CFI ≥ 3
Hα	1717	1720	1745	N21 W80	1N		$\Delta t \sim 1.5^h$
SWF	1710	1722	1755		2		(11)
Soft X-rays	1708	1720	1800		SIG		(159) no HXR data
5.0 GHz	1718.1	1718.6	1725	/11(2.1)	1.5 no lower frequency		
DS type IV		continues					(11)

1966	**April 15**	☉	McM 8262			Event 139	(−(2) 0 0)	CFI $\geqq$ 5
Hα		0956	1007	1106	N19 E40	2B		$\Delta t \sim 14^{\mathrm{h}}$
SWF		1003	1025	>1133		2	(11)	
Soft X-rays		0955	1020	1200		SIG	(159)	
Hard X-rays		1000	[1008 / 1015	1020		VSM	(161)	
3.0 GHz		0959	1007.2	1011	P0.6(2.0)	1.7	] no DS data	
260 MHz		1002	1005	1005		1.0		

1966	**May 4**	☉	McM 8278			Event 142	(−2 0 0)	CFI = 4
Hα		0150	0153	0220	N28 W67	SB		$\Delta t \sim 1^{\mathrm{h}}$
SWF		0153	0156	0223		1	(11)	
Soft X-rays		0150	0155	0300		SIG	(159)	
Hard X-rays			0153			VSM	(161)	
3.0 GHz		0151	0153	0156	P6(1.9)	1.9	(124)	
200 MHz		0152.5	0152.5	0154		2.9		
DS type III		0148		0154			(3, 11)	

1966	**May 28**	●	McM 8310			Event 145	(−(2) 0 0)	CFI = 10
Hα		1532	1622	1830	N15 W40	2B		$\Delta t \sim 0–1.5^{\mathrm{h}}$?
SWF		1614	1630	1730		3	(11)	
Soft X-rays		1530	[1550 / 1625	>1900		SM / SIG	(159) no HXR data	
3.0 GHz		1610	1622	1650	P4(1.7)	1.7	(124)	
200 MHz		1615	1627	1709		1.6		
DS type IV		1628		1855		2	(11)	
DS type III		1626–1628.7, 1633–1634					(4, 11)	

1966	**June 2**	☉	McM 8318			Appendix A		CFI = 0
Hα		1422	1424	1432	N04 W47	SB	(11) $\Delta t \sim 20^{\mathrm{m}}$	
SID		no event					(11)	
Soft X-rays		no event					(159) no HXR data	
3.0 GHz		1423	1424	1424	U1.7/3(0.7)	0.7	(124)	
111 MHz		1423.1	1423.4	1424.2		2.8		
DS type III		1423		1425			(11)	

1966	June 25	●	McM 8348			Event 146	(−2 0 0)	CFI = 11
Hα	1525	1543	1639		S24 W09	1B		$\Delta t \sim 0^m$?
SWF	1528		1655			3		(11)
SEA	1532	1539	1655			2−		(11)
Soft X-rays	1525	1540	~ 1900			SIG		(159)
Hard X-rays	1530	1535	1550			VSM		(161)
3.0 GHz	1525	1535	1714		U2P5(2.0)	2.0		(124)
200 MHz	1530	1542	1840			3.4		
DS type IV	1535		1940			2		(4, 11)
DS type II	1534.8		1607			3		(4, 11)
DS type III	1530–1532.5, 1536–1543, 1548–1556							(4, 11)
DS	1546		1555 unclassified activity					

1966	June 27	☉	McM 8344			Appendix A		CFI = 2
Hα	2358	2420	2520		N25 W58	2N		$\Delta t < 2^h$
SID	no event							(11)
X-rays	no event							(159, 163)
2.8 GHz	2400	2435	2545		?	0.7		the only bursts
500 MHz	2403.5	2409.5	2438			1.6		
DS type III	2409, 2415–2418							(11)

1966	July 7	●	McM 8362			Event 148	(1 2 1)	CFI = 13
Hα	0025	0040	0135		N35 W48	2B		$\Delta t \sim 15^m$
SWF	0025	0026	0329			3		(11)
Soft X-rays	0027	⎡ 0032 ⎣ 0042	> 0800			VLG		(159)
Hard X-rays	0026	⎡ 0031 ⎣ 0038.5	0105			VLG		(161)
3.0 GHz	0026	0038	0420		P10(4.1)	3.7		(124)
200 MHz	0030	0038.3	0206			2.9		
DS type IV	0042		0200					
DS type II	0038		0114					

1966	July 9	●	McM 8362			Event 149	(0 0 0)	CFI = 12
Hα	0310	0313	0435		N35 W75	2B		$\Delta t \sim 2^h$
SWF	0307	0309	0528			3		(11)
Soft X-rays	0230	⎡ 0237 ⎣ 0323	~ 0700			SIG LG		(159)
Hard X-rays	no event							(163)
3.0 GHz	0231	0414	> 0711		U4/9(2.6)	2.3		(124)
200 MHz	0315	0329	0435			2.1		
DS type IV	0330		0524					
DS type II	0321		0333					

1966	July 28	⊙	McM 8413			Event 152	(−1 0 0)	CFI = 10
Hα		2214	2245	2450	N36 E33	3B		$\Delta t \sim 18^h$
SPA		2230	2320	2415			(11)	
Soft X-rays		2215	⎡ 2230 ⎣ 2320	$> 29^d 06^h$		SM SIG	(164)	no HXR data
3.0 GHz		2214	2218	2400	P0.8(3.8)	2.1	(124)	
208 MHz		⎡ 2216 ⎣ 2239	2218 2254	2223 2400		2.1 1.7		
DS continuum		2255		$29^d 0440$				
DS type IV		2330		2448		3	(11)	
DS type II		2338		2351		2	(11)	
DS type III+V		2215		2219				

1966	August 28	●	McM 8461			Event 153	(1 2 0)	CFI = 16
Hα		1523	1529	1700	N22 E05	3B		$(\Delta t \sim 0^m)$ $\Delta t \sim 19^m$
SWF		1524	1530	1700		3+	(11)	
Soft X-rays		1522	1535	~2000		LG	(159, 164)	
Hard X-rays		1524.5	1530	1550		VLG	(161)	
3.0 GHz		1521	1527	1620	U2P7(3.5)	3.0	(124)	
200 MHz		1526.8	1527.4	1644		(4.8)		
DS type IV		1527		1749		3	(4, 11)	
DS type II		1530.3		1548		3	(4, 11)	
DS type III		1524–1532, 1534–1556, 1601–1609				very strong		
DS type V		1527		1532				
(Spectral diagram in (102).)								

1966	August 29	◐	McM 8461			Event 153	(1 2 0)	CFI = 8
Hα		1324	1335	1430	N21 W11	1N		
SWF		1332	1336	1430		2	(11)	
Soft X-rays		1323	~1340	1430		SM	(159)	
Hard X-rays		no event (particle flux)					(163)	
2.8 GHz		1323	1335	1413	U3P5(1.7)	1.4		
200 MHz		⎡ 1332.7 ⎢ 1401.0 ⎣ 1417.5	1332.8 1401.2	1334 1402 1417.9		(2.7) (3.2) (2.4)		
DS continuum		1325		1340 increased intensity			(11)	
DS type II		1319.5		1324.5		2	(11)	
DS type III		1326–1332, 1333–1337, 1341–1352, 1400–1402, 1417–1419						

1966	August 31	●	McM 8461		Appendix A		CFI = 9
Hα	0250	⌈0256 ⌊0353	0421	N22 W30	1N		$\Delta t = 1^h$
SWF	0040	0258	0440		3	(11)
SCNA	0040	0051	0449		2	(11)
Soft X-rays	0043	⌈0101 ⌊0355	0700		LG LG	(164)
Hard X-rays	no event					(163)
3.0 GHz	0154	0237	0454	P3(2.4)	2.4	(124)
500 MHz	⌈0200 ⌊0219	0202.8 0237	0211 0332		2.3 2.7	
200 MHz	0254.4	0254.6	0255		>3.3	
DS continuum		in progress since 30^d2330, end 0430				
DS type III	0215–0226, 0233–0255, 0318.5, 0349–0353					
DS type V	0233		0236			

1966	September 2	●	McM 8461		Event 154	(2 3 0)	CFI = 13
Hα	0542	0600	0800	N24 W56	3B	$\Delta t \sim 0^h$
SWF	0535	0543	0750		3	(11)
Soft X-rays	0530	0600	>1000		VLG	(159, 164)
Hard X-rays	no event (particle flux)					(163)
3.0 GHz	0547	0555	1045	/17(4.0)	3.4	(124) ⌉type IV deduced (spectral diagram
200 MHz	0552		0607		3.7	in (102))
DS type II	0554.5		0614 (herringbone structure)			⌋

1966	September 4	●	McM 8461		Event 155	((1) (0) 0)	CFI = 14
Hα	0407		>0519	N21 W87	3N	$\Delta t \sim 2.5^h$
SWF	0411	0412	0620		3	(11)
Soft X-rays	0410	0420	>0445		SIG	(159)
Hard X-rays	no event (particle flux)					(163)
3.0 GHz	0413	0418	>0820	/17(3.6)	3.0	(124)
200 MHz	⌈0413.5 ⌊0432	0502	0429 0518		>3.3 1.9	
DS type IV	0413.5		0458			
DS type II	0413.5		0441.5 (herringbone structure)			
DS type III	0412.5		0422			

1966	September 4	☉	McM 8461		Event 156	((1) (0) 0)	CFI = 1
Hα	1430	1440	1502	N22 W88	SN	(11) $\Delta t \sim 2.5^h$
SID	no event					(11)
X-rays	no event (particle flux)					(159, 163)
	no radio event at single frequencies					
DS type II	1452		1528		1	(11)

1966	September 11	☉	McM 8484		Appendix A		CFI = 0
Hα	0715	0735	0745	S22 W60	SB	(11) Δt ~ 15^m	
SID	no event					(11)	
Soft X-rays	0620	0710	1000		VSM	(159)	
Hard X-rays	no event					(163)	
3.0 GHz	no event above 111 MHz						
111 MHz	0733.1	0733.2	0737.5		3.7		
DS type III	0733		0737				

1966	September 12	☉	McM 8505		Event 157	(−2 0 0)	CFI ≧ 4
Hα	0925		1035	N12 E90	1N	Δt ~ ?	
SEA	0937	0950	1030		1	(11)	
SWF	0950	1035	1045		2	(11)	
Soft X-rays	0910	0955	2000		LG	(159)	
Hard X-rays	no event (particle flux)					(163)	
3.0 GHz	0925	0954	1418	P5(1.9)	1.9	(124) ⎤ no DS	
200 MHz	0929	0949	1044		1.7	⎦ data	

1966	September 14	●	McM 8484		Event 158	(0 0 0)	CFI ≧ 6
Hα	1014		1120	S21 W90	SN	Δt ~ 15^m	
SWF	1016	~ 1035	1059		1	(11)	
Soft X-rays	1010	1022	> 1130		LG	(159)	
Hard X-rays	no event (particle flux)					(163)	
3.0 GHz	1012	1014	1100	1.5\(1.8)	1.4	(124) ⎤ no DS	
234 MHz	1013	⎡ 1015	1116		4.3	⎥ data	
		⎣ 1042			3.7	⎦	

1966	September 17	●	McM 8496		Event 160	(−1 0 0)	CFI ≧ 4
Hα	0940	1005	1045	N24 W63	2N	Δt ~ 3^m?	
SPA	0950	1014	1215			(11)	
Soft X-rays	0940	1020	1230		SIG	(159)	
Hard X-rays	no event					(163)	
3.0 GHz	0954	1005	1018	P3(1.5)	1.5	(124) ⎤ no DS	
200 MHz	0954	1009	1054		1.3	⎦ data	

1966	September 20	●	McM 8505		Event 161	(−1 0 0)	CFI = 12
Hα	<1738	1805	2100	N06 W14		2B	Δt ~ 2^h
SWF	1710	1715	1920			3	(11)
Soft X-rays	~1630	1800	~2300			SIG	(159) (slow rise and fall)
Hard X-rays	1711	$\begin{bmatrix}1712\\1730\end{bmatrix}$	1830			SM SM	(161)
3.0 GHz	1641	1714	1800	U1P1.5(3.0)		2.4	(124)
200 MHz	1714	1714.2	1714.5			(3.0)	
DS continuum	1713		>2429				(11)
DS type III+V	1713		1715				

1966	October 20	●	McM 8546		Appendix A		CFI = 3
Hα	2152	2156	2208	N20 W50		SB	Δt ~ 30^m
SFD	2154	2155	2156				(11)
Soft X-rays	2152	2153	2205			VSM	(164)
Hard X-rays	no event						(163)
3.0 GHz	2154	2155	2157	P4(1.5)		1.5	(124)
200 MHz	2154	2154.3	2156			2.8	
DS type III+V	2154		2156.5				

1966	October 24	☉	McM 8546		Appendix A		CFI ≧ 0
Hα	0942	0946	0954	N15 W90		SB	Δt ~ 1^h
SID	no event						(11)
Soft X-rays	0930	$\begin{bmatrix}0935\\0945\end{bmatrix}$	1030			VSM VSM	(159)
Hard X-rays	no event						(163)
2.0 GHz	0950.5	0950.5	0953			0.2	(and 1.5 GHz only)
No DS data							

1966	November 1	☉	McM 8567		Appendix A		CFI ≧ 3
Hα	<1053		1118	N20 W15		2N	Δt ~ 2^h
SID	no event						
Soft X-rays	~1055	1105	1230			SM	(159, 164)
Hard X-rays	no event						(163)
3.0 GHz	1053	1100	1110	0.2\(1.2)		1.1	⎤ no DS data
200 MHz	1055	1129	1136			1.2	⎦

1966	November 2	●	McM 8567			Appendix A		CFI = 3
Hα		1707	1723	1755		N21 W35	SN	$\Delta t \sim 40^m$
SPA		1726	1738	1830				(11)
Soft X-rays		1703	1727	1840			VSM	(164)
Hard X-rays		no event						(163)
2.7 GHz		1721	1724	1733		0.5\(1.7)	1.4	
486 MHz		1722	1723	1725			1.7	no lower fre-
								quency
DS continuum		1736		1933				(11)
DS type III		1707, 1720–1727, 1743.5, 1746–1748						(4, 11)

1966	November 3	●	McM 8567			Appendix A		CFI = 0
Hα		1854	1900	1925		N21 W49	SN	(11) $\Delta t \lesssim 2.5^h$
SID		no event						(11)
Soft X-rays		1855	1902	1930			SM	(159, 164)
Hard X-rays		no event						(163)
3.0 GHz		1857	1900	1901		P3(0.9)	0.9	(124)
18 MHz		1902		1905 (no record between 18 and 1415 MHz)				
DS continuum		1905		1911				(11)
DS type III		1857–1859, 1903–1906						(4, 11)
DS type V			1904					

1966	November 4	☉	McM 8567			Appendix A		CFI = 0
Hα		1306	1313	1340		N23 W60	SN	$\Delta t \sim 20^m$
SID		no event						(11)
Soft X-rays		1314	1325	1410			VSM	(159)
Hard X-rays		no event						(163)
No radio event at single frequencies, no DS data								

1966	November 16	☉	McM 8573			Appendix A		CFI = 0
Hα		1753	1754	1806		N20 W54	SN	(11) $\Delta t \sim 25^m$
SID		no event						(11)
X-rays		no event						(163, 164)
2.8 GHz		1753	1754	1755		?	−0.1	(124)
18 MHz		1752		1754 (no other bursts)				
DS type III		1753.5		1753.7				(11)

1966	December 11	☉	McM 8612			Event 164	(−(2) 0 0)	CFI = 4
Hα		0537	0542	> 0609		N18 E77	1N	$\Delta t \sim 10^h$
SWF		0540	0550	0710			2	(11)
Soft X-rays		0520	0600	1300			LG	(159)
Hard X-rays		0535	0545	0555			SM	(161)
3.0 GHz		0523	0546	> 0648		P4(1.8)	1.8	(124)
		(no burst below 1.0 GHz)						
DS		no event						

1966	December 19	☉	McM 8613			Appendix A		CFI = 2
Hα		<1419	1424		1444	N13 W67	1N	$\Delta t \sim 30^m$
SID		no event						(11)
X-rays		no event						(159, 163)
3.0 GHz		1413	1418		1421	P3	1.2	
		(no burst below 1.4 GHz)						
DS		no event						

1966	December 30	●	McM 8629			Appendix A		CFI = 8
Hα		2231	2235		2254	S23 E32	1B	$\Delta t \sim 1.5^h$
SPA		2233						(11)
Soft X-rays		2224	2228		2240		SM	(159)
Hard X-rays		no event (particle flux)						(163)
2.7 GHz		2232	2235		2238	?	1.9	(no higher frequency)
200 MHz		2233	2235.4		2240		3.1	
DS type IV		2237			2257		doutful	(11)
DS type II		2236			2247			
DS type III		2230, 2233–2237						(4, 11)
DS type V		2234			2236			

1966	December 31	●	McM 8629			Appendix A		CFI ≧ 5
Hα		0842	0845		0900	S20 E27	1B	$\Delta t \sim 15^m$
SEA		0844	0853		0901		1—	(11)
Soft X-rays		0841	0848		0856		SM	(159) no HXR data
3.0 GHz		0843	0844		0846	0.2\(3.4)	1.4	⎫ no DS data
204 MHz		0841	0844		0846		3.4	⎭

1966	December 31	☉	McM 8629			Appendix A		CFI = 0
Hα		1516	1518		1525	S21 E22	SF	(11) $\Delta t \sim 15^m$
SID		no event						(11)
Soft X-rays		no event						(159) no HXR data
2.8 GHz		1516.5	1517.5		1519		0.1 no other burst	
DS type III		1510–1511, 1521						

1966	December 31	●	McM 8629			Appendix A		CFI = 2
Hα	1836	1842	1917	S23 E22	1B			$\Delta t < 15^{m}$
SPA	1838	1846	1915					(11)
Soft X-rays	1839	1845	1910					(159) no HXR data
2.8 GHz	1840	1841.7	1900	(P3)	1.4			
606 MHz	1840.8	1841.7	1843		1.8			
DS type III	1840–1845, 1850–1852							(4, 11)
DS type V		1842						

1967	January 3	☉	McM 8629			Appendix A		CFI ⩾ 4
Hα	1044	⎡ 1046 ⎣ 1102	1115	S23 W22	SN			$\Delta t \sim 0.75^{h}$
SID	no event							
Soft X-rays	1030	1050	1125		VSM			(229) no HXR data
9.1 GHz	1058	1059	1100		1			⎤ no DS data
200 MHz	1058	1058	1100		3.6			⎦

1967	January 11	☉	McM 8632			Event 168	(−1 0 0)	CFI = 5
Hα	0131	0259	0510	S26 W47	3B			$\Delta t \sim 1^{h}$
SID	no event							
Soft X-rays	0138	0148	> 0200		VSM			(229)
Hard X-rays	0240	⎡ 0248 ⎣ 0254	> 0300		SM			(161, 163)
3.0 GHz	0215	0310	0525	P7(1.2)	1.2			(124)
DS type II	0223		0231		2			(124)
DS type III	0149		0527 several					
DS uncl.	0442		0443		1			

1967	January 11	☉	McM 8631			Event 169	(−1 0 0)	CFI ⩾ 5
Hα	2016		> 2034	N16 W88	SB			(234)
								$\Delta t \sim 0.75^{h}$
SWF	2020	2024	2038		1			(11)
Soft X-rays	2015	⎡ 2025 ⎣ 2055	2200		SIG			(229)
Hard X-rays	no event							(161, 163)
2.8 GHz	2017	2046	2052		1.5			
600 MHz	2020	2040	2055		1.6			
DS type III	2020–2021, 2033–2034				2			
DS type IV	2119		2205		1			
DS cont	2033		2100		2			
DS type II	2054		2104		2			

1967	January 14	⊙	McM 8632			Appendix A		CFI = 0
Hα		0118	0126	0133		S31 W89	SN	$(234) \cdot \Delta t \sim 1^h$
SID		no event						
Soft X-rays		no event						
Hard X-rays		no event						(161, 163)
DS type III		0117–0117, 0118–0121, 0121–0122						

1967	February 13	●	McM 8687			Event 176	(−1 1 0)	CFI ⩾ 8
Hα		1747	1820	> 2130		N21 W11	3B	$\Delta t \sim 1^h$
SWF		1804	1825	1910			1+	(11)
Soft X-rays		1755	1825	2300			SIG	(229)
Hard X-rays		1808	1809	1811			VSM	(161, 163)
3.0 GHz		1753	1804	2045	3–9(1.7)		1.7	(124)
600 MHz		1754	1810	1829			1.6	
DS cont.		1801		1829			3	(124)
DS type III(+V)		1819		1831			3	(124)
DS type II		1803		1820			3	(124)
DS type IV		1829		2438			2	(124)

1967	February 27	●	McM 8704			Event 177	(0 0 0)	CFI = 12
Hα		1637	1644	1830		N27 E02	2N	$\Delta t \sim 0.5^h$
SWF		1640	1655	1752			2+	(11)
Soft X-rays		164.	1655	1830			LG	(229)
Hard X-rays		1640	1650	1703			LG	(161, 163)
3.0 GHz		1637	1650	1800	/10(0.7)		2.7	(124)
200 MHz		1640		1701			> 3.5	
DS type II		1640		1703			3	(124)
DS type IV		1641		1827			3	(124)
DS type III		1641–1646, 1652–1658					3	

1967	April 1	⊙	McM 8740			Appendix A		CFI ⩾ 4
Hα		0835		0855		N19 W74	SB	(233)
								$\Delta t \sim 0.5^h$
SWF		0834	0841	0848			1−	(11)
Soft X-rays		0833	0838	0845			SIG	(229)
Hard X-rays		0834	0835	0840			SM	(161, 163)
2.0 GHz		0847	0848	0849			0.5	DS no data
200 MHz		0844	0851	0853			3	

1967	April 1	●	McM 8740		Appendix A		CFI = 7
Hα	1410		> 1433	N21 W80	1B		$\Delta t \sim 1^{\mathrm{h}}$
SWF	1412	1416	1430		1+	(11)	
Soft X-rays	1410	1415	1426		SM	(229)	
Hard X-rays	1410	1415	1417		LG	(161, 163)	
3.0 GHz	1410	1411	1415	/10(2.8)	2.5	(124)	
200 MHz	1410	1411	1417		3.8		
DS type III(+V)	1410		1418		3	(124)	

1967	April 14	☉	McM 8760		Event 181	(−2 0 0)	CFI = 5
Hα	1703	1715	1737	N24 W71	2N		$\Delta t \geqslant 2^{\mathrm{h}}$
SWF	1708	1718	1745		1	(11)	
Soft X-rays	1705	1715	1740		SM	(229)	
Hard X-rays	1724	1724	1747		EB	(231)	
2.8 GHz	1707	1709	1710	0.6\	0.6	(124)	
600 MHz	1708	1709	1709		1.5		
DS type III	1707–1718, 1726–1727						
DS type II	1712		1724		1	(124)	
DS cont.	1724		1738		2	(124)	

1967	May 21	●	McM 8818		Event 182	(−2 0 0)	CFI = 12
Hα	1919	1926	2025	N24 E39	2N		$\Delta t \sim 1.5^{\mathrm{h}}$
SWF	1924	1932	2010		2	(11)	
Soft X-rays	1919	1927	2050		LG	(229)	
Hard X-rays	1921	1925	1940		VLG	(161, 163)	
2.8 GHz	1922	1924	1948		2.9		
600 MHz	1922	1923	2017		3.3		
DS type III	1922–1926 (also type V), 1924–1951, 1933–1947						
DS type IV	1922		2007		3		
DS type II	1923		1945		3		
White light	1920	1925				(230)	

1967	May 23	●	McM 8818		Event 183	(3 3 0)	CFI = 16
Hα	⌈1802	⌈1814	> 1930	N28 E25			
	⌊1844				3B	(121) $\Delta t < 1^{\mathrm{h}}$	
	⌊1932	1947	2200	N28 E28			
SWF	⌈1809	1815	2049		3	(11)	
	⌊1834		2230		3+	(11)	
		⌈1820			SM	(229)	
Soft X-rays	1805	1850	24ᵈ0200		LG	(229)	
		⌊1950			LG	(229)	
Hard X-rays	⌈1808	1810	1818		SM	(161, 163)	
	1835	1842	1915		VLG	(161, 163)	
	⌊1935	1941	2015		LG	(161, 163)	
2.8 GHz	1836	1952	2218	U3/9	3.9	(124)	
200 MHz	⌈1836	1838	1842		4.1		
	⌊1848	1923	> 1937		3.6		

(*continues on next page*)

1967 **May 23** (continued)

DS type III	1852–1900, 1937–1940				2	
DS type II	1838		1900		3	(124)
DS type IV	1839		2400		3	(124)
White light	1835	1841				(230)

1967 **May 28** ● McM 8818 Event 184 (2 2 0) CFI = 12

Hα	<0529	0543	0700	N28 W32	3B	(233) $\Delta t \sim 0.25^{\rm h}$
SWF	0530	0533	0640		3	(11)
Soft X-rays	0530	0550	0720		LG	(229)
Hard X-rays	<0548	<0548	>0623		LG	(231)
2.8 GHz	0529	0542	0610	U2P10	3	(124)
200 MHz	0537	0540	0800		3.3	
DS type II	0539		0552		2	(124)
DS type III	0532		0532			

1967 **July 28** ⊙ McM 8905 Appendix A CFI = 2

Hα	0053	⌈ 0115 ⌊ 0127	0145	N25 E08	SF	(233) $\Delta t \sim 1^{\rm h}$
SWF	0117	0129	0139		1	(11)
Soft X-rays	0100	0130	0200		SM	(229)
Hard X-rays	0124	0127	0132		SM	(161, 163)
2.7 GHz	0052	0052	0053		0.4	
500 MHz	0052	0053	0057		2.8	
DS cont.	0000		0200			
DS type III	0000		0410 intermittent			

1967 **July 30** ● McM 8905 Event 194 (−1 0 0) CFI = 6

Hα	0508	0513	0532	N24 W27	1B	$\Delta t \sim 0.2^{\rm h}$
SWF	0508	0513	0532		1−	(11)
Soft X-rays	0508	0515	0530		VSM	(229)
Hard X-rays	no event					(231)
3.0 GHz	0509	0511	0513		1.8	
200 MHz	0508	0509	0510		3.3	
DS type III	0507		0518			
DS type V	0508		0510			

1967	July 30	☉	McM 8907		Appendix A		CFI = 5
Hα	1555	1601	1615	N13 W07	SB	(233)	
							$\Delta t \sim 0.5^{\mathrm{h}}$
SWF	1600	1604	1614		1−	(11)	
Soft X-rays	no data						
Hard X-rays	< 1601	1603	> 1613		EB	(231)	
2.8 GHz	1600	1601	1605		1.5		
200 MHz	1600	1601	1602		3		
DS type III(+V)	1558		1605		3		

1967	July 30	☉	McM 8905		Appendix A		CFI = 3
Hα	1612	1635	> 1654	N26 W36	1N		$\Delta t \sim 0.3^{\mathrm{h}}$
SID	no event						
Soft X-rays	1630	1630	1700		VSM	(229)	
Hard X-rays	no event					(161, 163)	
2.8 GHz	1625	1626	1639		1.4		
600 MHz	1627		1628		1.3		
DS type III(+V)	1623		1938		3		

1967	July 30	☉	McM 8907 + 8905		Appendix A		CFI = 2
Hα	1945	1948	2029	N19 W10	SF	(11)	
or							$\Delta t \sim 0.25^{\mathrm{h}}$
Hα	1946	1955	2021	N30 W33	SN		
SWF	2000	2003	2034		1	(11) also SPA	
Soft X-rays	1950	1958	2010		VSM	(229)	
Hard X-rays	no event					(161, 163)	
2.8 GHz	1947	1950	1954		1.6		
DS type III	1937–1955, 2006–2009				3		

1967	July 31	☉	McM 8905		Event 195	(−2 0 0)	CFI ⩾ 4
Hα	0808		0850	N25 W42	1N		$\Delta t > 1^{\mathrm{h}}$
SWF	0814	0828	0835		1−	(11)	
Soft X-rays	0806	0827	0850		SM	(229)	
Hard X-rays		0808			VSM	(231)	
3 GHz	0813	0815	0817		2.4		DS no data
100 MHz	0813	0813	0834		3		

1967	July 31	☉	McM 8905		Appendix A		CFI ⩾ 2
Hα	2047	2114	2140	N23 W50	1B		$\Delta t = 1^{\mathrm{h}}$
SPA	2052	2056	2232			(11)	
Soft X-rays	no data						
Hard X-rays	2056	2058	2100		VSM	(161, 163)	
5 GHz	2054	2056	2059		1.3		
600 MHz	2054	2056	2059		2.2		
DS type III	2047–2051, 2054–2058				3		
	2100–2101, 2104–2107				2		

1967	August 1	⊙	McM 8907		Event 196	(−1 0 0)	CFI =5
Hα	0120	0121	0130	N13 W25	SB		$\Delta t \sim 8^h$
SWF	0120	0121	0129		1−	(11)	
Soft X-rays	0118	0121	0126		SIG	(229)	
Hard X-rays	0117	0120	0123		SM	(161, 163)	
3.7 GHz	0117	0118	0119		2		
500 MHz	0120	0121	0121		2.1		
DS type II	0120–0123, 0126–0130						
DS type III(+V)	0118		0119		3		

1967	August 1	⊙	McM 8905		Event 196	(−1 0 0)	CFI = 4
Hα	<0535	⌜0555 ⌞0641	0655	N25 W53	1N		$\Delta t \sim 4^h$
SWF	0635	0645	0648		1	(11)	
Soft X-rays	no data						
Hard X-rays	no data						
3.1 GHz	0637	0640	0642		0.8		
200 MHz	0638	0638	0646		2.7		
DS type III	0530–0531, 0636						

1967	August 1	●	McM 8905		Event 197	(0 0 0)	CFI = 9
Hα	1721	1738	1810	N27 W62	2B		$\Delta t \sim 0.5^h$
SWF	1731	1740	1800		2	(11)	
Soft X-rays	1718	1738	1810		SIG	(229)	
Hard X-rays		1722			VSM	(231)	
3 GHz	1730	1731	1738	P5	2.2		
200 MHz	1729	1731	1737		1.4		
DS cont.	1730		1808		2		
DS type V	1730		1738		3		
DS type III	1730–1738, 1740–1800				2		
DS type II	1735		1738		2		

1967	August 2	⊙	McM 8905		Appendix A		CFI = 1
Hα	1726	1731	1748	N26 W76	SN	(233)	
						$\Delta t \sim 0.25^h$	
SES	1728	1733	1828		1	(11) no SWF	
Soft X-rays	no data						
Hard X-rays	no event				(161, 163)		
2.8 GHz	1728	1729	1730		1		
DS type III	1729		1732		2		

1967	August 3	☉	McM 8905		Event 199	(−1 0 0)	CFI ⩾ 4
Hα		0918	⎡ 0920 ⎣ 0930	0950	N27 W85	1B	$\Delta t \sim 0.2^{\mathrm{h}}$
SWF		0923		1000		2−	(11)
Soft X-rays		0920	0930	1045		SIG	(229)
Hard X-rays		⎡ 0919 ⎣ 0925	0919 0929	0922 0935		LG LG	(161, 163) (161, 163)
3 GHz		0919	0919	0943	P5	1.8	⎤ DS no data
100 MHz		0918		0921		2.2	⎦

1967	August 4	☉	McM 8905		Appendix A		CFI = 3
Hα		1511	1515	1550	N31 W87	1N	$\Delta t \sim 0.5^{\mathrm{h}}$
SWF		1516	1518	1550		1+	(11)
Soft X-rays		1440	1520	1600		SM	(229)
Hard X-rays		< 1457	> 1504	> 1533		LG	(231)
2.8 GHz		1514	1514	1516		1.8	
DS		no event					

1967	August 9	☉	McM 8926		Event 203	(0 0 0)	CFI = 4
Hα		1758	⎡ 1813 ⎣ 1830	1930	S24 E32	2B	$\Delta t \sim 0.5^{\mathrm{h}}$
SWF		1824	1830	1853		1+	(11)
Soft X-rays		no data					
Hard X-rays		< 1752	1801	> 1804		EB	(231)
2.7 GHz		⎡ 1808 ⎣ 1824	1809 1827	1812 1932	P1	0.6 1.9	
600 MHz		1825	1827	1828		1.3	
DS type III		1825		1827		1	

1967	August 18	☉	McM 8942		Appendix A		CFI = 4
Hα		0239	0248	0305	N24 E89	SB	(11) $\Delta t \sim 8^{\mathrm{h}}$
SWF		0221	0230	0307		2−	(11)
Soft X-rays		0150	0237	0900		SIG	(229)
Hard X-rays		< 0220	> 0227	0319		LG	(231)
2.0 GHz		0223	0223	0236		2.1	DS no data
200 MHz		no event					

1967	August 26	●	McM 8949			Appendix A		CFI = 9
Hα	0014	0023	0108	S19 E00	1B	$\Delta t = 2^h$
SWF	0018	0022	0100		1+	(11)
Soft X-rays	0015	0025	0110		SM	(229)
Hard X-rays	<0002	0023	>0047		EB	(231)
2.7 GHz	0014	0017	0025	P0.5(4)	2.2	
200 MHz	0016	0028	0130		2.5	
DS type III	0019		0019		2	
DS type II	0017		0034		3	
DS type IV	0017		0025		3	

1967	September 10	⊙	McM 8962			Appendix A		CFI ⩾ 1
Hα	0852	0857	0907	S22 W46	SB	$\Delta t \sim 0.5^h$
SID	no event					
Soft X-rays	0850	0859	0910		VSM	(229)
Hard X-rays	no event					(161, 163)
2.0 GHz	0854	0856	0859		0.6	⎤ DS no data
300 MHz	0853	0857	0857		1.5	⎦

1967	September 18	●	McM 8973			Event 205	(−2(1) 0)	CFI = 8
Hα	2316	2345	19^{d}0145	N16 W60	2B	$\Delta t \sim 2.5^h$
SWF	2328	19^{d}0005	0020		1	(11)
Soft X-rays	no data					
Hard X-rays	no data					
2.7 GHz	2315	2345	19^{d}0320	U3/9	1.3	
200 MHz	2312	2331	2353		1	
DS type III	2334		2336			
DS type IV	2335		19^{d}0037		3	
DS type II	2338		2345		3	
DS unclassified	2337		19^{d}0016			

1967	October 7	⊙	McM 9004			Event 208	(−2 0 0)	CFI = 3
Hα	2044	2050	2110	S18 W62	2B	$\Delta t \sim 0^h$
SPA	2046	2049	2051			(11)
Soft X-rays	no event					
Hard X-rays	no data					
2.8 GHz	2049	2049	2050		1	
600 MHz	2049	2050	2050		0.1	
DS type III	2047		2051		3	

1967	October 25	⊙	McM 9034		Appendix A		CFI = 5
Hα	< 1327	⌈ 1350 ⌊ 1412	1445	N09 W24	1N		$\Delta t \sim 1^h$
SID	no event						
Soft X-rays	1325	1400	> 1600		SIG	(229)	
Hard X-rays	< 1338	1353	> 1407		EB	(231)	
2.8 GHz	1325		1340		0.9		
200 MHz	1348	1349	1354		1.5		
DS type IV	1338–1405, 1420–1443				3		

1967	October 25	⊙	McM 9034		Appendix A		CFI = 1
Hα	2129	⌈ 2131 ⌊ 2204	> 2205	N10 W34	SB		$\Delta t \sim 0.3^h$
SWF	2150	2200	2215		1	(11)	
Soft X-rays	⌈ 2129 ⌊ < 2150	2133 2202	> 2150 > 2245		SM SIG	(229) (229)	
Hard X-rays	⌈ 2126 ⌊ 2147	2132 2205	2139 > 2205		EB EB	(231) (231)	
2.7 GHz	2150		2200		0.9		
200 MHz	no event						
DS	no event						

1967	October 25	⊙	McM 9034		Appendix A		CFI = 3
Hα	< 2312	⌈ 2328 ⌊ 2347	2400	N10 W28	1B		$\Delta t = 1^h$
SWF	2315	2329	2351		1	(11)	
Soft X-rays	2306	2330	> 2400		SM	(159)	
Hard X-rays	< 2301	> 2319	> 2336		EB	(231)	
2 GHz	2316	2319	2331		1		
500 MHz	2318	2320	2325		1.9		
DS type III	2321		2326				

1967	October 26	●	McM 9034		Event 209	(−2 0 0)	CFI = 6
Hα	0608	0614	0640	N10 W38	1B		$\Delta t \sim 0.5^h$
SWF	0610	0620	0650		1	(11)	
Soft X-rays	0609	0616	0645		SIG	(229)	
Hard X-rays	no event					(161, 163)	
3.1 GHz	0609	0610	0624		1.5		
200 MHz	0610	0610	0612		2		
DS type III(+V)	0609		0610		2		
DS type II	0612		0617		2		

1967	October 27	⊙	McM 9034		Event 210	(−2 0 0)	CFI = 4
Hα	1107	1110	1130	N09 W46	SN	(233) $\Delta t \sim 0.5^h$	
SEA	1113	1117	1157		1+	(11)	
Soft X-rays	1105	1110	1120		VSM	(229)	
Hard X-rays	no event					(161, 163)	
3 GHz	1107	1108	1110		1.3		
200 MHz	1107	1109	1111		3		
DS type III(+V)	1107		1110		1		

1967	October 29	⊙	McM 9034		Event 211	(−1 0 0)	CFI = 4
Hα	0258	0302	0307	N10 W80	1N	$\Delta t \sim 1^h$	
SWF	0305		0400		2+	(11)	
Soft X-rays	0255	0325	0440		LG	(229)	
Hard X-rays	< 0322	0322?	> 0409		EB	(231)	
2 GHz	0257	0300	0343	U1P9(2.3)	1.7		
700 MHz	0256		0334		1.7		
DS unclassified	0258						

1967	October 29	●	McM 9034		Event 212	(−1 0 0)	CFI = 9
Hα	2347	[2351 30^d0014	30^d0100	N10 W90	2B	$\Delta t \sim 0.25^h$	
SWF	2342	2346	30^d0108		2	(11)	
Soft X-rays	2342	30^d0000	0130		LG	(229)	
Hard X-rays	< 2352	2352	30^d0049		LG	(231)	
2 GHz	2343	2347	30^d0018	P10	3.6	(124)	
200 MHz	2343	2358	30^d0030		2.4	also uncl. 2358	
DS type III(+V)	2345		2346		2	(124)	

1967	November 2	●	McM 9047		Event 213	(0 1 0)	CFI = 12
Hα	0852	0856	0914	S18 W02	2B	$\Delta t < 0.5^h$	
SWF	0857		0920		2+	(11)	
Soft X-rays	0904	0906	0923		SIG	(229)	
Hard X-rays	0853	0856	0900		SM	(161, 163)	
2 GHz	0854	0856	0915	U2P7	2.7	(124)	
200 MHz	0854	0856	0858		3.8		
DS type IV	0854		0910		1	(124) also type III 0853	

1967	November 4	●	McM 9047			Event 214	(−1 0 0)	CFI = 7
Hα		1151	1154	1220		S18 W33	1B	$\Delta t = 0.3^h$
SWF		1157		1232			1−	(11)
Soft X-rays		1152	1158	1230			SIG	(229)
Hard X-rays		<1208	<1208	>1218			SIG	(231)
2 GHz		1152	1154	1205		P3	2	(124)
200 MHz		1153	1154	1158			2.6	
Type IV		1152		1156			2	(124)
DS type II		1159		1208				(124)
DS type III		1155		1156				

1967	November 7	⦸	McM 9047		.	Appendix A		CFI = 0
Hα		2142	⌈2147 ⌊2200	2210		S22 W60	SN	$\Delta t \sim 1^h$
SID		no event						
Soft X-rays		2140	2200	2330			SM	(229)
Hard X-rays		2139	2149	>2158			EB	(231)
2 GHz		no event						
DS		no event						

1967	November 7	☉	McM 9048			Appendix A		CFI = 1
Hα		2218	2243	2325		N25 W63	1F	$\Delta t = 0.5^h$
SID		no event						
Soft X-rays		no event						
Hard X-rays		<2200	<2200	>2212			SM	(231)
2 GHz		no event						
DS		no event						

1967	December 1	☉	McM 9091			Appendix A		CFI ⩾ 3
Hα		<0335	0345	0416		S27 W70	1N	$\Delta t \sim 0.5^h$
SWF		0315	0337	0415			2	(11)
Soft X-rays		0215	0338	0440			LG	(229)
Hard X-rays		<0311	0315?	>0400			LG	(231)
2 GHz		0305	0330	0343		P9	1	
DS		no event						

1967	December 1	☉	McM 9091			Appendix A		CFI = 0
Hα		0942	0944	1003		S31 W69	SN	(11) $\Delta t = 1^h$
SPA		0942	0949	1018				(11)
Soft X-rays		no event						
Hard X-rays			<0959				SM	(231)
3 GHz		0940	0942	0942			0.8	
DS type III		0947–0948, 0956–0957					1	

1967	December 2	⊙	McM 9091		Appendix A		CFI = 2
Hα	⌈0535 ⌊0601	0546 0606	0600 0619	S27 W87 S29 W90	SN SN	(11) $\Delta t \sim 1^h$ (11) $\Delta t \sim 0.5^h$
SWF	0514	0523	0605		1	(11)
Soft X-rays	0505	⌈0508 0522 ⌊0540	0930		LG	(229)
Hard X-rays	⌈0513 ⌊0533	0518 0543	0528 0600		SIG SIG	(161, 163) (161, 163)
2 GHz	0513	0555	0603		1.3	
DS	no event					

1967	December 11	●	McM 9108		Event 224	(−1 0 0)	CFI = 8
Hα	2347	2358	12^d0110	S22 W17	2B	(233) $\Delta t < 3^h$
SWF	2356	12^d0007	0207		1	(11)
Soft X-rays	2346	12^d0012	>0100		SIG	(229)
Hard X-rays	12^d0014	<0014	>0111		LG	(231)
2 GHz	2346	2355	12^d0014	P2	2.5	
200 MHz	2352	2353	2356		3.9	
DS type III(+V)	2353		2355		2	
DS type II	2357	12^d0018		(m+Dkm)		

1967	December 16	●	McM 9118		Event 225	(0 1 0)	CFI = 9
Hα	0247	0255	0430	N23 E66	3N	$\Delta t \sim 2^h$
SWF	0242	0318	0410		1	(11)
Soft X-rays	0246	0305	>0400		SM	(229)
Hard X-rays	0250	0255	>0330		LG	(161, 163)
2 GHz	0246	0252	0527	P1	2.9	
200 MHz	0245	0300	0345		3.1	
DS type III(+V)	0244		0253		1	
DS type II	0253		0326		2	

1967	December 28	●	McM 9128		Appendix A		CFI = 4
Hα	1335		1530	S21 W19	1N	$\Delta t \sim 1^h$
SID	no event					
Soft X-rays	<1320	1430	1630		SM	(229)
Hard X-rays	<1332	>1403	>1403		LG	(231)
2.8 GHz	1358	1435	1538		1.2	
600 MHz	1334	1505	1811		1.1	
DS cont.	1420		1901		3	

1967	December 29	☉	McM 9120			Event 229	(−1 0 0)	CFI = 3
Hα		0047	0050	0100		S27 W78	1N	$\Delta t \sim 0.2^{\rm h}$
SWF		0049	0052	0111			1−	(11)
Soft X-rays		0045	0050	0145			SIG	(229)
Hard X-rays		0047	0049	0052			SM	(161, 163)
2 GHz		0047	0048	0050		U1P9	1.1	
DS type III(+V)		0046		0049			2	

1967	December 29	◐	McM 9128			Event 229	(−1 0 0)	CFI = 3
Hα		0106	0110	0140		S15 W22	1B	
SWF		0111	0114	0126			1−	(11)
Soft X-rays		no data						
Hard X-rays		0100	0102	>0108			SM	(161, 163)
3.7 GHz		0105	0109	0130		P9(2.2)	1.5	
DS type III(+V)		0130		0138			2	

1967	December 29	☉	McM 9128			Appendix A		CFI = 0
Hα		1859		1911		S18 W36	SN	$\Delta t \sim 0.3^{\rm h}$
SID		no event						
Soft X-rays		1829	1838	1945			VSM	(229)
Hard X-rays		no data						
5 GHz		1857	1905	1908			0.7	
DS type III		1857		1902			3	

1968	January 5	●	McM 9146			Event 232	(−1 0 0)	CFI = 8
Hα		0458	0459	0504		N12 E72	2B	$\Delta t \sim 9^{\rm h}$
SWF		0458	0500	0510			1−	(11)
Soft X-rays		0458	0501	0510			SIG	(229)
Hard X-rays		0456	0500	>0511			EB	(231)
2.0 GHz		0456	0458	0500		U1/9(2.7)	2.2	
200 MHz		0457	0458	0502			2.8	
DS type III(+V)		0457		0459			3	
DS type II		0459–0505, 0501–0506						

1968	January 9	☉	McM 9146			Appendix A		CFI = 7
Hα		0507	0511	0530		N09 E26	2N	$\Delta t \sim 2^{\rm h}$
SWF		0508	0510	0524			1−	(11)
Soft X-rays		0504	0511	>0550			SIG	(229)
Hard X-rays		<0522	<0522	>0617			EB	(231)
2 GHz		0508	0511	0515			2.5	
200 MHz		0512	0512	0516			2.4	
DS		no event						

1968	January 10	⊘	McM 9145			Appendix A		CFI ⩾ 3
Hα	2145	2149	2213		S26 W26	1B		
SES	2146	2148	2152			1−	(11)	
Soft X-rays	2140	2151	2210			SM	(229)	
Hard X-rays	2140	> 2152	> 2152			EB	(231)	
2 GHz	2145	2148	2151			1.7		
DS type III	2146		2151			3	also uncl.	
DS type II	2152		2158			1	2157	

1968	January 11	●	McM 9145			Event 233	(−1(1)0)	CFI = 9
Hα	1659	⌈ 1701 ⌊ 1708	1728		S25 W38	1B		$\Delta t \sim 1^h$
SWF	1701	1707	1732			1+	(11)	
Soft X-rays	< 1659	1702	1740			SIG	(229)	
Hard X-rays	no data							
2.8 GHz	1659	1701	1711		P7	2.3	(124)	
200 MHz	1702	1703	1706			> 2.5		
DS type III(+V)	1658		1702			3	(124)	
DS type II	1703		1717			2	(124)	
DS type IV	1705		1845			2	(124)	

1968	January 12	☉	McM 9146			Event 234	(0(1)0)	CFI = 0
Hα	0019	0021	0032		N11 W24	SN	(11) $\Delta t \sim 1^h$	
SID	no event							
Soft X-rays	no event							
Hard X-rays	< 0015	0023	> 0025			EB	(231)	
2 GHz	no event						no event on 200 MHz	
DS type III	0020		0022					

1968	January 12	●	McM 9145			Event 235	(0 0 0)	CFI ⩾ 6
Hα	1807	1811	1832		S25 W53	2B		$\Delta t \sim 0.5^h$
SWF	1810	1814	1835			1+	(11)	
Soft X-rays	1805	1812	1830			LG	(229)	
Hard X-rays	1801	> 1814	> 1814			EB	(231)	
2.8 GHz	1806	1810	1812		P7(2.3)	2		
600 MHz	1807	1810	1814			1.4		
DS type III	1806		1811			3		
DS type II	1812–1816, 1821–1823					2		

1968	January 14	☉	McM 9153			Event 236	(−2 0 0)	CFI ⩾ 6
Hα		0725	⌈0729 ⌊0734	0751		N24 W09	1B	Δt ~ 6ʰ
SWF		0726	0729	0746			1−	(11)
Soft X-rays		0721	0732	1000			LG	(229)
Hard X-rays		no data						
3 GHz		0725	0731	0735		P9(2.5)	2	⌉ DS no data
200 MHz		0732	0732	0733			2.8	⌋

1968	January 17	●	McM 9146			Event 237	(−1 0 0)	CFI ⩾ 5
Hα		0534	0543	0558		N08 W90	1N	Δt ~ 1.5ʰ
SWF		0501	0513	0612			2	(11)
Soft X-rays		0502	0519	>0525			LG	(229)
Hard X-rays		<0505	>0511	>0524			LG	(231)
2 GHz		0538	0541	0543			1	
200 MHz		no event						
DS type II		0557		0602			1	

1968	January 29	●	McM 9184			Appendix A		CFI = 10
Hα		1537	1540	1558		N14 E28	1B	Δt ~ 16ʰ
SWF		1538	1544	1559			1+	(11)
Soft X-rays		1510	1516	1525			VSM	(229)
Hard X-rays		no data						
2.8 GHz		1537	1539	1542		P5(2.9)	2.6	
200 MHz		1537	1542	1543			4.5	
DS type III		1538		1539			3	
DS type II		1540		1547			3	

1968	January 31	◐	McM 9184			Event 238	(−1 0 0)	CFI = 4
Hα		2129	⌈2132 ⌊2145	2200		N15 W18	2N	Δt ~ 12ʰ
SWF		2129	2157	2133			1	(11)
X-rays		no data						
2.8 GHz		no event						
200 MHz		no event						
DS type I(+cont.)		2000		2400			2	

1968	February 1	⊙	McM 9193+9184		Event 239	(−1 0 0)	CFI ⩾ 4
Hα		1756	⌈ 1801 ⌊ 1810	1830	S15 E48	SN	
	or						Δt ~ 0.25^h
Hα		1802	⌈ 1804 ⌊ 1811	1835	N09 W35 N17 W20	SN	
SWF		1759	1805	1830		1+	(11)
Soft X-rays		1800	~ 1804	1840		SIG	(229)
Hard X-rays		no data					
2.8 GHz		1802	1803	1810	U10/15(2.9)	2.3	
300 MHz		1801	1803	1807		1.9	
DS type III(+V)		1802		1804		3	
DS uncl.		1804		1810		2	
DS type III		1811–1815, 1820–1823					
DS type II		1820		1824		2	

1968	February 1	●	McM 9184		Appendix A		CFI = 4
Hα	1915	1920	2005	N16 W16	1N		Δt < 0.5^h
SWF	1917	1923	1936		1		(11)
Soft X-rays	1917	1920	1935		SIG		(229)
Hard X-rays	no data						
2.7 GHz	1916	1918	1925	P10(2.3)	1.8		
300 MHz	1917	1918	1921		2.1		
DS type III(+V)	1917		1924		3		
DS type I	1917		2000		1		
DS type II	1921		1923		3		
DS type III	1924		1928	Dkm			

1968	February 2	●	McM 9184		Event 240	(−1 0 0)	CFI = 8
Hα	0541		0605	N15 W24	1B		Δt ~ 0.3^h
SWF	0542	0546	0605		1−		(11)
Soft X-rays	0540	0546	0630		LG		(229)
Hard X-rays	no data						
2 GHz	0540	0543	0547	P9(2.5)	2.3		
200 MHz	0541	0547	0549		3.1		
DS type III(+V)	0542		0545		2		
DS type II	0545		0548		2		

1968	February 17	●	McM 9204		Event 245	(0 1 0)	CFI = 7
Hα	0252	0254	0313	N17 W47	1B		Δt ~ 0.3^h
SWF	0252	0255	0311		1		(11)
Soft X-rays	0251	0256	>0320		LG		(229)
Hard X-rays	0251	0256	>0309		LG		(231)
2 GHz	0251	0256	0323	P10 ·	2		(124)
200 MHz	0251	0253	0258		2.5		
DS type III(+V)	0251		0257		2		(124)
DS type II	0259		0307		2		(124)

1968	February 18	●	McM 9206			Appendix A		CFI ≥ 1
Hα		0851		1055	S16 W47	1N		$\Delta t \sim 0.5^{\text{h}}$
SID		no event						
Soft X-rays		0855	0935	> 1100			SM	(229)
Hard X-rays		< 0857	0906	> 0909			SM	(231)
2 GHz		no event						DS no data
200 MHz		no event						

1968	March 21	☉	McM 9266			Event 253	(−2 0 0)	CFI = 4
Hα		1421		1515	N17 W54	1B		$\Delta t \sim 1^{\text{h}}$
SWF		1427		1456		1		(11)
Soft X-rays		1421	1428	> 1443		SIG		(229)
Hard X-rays		no data						
2.7 GHz		1417	1431	1436		1.7		
600 MHz		1427	1432	1439		1.3		
DS type II		1423–1440, 1439–1454				3		
DS type III		1426		1438		2		

1968	March 23	☉	McM 9279			Appendix A		CFI = 3
Hα		0935	0940	0953	S23 E33	1B		$\Delta t \sim 2.5^{\text{h}}$
SWF		0938		1010		1		(11)
Soft X-rays		~ 0936	0940	0950		SM		(229)
Hard X-rays		no data						
3 GHz		0937	0940	0942		0.9		
200 MHz		0939	0940	0941		1.6		
DS type III		0940–0941				1		

1968	March 27	◐	McM 9273			Event 255	(−2 0 0)	CFI = 7
Hα		< 1757	⌈ 1802 ⌊ 1814	> 1844	S12 W42	2B		
SWF		1801	1820	1839		2−		(11)
Soft X-rays		1754	1758	> 1800		SM		(229)
Hard X-rays		1756	1805	> 1809		EB		(231)
2.8 GHz		⌈ 1758 ⌊ 1807	1800 1816	1806 1822	P5(2)	1.7 1.5		
600 MHz		1814	1815	1820		1.9		
DS type I(+cont.)		1740–1800				2		
DS type III		1823				1		

1968	March 28	☉	McM 9273		Appendix A		CFI = 4
Hα	0321	0329	0400	S15 W51	1N		Δt ~ 1.5^h
SWF	0323	0328	0351		1		(11)
Soft X-rays	0320	0332	>0335		SM		(229)
Hard X-rays	<0320	0332	>0345		EB		(231)
2 GHz	0321	0324	0351	P1(2.4)	1.4		
700 MHz	0321		0333		2.4		
DS type III(+V)	0321		0331		2		also uncl.
DS type III	0331		0341		2		
DS type II	0331		0341		2		

1968	March 30	☉	McM 9273		Event 256	(−1 0 0)	CFI ≥ 1
Hα	0025	0035	0045	S15 W65	SF		(11) Δt ~ 0.5^h
SID	no event						
Soft X-rays	no event						
Hard X-rays	no data						
2.7 GHz	0034	0036	0039		1.3		
200 MHz	no data						
DS type I	0000		0038		1		
DS type III(+V)	0035		0037		2		

1968	April 19	☉	McM 9313		Event 262	(−2 0 0)	CFI = 5
Hα	<1640	[1648 / 1658	1738	N20 W62	SN		Δt ~ 1.5^h
SID	no event						
Soft X-rays	~1640	1700	~1830		SIG		(229)
Hard X-rays	1609	1613	1618		SIG		(231) no data after 1636
2 GHz	1627	1642	1646	U3	2		
600 MHz	[1607 / 1627	1611 / 1629	1611 / 1631		2.8 / 0.6		
DS type III	1610		1613		3		
DS type IV	1650		1818		3		

1968	April 27	☉	McM 9337		Appendix A		CFI = 0
Hα	1729	1732	1741	N08 W65	SN		(11) Δt < 0.5^h
SID	no event						
Soft X-rays	1742	~1750	>1750		VSM		(229)
Hard X-rays		1725			VSM		(231)
2 GHz	no event						
200 MHz	no event						
DS type III	1730		1732		2		

1968	May 3	●	McM 9372		Event 268	(−2 0 0)	CFI = 9
Hα	2123		2150	N18 E49	1B		$\Delta t \sim 2.5^{\mathrm{h}}$
SWF	2125	2132	2153		3		(11)
Soft X-rays	2120	⎡2121 ⎣2130	>2150		SM VLG		(229) (229)
Hard X-rays	2123	2129	2149		VLG		(229)
2.8 GHz	2125	2128	2132		2		
200 MHz	2128	2130	2134		2.3		
DS type II	2127		2143		3		
DS type III	2128		2147		2		

1968	May 4	⊙	McM 9364		Event 269	((−2) 0 0)	CFI = 7
Hα	0427	0428	0453	N20 E18	1B		$\Delta t \sim 9^{\mathrm{h}}$
SWF	0428	0431	0436		1−		(11)
Soft X-rays	0427	0430	>0435		SM		(229)
Hard X-rays	no data						
2 GHz	0427	0429	0437	P9(2.4)	1.9		
200 MHz	0428	0428	0434		3.3		
DS type III(+V)	0428		0430		2		
DS unclassified	0429				2		

1968	May 8	⊙	McM 9382		Appendix A		CFI = 5
Hα	<1415		>1419	N23 E58	SF		(11) $\Delta t \sim 2.5^{\mathrm{h}}$
SPA	1417	1425	1436				(11)
X-rays	no data						
2.8 GHz	1415	1418	1420	U15/3	2.3		
200 MHz	1414	1415	1416		2.8		
DS type II	1416		1426		2		
DS type III	1406				1		

1968	May 10	⊙	McM 9364		Appendix A		CFI = 5
Hα	0840	0845	0854	N20 W70	1N		$\Delta t \sim 0.2^{\mathrm{h}}$
SID	no event						(11)
X-rays	no data						
3 GHz	0842	0843	0852		0.9		
200 MHz	0846	0852	0853		3.0		
DS type III	⎡0840–0844, 0846–0847, 0850–0950 ⎣0852–0853				2 3		

1968	May 21	⦰		McM 9410		Event 271	(−2 0 0)	CFI ⩾ 3
Hα		1952	1957	2003	N19 E90	SN	(11)	$\Delta t \sim 6^{h}$
SWF		1952	2020	2050		1−	(11)	
Soft X-rays		1948	2020	2230		LG	(229)	
Hard X-rays		1945	2018	2050		LG	(229)	
2.8 GHz		2001	2010	2028		1.3		
600 MHz		1957	1958	2001		1.2		
DS type III		1955		2004		3		
DS type II		2024		2036		3		
DS unclassified		2004		2012				

1968	June 7	⊙		McM 9429		Event 273	(−1 0 0)	CFI = 4
Hα		1226	1250	1400	S13 E16	1N		$\Delta t \sim 1.5^{h}$
SWF		1231	1240	1400		1		
Soft X-rays		<1228	1236	>1246			(11)	
Hard X-rays		no data					(229)	
2.8 GHz		1226	1230	1238		1		
DS type I		1231		1410		1		intermittent
DS cont.		<1235		1435		2		
DS type III		1304		1305				

1968	June 7	⦰		McM 9423		Event 273	(−1 0 0)	CFI = 1
Hα		<1610		>1640	N21 W35	1N		
SWF		no event						
X-rays		no data						
2.8 GHz		no event						
200 MHz		no event						
DS type III		1554–1556, 1611–1611				1		
DS type III(Dkm)		1619–1620				2		
DS cont.		1545		1653		2		

1968	June 9	⊙		McM 9443		Appendix A		CFI = 6
Hα		0025	0036	0155	N14 E38	2N		$\Delta t \sim 1.5^{h}$
SWF		0034	0050	0058		1		(11)
Soft X-rays		0025	0034	>0101		LG		(229)
Hard X-rays		no data						
2.7 GHz		0026	0036	0054		2		
500 MHz		0028	0038	0117		1.4		
DS type III		0028–0035, 0041–0054				3		
DS cont.		0028		0131				
DS type I		0029		0053				
DS type II		0107		0122		1		

1968 June 9	●	McM 9429		Event 274	(2 3 0)	CFI = 14
Hα	0830	0854	1030	S14 W09	3B	$\Delta t \sim 1^h$
SWF	0840		1000		3	(11)
Soft X-rays	0828	⎡0831 0835 0844 ⎣0854	> 1025		VLG	(229)
Hard X-rays	~ 0834		0845	·	3	(232)
3 GHz	0839	0851	1130	P7(3.1)	2.9	(124)
200 MHz	0847	0848	0854		3.8	
DS type IV(dm)	0839		0940		2?	(124)
DS type IV(m)	0900		0910			(124)
DS type III	0839		0846		2	

1968 June 17	☉	McM 9459		Appendix A		CFI = 3
Hα	< 0255		0410	N26 E41	1N	$\Delta t \sim 6^h$
SID	no event					
Soft X-rays	0248	⎡0255 ⎣0312	> 0340	·	SIG	(229)
Hard X-rays	0249	0305	0340		SIG	(229)
2 GHz	0250	0258	> 0340		1	
600 MHz	0255	0300	0602		1.3	
DS type II	0301		0326		1	
DS type III	0324		0325		1	
DS cont.	0324		0325		1	

1968 June 26	☉	McM 9462		Appendix A		CFI = 5
Hα	< 0509	0513	0532	N14 W68	SB	$\Delta t > 0.5^h$
SWF	0512	0516	0530		1−	(11)
X-rays	no data					
2 GHz	0512	0512	0513		1	
200 MHz	0512	0512	0513		3.1	
DS type III(+V)	0511		0512		3	
DS type III	0511		0524		3	

1968 July 6	■	McM 9503		Event 276	(0 1 0)	CFI > 8
Hα	0943	⎡0945 ⎣0956	1033	N13 E89	1N	$\Delta t \sim 0.3^h$
SWF	0948	1010	1020		2+	(11)
Soft X-rays	0941	0950	1220		VLG	(229)
Hard X-rays	0942	0942	0943		3	(232)
3 GHz	0941	0949	1131	U1/70	2 7	(124) ⎤ DS no
200 MHz	0945	0946	1000		> 4.1	⎦ data
Type IV?	0944		1000		3	(124)

1968	July 8	●	McM 9503		Event 277	(1 1 0)	CFI = 17
Hα	<1708	1715 / 1725	1825	N13 E58	3B		Δt ~ 1^h
SWF	1706	1710	2031		3+	(11)	
Soft X-rays	1705	~1725	2200		VLG	(229)	
Hard X-rays	1655	1710	>1900		VLG	(229)	
3 GHz	1704	1712	1837	P15	3.2	(124)	
200 MHz	1709	1711	1748		5.7		
DS type II	1709		1752		3	(124)	
DS type IV	1709		1830		3	(124)	
DS type III	1712		1721		3		
White light	1707	1710				(230) possible	

1968	July 9	⊘	McM 9503		Event 277	(1 1 0)	CFI = 7
Hα	1809	1817	1915	N13 E40	2B	
SWF	1812	1820	1913		3−	(11)
Soft X-rays	1807	1820	~2400		LG	(229)
Hard X-rays	no data					
2.8 GHz	1807	1819	1835	P5	2.3	
600 MHz	1814	1819	1831		1.8	
DS type III	>1805		1806		3	
DS type I	1840		2153		1	

1968	July 12	⊙	McM 9503		Event 278	(0 1 0)	CFI = 5
Hα	<0000	0009	0055	N12 E10	2N		Δt ~ 5^h
SWF	0004	0008	0045		1+	(11)	
Soft X-rays	11^{d}2358	0014	0800		LG	(229)	
Hard X-rays	11^{d}2359	0007?	>0030		LG	(229)	
2 GHz	0000	0011	0023		1.9		
600 MHz	0003	0003	0021		2.2		
DS type III	0054		0055		1		
DS type I	0000		0146		1		

1968	July 12	⊙	McM 9499		Event 279	(1 2 0)	CFI = 8
Hα	<1348		1530	N11 W20	2N		Δt ~ 0.25^h
SWF	1342	1420	1440		1+	(11)	
					SM	(229)	
Soft X-rays	1256	1310 / ~1415	2000		SIG	(229)	
Hard X-rays	no data						
2.7 GHz	1343	1401	1510	P1	2.4	(124)	
600 MHz	1343	1414	1432		3.3		
DS type IV	1354		1455		3	(124)	
DS cont.	<1354		1455		2		
DS type III	1342		1355		2		

1968	August 3	●	McM 9567		Event 281	(−2 0 0)	CFI = 9
Hα	0313	0320	0442	N10 E75	2N		$\Delta t \sim 3^{\mathrm{h}}$
SWF	0311	0315	0445		2+	(11)	
Soft X-rays	0310	0322	0700		LG	(229)	
Hard X-rays	no data						
3 GHz	0312	0317	0507	P4	2.8	(124)	
200 MHz	0315	0319	0335		2.3		
DS type II	0316		0406		2	(124)	
DS type III	0330		0333		1		
DS type I	0358		0404		1		

1968	August 6	☉	McM 9545+9567		Event 283	(−1 0 0)	CFI = 3
Hα	1318	1321	>1339	N15 W88	SN	(11)	$\Delta t \sim 6^{\mathrm{h}}$
Hα or	1315	1315	>1320	N07 E31	SB	(11)	
	<1344	1347	1418	N13 E32	SN	(11)	
SWF	1321		1332		2	(11)	
Soft X-rays	<1330	1333	1540		SIG	(229)	
Hard X-rays	no data						
2.7 GHz	1313	1323	1447		1		
DS type I	1246		1330		1		

1968	August 14	●	McM 9567		Event 284	(−1 0 0)	CFI = 7
Hα	<1327	1332	1354	N13 W80	1B		$\Delta t < 0.5^{\mathrm{h}}$
	1400	1401	1425	N13 W82	1N		
SWF	1328	1344	1500		3	(11)	
Soft X-rays	∼1325	<1400	1540		LG	(229)	
Hard X-rays	1325	1326	∼1327		2	(232)	
2.8 GHz	1326	1339	1447	P7	2.1	(124)	
600 MHz	1326	1333	1346		1.1		
DS type II	1339		1400		2	(124)	
DS cont.	1328		1530			(124)	
DS type III	1327		1330		1		
DS type I	1259		1448				

1968	August 21	●	McM 9593		Appendix A		CFI = 8
Hα	0146	0149	>0158	S16 W43	1N		$\Delta t \sim 11^{\mathrm{h}}$
SWF	0148	0148	0300		2	(11)	
Soft X-rays	<0140	>0215	0400		SIG	(229)	
		0144			SM	(229)	
Hard X-rays	0143	0148	0225		LG	(229)	
		0150			LG	(229)	
2 GHz	0147	0148	0155		2.4		
200 MHz	0147	0148	0149		2.7		
DS type III(+V)	0147		0150		2		
DS type II	0154		0200		1		
DS type I	0000		0632 storm		2		

1968	September 1	●	McM 9611			Appendix A		CFI = 4
Hα		1622	1634	1645	N16 W90	SN	$\Delta t \sim 0.25^h$
SWF		1613	1618	1654		1	(11)
Soft X-rays		< 1610	1625	1800		SIG	(229)
Hard X-rays		1607	⌈1614 ⌊1626	1655		SIG SIG	(229) (229)
2.8 GHz		1623	1623	1624		0.5	
DS unclassified		1621		1642		1	
DS type II		1626		1636		3	
DS type III		1614		1616		2	
DS type IV		1636		1638		3	

1968	September 3	☉	McM 9634			Appendix A		CFI ⩾ 4
Hα		⌈1553 ⌊1715	1625 1723	> 1706 1743	S15 E07	1N	$\Delta t < 3^h$
SWF		⌈1620 ⌊1715	1628 1729	1638 1752		1− 1	(11) (11)
Soft X-rays		no event					
Hard X-rays		no data					
2.8 GHz		1620	1625	1710		0.3	
DS cont.		1601		1905		2	
DS type III		1609		1944			
DS type I		1540		1836		2	

1968	September 4	●	McM 9630			Appendix A		CFI = 9
Hα		< 0031	0042	0200	N13 W14	1N	$\Delta t \sim 2^h$
SWF		0040	0045	0140		1+	(11)
Soft X-rays		0028	~ 0050	0400		LG	(229)
Hard X-rays		no data					
2.7 GHz		0029	0040	0057		1.2	
500 MHz		0032	0042	0126		2.3	
DS type IV		0033		0054		2	
DS type III		0033–0045, 0045–0055				3	
DS type II		0041		0054		2	
DS type I		0000		0054		1	

1968	September 26	●	McM 9687			Event 287	(1 1 0)	CFI = 9
Hα		0026	0031	0108	N14 E34	2B	$\Delta t \sim 1^h$
SWF		0028	0040	0140		2	(11)
Soft X-rays		0025	0030	0130		LG	(229)
Hard X-rays		no data					
2.7 GHz		0025	0031	0140		2.4	
200 MHz		0027	0030	0039		2.3	
DS type III(+V)		0026		0033		2	
DS type II		0034		0047		2	
DS type III		0038–0046, 0055–0104				2	

1968	September 28	●	McM 9687		Appendix A		CFI = 7
Hα	<0155	0202	>0230	N13 E06	1B		$\Delta t \sim 5^h$
SWF	0153	0159	0343		2+	(11)	
Soft X-rays	0150	0205	0420		LG	(229)	
Hard X-rays	0153					(232)	gradual
3.8 GHz	0147	0201	0340	P2	2.6	(124)	
200 MHz	0152	0157	0217		2.5		
DS type III	0156		0204		2		
DS uncl.	0153		0204				

1968	September 28	●	McM 9692		Event 288	(1 1 0)	CFI ≥ 8
Hα	0721		>0930	S18 E39	2B		$\Delta t \sim 2^h$
SWF	0735	0830	1005		2	(11)	
Soft X-rays	0720	0830	1400		LG	(229)	
Hard X-rays	0716	0815	>0845		LG	(229)	
2.8 GHz	0735	0752	1045	P3	2.8	(124)	]DS no
200 MHz	0738	0739	0753		2.3		]data

1968	September 29	●	McM 9687		Appendix A		CFI ≥ 9
Hα	<0920	0940	1050	N13 W13	2B		$\Delta t \sim 0.25^h$
SWF	0931	0937	1045		2+	(11)	
Soft X-rays	0920	0942	>1110		LG	(229)	
Hard X-rays	0932	0932	>0933		>2	(232)	
2.8 GHz	0929	0933	1020		2.8	(124)	
200 MHz	0931	0932	0955		3.3		]DS no
Type IV?	0931		0955		2	(124)	]data

1968	September 29	●	McM 9678		Event 289	(1 2 1)	CFI = 12
Hα	1617	1623	1700	N17 W51	2B		$\Delta t \sim 0.5^h$
SWF	1619	1621	1721		2	(11)	
Soft X-rays	1616	1624	1900		LG	(229)	
Hard X-rays	~1617				>2	(232)	
3 GHz	1616	1621	1702	P7	2.9	(124)	
200 MHz	1619	1620	1621		3.8		
DS type II	1626		1644		3	(124)	
DS type II	1619		1639		3		
DS type IV	1636		1650		3	(124)	
DS type III	1619		1627		3		

1968	October 3	⊙	McM 9687		Appendix A		CFI = 0
Hα	1239		1310		N14 W68	SN	$\Delta t < 0.5^h$
SID	no event						
Soft X-rays	1232		~ 1400				(229)
Hard X-rays	no data						
3 GHz	1239	1300	1343			0.4	
DS	no event						

1968	October 3	●	McM 9692		Event 290	(1 2 0)	CFI = 9
Hα	2343		4^d0200		S17 W36	2B	$\Delta t \sim 1^h$
SWF	2347	2354	4^d0240			2−	(11)
Soft X-rays	2330	4^d0015	0600			SIG	(229)
Hard X-rays	no data						
2.8 GHz	2343	4^d0020	0313		P7	1.8	(124)
200 MHz	2347	4^d0004	0151			2	
DS type II	2359		4^d0027			1	(124)
DS cont.	2359		4^d0015			2	(124)

1968	October 23	●	McM 9740		Event 291	(0 0 0)	CFI = 11
Hα	2352	$\lceil 24^d0004$ / $\lfloor\ \ 0021$	0140		S13 E59	2N	
							$\Delta t \sim 2^h$
SWF	2355	2358	24^d0055			2+	(11)
Soft X-rays	2340	24^d0012	0500			LG	(229)
Hard X-rays	2348	24^d0006	0124			LG	(229)
2 GHz	2354	24^d0004	0054			2.5	
200 MHz	2354	24^d0001	0010			2.2	
DS type III(+V)	2356		2357			2	
DS type II	2358		24^d0014			2	
DS type III	2353		24^d0030			1	
DS type IV	24^d0003		0055			1	

1968	October 26	⊙	McM 9740		Event 292	(−1 0 0)	CFI = 5
Hα	0046	0120	0218		S20 E32	1N	$\Delta t \sim 6.5^h$
SID	no event						
Soft X-rays	no event						
Hard X-rays	no data						
3 GHz	0104	0132	0234		P0.7	1.7	(124)
200 MHz	0100	0156	0210			2.3	
DS type III	0045–0048, 0059–0118					1	
DS type I	0120		0215				(124)
DS type IV(dm)	0120		0208				(124)

1968	October 27	●	McM 9740			Event 293	(−1 0 0)	CFI = 12
Hα		[1232	1237	> 1306		S17 E17	1B	Δt ~ 1.5[h]
		[< 1318	1322	1500			2N	
SWF		1235	1335	1545			3+	(11)
Soft X-rays		[1230	1239	> 1305			VLG	(229)
		[< 1305	1330	1800			VLG	(229)
Hard X-rays			< 1406	> 1438				(229)
2.8 GHz		[1233	1236	1255		P7	3	(124)
		[1306	1319	1635		P4	2.9	(124)
200 MHz		1305	1313				2.8	
DS cont.		1230		1630				(124)
DS type IV		1307		1830			3	(124)
DS type I		1330		1800			3	

1968	October 29	☉	McM 9740			Event 294	(0 0 0)	CFI = 14
Hα		< 1222	[1223	> 1300		S16 W12	2B	Δt ~ 3[h]
			[1234					
SWF		1218	1230	1328			3−	(11)
Soft X-rays		1210	~ 1233	> 1400			VLG	(229)
Hard X-rays		< 1211	[1217	> 1319			VLG	(229)
			[1227					
2.8 GHz		1219	1226	1307		P9	2.4	(124)
200 MHz		1221	> 1314	> 1314			4.7	
DS type IV		1225		1540			3?	(124)
DS type III(+V)		1245		1252			3?	(124)
DS type I		< 1328		2120 continuum			2	

1968	October 29	☉	McM 9735 + 9740			Event 294	(0 0 0)	CFI = 9
Hα		[1515	1518	> 1525		N15 W82	SN	(11) Δt ~ 0[h]
		[1515	1521	1558		S14 W19	1N	(11)
SID		no event						
Soft X-rays		no distinct event						
Hard X-rays		no data						
2.8 GHz		1516	1523	1615		P0.6	3.6	(124)
200 MHz		1518	1525				3.8	
DS type IV		1516		1538			3	(124)
DS cont.		1524		2125			3	(124)

1968	October 30	●	McM 9740		Event 295	(0 0 0)	CFI = 9
Hα	<1235	⎡1252 ⎣1346	1445	S18 W26	2B		$\Delta t \sim 0^h$
SWF	1243	1250	1452		3+	(11)
Soft X-rays	1222	1255	1305		LG	(229)
Hard X-rays	1232	1233	1234		1	(232)
2.8 GHz	1235	1250	1347	P3	1.5	(124)
300 MHz	1334	1337	1443		2	
DS type IV	1255		1545		2	(124)
DS type I	<1328		1600		3	
DS type III	1337–1337, 1358–1432				2	
DS cont.	<1400		1740		3	

1968	October 30	●	McM 9740		Event 296	(2 3 0)	CFI = 13
Hα	2340	⎡2358 ⎣2412	31^d0135	S14 W37	3B		$\Delta t \sim 0.2^h$
SWF	2343	2355	31^d0301		2+	(124)
Soft X-rays	2338	31^d0012	>0600		VLG	(229)
Hard X-rays	2343	2344	2345		2	(232)
3.7 GHz	2340	31^d0011	0440	P6	3.3	(124)
200 MHz	2339	31^d0009	0039		2.9	
DS type II	2359		2405		2	(124)
DS type III(+V)	31^d0009		0010		3	(124)
DS type IV	31^d0002		0035		1	(124)
DS type III	2351–2428, 2358–2358					
DS unclassified	2344		2355			

1968	November 1	●	McM 9740		Event 297	(2 3 0)	CFI = 10
Hα	0814	⎡0843 ⎣0903	>0930	S16 W47	2N		$\Delta t \sim 4^h$
SWF	0828	0858	1012		2+	(11)
Soft X-rays	<0810	0908	>1015		LG	(229)
Hard X-rays	no data					
3 GHz	0800	0912	1052	P5	3.4	(124)
200 MHz	<0830	0841	1007		2.2	
DS type II	0851		0901		2	(124)
DS type IV	0853		0952		2?	(124)
DS type I	0750		1508 storm		2	

1968	November 4	■	McM 9740		Event 298	(1 2 0)	CFI ⩾ 6
Hα	< 0524	0529	> 0606	S15 W90	1B		Δt ~ 0.25ʰ
SWF	0515	0516	0628		2	(11)	
Soft X-rays	< 0517	~ 0525	1200		VLG	(229)	
Hard X-rays	no data						
2.8 GHz	0513	0523	> 0710	/9	3.3	(124)	
600 MHz	0515	0517	0524		2.3		
DS type II	0517		0550		3	(124)	
DS cont.	0518		0522			(124)	
DS type III(+V)	0514		0516				

1968	November 18	■	McM 9760		Event 301	(2 3 3)	CFI = 9
Hα	< 1026	1035	1127	N21 W87	1B		Δt ~ 0ʰ
SWF	1028	1029	1154		3	(11)	
Soft X-rays	1024	⌈ 1029 ⌊< 1038	1800		VLG VLG	(229) (229)	
Hard X-rays	1025	1026	1027		⩾ 3	(232)	
2.8 GHz	1026	1030	1242	P10	3.1	(124)	
200 MHz	1027	1032	1036		3.5		
DS type IV	1026		1050		3?	(124)	
DS type II	1026		1105		3?	(124)	
DS type III	1045		1104		2		

1968	December 2	●	McM 9802		Event 304	(1 2 0)	CFI ⩾ 9
Hα	⌈ 2115 ⌊< 2202		> 2130 2400	N20 E89 N19 E80	1N 3N		Δt ~ 5.5ʰ
SWF	2113	2118	2126		1−	(11)	
Soft X-rays	2100	2120	3ᵈ0600		LG	(229)	
Hard X-rays	no data						
2.8 GHz	2105	2116	> 2247		2.4	(124)	
400 MHz	2105	2119	2140		1.7		
DS type II	2114		2141		3	(124)	
DS type IV	2117		2152		3	(124)	
DS type III	2115		2117		3		

1968	December 24	☉	McM 9842		Event 306	(−1 0 0)	CFI = 7
Hα	2227	2230	> 2242	N19 E29	1B		Δt ~ 0ʰ
SWF	2227	2230	2249		1−	(11)	
Soft X-rays	2226	2230	2245		SIG	(229)	
Hard X-rays	no data						
2.8 GHz	2227	2229	2236		2.2		
200 MHz	2227	2228	2237		2.4		
DS type III(+V)	2227–2230, 2235–2247				3		
DS type IV	2229		2248		3		
DS unclassified	2230		2233		1		
DS type II	2224		2224		1		
DS type II	2230		2242		2		

1968	December 27	●	McM 9842		Event 307	(−1 0 0)	CFI = 10
Hα	<1050	1056	1150	N16 E03	2B		Δt ~ 2^h
SWF	1055	1056	1128		2+	(11)	
Soft X-rays	1045	1058	1400		VLG	(229)	
		⌐1049			SM	(229)	
		1052			SIG	(229)	
Hard X-rays	1048	1055			VLG	(229)	
		└1103	1150		LG	(229)	
3 GHz	1053	1056	1245	/10	2.9	(124)	
200 MHz	1056	1057	1101		3.3		
DS type III	1054		1101		3		
DS type II	1055		1115		3	(124)	
DS type IV	1100		1400		?	(124)	

1968	December 28	●	McM 9838		Appendix A		CFI = 1
Hα	0957	⌐0959	1027	S16 W38	SN		Δt ~ 0.5^h
		└1006					
SWF	no event.						
Soft X-rays	no distinct event						
Hard X-rays	no data						
3 GHz	1007	1008	1010		1.3		
200 MHz	no event						
DS type III	0957–0958, 1003–1008						

1968	December 30	☉	McM 9847		Appendix A		CFI = 5
Hα	1424	1427	1444	N13 E26	SN		Δt ~ 4.5^h
SWF	no event						
Soft X-rays	1424	1427	1435		VSM	(229)	
Hard X-rays	no data						
3 GHz	1425	1426	1427		1.7		
200 MHz	1424	1425	1427		3.1		
DS type III	1424		1426				
DS type III(+V)	1424		1428		3		
DS type I	1415		2343		1		
DS type II	1433		1437		2		

1969	January 2	☉	McM 9841		Appendix A		CFI ≥ 3
Hα	1140	1143	1203	S22 W83	1N		Δt < 1^h
SWF	no event						
Soft X-rays	1123	⌐1130	>1230		SM	(229)	
		└1143			SIG	(229)	
Hard X-rays	no data						
2.8 GHz	no event						DS no data
500 MHz	1143	1143	1145		2		

1969	January 3	∅	McM 9855		Appendix A ᴀ		CFI ⩾ 4
Hα		0439	⌈ 0443 ⌊ 0503	0535	N26 E72	1N	
SWF		0442	0504	0607		2	(11)
Soft X-rays		0430	0450	0700		LG	(229)
Hard X-rays		no data					
2 GHz		0430	0446	0540		0.8	DS no data

1969	January 17	⊙	McM 9873		Event 308	(0 0 0)	CFI ⩾ 5
Hα		< 1242	1247	1300	N16 E48	SB	$\Delta t \sim 1.5^h$
SWF		1247	1248	1310		1+	(11)
Soft X-rays		1240	1248	~ 1300		LG	(229)
Hard X-rays		no data					
3 GHz		1242	1247	1248	P15	2.2	
200 MHz		1242	1246	1247		2.8	
DS type III		1246		1247		1	

1969	January 17	⦸	McM 9873		Event 308	(0 0 0)	CFI ⩾ 6
Hα		1703	1705	1717	N16 E45	SB	
SWF		1705	1708	1718		1	(11)
Soft X-rays		1702	1705	1720		LG	(229)
Hard X-rays		no data					
2.8 GHz		1704	1705	1712	P35	2.7	
400 MHz		1703	1704	1720		3	
DS type III		1703		1705		3	
DS type III(+V)		1704		1714		3	
DS type IV		1705		1737		3	
DS type II		1705		1720		3	

1969	January 24	●	McM 9879		Event 309	(0 1 0)	CFI ⩾ 7
Hα		< 0706	0728	0930	N20 W08	3B	$\Delta t \sim 0.5^h$
SWF		0718	0729	0805		1	(11)
Soft X-rays		0700	0733	1200		VLG	(229)
Hard X-rays		0704	0727	~ 0840		LG	(229)
2.7 GHz		0705	0721	0845	P4	2.2	(124) ⌉ DS no
200 MHz		0713	0722	0736		2	⌋ data
Type II or III?		0716		0736		1	(124)

1969	February 23	●	McM 9946		Event 310	(−1 0 0)	CFI = 7
Hα	0442			0509	N12 W09	1N	$\Delta t \sim 0.25^h$
SWF	0445	0446		0518		1	(11)
Soft X-rays	<0442	0450		0520		LG	(229)
Hard X-rays	no data						
3.7 GHz	0442	0445		0504		2.3	
200 MHz	0439	0449		0517		2.9	
DS type III(+V)	0440–0447, 0451–0452					1	
DS type II	0447			0456		3	
DS type III	0451			0458		1	

1969	February 24	●	McM 9946		Appendix A		CFI = 4
Hα	1431	⌈1433 ⌊1510		1542	N11 W27	SN	$\Delta t \sim 0.3^h$
SWF	1508	1511		1518		1−	(11)
Soft X-rays	1430	1432		>1445		VSM	(229)
Hard X-rays	no data						
3 GHz	1427	1433		1440		1.4	
200 MHz	1427	1432		1434		2.7	
DS type III	1432			1434		2	
DS type I	1351			1501		1	

1969	February 24	●	McM 9946		Event 311	(−1 0 0)	CFI = 12
Hα	2305	2315		2400	N12 W31	2B	$\Delta t \sim 0.5^h$
SWF	2310	2325		2440		2+	(11)
Soft X-rays	2305	2321		25^d0145		LG	(229)
Hard X-rays	2307	2308		2310		2	(232)
3.7 GHz	2307	2314		2354	P6	2.9	(124)
200 MHz	2306	2313		2322		2.6	
DS type III(+V)	2307			2308		2	(124)
DS type II	2316			2322		3	(124)
DS type IV	2308			2430		3	(124)
DS unclassified	2308–2322, 2328–2329					2	
DS type III	2310, 2316–2319					3	

1969	February 25	●	McM 9946			Event 312	(1 2 3)	CFI = 13
Hα	0900	⌈ 0913 ⌊ 1001	1040		N13 W37	2B		$\Delta t \sim 0.25^{\mathrm{h}}$
SWF	0910	0916	0952			1	(11)
Soft X-rays	0858	⌈ 0906 ⌊ 0918	1110			VLG	(229)
Hard X-rays	⌈ 0901	0902	0903			2	(232)
	| 0905	0905	0906			3	(232)
	⌊~ 0909					≫ 3	(232)
2.8 GHz	0905	0912	0959		U3P10	3.3	(124)
200 MHz	0904	0910	0920			4.1	
DS type IV	0904		1130			3	(124)
DS type III	0901–0903, 0905–0906					2	
DS type I	0650		1540			1	
White light	0909	0912					(230)

1969	February 26	●	McM 9946			Event 313	(1 1 0)	CFI = 12
Hα	0418	⌈ 0427 ⌊ 0459	0525		N13 W46	2B		$\Delta t \sim 0.1^{\mathrm{h}}$
SWF	0419	0429	0545			2	(11)
Soft X-rays	0415	⌈ 0420 ⌊ 0430	0545			VLG	(229)
Hard X-rays	saturation						(232)
2 GHz	0416	0425	0510		P10	3.1	(124)
200 MHz	0423	0425	0740			>3.1	(124)
DS type III	0423		0426			3	(124)
DS type II	0425		0444			2	(124)
DS cont.	0425		0555			2	(124)

1969	February 27	●	McM 9946			Event 314	(1 1 0)	CFI = 12
Hα	1348	1413	1505		N13 W65	2B		$\Delta t \sim 0.5^{\mathrm{h}}$
SWF	1404	1410	1454			2	(11)
Soft X-rays	1356	1414	1600			VLG	(229)
Hard X-rays	1356	1357	1358			3	(232) complex
2.8 GHz	1400	1408	1530		P10	3.1	(124)
200 MHz	1404	1405	1417			3.6	
DS type II	1400–1405, 1405–1514					2	(124)
DS type IV	1405		1531			2	(124)
DS type III	1403		1416			2	

1969	March 12	●	McM 9966		Event 316	(0 1 0)	CFI = 14
Hα	1739	1742	1809	N12 W80	2B		$\Delta t \sim 0.5^{h}$
SWF	1739	1740	1945		3+		(11)
Soft X-rays	1739	1744	> 2000		VLG		(229)
Hard X-rays	1738				≫ 3		(232)
2.8 GHz	1738	1740	1932	P15	3.2		(124)
200 MHz	1740	1741	1838		3.2		
DS type II	1742		1800		3		(124)
DS type IV	1741		1748		2		(124)
DS type IV	1800		2035		3		(124)
DS type III	1745–1746, 1750–1756				3		
DS type I	1748		1920		2		
White light	1738	1741					(230)

1969	March 21	●	McM 9994		Event 319	(0 1 0)	CFI = 12
Hα	0139	⌈0149 ⌊0159	0330	N20 E17	2B		$\Delta t \sim 3^{h}$
SWF	0127	0147	0515		3		(11)
Soft X-rays	< 0146	0207	0600		VLG		(229)
Hard X-rays	⌈0147	0148	0148		3		(232)
	⌊0149	0150			3		(232)
3.7 GHz	0121	0153	0550	P7	3		(124)
200 MHz	0100	0149	0335		2.6		
DS type II	0150		0210		2		(124)
DS cont.	0150		0330		2		(124)
DS type III	0142		0152		2		
DS unclassified	0142		0152		2		

1969	March 21	⊘	McM 9994		Event 319	(0 1 0)	CFI = 14
Hα	1312	⌈1334 ⌊1405	1430	N19 E09	2B		
SWF	1330	1338	1500		3		(11)
Soft X-rays	1322	1341	1700		VLG		(229)
Hard X-rays	no event						(232)
2.8 GHz	1316	1334	1356		3.3		(124)
200 MHz	1328	1335	1350		3.1		
DS type IV	1315		1840		3		(124)
DS type II	1319		1346		2		(124)
DS type III	1320–1345, 1458–1459				2		

1969	March 24	⊙	McM 9994		Appendix A		CFI = 5
Hα	1449	1450	1510	N18 W42	1N		$\Delta t \sim 3^{h}$
SWF	1449	1455	1505		3–		(11)
Soft X-rays	1447	1451	> 1455		VSM		(229)
Hard X-rays	no data						
3 GHz	1449	1450	1451		1.3		
DS type I	1350		2400		1		

1969	March 27	●	McM 9994			Event 320	(−1 0 0)	CFI = 12
Hα		1315	⌈1327 ⌊1342	1430		N20 W69	2B	Δt ~ 0.75^h
SWF		1318	1334	1400			2−	(11)
Soft X-rays		1319	1335	1630			VLG	(229)
Hard X-rays		no data						
3 GHz		1318	1327	1530		P20	3	(124)
200 MHz ,		1324	1333	1333			3.5	
DS type IV		1336		1515			3	(124)
DS type II		1331		1339			2	(124)
DS type III		1336–1339, 1326–1328					3	

1969	March 30	⊡	McM 9994			Event 321	(1 1 2)	CFI = 10
Hα		< 0332		0400		N19 W90	1N	(11) Δt ~ 0^h
SWF		0248	0249	0545			2−	(11)
Soft X-rays		0247	⌈0248 ⌊0300	0400			LG	(229)
Hard X-rays		0247	0249				> 3	(232)
2.8 GHz		0247	0249	0617		P5	4.5	(124)
200 MHz		0248	0253	0318			2.9	
DS type II		0250		0318				(124)
DS type III		0248		0251			1	

1969	April 10	■	McM 10035			Event 322	(3 3 0)	CFI = 8
Hα		0410		0445		N11 E90	1N	(11) Δt ~ 20^h
SWF		0357	0402	0422			2+	(11)
Soft X-rays		0355	0402	0630			VLG	(229)
Hard X-rays		0355	0356	0357			3	(232)
2.8 GHz		0354	0403	0618		P7	2.6	(124)
200 MHz		0356	0358	0410			2.9	
DS type II		0356		0422			2	(124)
DS type III		0403		0406			2	

1969	May 2	●	McM 10057			Event 326	(−2 0 0)	CFI = 10
Hα		1745	1752	1900		N09 W40	1B	Δt ~ 1.3^h
SWF		1746	1750	1845			2	(11)
Soft X-rays		1744	1757	> 1850			LG	(229)
Hard X-rays		no data						
3 GHz		1748	1751	1756		U3P10	2	
200 MHz		1746	1754	2105			2.5	
DS type III		1754–1758, 1811–1815						
DS type I		1800		1809			2	
DS type II		1752		1806			3	
DS type IV		1748		1834			2	

1969	May 5	●	McM 10057			Event 327	(−2 0 0)	CFI = 5
Hα		1237	1244	1250	N08 W77	SN	$\Delta t \sim 0.2^{\text{h}}$
SWF		1243	1252	1310		2−	(11)
Soft X-rays		1238	1249	1320		SIG	(229)
Hard X-rays		no data					
3 GHz		1235	1238	1247		1.5	
200 MHz		1235	1242	1249		2.6	
DS type IV		1236		1244		1	
DS type III		1232		1246		2	

1969	May 28	☉	McM 10109			Appendix A		CFI = 2
Hα		0737		0806	N11 W56	SN	$\Delta t \sim 0.1^{\text{h}}$
SID		no event					
Soft X-rays		no event					
Hard X-rays		no data					
2 GHz		0738	0739	0739		0.3	
200 MHz		0739	0739	0739		2.1	
DS type V		0736		0739		2	
DS type III		0736		0738		2	

1969	May 28	☉	McM 10109			Appendix A		CFI = 3
Hα		1038	1046	1110	N10 W58	1N	$\Delta t \sim 0.3^{\text{h}}$
SID		no event					
X-rays		no data					
3 GHz		1046	1047	1047		0.7	
200 MHz		1045	1045	1047		2.2	
DS type III		1036		1037		1	

1969	May 28	●	McM 10109			Event 329	(−2 0 0)	CFI = 6
Hα		1241	1258	1320	N10 W59	1B	$\Delta t \sim 0.5^{\text{h}}$
SWF		1256	1302	1330		1	(11)
X-rays		no data					
3 GHz		1255	1258	1302		1.5	
200 MHz		1255	1311	1314		2.5	
DS type III		1254–1258, 1309–1310				2	
DS type III(+V)		1311		1313		2	
DS type II		1305		1313		2	

1969	May 28	· ☉	McM 10109			Appendix A		CFI = 1
Hα		2313		2343	N12 W64	SN	$\Delta t \sim 0.3^{\text{h}}$
SWF		2313	2317	2325		1−	(11)
X-rays		no data					
2.8 GHz		2314	2314	2315		0.9	
200 MHz		no event					
DS type III		2312		2324		2	

1969	May 29	⊙	McM 10109		Event 330	(−2 0 0)	CFI = 6
Hα		0020	0022	0047	N11 W64	1B	$\Delta t \sim 0.3^h$
SWF		0021	0033	0105		1−	(11)
Soft X-rays		0019	0025	0130		SIG	(229)
Hard X-rays		no data					
2.7 GHz		0020	0022	0026		1.5	
200 MHz		0018	0021	0025		2.5	
DS type III		0018		0025		2	
DS type II		0027		0038		2	

1969	May 29	●	McM 10109		Appendix A		CFI = 5
Hα		0406	0411	0443	N12 W67	1N	$\Delta t \sim 0.3^h$
SWF		0408	0410	0442		2−	(11)
Soft X-rays		0406	0412	0430		SIG	(229)
Hard X-rays		no data					
3 GHz		0406	0407	0412		1.6	
200 MHz		no event					
DS type III(+V)		0407		0410		2	
DS type II		0411		0422		1	

1969	May 29	⊙	McM 10109		Appendix A		CFI = 7
Hα		< 1442	1454	1516	N11 W73	1B	$\Delta t < 0.2^h$
SWF		1450	1458	1615		2−	(11)
Soft X-rays		1440	[1443 / 1459	> 1620		VSM / LG	(229) / (229)
Hard X-rays		no data					
3 GHz		1451	1453	1455		1.4	
200 MHz		1451	1452	1453		2.3	
DS type III		1450		1453		2	
DS type II		1508		1515		1	

1969	May 29	⊙	McM 10109		Event 331	(−2 0 0)	CFI = 6
Hα		1939	1944	2010	N10 W76	1B	$\Delta t \sim 0.3^h$
SWF		1940	1950	2035		2	(11)
Soft X-rays		< 1938	1942	~ 2045		LG	(229)
Hard X-rays		no data					
2.8 GHz		1938	1942	1946	P10	1.9	
200 MHz		1939	1941	1945		2.4	
DS type III		1938		1943		3	

1969	May 30	⊙	McM 10108			Appendix A		CFI = 1
Hα		0420	0422	0428		S16 W90	1N	$\Delta t \sim 0.1^{\text{h}}$
SID		no event						
Soft X-rays		0419		<0905			SM?	(229)
Hard X-rays		no data						
3.7 GHz		0421	0422	0425			0.2	DS no event
200 MHz		no event						

1969	June 5	⊙	McM 10134			Appendix A		CFI = 14
Hα		0952	⌈1001 ⌊1010	1050		N12 E64	2B	$\Delta t \sim 10^{\text{h}}$
SWF		0956	1000	1220			3+	(11)
Soft X-rays		0929	<1030	1230			VLG	(229)
Hard X-rays		no data						
3 GHz		0954	1008	1028		U0.6/P10	3.2	
200 MHz		0954	0958	1023			3.3	
DS type IV		0953		1025			3	
DS type II		0957		1014			3	
DS type III		0954		1014			3	

1969	June 5	⊘	McM 10134			Appendix A		CFI = 11
Hα		<1442	⌈1455 ⌊1557	>1703		N09 E53	3B	
SWF		1455	1505	1510			1−	(11)
Soft X-rays		1445	<1515	1900			LG	(229)
Hard X-rays		no data						
3 GHz		1442	1500	1529			2.8	
200 MHz		1455	1501	1516			2.4	
DS cont.		1428		1823			3	
DS type IV		1451		1515			2	
DS unclassified		1507		1516			2	

1969	June 7	●	McM 10134			Event 332	(1 1 0)	CFI = 7
Hα		<0945	⌈0955 ⌊1006	1050		N11 E34	1N	$\Delta t \sim 5^{\text{h}}$
SWF		0951	0955	1110			3	(11)
Soft X-rays		<0941	>1010	1150			LG	(229)
Hard X-rays		no data						
2.8 GHz		0948	0956	0959			2.2	
200 MHz		0949	0955	0959			1.7	
DS type III		0954		0955			1	

1969	June 11	●	McM 10134			Appendix A		CFI = 12
Hα		1615	1627	1730	N10 W20	2B		$\Delta t \sim 0.5^{\mathrm{h}}$
SWF		1620	1640	1730		2+		(11)
Soft X-rays		1616	1627	1830		VLG		(229)
Hard X-rays		1623	1624	1629	20–50 keV	LG		(229)
3 GHz		1614	1623	1638		2.5		
200 MHz		1620	1621	1628		3.4		
DS type III(+V)		1619		1625		3		
DS type IV		1619		1628		2		
DS type II		1622		1640		3		
DS type III		1614–1615, 1621–1622				3		

1969	July 13	●	McM 10181			Appendix A		CFI = 1
Hα		0647	0647	0653	S15 W90	SN		$\Delta t \sim 0.25^{\mathrm{h}}$
SWF		no event						
Soft X-rays		0647	0650	0710		VSM		(229)
Hard X-rays		no data						
2.8 GHz		0649	0649	0650		0.1		
200 MHz		0649	0650	0650		1.9		
DS type II(+V)		0649		0650		2		

1969	August 7	☉	McM 10253			Appendix A		CFI ≥ 4
Hα		0901	0911	0955	N21 E28	2N		$\Delta t \sim 3^{\mathrm{h}}$
SWF		0907	0910	0945		1		(11)
Soft X-rays		0900	0914	1020		SIG		(229)
Hard X-rays		no event			20–50 keV			(229)
2.9 GHz		0855	0922	1020		0.9		
200 MHz		no event						
DS type II		0910		0917		1		

1969	August 18	☉	McM 10262			Appendix A		CFI ≥ 2
Hα		2201	⌈ 2210	2300	S15 W53	1B		$\Delta t \sim 4^{\mathrm{h}}$
			⌊ 2221					
SID		no event						
Soft X-rays		2207	2217	2400		LG		(229)
Hard X-rays		no event			20–50 keV			(229)
DS type II		2214–2220, 2217–2222				2		⌉ no other
DS type III		2215		2220		3		⌋ radio report

1969	September 17	⊙	McM 10317			Appendix A		CFI = 6
Hα		0836	0839	0855		S04 W06	SN	$\Delta t \sim 5.5^{\mathrm{h}}$
SWF		0839	0844	0850			2+	(11)
Soft X-rays		0837	⌈0838 ⌊0842	0848			SIG SIG	(229) (229)
Hard X-rays		no data						
3 GHz		0837	0838	0842			1.8	
200 MHz		0835	0837	0839			3.4	
DS type V		0836		0838			2	
DS type III		0830		0842			2	

1969	September 17	⊙	McM 10309			Appendix A		CFI ⩾ 4
Hα		0935	⌈0945 ⌊0954	1030		S18 W27	1N	$\Delta t \sim 4.5^{\mathrm{h}}$
SWF		0939	0950	1020			2−	(11)
Soft X-rays		>0920	<0950	1035			LG	(229)
Hard X-rays		no data						
2.8 GHz		0936	0945	0948			1.4	
DS type III		0959–0959, 1001–1009						

1969	September 17	⊙	McM 10317			Appendix A		CFI = 5
Hα		1012	1016	1030		S04 W06	SN	$\Delta t \sim 4^{\mathrm{h}}$
SWF		no event						
			⌈1014				SM	(229) in decay of pre-
Soft X-rays		1012		1022				vious
			⌊1016				SM	(229) event
Hard X-rays		no data						
3 GHz		1013	1014	1021			1.5	
200 MHz		1013	1014	1020			3.3	
DS type V		1013		1020			3	
DS type II		1014		1015			3	
DS type III		1012–1017, 1017–1020						3

1969	September 17	⊙	McM 10317			Appendix A		CFI = 3
Hα		1207	1215	1230		S04 W06	SB	$\Delta t \sim 2^{\mathrm{h}}$
SWF		no event						
Soft X-rays		no event						
Hard X-rays		no event			20–50 keV			(229)
3 GHz		1211	1211	1211			0.9	
200 MHz		1207	1211	1220			2.7	
DS type II		1211		1212			2	
DS type III		1207		1220			3	

1969	September 25	●	McM 10326		Event 333	(1 1 0)	CFI = 5
Hα	<0658		>0900		N13 W15	3N	$\Delta t \sim 0.5^h$
SWF	no event						
Soft X-rays	<0640	0730	>1000			SIG	(229)
Hard X-rays	no data						
3 GHz	0713	0753	0900		P3	1.5	(124)
200 MHz	0700	0720	0745			1.9	
DS type I	0800		0930			1	(124)
DS type III	0657–0658, 0746–0754					1	

1969	September 25	●	McM 10325		Appendix A		CFI = 6
Hα	1414	⌈1425 ⌊1434	1505		N05 W39	1N	$\Delta t \sim 30^h$
SWF	1426	1438	1505			1	(11)
Soft X-rays	1414	1430	1700			SM	(229)
Hard X-rays	no event				20–50 keV		(229)
2.8 GHz	1410	1425	1430			2	
200 MHz	1423	1425	1435			2.1	
DS type I	1406		1536			1	
DS type III	1421		1429			3	
DS uncl.	1424		1427			2	

1969	September 27	●	McM 10333		Event 334	(−1 1 0)	CFI = 12
Hα	<0350	⌈0357 ⌊0412	0558		N09 E02	3B	$\Delta t \sim 4^h$
SWF	0353	0358	0541			2+	(11)
Soft X-rays	0346	0418	0930			LG	(229)
Hard X-rays	no data						
3.7 GHz	0325	0425	0725		P7	1.8	(124)
200 MHz	0345		0540			2.1	
DS type II	0402		0420			2	(124)
DS type IV	0359		0808			1	(124)
DS type III	0355–0356, 0359–0401					1	

1969	October 14	●	McM 10352		Event 337	(−1 0 0)	CFI = 6
Hα	0539	0544	0616		N25 W71	2N	$\Delta t \sim 0.5^h$
SWF	no event						
Soft X-rays	0539	0554	0640			SIG	(229)
Hard X-rays	no event						(232)
3.7 GHz	0540	0554	0640			0.8	(124)
600 MHz	0530	0547	0606			1	
DS type II	0540		0558			3	(124)

1969	October 20	⊙	McM 10385		Appendix A		CFI = 5
Hα	<0326	0336	>0355	N08 E80	1F		Δt ~ 10^h
SWF	0302	0309	0543		3		(11)
Soft X-rays	0240	⌐0259 0303 0309 └0328	>0620		VSM SM LG LG		(229) (229) (229) (229)
Hard X-rays	⌐0300 └<0304	0303 0307	0304 0311	20–50 keV 20–50 keV	SIG SM		(229) (229)
3.7 GHz	0321	0324	0336		0.8		DS no event
200 MHz	no event						

1969	November 2	■	McM 10385		Event 340	(3 3 0)	CFI = 11
Hα	<1102	1138 1146	1220	N14 W90	1N		Δt ~ 0^h
SWF	1017	1122	1218		3+		(11)
Soft X-rays	0940	1045	1700		VLG		(229)
Hard X-rays	no event						(232)
3 GHz	1008	1040	1530	P3	3.1		(124)
200 MHz	1000	1010	1035		2.5		
DS type III	⌐1000 └1043		1037 also continuum 1050		1 1		(124)

1969	November 7	⊙	McM 10406		Event 341	(0 1 0)	CFI ⩾ 2
Hα	0322	0345	0459	N13 E11	2N		Δt ~ 4.5^h
SWF	no event						
Soft X-rays	0330	<0410	0900		SM		(229)
Hard X-rays	no data						
2.8 GHz	no event						DS no event

1969	November 18	●	McM 10432		Event 345	(−1 0 0)	CFI = 14
Hα	1633	1654	1756	N14 E40	2B		Δt ~ 0^h
SWF	1638	1644	1844		3+		(11)
Soft X-rays	1622	⌐1630 └1655	2000		VSM VLG		(229) (229)
Hard X-rays	1645	1650	>1705	50–150 keV	LG		(229)
2.8 GHz	1635	1655	1710		3		
200 MHz	1644	1651	1844		3.7		
DS type III	1644–1646, 1655–1657						
DS type IV	1657		1841		3		
DS type II	1649		1709		2		

1969	November 20	⊙	McM 10432			Event 346	(−2 0 0)	CFI = 5
Hα		1619	1624	1700		N07 E07	2B	$\Delta t \sim 7^{\mathrm{h}}$
SWF		1621	1628	1730			1+	(11)
Soft X-rays		1619	1626	1700			LG	(229)
Hard X-rays		·no data						
2.8 GHz		1619	1621	1626			2.6	DS no even

1969	November 23	●	McM 10432			Appendix A		CFI ⩾ 7
Hα		0958	⌈1004 ⌊1019	1125		N15 W19	1B	$\Delta t \sim 4^{\mathrm{h}}$
SWF		1015	1022	1125			3	(11)
Soft X-rays		0955	⌈1005 ⌊1020	1220			SM VLG	(229) (229)
Hard X-rays		no data						
3 GHz		1015	1018	1023			2.5	⌉ DS no data
600 MHz		1016	1017	1024			2	⌋

1969	November 24	●	McM 10432			Event 348	(0 1 0)	CFI ⩾ 12
Hα		0914	0919	1030		N15 W31	2B	$\Delta t \sim 0.5^{\mathrm{h}}$
SWF		0915	0917	1300			3+	(11)
Soft X-rays		0914	∼0925	>1130			VLG	(229)
Hard X-rays		0913					>3	(232)
3 GHz		0914	0918	1125			3.4	(124) ⌉ DS no
200 MHz		0917	0920	0934			4	⌋ data
Type II or IV?		0917		0934			3	(124)

1969	December 30	●	McM 10491			Event 352	(0 1 0)	CFI = 7
Hα		1927	1934	2004		S14 W85	1N	(11)
								$\Delta t \sim 0.75^{\mathrm{h}}$
SWF		<1916	1924	2007			2	(11)
Soft X-rays		1903	1929	2200			SIG	(229)
Hard X-rays		no data						
2.8 GHz		1914	1925	1942		P4	2	(124)
200 MHz		1914	1924	2051			1.7	
DS type II		1930		1938			?	(124)
DS unclassified		1923		1931			1	(124)
DS type I		1923–1931, 1932–2035					1	

REFERENCES TO PARTS 1 AND 2 AND APPENDIX A

1   Bezprozvannaya, A. S.: 1962, *J. Phys. Soc. Japan* 17, Suppl. A-I, 146.
2   Bailey, D. K.: 1964, *Planetary Space Sci.* 12, 495.
3   Hakura, Y.: 1968, NASA Washington D. C., NASA TN D-4473.
4   *Quarterly Bulletin on Solar Activity,* published by the Eidgen. Sternwarte in Zürich.
5   Mac Math-Hulbert Observatory, Pontiac, unpublished data.
6   Collins, C., Jelly, D. H., and Matthews, A. G.: 1961, *Can. J. Phys.* 39, 35.
7   Notuki, M., Hatanaka, T., and Unno, W.: 1956, *Publ. Astron. Soc. Japan* 8, 52.
8   Obayashi, T.: 1962, *J. Geophys. Res.* 67, 2039.
9   Fritzová, L. and Švestka, Z.: 1966, *Bull. Astron. Inst. Czech.* 17, 249.
10 Jelly, D. H. and Collins, C.: 1962, *Can. J. Phys.* 40, 706.
11 *Solar Geophysical Data,* CRPL, ESSA, Boulder.
12 Reid, G. C. and Leinbach, H.: 1959, *J. Geophys. Res.* 64, 1801.
13 Warwick, C. S. and Haurwitz, M.W.: 1962, *J. Geophys. Res.* 67, 1317.
14 Hill, G. E.: 1962, *J. Phys. Soc. Japan* 17, Suppl. A-I, 97.
15 Obayashi, T.: 1967, *Table of Solar-Geophysical Events* (Jan. 1957–Dec. 1963), Ionosphere Research Laboratory, Kyoto University.
16 Dvoryashin, A. S., Levitzky, L. S., and Pankratov, A. K.: 1961, *Izv. Krymsk. Astrofiz. Obs.* 26, 90.
17 Dodson, H. W. and Hedeman, E. R.: IGY Solar Activity Report, Series No. 12, 15, 18, 21, and 25.
18 Kahle, A. B.: 1962, Geophys. Inst. Univ. Alaska, Sci. Rept. No. 2, AFCRL-62-708.
19 Bookin, G. V.: 1962, *J. Phys. Soc. Japan* 17, Suppl. A-I, 150.
20 Hakura, Y. and Goh, T.: 1959, *J. Radio Res. Lab. Japan* 6, 635.
21 Anderson, K. A.: 1964, *J. Geophys. Res.* 69, 1743.
22 Leinbach, H.: 1962, Geophys. Inst. Univ. Alaska, Sci. Rept. No. 3, UAG-R127.
23 Piggott, W. R. and Shapley, A. H.: 1962, Am. Geophys. Un., Antarctic Res. Geophys. Monograph No. 7, 111.
24 Leinbach, H.: 1962, Geophys. Inst. Univ. Alaska, Sci. Rept. No. 2, NSF Grant G-14133.
25 Maxwell, A., Hughes, M. P., and Thompson, A. R.: 1963, *J. Geophys. Res.* 68, 1347.
26 Egeland, A., Hultqvist, B., and Ortner, J.: 1963, in G.J. Gassmann (ed.), *The Effect of Disturbances of Solar Origin on Communications,* AGARDograph No. 59, Pergamon Press, Oxford, p. 79.
27 Winckler, J. R. and Bhavsar, P. D.: 1960, *J. Geophys. Res.* 65, 2637.
28 University of Michigan: 'Solar Dynamic Spectral Data', Aug. 1957–July 1961.
29 Winckler, J. R., Bhavsar, P. D., and Peterson, L.: 1961, *J. Geophys. Res.* 66, 995.
30 Sinno, K.: 1961, *J. Geomagn. Geoelec.* 13, 1.
31 Noyes, J. C.: 1962, *J. Phys. Soc. Japan* 17, Suppl. A-II, 275.
32 Fan, C. Y., Meyer, P., and Simpson, J. A.: 1960, in H. Kallmann (ed.), *Space Research* I, North-Holland, Amsterdam, p. 951.
33 Bruzek, A.: 1964, *Astrophys. J.* 140, 746.
34 Kursonova, L. V., Razorenov, L. A., and Fradkin, M. I.: 1962, *J. Phys. Soc. Japan* 17, Suppl. A-II, 315.
35 Gregory, J. B.: 1963, *J. Geophys. Res.* 68, 3097.
36 Van Allen, J. A. and Lin, W. C.: 1960, *J. Geophys. Res.* 65, 2998.
37 Arnoldy, R. L., Hoffman, R. A., and Winckler, J. R.: 1960, *J. Geophys. Res.* 65, 3005.
38 Masley, A. J., May, T. C., and Winckler, J. R.: 1962, *J. Geophys. Res.* 67, 3243.
39 Jenkins, R. W. and Paghis, J.: 1963, *Can. J. Phys.* 41, 1056.
40 Malitson, H. H.: 1963, in F. B. McDonald (ed.), *Solar Proton Manual,* NASA Rept. TRR-169, p. 109.
41 Smith, E. v. P.: 1968, in Y. Öhman (ed.), 'Mass Motions in Solar Flares and Related Phenomena', Nobel Symposium 9, Almqvist and Wiksells, Stockholm, p. 137.
42 Lin, W. C. and Van Allen, J. A.: 1964, *Proc. of the International School of Physics Enrico*

Fermi, Course 24, Space Exploration and the Solar System, Academic Press, New York, p. 194.

43 Vernov, S. N.: 1972, 'U.S.S.R. Observations of Proton Increases in Stratosphere and Interplanetary Space', unpublished, special report prepared for the present publication.

44 Bryant, D. A., Cline, T.L., Desai, U.D., and McDonald, F.B.: 1965, *Phys. Rev. Letters* 14, 481.

45 Bryant, D. A., Cline, T. L., Desai, U. D., and McDonald, F. B.: 1962, *J. Geophys. Res.* 67, 4983.

46 Anderson, K. A. and Winckler, J. R.: 1962, *J. Geophys. Res.* 67, 4103.

47 Boorman, J. A., McLean, D. J., Sheridan, K. V., and Wild, J. P.: 1961, *Monthly Notices Roy.. Astron. Soc.* 123, 87.

48 Rothwell, P. and McIlwain, C.: 1959, *Nature* 184, 138.

49 Anderson, K. A. and Enemark, D. C.: 1960, *J. Geophys. Res.* 65, 2657.

50 Fan, C. Y., Glockler, G., and Simpson, J. A.: 1966, *Proc. of the 9th International Conference on Cosmic Rays*, London, The Institutes of Physics and the Physical Society, Vol. 1, p. 109.

51 Tokyo Astronomical Observatory Bulletin on Solar Phenomena.

52 Jelly, D. H.: 1963, *J. Geophys. Res.* 68, 1705.

53 Dvoryashin, A. S.: 1962, *Izv. Krymsk. Astrofiz. Obs.* 28, 293.

54 Winckler, J. R. and Bhavsar, P. D.: 1963, *J. Geophys. Res.* 68, 2099.

55 Fichtel, C. E., Kniffen, D. A., and Ogilvie, K. W.: 1962, *J. Geophys. Res.* 67, 3669.

56 Yoshida, S., Nagashima, K., Kawabata, K., and Morimoto, M.: 1963, in W. Priester (ed.), *Space Research* III, North-Holland, Amsterdam, p. 608.

57 Hofmann, D. J. and Winckler, J. R.: 1963, *J. Geophys. Res.* 68, 2067.

58 Keppler, E., Ehmert, A., and Pfotzer, G.: 1963, in W. Priester (ed.), *Space Research* III, North-Holland, Amsterdam, p. 676.

59 Pieper, G. F., Zmuda, A. J., Bostrom, C. O., and O'Brien, B. J.: 1962, *J. Geophys. Res.* 67, 4959.

60 Gregory, J. B. and Newdick, R. E.: 1964, *J. Geophys. Res.* 69, 2383.

61 Krimigis, S. M. and Van Allen, J. A.: 1963, *Trans. Am. Geophys. Un.* 44, 882.

62 Skerjanec, R. E., Wightman, D. W., and Warwick, J. W.: 1963, *Inf. Bull. Solar Radio Obs.* 13, 5.

63 Bryant, D. A., Cline, T. L., Desai, U. D., and McDonald, F. B.: 1965, *Astrophys. J.* 141, 478.

64 Zmuda, A. J., Pieper, G. F., and Bostrom, C. O.: 1963, *J. Geophys. Res.* 68, 1160.

65 Kinsey, J. H.: 1969, NASA Document X-611-69-396, Sept. 1969.

66 Lin, R. P. and Anderson, K. A.: 1967, *Solar Phys.* 1, 446.

67 Cline, T. L. and McDonald, F. B.: 1968, *Solar Phys.* 5, 507.

68 Simpson, J. A.: 1972, unpublished data from IMP-1, deposited in WDC-A.

69 Lin, R. P.: 1970, *Solar Phys.* 12, 266.

70 Fan, C. Y., Pick, M., Pyle, R., Simpson, J. A., and Smith, D. R.: 1968, *J. Geophys. Res.* 73, 1555.

71 McDonald, F. B.: 1967, private communication to Lin and Anderson.

72 McDonald, F. B.: 1966, private communication.

73 Bhavsar, P. D.: 1962, *J. Phys. Soc. Japan* 17, Suppl. A-II, 329.

74 Basler, R. P. and Owren, L.: 1964, Geophys. Inst. Univ. Alaska, Sci. Rept., UAG-R152.

75 Goedeke, A. D. and Masley, A. J.: 1964, *J. Geophys. Res.* 69, 4166.

76 Simpson, J. A.: 1972, unpublished data from IMP-2, deposited in WDC-A.

77 O'Gallagher, J. J. and Simpson, J. A.: 1966, *Phys. Rev. Letters* 16, 1212.

78 Krimigis, S. M. and Van Allen, J. A.: 1967, *J. Geophys. Res.* 72, 4471.

79 Krimigis, S. M. and Van Allen, J. A.: 1966, *Phys. Rev. Letters* 16, 419.

80 Simpson, J. A.: 1972, unpublished data from IMP-3, deposited in WDC-A.

81 Van Allen, J. A. and Krimigis, S. M.: 1965, *J. Geophys. Res.* 70, 5737.

82 Kahler, S. W. and Anderson, K. A.: 1972, unpublished event list from OGO data, obtained from NASA NSSDC.

83 Hultqvist, B.: 1969, in C. de Jager and Z. Švestka (eds.), *Solar Flares and Space Research*, North-Holland, Amsterdam, p. 215.

84 McCracken, K. G., Rao, U. R. and Bukata, R. P.: 1967, *J. Geophys. Res.* 72, 4293.

85 Stone, E. C.: 1964, *J. Geophys. Res.* 69, 3577.

86 Goedeke, A. D., Masley, A. J., and Adams, G. W.: 1967, *Solar Phys.* 1, 285.

87 Krimigis, S. M., Van Allen, J. A., and Armstrong, T. P.: 1969, *Ann. IQSY* 3, 395.

88 Bailey, D. K.: 1962, *J. Phys. Soc. Japan* 17, Suppl. A-I, 106.

89 Winckler, J. R.: 1960, *J. Geophys. Res.* 65, 1331.

90 Freier, P. S., Ney, E. P., and Winckler, J. R.: 1959, *J. Geophys. Res.* 64, 685.

91 Ney, E. P., Winckler, J. R., and Freier, P. S.: 1959, *Phys. Rev. Letters* 3, 183.

92 Anderson, K. A., Arnoldy, R., Hoffman, R., Peterson, L., and Winckler, J. R.: 1959, *J. Geophys. Res.* **64**, 1133.

93 Brown, R. R. and D'Arcy, R. G: 1959, *Phys. Rev. Letters* **3**, 390.

94 Brown, R. R. and D'Arcy, R. G., 1961, *J. Geophys. Res.* **66**, 2516 (Abstract).

95 Leinbach, H., Venkatesan, D., and Parthasarathy, R.: 1965, *Planetary Space Sci.* **13**, 1075.

96 Jelly, D.: 1973, private communication.

97 Charakhchyan, A. N., Tulinov, V. F., and Charakhchyań, T. N.: 1960, in H. Kallmann (ed.), *Space Research* I, North-Holland, Amsterdam, p. 649.

98 Pick-Gutmann, M.: 1961, *Ann. Astrophys.* **24**, 183.

99 Švestka, Z. and Olmr, J.: 1966, *Bull. Astron. Inst. Czech.* **17**, 4.

100 Jonah, F. C., Dodson, H. W., and Hedeman, E. R.: 1965, 'Solar Activity Catalogue', NASA Reports 00.538 and 00.594, NASA, Manned Spacecraft Center, Houston.

101 Vakulov, P. V., Vernov, S. N., Gorchakov, E. B., Logachev, Yu. I., Charakhchyan, A. N., Charakhchyan, T. N., and Chudakov, A. E.: 1964, in P. Muller (ed.), *Space Research* IV, North-Holland, Amsterdam, p. 26.

102 Krüger, A., Fürstenberg, F., and Fricke, K. H.: 1971, *HHI Suppl. Series of Solar Data* 2, No. 5.

103 Vernov, S. N., Gorchakov, E. V., Logachev, Yu. I., Lyubinov, G. B., Pereslegina, N. V., Tverskoy, B. A., and Chudakov, A. E.: 1968, *Can. J. Phys.* **46**, S812.

104 Simpson, J. A. and Fan, C. Y.: unpublished data from Pioneer 6, deposited in WDC-A.

105 Kahler, S.W., Primbsch, J. H., and Anderson, K. A.: 1967, *Solar Phys.* **2**, 179.

106 Shea, M. A. and Smart, D. F.: 1973, private communication.

107 Simpson, J. A.: unpublished data from IMP-4, deposited in WDC-A.

108 Cline, T. L. and Hones, E. W.: 1967, presented at the OGO-1 and OGO-3 Working Groups Meeting March 1967, unpublished NSSDS document.

109 Ahluwalia, H. S., Sud, L. V., and Schreier, M.: 1969, *Ann. IQSY* 3, 254.

110 Heristchi, Dj., Kangas, J., Kremser, G., Legrand, J. P., Masse, P., Palous, M., Pfotzer, G., Riedler, W., and Wilhelm, K.: 1969, *Ann. IQSY* 3, 267.

111 Hakura, Y.: 1969, *Ann. IQSY* 3, 337.

112 Masley, A. J. and Goedeke, A. D.: 1969, *Ann. IQSY* 3, 353.

113 Stickland, A. C. (ed.): 1969, *The Proton Flare Project* (The July 1966 Event), *Ann. IQSY* 3, The MIT Press, Cambridge (U.S.A.) and London.

114 Lin, R. P., Kahler, S. W. and Roelof, E. C.: 1968, *Solar Phys.* **4**, 338.

115 Švestka, Z.: 1968, *Solar Phys.* **4**, 361.

116 Dodson, H. W., Hedeman, E. R., Kahler, S. W., and Lin, R. P.: 1969, *Solar Phys.* **6**, 294.

117 Anderson, K. A.: 1969, *Solar Phys.* **6**, 111.

118 Simnett, G. N., Cline, T. L., Holt, S. S., and McDonald, F. B.: 1970, *Acta Phys. Acad. Sci. Hungaricae* **29**, Suppl. 2, 649.

119 Simpson, J. A. and Fan, C. Y.: unpublished data from Pioneer 7, deposited in WDC-A.

120 Švestka, Z. and Simon, P.: 1969, *Solar Phys.* **10**, 3.

121 Dodson, H. W. and Hedeman, E. R.: 1969, in J. V. Lincoln (ed.), WDC-A Report UAG-5, ESSA, Boulder, p. 7.

122 Fokker, A. D.: 1966, *Utrechtse Sterrekundige Overdrukken*, No. 23.

123 Roosen, J.: 1966, *Inf. Bull. Solar Radio Obs.* **20**, 8.

124 Tanaka, H.: unpublished list of radio events, specially prepared for the IUCSTP W.G. 2.

125 McCracken, K. G., Rao, U. R., and Bukata, R. P.: unpublished Pioneer 6 and 7 data records, now deposited in WDC-A.

126 Collins, C. and Jelly, D. H.: 1961, *Nature* **189**, 128.

127 Boischot, A. and Fokker, A. D.: 1959, in R. N. Bracewell (ed.), 'Paris Symposium on Radio Astronomy', *IAU Symp.* 9 and *URSI Symp.* 1, held from 30 July to 6 August 1958, Stanford University Press, Stanford, p. 263.

128 Ortner, J., Hultqvist, B., Brown, R. R., Hartz T. R., Holt, O., Landmark, B., Hook, J. L., and Leinbach, H.: 1962, *J. Geophys. Res.* **67**, 4169.

129 Ney, E. P. and Stein, W. A.: 1962, *J. Geophys. Res.* **67**, 2087.

130 Charakhchyan, A. N., Tulinov, V. F., and Charakhchyan, T. N.: 1962, *J. Phys. Soc. Japan* **17**, Suppl. A-II, 360.

131 Ogilvie, K. W., Bryant, D. A., and Davis, L. R.: 1962, *J. Geophys. Res.* **67**, 929.

132 Ortner, J., Egeland, A., and Hultqvist, B.: 1961, in H. C. van de Hulst, C. de Jager and A. F. Moore (eds.), *Space Research* II, North-Holland, Amsterdam, p. 722.

133 Campbell, W. S. and Hubert, P. L.: 1961, *Can. J. Phys.* **39**, 614.

134 Vogan, E. L. and Hartz, T. R.: 1961, *Can. J. Phys.* **39**, 630.

135 Van Allen, J. A., Lin, W. C., and Leinbach, H.: 1964, *J. Geophys. Res.* **69**, 4481.

136   Axford, W. I. and Reid, G. C.: 1963, *J. Geophys. Res.* **68**, 1793.
137   Takakura, T.: 1963, *Publ. Astron. Soc. Japan* **15**, 327.
138   Guss, D. E. and Waddington, C. J.: 1963, *J. Geophys. Res.* **68**, 2619.
139   Becker, U.: 1958, *Z. Astrophys.* **46**, 168.
140   Freier, P. S. and Webber, W. R.: 1963, *J. Geophys. Res.* **68**, 1605.
141   Shea, M. A. and Smart, D. F.: 1973, *Proc. of the 13th International Conference on Cosmic Rays,* Denver, Conference Papers Vol. 2, p. 1548.
142   Bryant, D. A., Cline, T. L., Desai, U. D., and McDonald, F. B.: 1963, *Phys. Rev. Letters* **11**, 144.
143   Van Allen, J. A. and Whelpley, W. H.: 1962, *J. Geophys. Res.* **67**, 1660 (Abstract).
144   Bates, H. F.: 1962, *J. Geophys. Res.* **67**, 2745.
145   Banks, P. M. and Sechrist, C. F. Jr.: 1967, *J. Geophys. Res.* **72**, 2275.
146   Cormier, R. J.: 1973, 'Air Force Surveys in Geophysics', No. 255, AFCRL TR-73-0060.
147   Van Allen, J. A., Frank, L. A., and Venkatesan, D.: 1964, *Trans. Am. Geophys. Un.* **45**, 80.
148   Anderson, H. R.: 1966, in R. J. Mackin, Jr. and M. Neugebauer (eds.), *The Solar Wind,* Pergamon Press, New York, p. 53.
149   Anderson, H. R.: 1963, *Science* **139**, 42.
150   Franck, L. A.: 1965, *J. Geophys. Res.* **70**, 1593.
151   Masley, A. J. and Goedeke, A. D.: 1964, *Proc. of the 8th International Conference on Cosmic Rays,* Jaipur, Tata Institutes of Fundamental Research, Bombay, Vol. 1, p. 166.
152   Masley, A. J. and Goedeke, A. D.: 1968, *Can. J. Phys.* **46**, S766.
153   Maehlum, B. and O'Brien, B. J.: 1962, *J. Geophys. Res.* **67**, 3269.
154   Fan, C. Y., Lamport, J. E., Simpson, J. A., and Smith, D. R.: 1966, *J. Geophys. Res.* **71**, 3289.
155   Leinbach, H.: 1973, private communication.
156   Burrows, J. R.: 1973, unpublished Alouette 1 data.
157   Burrows, J. R., McDiarmid, I. B., and Wilson, M. D.: 1965, in B. Maehlum (ed.), *Proc. of the Symposium on High Latitude Particles and the Ionosphere,* Alpbach 1964, Logos Press and Academic Press, London and New York, p. 119.
158   Chivers, H. J. A. and Burrows, J. R.: 1966, *Planetary Space Sci.* **14**, 131.
159   Conner, J. P.: 1972, unpublished data from Vela 3A and 3B.
160   Kreplin, R. W.: 1966, 'Final Data and Calibrations for the Explorer 30 (NRL Solrad 8) X-Ray Monotoring Experiment', E. O. Hulburt Center for Space Research, U.S. Naval Research Laboratory, Washington D.C.
161   Arnoldy, R. L., Kane, S. R., and Winckler, J. R.: 1968, University of Minnesota Technical Report, C. R.-108, Jan. 1968.
162   Kane, S. R. and Winckler, J. R.: 1969, University of Minnesota Technical Report, C. R.-134.
163   Kane, S. R. and Winckler, J. R.: 1969, University of Minnesota Technical Report, C. R.-135.
164   Wende, Ch.: 1972, unpublished data from Explorer 33, deposited in NSSDC.
165   McCracken, K. G. and Rao, U. R.: 1967, unnumbered report data from SCAS experiment on Pioneer 6 and 7 deep space probes, Vol. 1 and 2.
166   Simnett, G. M.: 1972, *Solar Phys.* **22**, 189.
167   Potemra, T. A. and Lanzerotti, L. J.: 1971, *J. Geophys. Res.* **76**, 5244.
168   Singer, S.: 1972, private communication, unpublished data from the Vela 4 satellite, Los Alamos Scientific Laboratories.
169   Masley, A. J.: 1971, private communication.
170   Innanen, W. G.: 1968, University of Iowa, unpublished thesis.
171   Barcus, J. R.: 1969, *Solar Phys.* **8**, 186.
172   Kahler, S. W.: 1969, *Solar Phys.* **8**, 166.
173   Bostrom, C. O., Williams, D. J., and Arens, J. F.: 'Solar Proton Monitoring Experiment Explorer 34', *Solar Geophysical Data,* NOAA Boulder, and NSSDC Data Set 67-051A-07D.
174   Rao, U. R., McCracken, K. G., Allum, F. R., Palmeira, R. A. R., Bartley, W. C., and Palmer, I.: 1971, *Solar Phys.* **19**, 209.
175   Graedel, T. E. and Lanzerotti, L. J.: 1972, *J. Geophys. Res.* **76**, 6932.
176   Allum, F. R., Palmeira, R. A. R., Rao, U. R., McCracken, K. G., Harries, J. R., and Palmer, I.: 1971, *Solar Phys.* **17**, 241.
177   Vernov, S. N., Chudakov, A. E., Vakulov, P. V., Gorchakov, E. V., Kontor, N. N., Logachev, Yu. I., Lyubimov, G. P., Pereslegina, N. V., and Timofeev, G. A.: 1970, in V. Manno and D. E. Page (eds.), *Intercorrelated Satellite Observations Related to Solar Events,* D. Reidel, Dordrecht-Holland, p. 33.
178   Wang, J. R., Fisk, L. A., and Lin, R. P.: 1971, *Proc. of the 12th International Conference on Cosmic Rays,* Hobart, University of Tasmania, Conference Papers Vol. 2, p. 438.
179   Simnett, G.M.: 1974, *Space Sci. Rev.* **16**, 257.

180   Evans, L. C. and Stone, E. C.: 1969, *J. Geophys. Res.* **74**, 5127.

181   Bostrom, C. O., Williams, D. J., and Arens, J. R.: 'Solar Proton Monitoring Experiment
       Explorer 34', NSSDC Data Set 67-051A-07C.

182   Van Hollebeke, M., Wang, J. R., and McDonald, F. B.: 1973, private communication,
       unpublished data from GSFC Experiment on IMP 4.

183   Webber, W. R.: 1972, private communication, unpublished Pioneer 8 data, NSSDC data set
       67-123A-06A.

184   Masley, A. J. and Satterblom, P. R.: 1970, *Acta Phys. Acad. Sci. Hungaricae* **29**, Suppl. 2, 513.

185   McKibben, R. B.: 1972, *J. Geophys. Res.* **77**, 3957.

186   Datlowe, D.: 1971, *Solar Phys.* **17**, 436.

187   Lanzerotti, L. J., Venkatesan, D., and Wibberenz, G.: 1973, J. Geophys. Res. **78**, 7986.

188   Rao, U. R.: 1973, private communication, particulars of 56 solar flare proton events which
       occured during 1965–1968.

189   Kane, S. R. and Lin, R. P.: 1972, *Solar Phys.* **23**, 457.

190   Simnett, G. M.: 1973, in M. J. Rycroft and S. K. Runcorn (eds.), *Space Research* **XIII**,
       Akademie-Verlag, Berlin, Vol. 2, p. 745.

191   Simnett, G. M. and Holt, S. S.: 1971, *Solar Phys.* **16**, 208.

192   Westerlund, S.: 1972,'Kiruna Geophysical Observatory Working List of PCA's', Internal
       Document No. 6.

193   Fennell, J. F.: 1973, *J. Geophys. Res.* **78**, 1036.

194   Heristchi, Dj., Legrand, J. P., and Petrou, D.: 1971, *Solar Phys.* **18**, 321.

195   *Kiruna Geophysical Data,* 1969, Suppl. 69-1, PCA events observed at Kiruna in the year 1968.

196   Lanzerotti, L. J.: 1970, in J. V. Lincoln (ed.), WDC-A Report UAG-9, ESSA, Boulder, p. 34.

197   Venkataragan, P., Venkatesan, D., and Van Allen, J. A.: 1970, *Acta Phys. Acad. Sci. Hungaricae*
       29, Suppl. 2, 409.

198   Sarris, E. T. and Shawan, S. D.: 1973, *Solar Phys.* **28**, 519.

199   McCracken, K. G., Rao, U. R., Bukata, R. P., and Keath, E. P.: 1971, *Solar Phys.* **18**, 100.

200   Bostrom, C. O., Williams, D. J., and Arens, J. F.: 'Solar Proton Monitoring Experiment
       Explorer 41', *Solar Geophysical Data,* NOAA Boulder, and NSSDC Data Set 69-053A-07C.

201   Barouch, E., Gros, M., and Masse, P.: 1971, *Solar Phys.* **19**, 483.

202   Gros, M.: 1973, private communication,\unpublished data from Centre d'Etudes Nucléaires
       Saclay, Experiment S. 72 on Heos 1.

203   Piggott, W. R.: 1973, private communication, to be published in *INAG Bull.,* 1974.

204   Dilworth, C., Maccagni, D., Perotti, F., Tanzi, E. G., Mercier, J. P., Raviart, A., Treguer, L., and
       Gros, M.: 1972, *Solar Phys.* **23**, 487.

205   Engel, A. P., Balogh, A., Elliot, H., Hynds, R. J., and Quenby, J. J.: 1970, *Acta Phys. Acad. Sci.
       Hungaricae* **29**, Suppl. 2, 439.

206   Barouch, E., Engelmann, J., Gros, M., Koch, L., and Masse, P.: 1970, *Acta Phys. Acad. Sci.
       Hungaricae* **29**, Suppl. 2, 493.

207   Keath, E. P., Bukata, R. P., McCracken, K. G., and Rao, U. R.: 1971, *Solar Phys.* **18**, 503.

208   Gros, M., Masse, P., Engelmann, J., and Barouch, E.: 1972, in J. C. Ulwick (ed.), *Proc. of the
       COSPAR Symposium on Solar Particle Event of November 1969,* Boston 1971, Air Force
       Cambridge Laboratories, Bedford, Special Rept. No. 144, AFCRL-72-0474, p. 115.

209   Vernov, S. N., Chuchkov, E. A., Kontor, N. N., Lyubimov, G. P., Nikolaev, A. G., and
       Pereslegina, N. V.: 1971, in K. Ya. Kondratyev, M. J. Rycroft and C. Sagan (eds.), *Space
       Research* **XI**, Akademie-Verlag, Berlin, p. 1213.

210   Innanen, W. G. and Van Allen, J. A.: 1973, *J. Geophys. Res.* **78**, 1019.

211   Murray, S. S., Stone, E. C., and Vogt, R. E.: 1971, *Phys. Rev. Letters* **26**, 663.

212   Bostrom, C. O., Williams, D. J., and Arens, J. F.: 'Solar Proton Monitoring Experiment
       Explorer 41', NSSDC Data Set 69-053A-07B.

213   Asbridge, J. R.: 1973, private communication, unpublished data from Vela satellites.

214   Baker, M. B., Masley, A. J., and Satterblom, P. R.: 1973, *Proc. of the 13th International
       Conference on Cosmic Rays,* Denver, Conference Papers, Vol. 2, p. 1440.

215   Lanzerotti, L. J. and Maclennan, C. G.: 1972, in J. C. Ulwick (ed.), *Proc. of COSPAR
       Symposium on Solar Particle Event of November 1969,* Boston 1971, Air Force Cambridge
       Laboratories, Bedford, Special Rept. No. 144, AFCRL-72-0474, p. 85.

216   Turtle, J. P., Oelbermann, E. J., Jr., Blake, J. B, Lanzerotti, L. J., Vampola, A. L., and Yates,
       G. K.: 1972, *J. Geophys. Res.* **77**, 730.

217   Webber, W. R.: 1973, private communication, unpublished Pioneer 9 data, and NSSDC Data
       Set 68-100A-06.

218   Balogh, A., Hedgecock, P. C., Hynds, R. J., and Sears, J.: 1971, *Proc. of the 12th International*

Conference on Cosmic Rays, Hobart, Univ. of Tasmania, Conference Papers, Vol. 5, p. 1837.

219   Baird, G. A., Bell, G. G., Duggal, S. P., and Pomerantz, M. A.: 1967, *Solar Phys.* **2**, 491.

220   Lockwood, J. A.: 1968, *J. Geophys. Res.* **73**, 4247.

221   Dodson, H. W. and Hedeman, E. R.: 1969, *Solar Phys.* **9**, 278.

222   Bukata, R. P., Gronstal, P. T., Palmeira, R. A. R., McCracken, K. G., and Rao, U. R.: 1969, *Solar Phys.* **10**, 198. -

223   Masley, A. J.: 1968, in C. de Jager and Z. Švestka (eds.), *Solar Flares and Space Research,* North-Holland, Amsterdam, p. 279.

224   Hakura, Y.: 1968, 'Table of Outstanding Solar Terrestrial Events in 1954 through 1967', NASA Preprint X-641-68-62.

225   Lanzerotti, L. J. and Robbins, M. F.: 1969, *Solar Phys.* **10**, 212.

226   Axisa, F.: 1972, list of electron events prepared for this catalogue, according to data supplied by R. P. Lin.

227   Masley, A. J. and Satterblom, P. R.: 1971, *Proc. of the 12th International Conference on Cosmic Rays,* Hobart, Univ. of Tasmania, Conference Papers, Vol. 5, p. 1849.

228   Masley, A. J., McDonough, J. W., and Satterblom, P. R.: 1970, *Antarctic J. of the U.S.* **5**, 172.

229   Kreplin, R. W.: 1974, unpublished data from Solrad 9, Explorer 35, OGO-4, OGO-5 and OGO-6.

230   McIntosh, P. S. and Donnelly, R. F.: 1972, *Solar Phys.* **23**, 444.

231   Hudson, H.: 1973, private communication, catalogue of OSO-3 data.

232   Kane, S. R.: 1973, private communication, unpublished data from OGO-5.

233   Dodson, H. W., Hedeman, E. R., and de Miceli, M. R.: 1972, WDC-A Report UAG-19, NOAA, Boulder.

234   Dodson, H. W. and Hedeman, E. R.: unpublished list of 1967 subflares.

PART 3

MAPS OF SELECTED ACTIVE REGIONS WHICH WERE SOURCES OF PARTICLE EVENTS

SUMMARY OF PART 3

McMath plage	Particle events	McMath plage	Particle events
3400	2	8105	125
3432	3	8131	127*, 129*
3643	5*	8207	130, 131*, 133, 135*
3820	7*	8310	145
3907	9	8348	146*
4039	16*	8362	148*, 149*
4070	17*	8461	153*, 154*, 155
4124	23*	8484	158
4151	26	8496	160
4159	28*	8505	161*
4189	29	8687	176*
4321	32	8704	177
4400	33*	8818	182, 183*, 184
4476	37*	8905	194, 197
4578	41*	8973	205*
4634	43	9034	209, 212*
4659	44	9047	213*, 214
4686	45*	9108	224
4708	47, 48	9118	225*
5148	52	9145	233*, 235
5204	53	9146	232, 237
5265	55, 56, 57	9184	240
5323	58*	9204	245
5527	61*	9372	268
5615	64*, 65, 66*, 67, 68	9429	274*
5642	70, 71	9503	276*, 277*
5645	69	9567	281, 284
5653	72*	9678	289
5654	74*	9687	287
5669	76*	9692	288, 290
5680	77*	9740	291, 293*, 295, 296*, 297, 298*
5837	83		
5858	84	9760	301*
5909	86	9802	304*
5925	⌈87*, 88, 89, 90* ⌊91, 93	9842	307*
		9879	309*
5929	94	9946	310, 311*, 312*, 313, 314*
6171	96, 97*, 98, 99*, 100*		
6178	101	9966	316*
6212	103	9994	319, 320
6235	104	10057	326, 327
6264	106*	10109	329
6908	116	10134	332
6964	117, 118, 119, 120*	10326	333*
7003	121	10333	334
7182	122*	10352	337
7661	123	10432	345, 348
8012	124*	10491	352

* Hα flare picture.

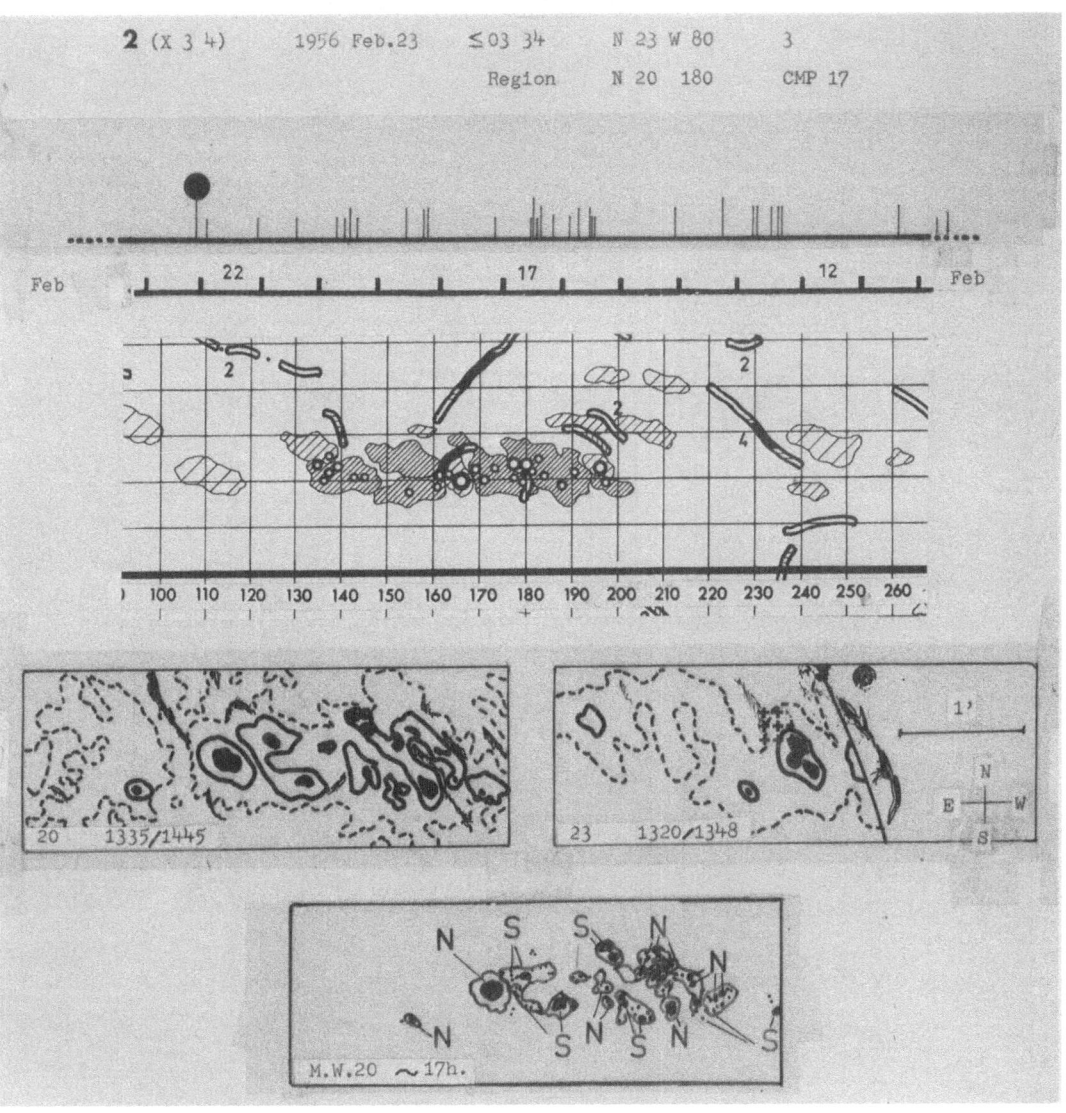

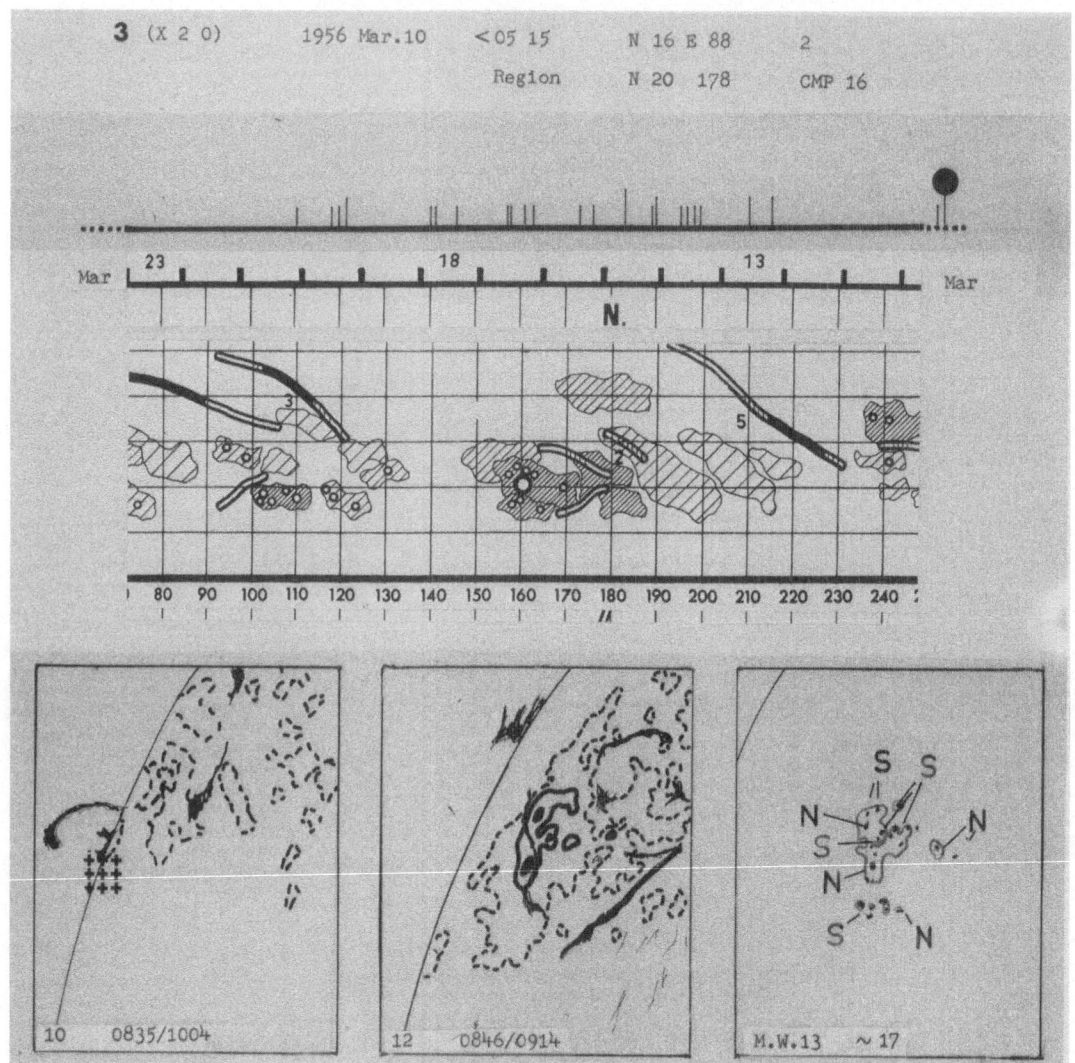

3 (X 2 0)   1956 Mar.10   < 05 15   N 16 E 88   2
                Region   N 20 178   CMP 16

Mar   23   18   13   Mar

N.

80 90 100 110 120 130 140 150 160 170 180 190 200 210 220 230 240

10   0835/1004

12   0846/0914

M.W.13   ~ 17

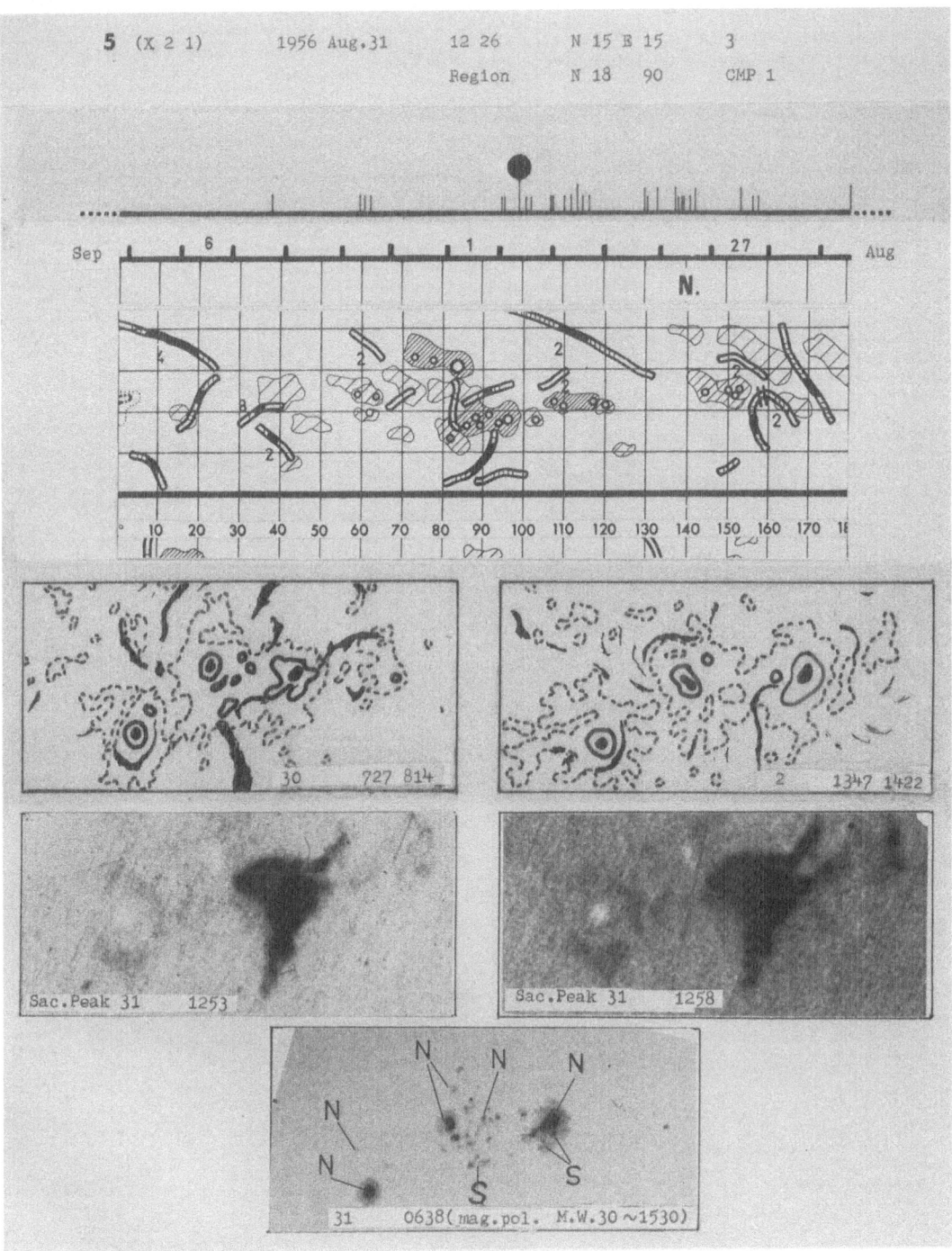

5 (X 2 1) 1956 Aug.31 12 26 N 15 E 15 3
Region N 18 90 CMP 1

30 727 814

2 1347 1422

Sac.Peak 31 1253

Sac.Peak 31 1258

31 0638(mag.pol. M.W.30 ~1530)

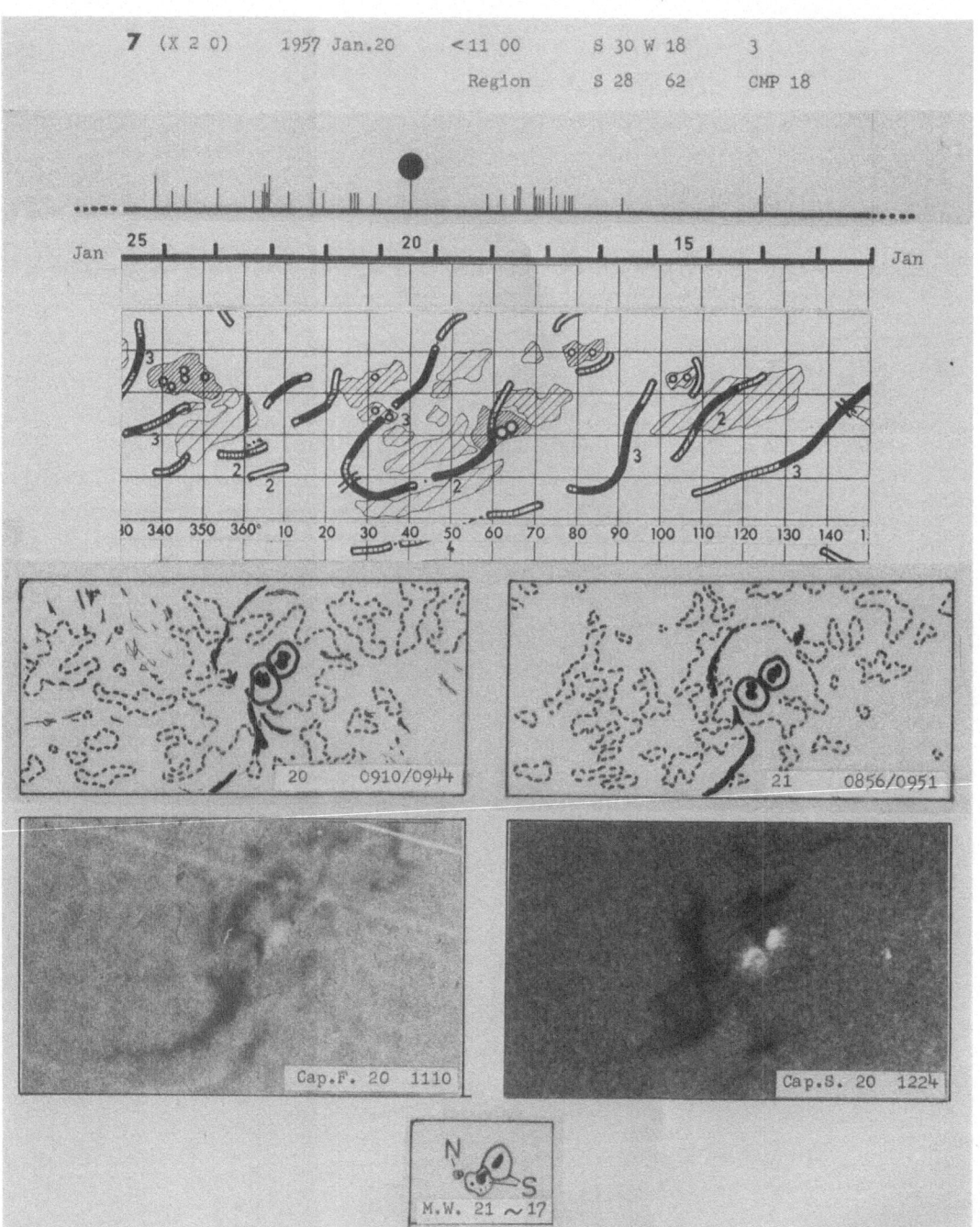

7 (X 2 0)   1957 Jan.20   < 11 00   S 30 W 18   3
                          Region   S 28   62   CMP 18

Jan   25   20   15   Jan

30 340 350 360° 10 20 30 40 50 60 70 80 90 100 110 120 130 140 1.

20   0910/0944

21   0856/0951

Cap.F. 20 1110

Cap.S. 20 1224

M.W. 21 ~ 17

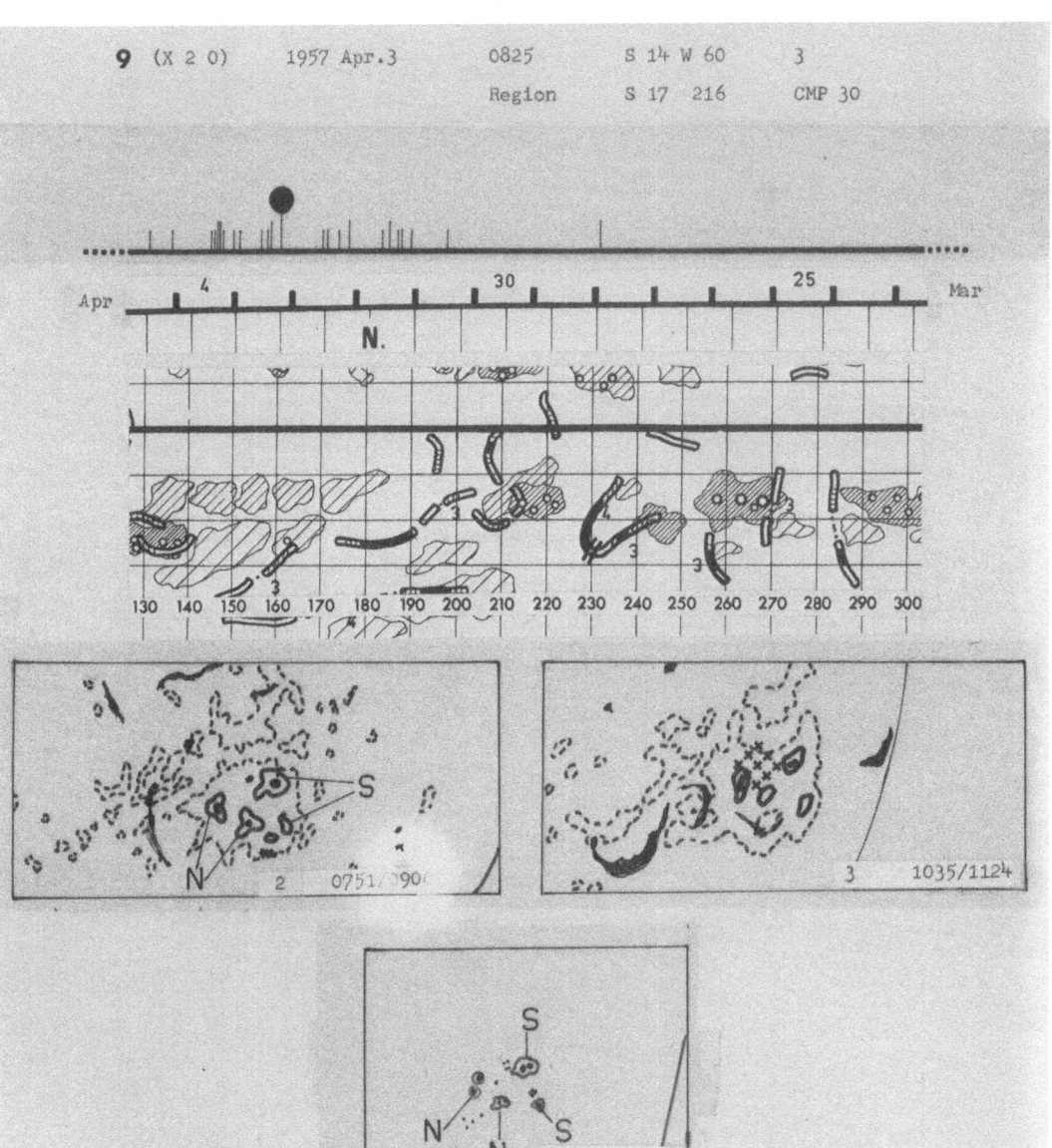

9 (X 2 0)   1957 Apr.3   0825   S 14 W 60   3
                            Region S 17 216   CMP 30

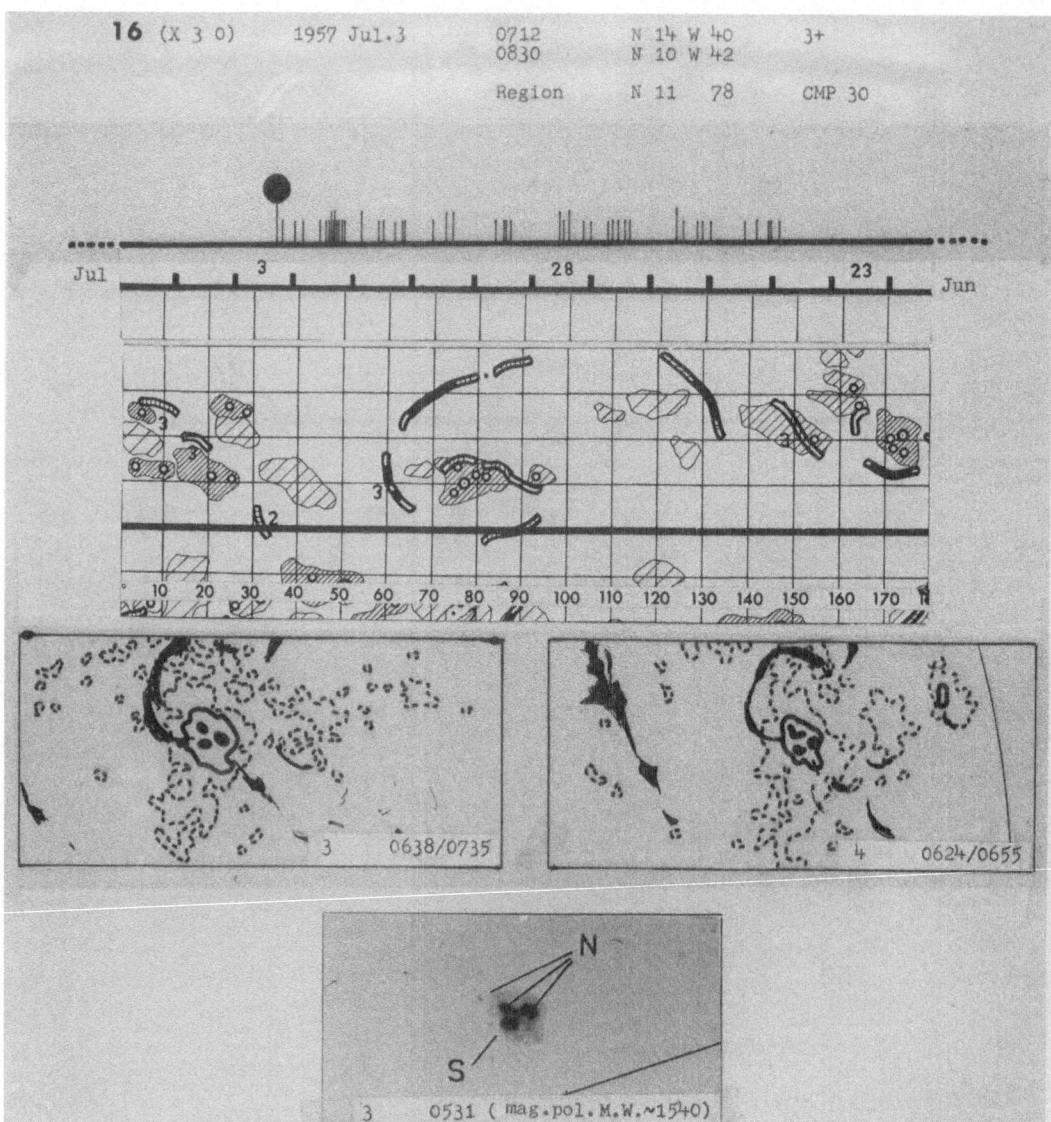

16 (X 3 0)   1957 Jul.3   0712   N 14 W 40   3+
                              0830   N 10 W 42

                              Region   N 11   78   CMP 30

3   0638/0735

4   0624/0655

3   0531 (mag.pol. M.W. ~1540)

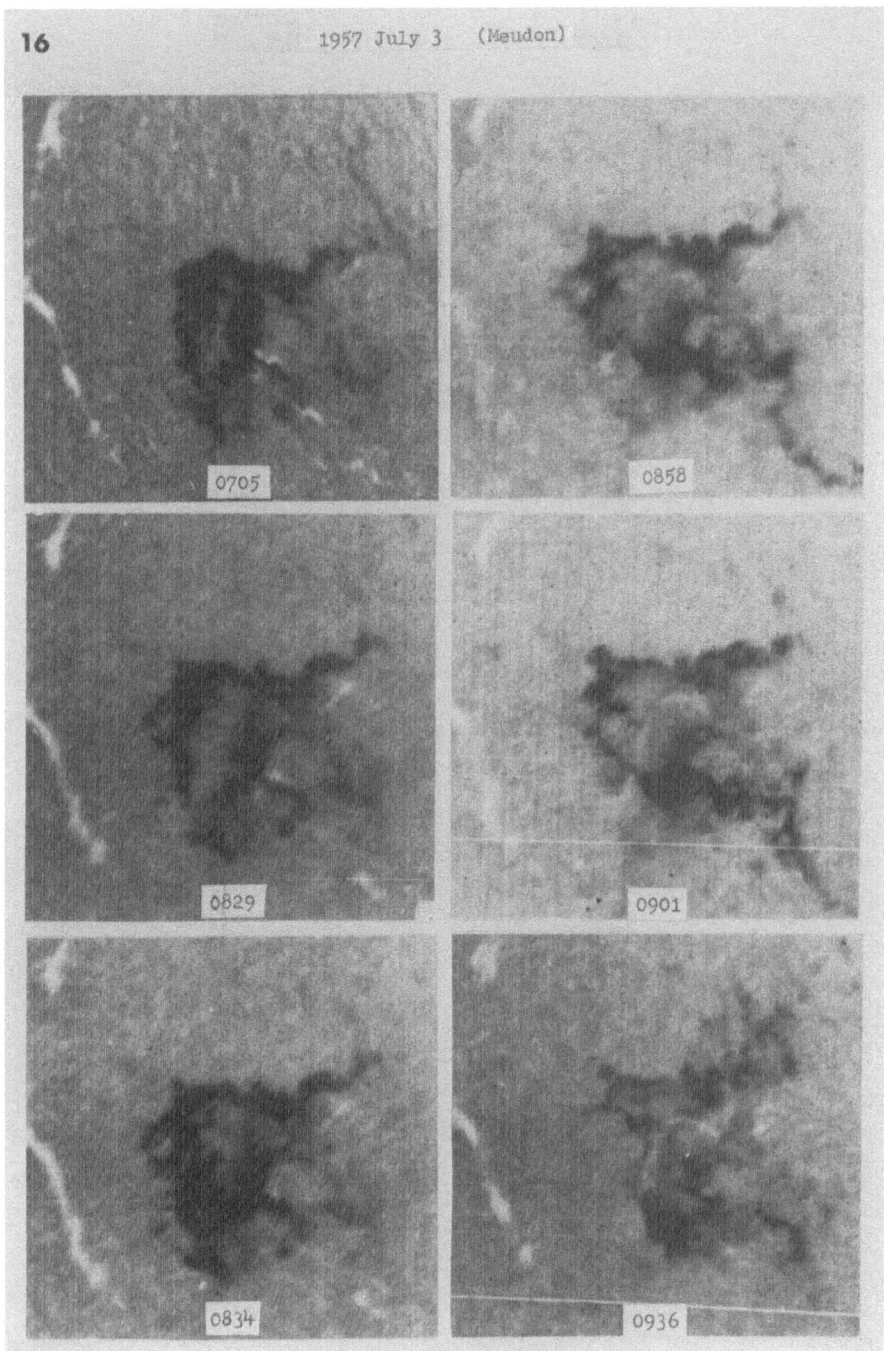

16   1957 July 3   (Meudon)

0705

0858

0829

0901

0834

0936

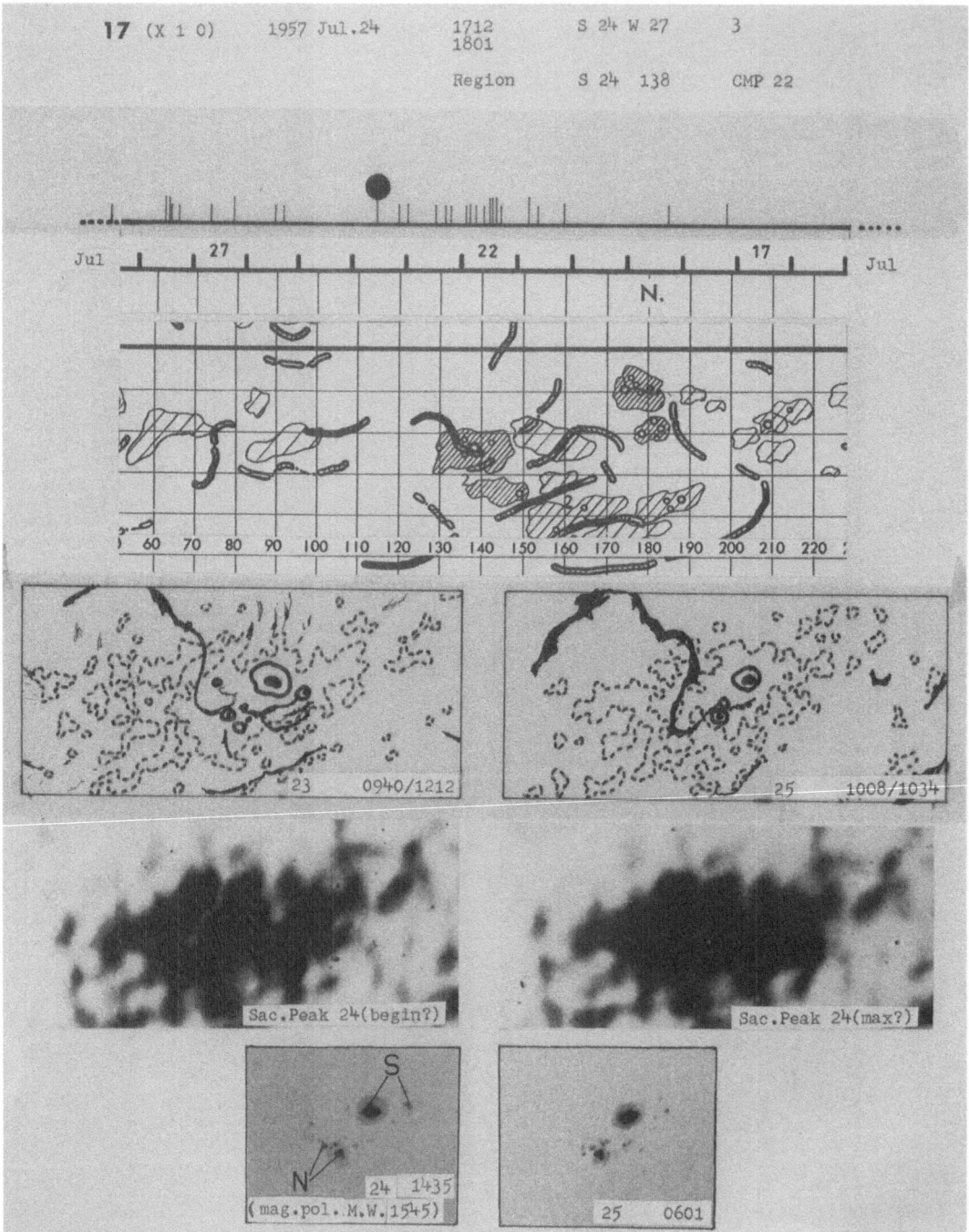

17 (X 1 0)   1957 Jul.24   1712   S 24 W 27   3
                               1801

                            Region   S 24   138   CMP 22

23   0940/1212

25   1008/1034

Sac.Peak 24(begin?)

Sac.Peak 24(max?)

24   1435
(mag.pol.M.W.1545)

25   0601

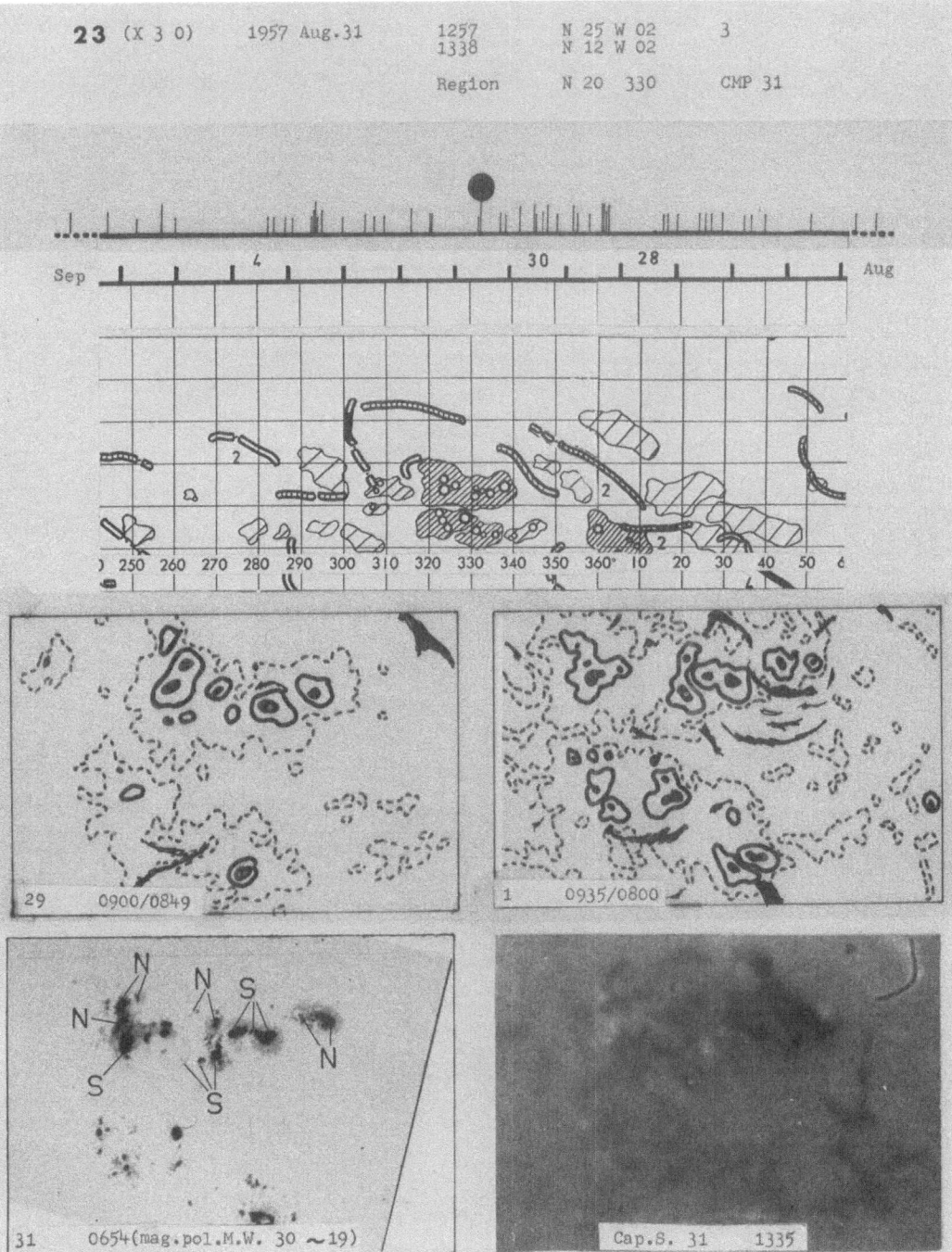

23 (X 3 0)   1957 Aug.31   1257   N 25 W 02   3
                           1338   N 12 W 02

                           Region   N 20 330   CMP 31

29   0900/0849

1   0935/0800

31   0654(mag.pol.M.W. 30 ~19)

Cap.S. 31   1335

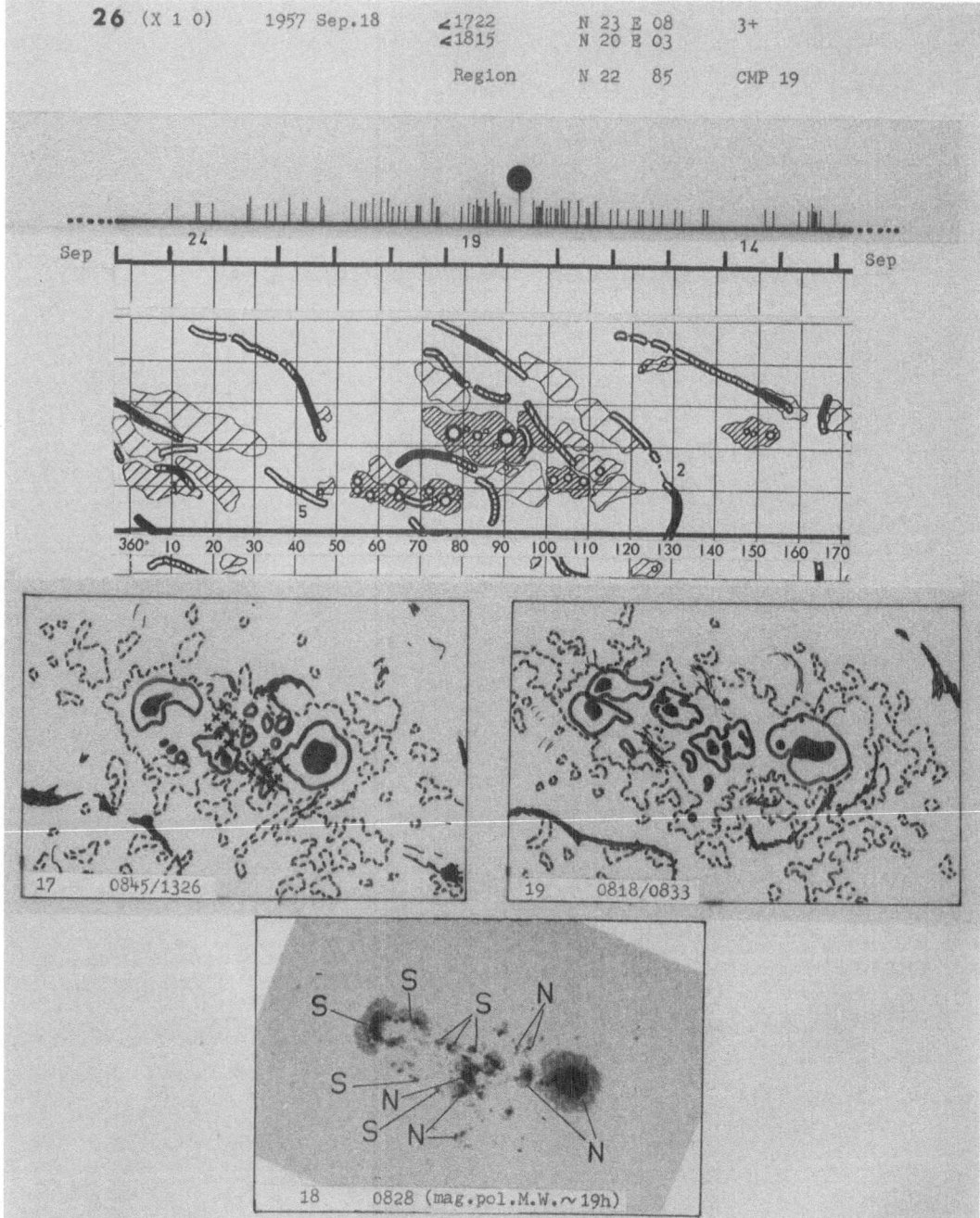

26 (X 1 0)   1957 Sep.18   < 1722   N 23 E 08   3+
                                 < 1815   N 20 E 03

                                 Region   N 22   85   CMP 19

17   0845/1326

19   0818/0833

18   0828 (mag.pol.M.W.∼19h)

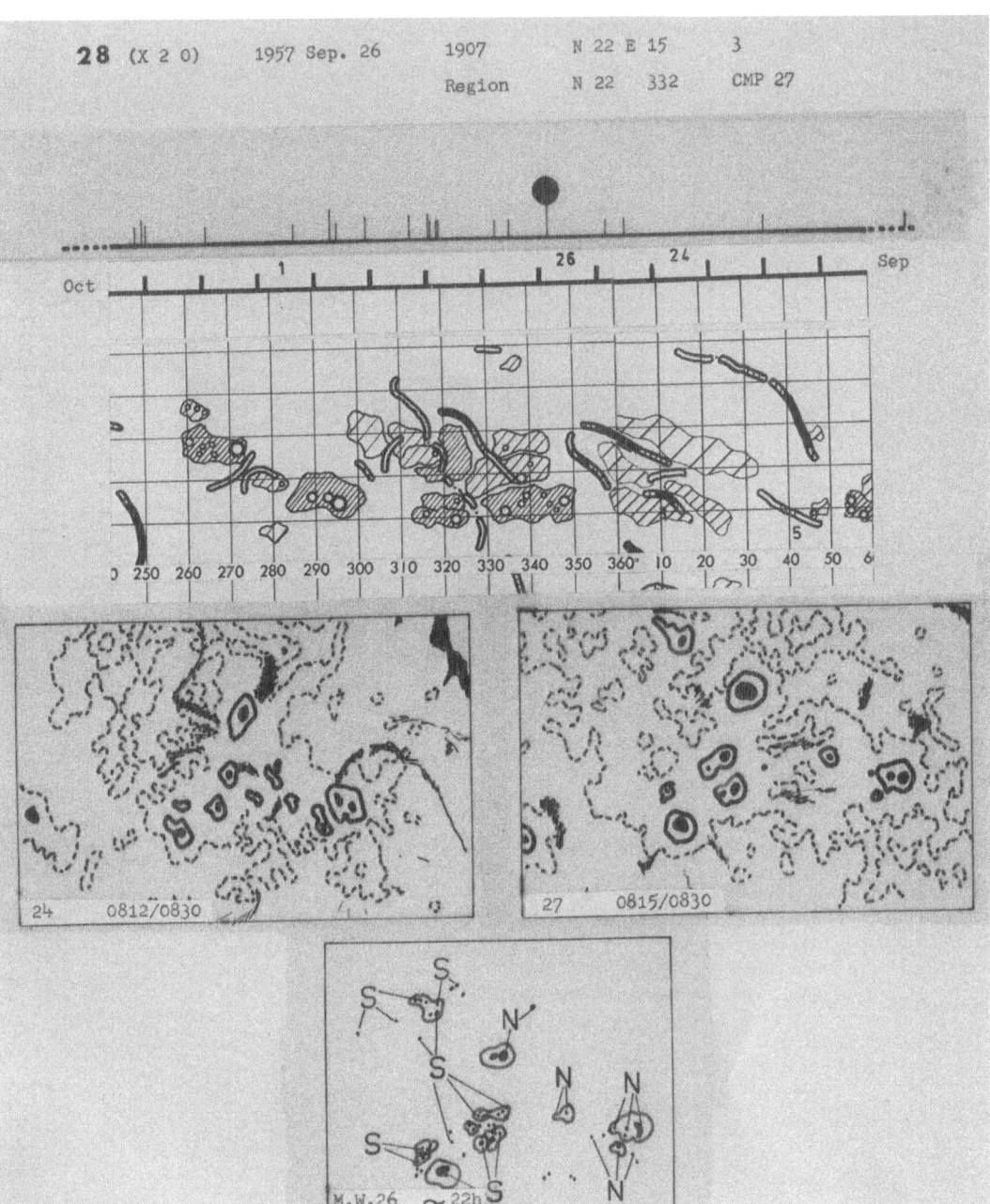

28 (X 2 0)   1957 Sep. 26   1907   N 22 E 15   3
                                 Region   N 22   332   CMP 27

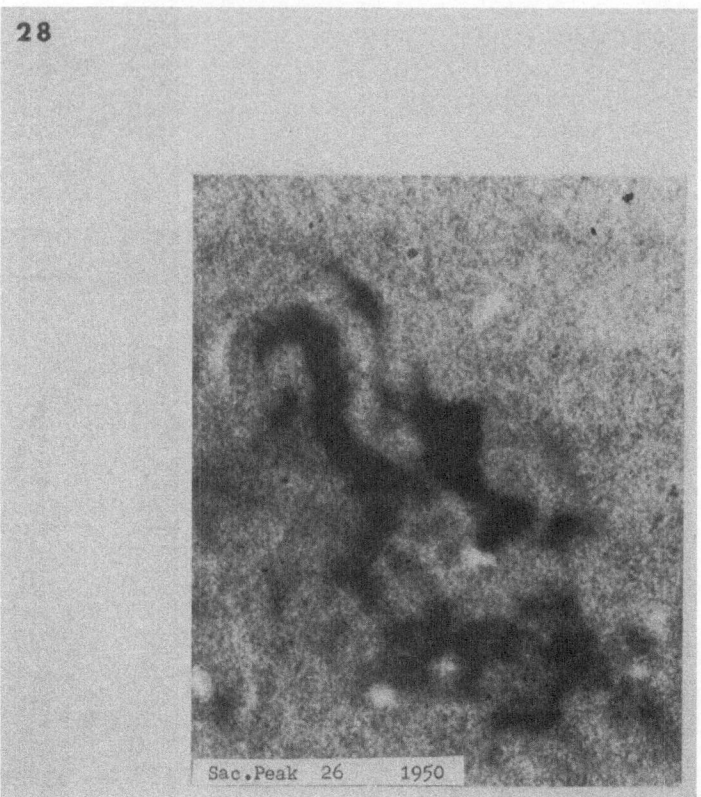

28

Sac.Peak   26   1950

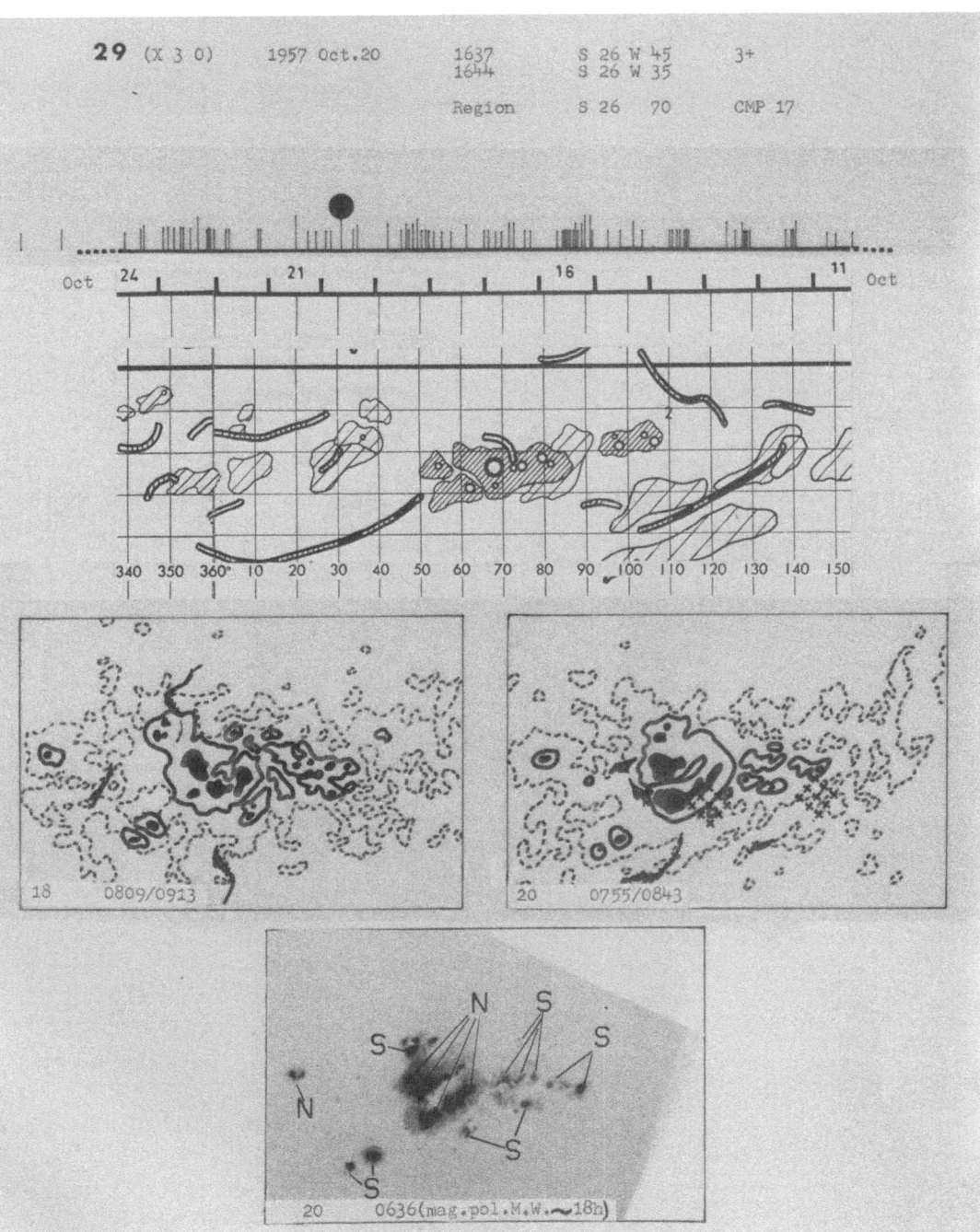

29 (X 3 0)   1957 Oct.20   1637   S 26 W 45   3+
1644   S 26 W 35

Region   S 26   70   CMP 17

18   0809/0913

20   0755/0843

20   0636(mag.pol.M.W.∼18h)

32 (X 1 0) 1957 Dec.28 2229 N 25 W 50 2

Region N 25 250 CMP 24

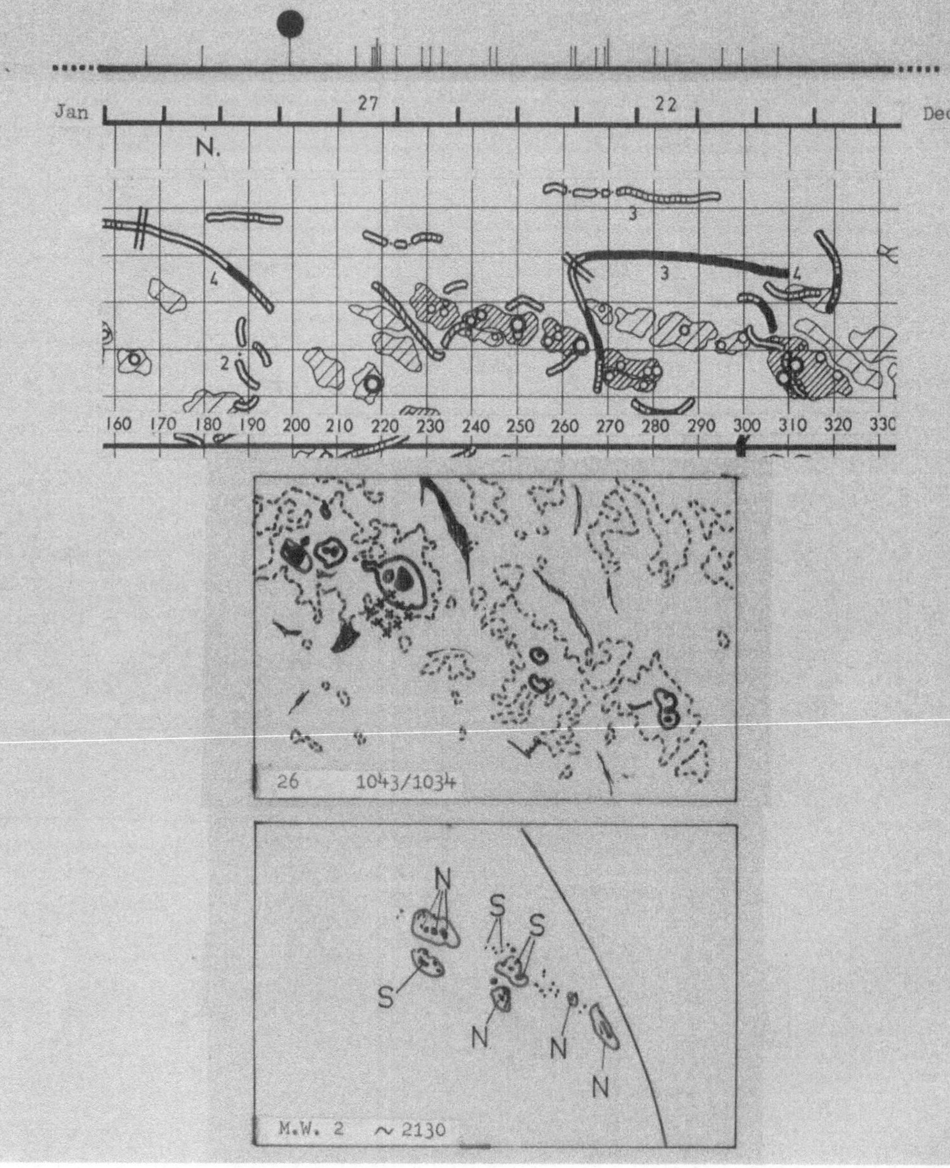

26 10⁴3/10³⁴

M.W. 2 ∼ 2130

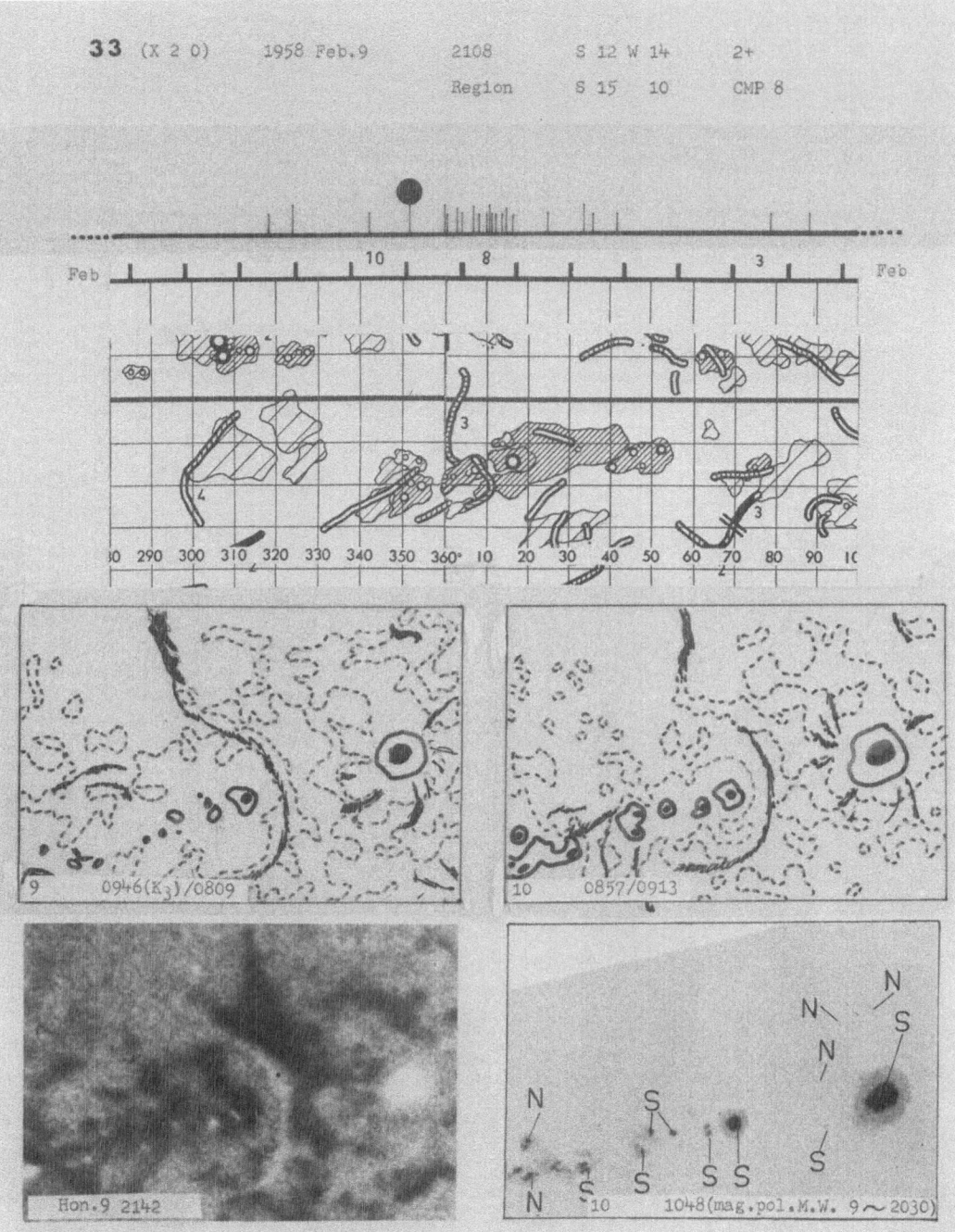

33 (X 2 0)   1958 Feb.9   2108   S 12 W 14   2+

Region   S 15   10   CMP 8

9   0946(K₃)/0809

10   0857/0913

Hon.9 2142

10   1048(mag.pol.M.W. 9 ～ 2030)

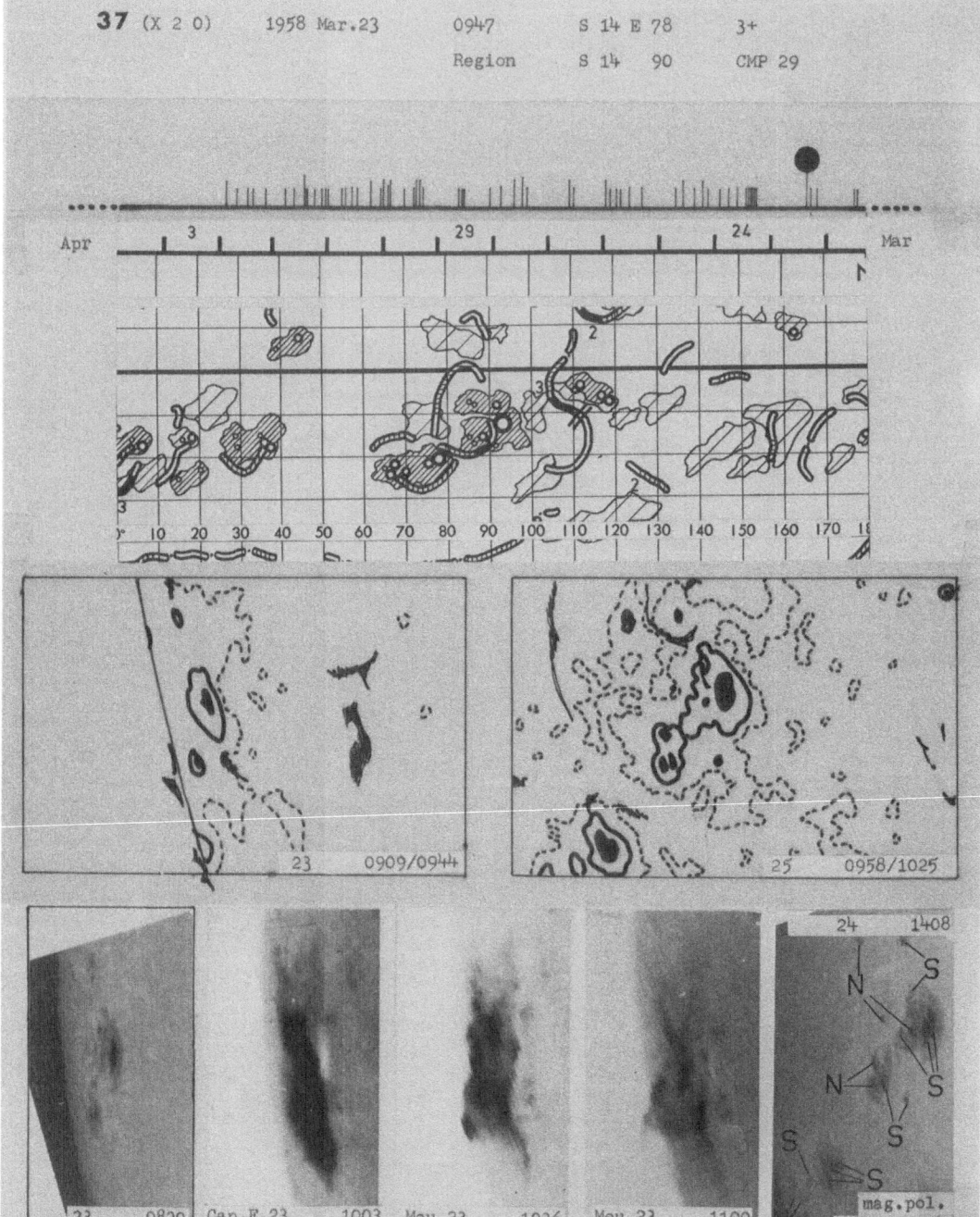

37 (X 2 0)   1958 Mar.23   0947   S 14 E 78   3+

Region   S 14   90   CMP 29

Apr   3   29   24   Mar

10 20 30 40 50 60 70 80 90 100 110 120 130 140 150 160 170 18

23   0909/0944

25   0958/1025

23   0820   Cap.F.23   1003   Meu.23   1036   Meu.23   1109   24   1408

mag.pol.
M.W. ~2030)

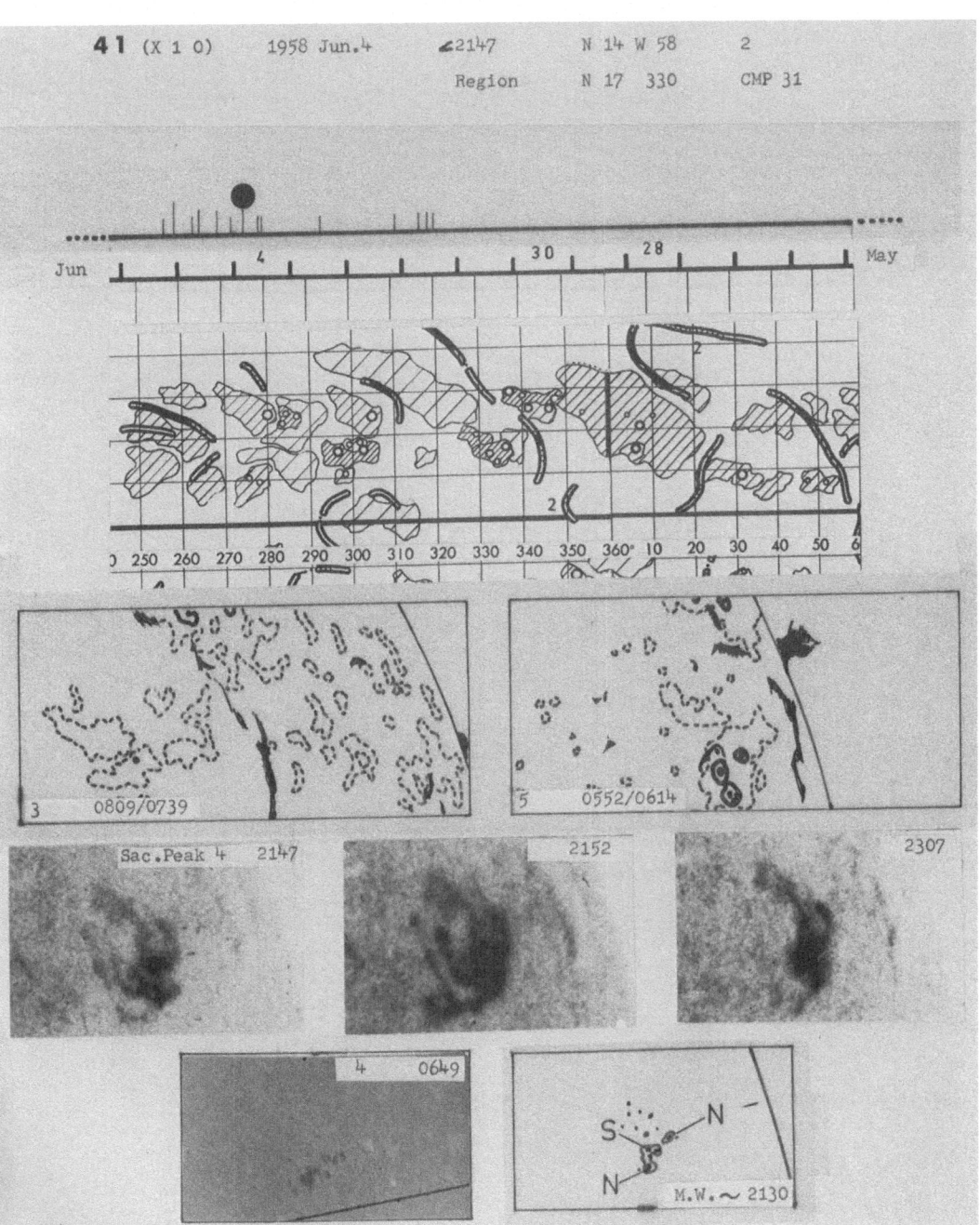

41 (X 1 0)   1958 Jun.4   ∠2147   N 14 W 58   2
                            Region   N 17 330   CMP 31

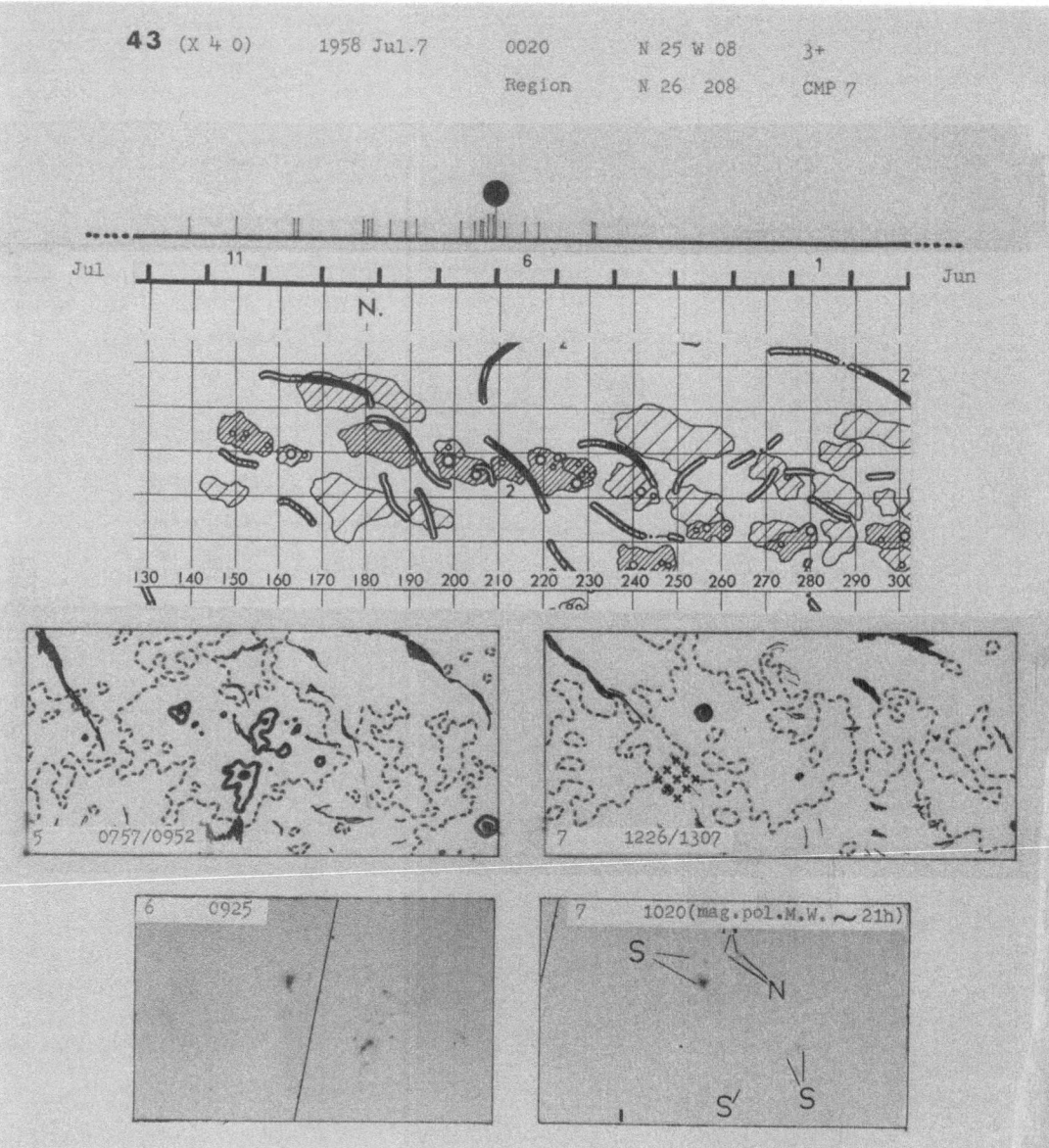

43 (X 4 0) 1958 Jul.7 0020 N 25 W 08 3+

Region N 26 208 CMP 7

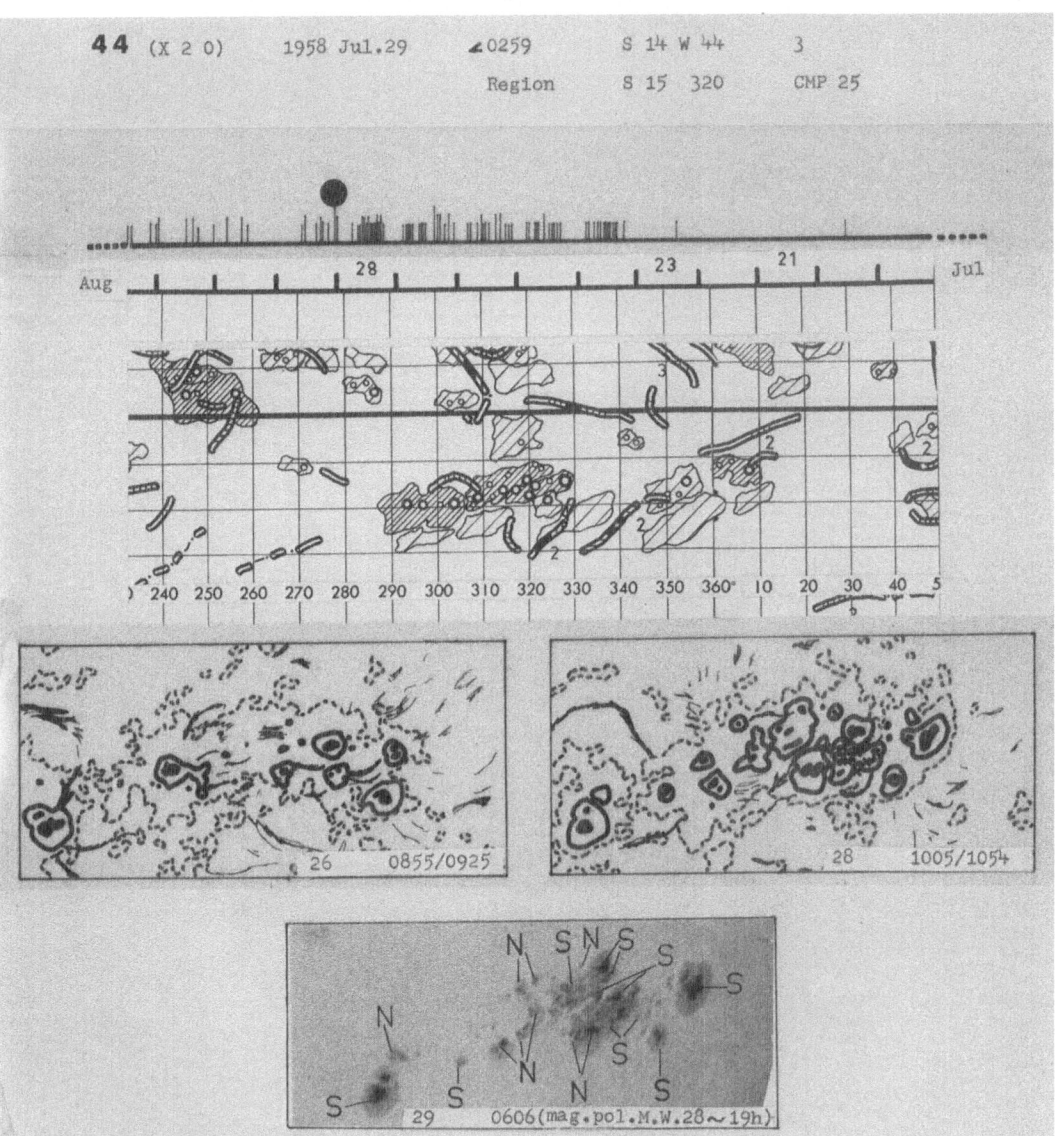

44 (X 2 0)   1958 Jul.29   ∠0259   S 14 W 44   3
Region   S 15 320   CMP 25

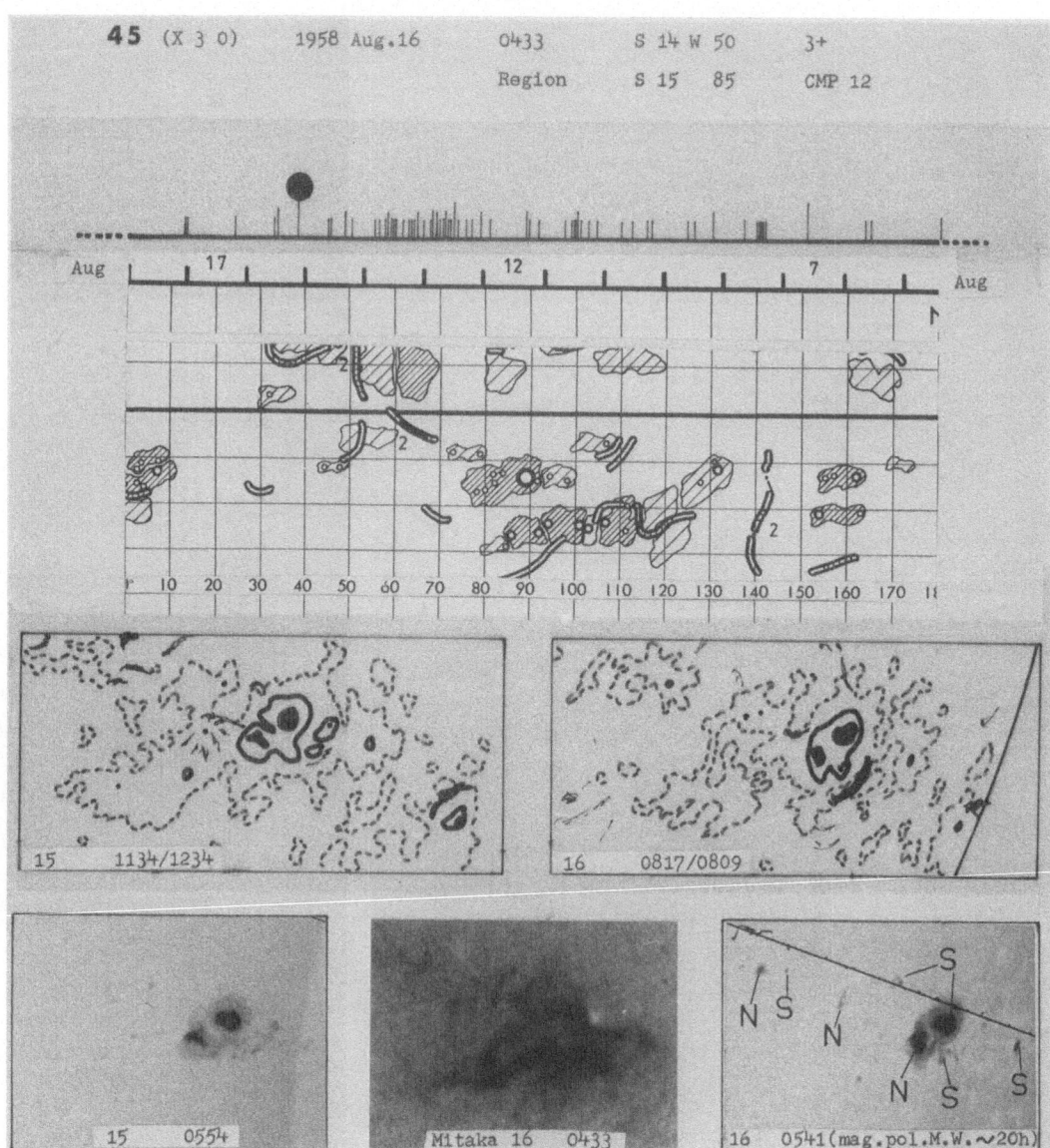

45 (X 3 0)   1958 Aug.16   0433   S 14 W 50   3+

Region   S 15   85   CMP 12

15   1134/1234

16   0817/0809

15   0554

Mitaka 16   0433

16   0541(mag.pol.M.W.~20h)

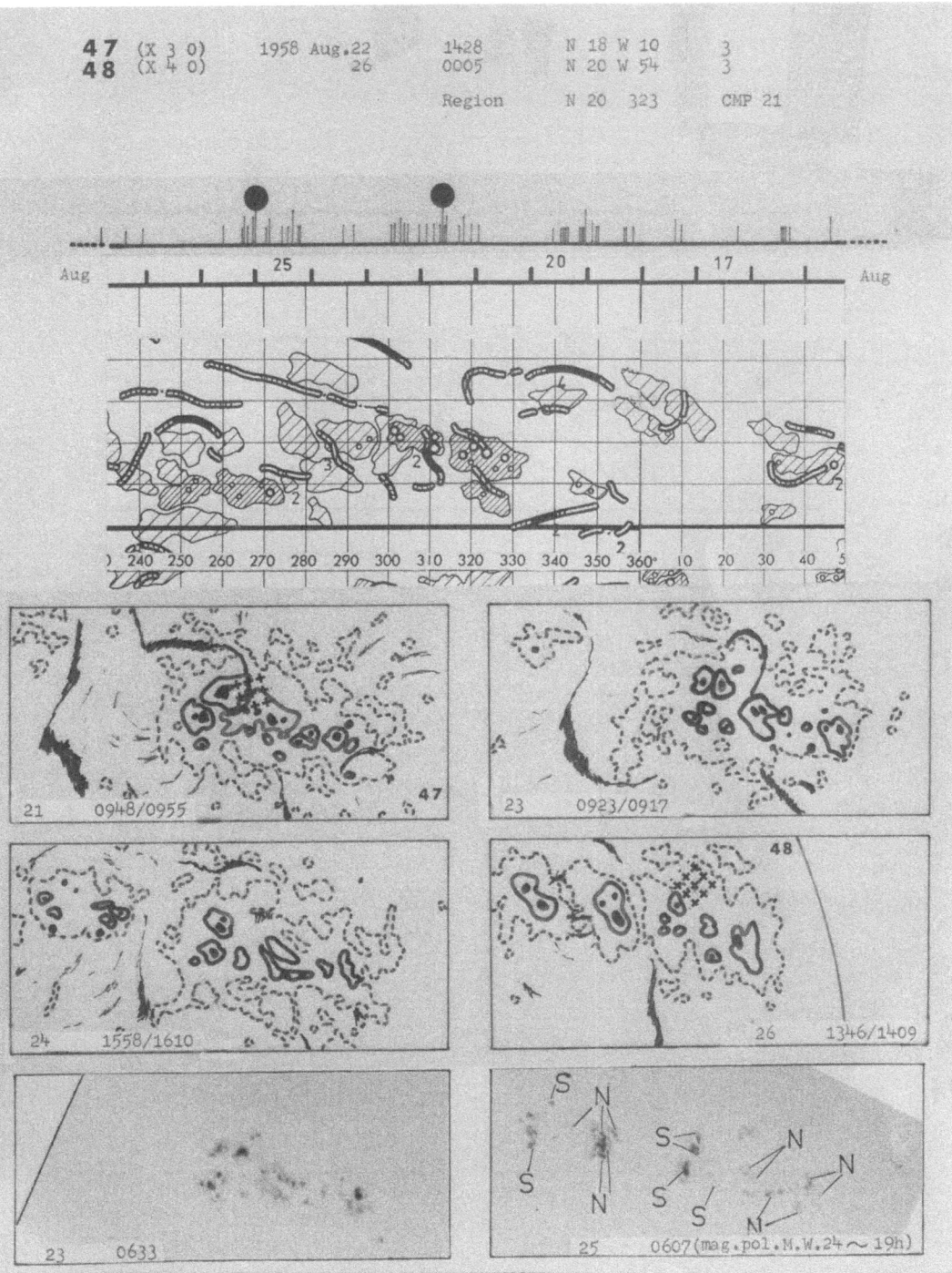

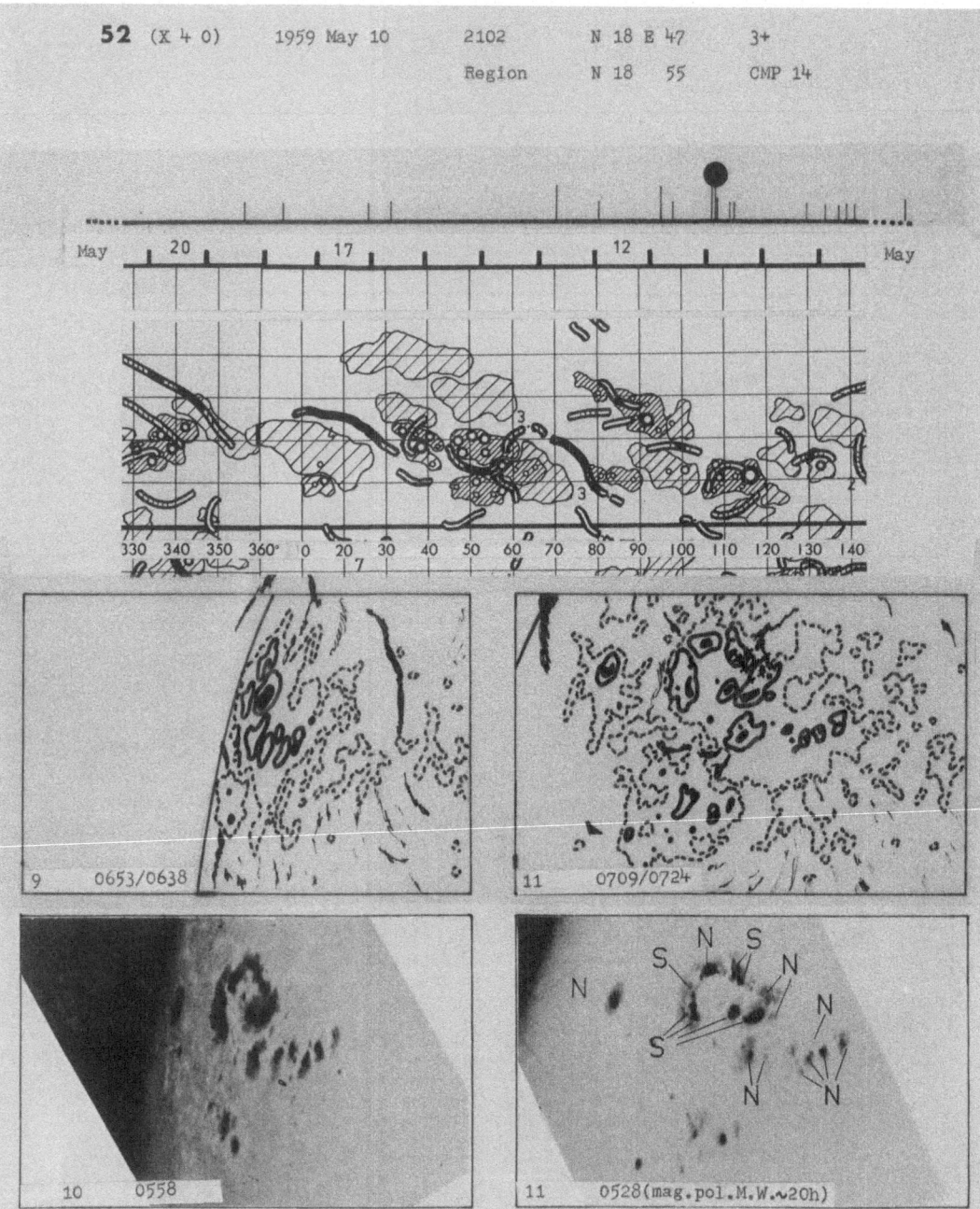

52 (X 4 0)   1959 May 10   2102   N 18 E 47   3+
                             Region N 18 55   CMP 14

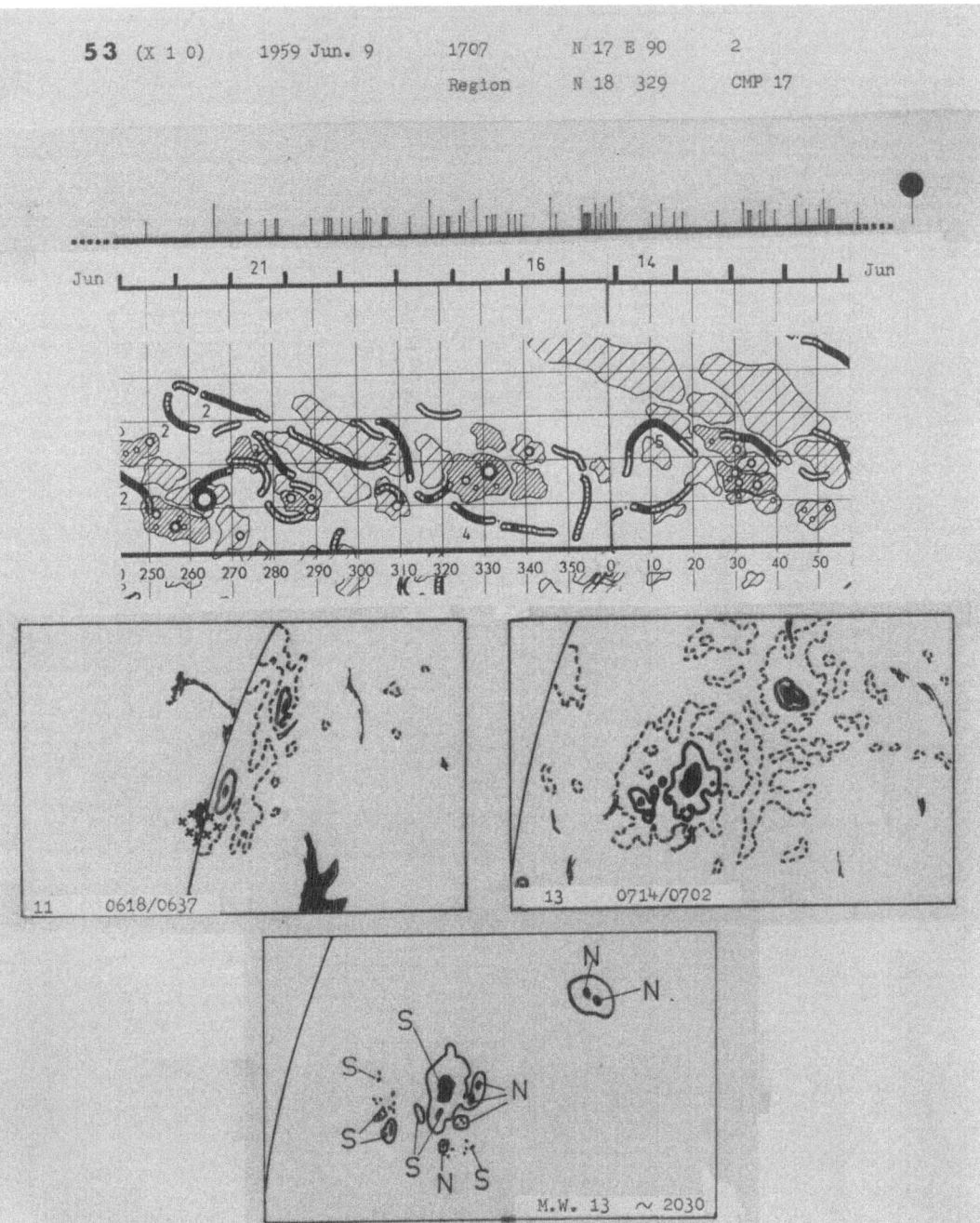

53 (X 1 0)   1959 Jun. 9   1707   N 17 E 90   2

Region   N 18 329   CMP 17

11   0618/0637

13   0714/0702

M.W. 13   ~ 2030

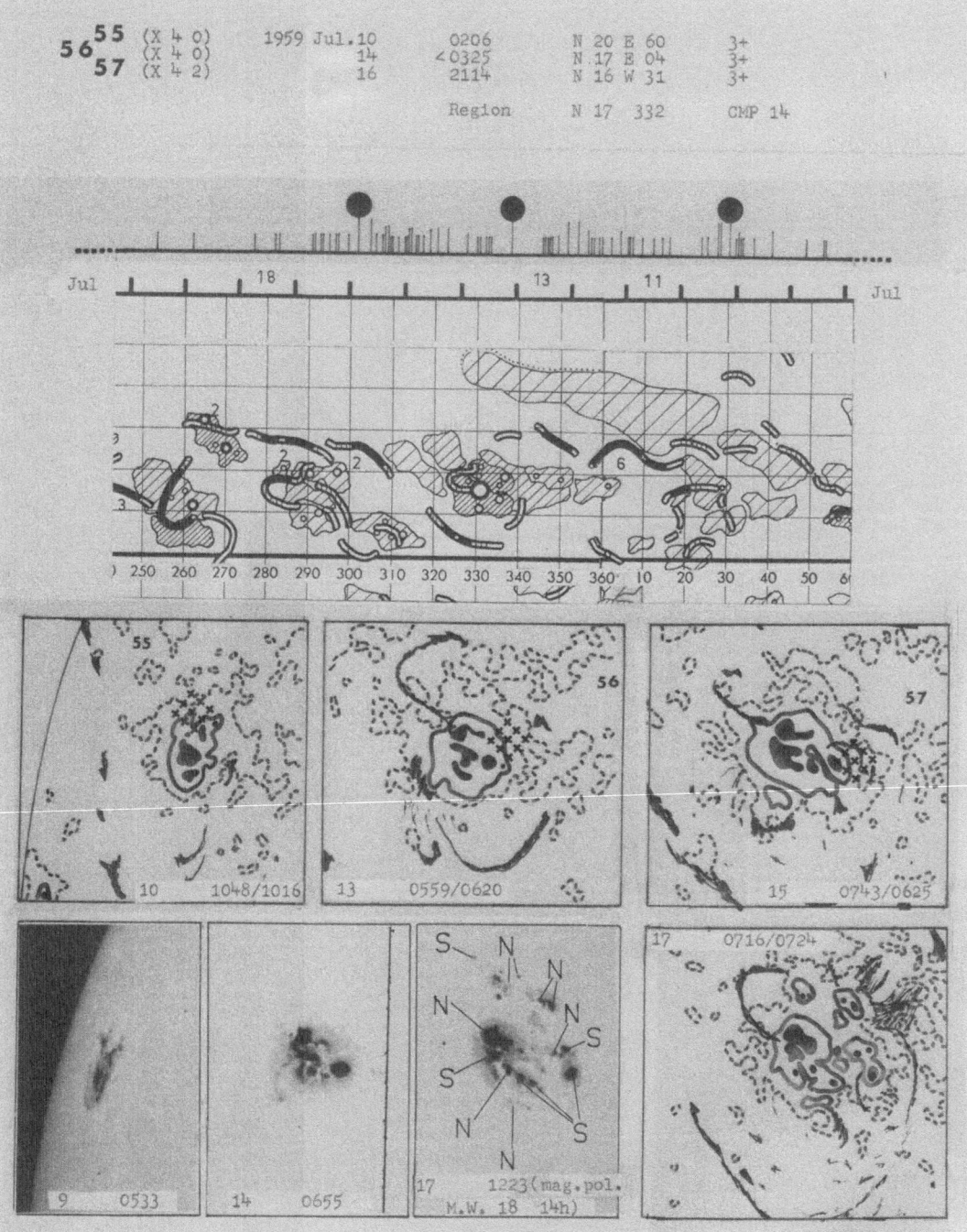

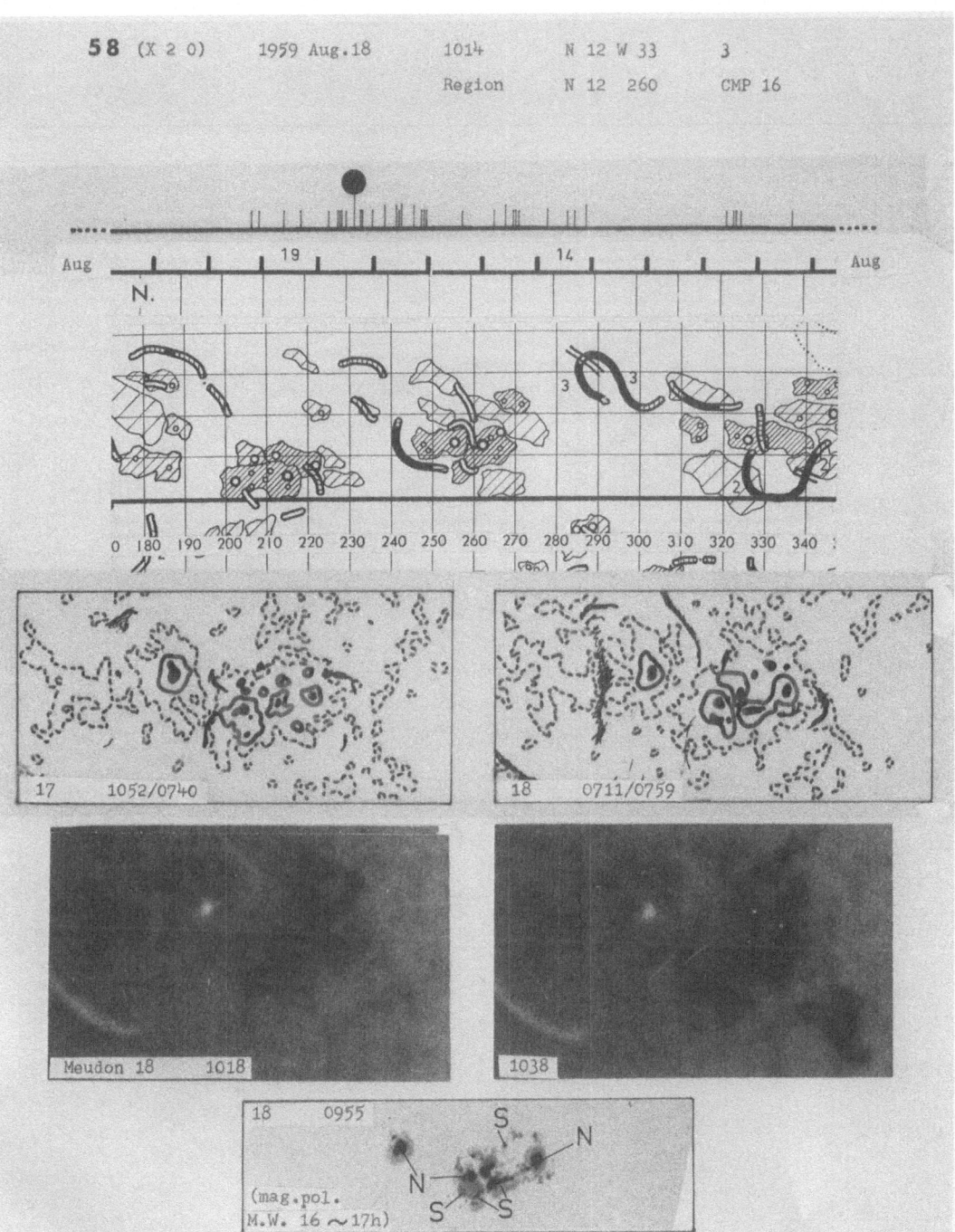

58 (X 2 0)   1959 Aug.18   1014   N 12 W 33   3
                  Region   N 12   260   CMP 16

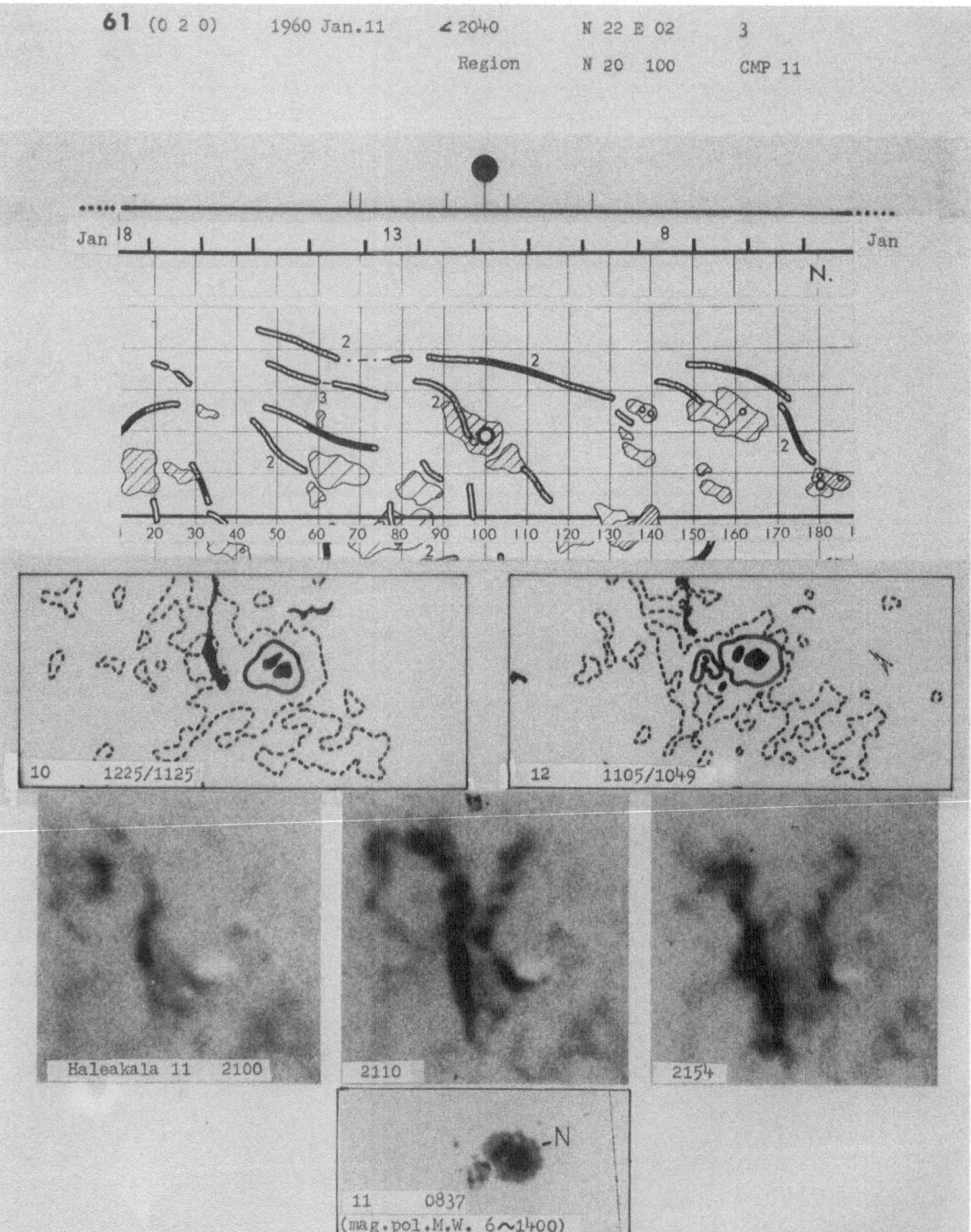

61 (0 2 0)   1960 Jan.11   < 2040   N 22 E 02   3
                                 Region   N 20   100   CMP 11

10   1225/1125

12   1105/1049

Haleakala 11   2100   2110   2154

11   0837
(mag.pol.M.W. 6∿1400)

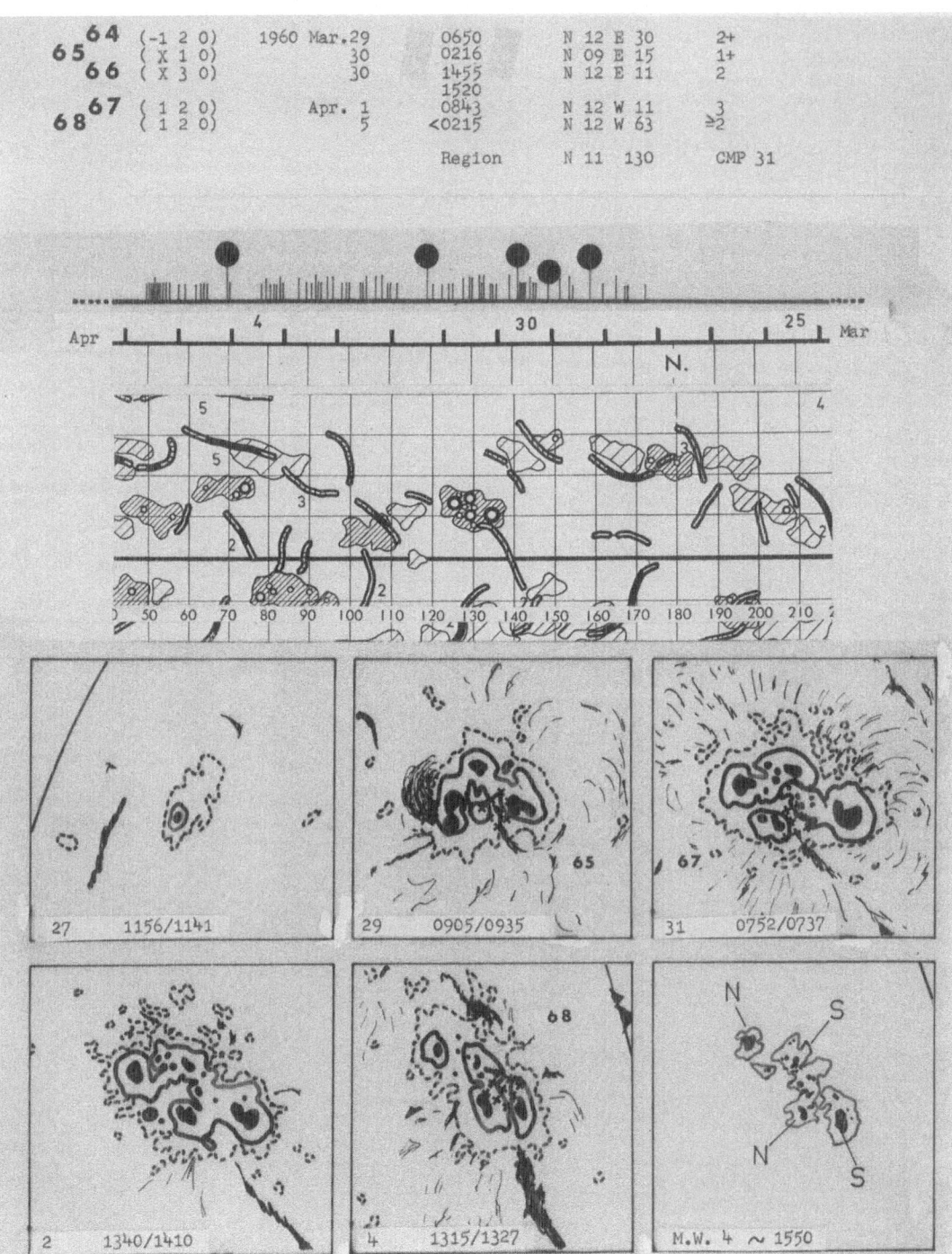

65	64	(-1 2 0)	1960 Mar.29	0650	N 12 E 30	2+
		(X 1 0)	30	0216	N 09 E 15	1+
	66	(X 3 0)	30	1455	N 12 E 11	2
				1520		
68	67	(1 2 0)	Apr. 1	0843	N 12 W 11	3
		(1 2 0)	5	<0215	N 12 W 63	≧2
			Region		N 11 130	CMP 31

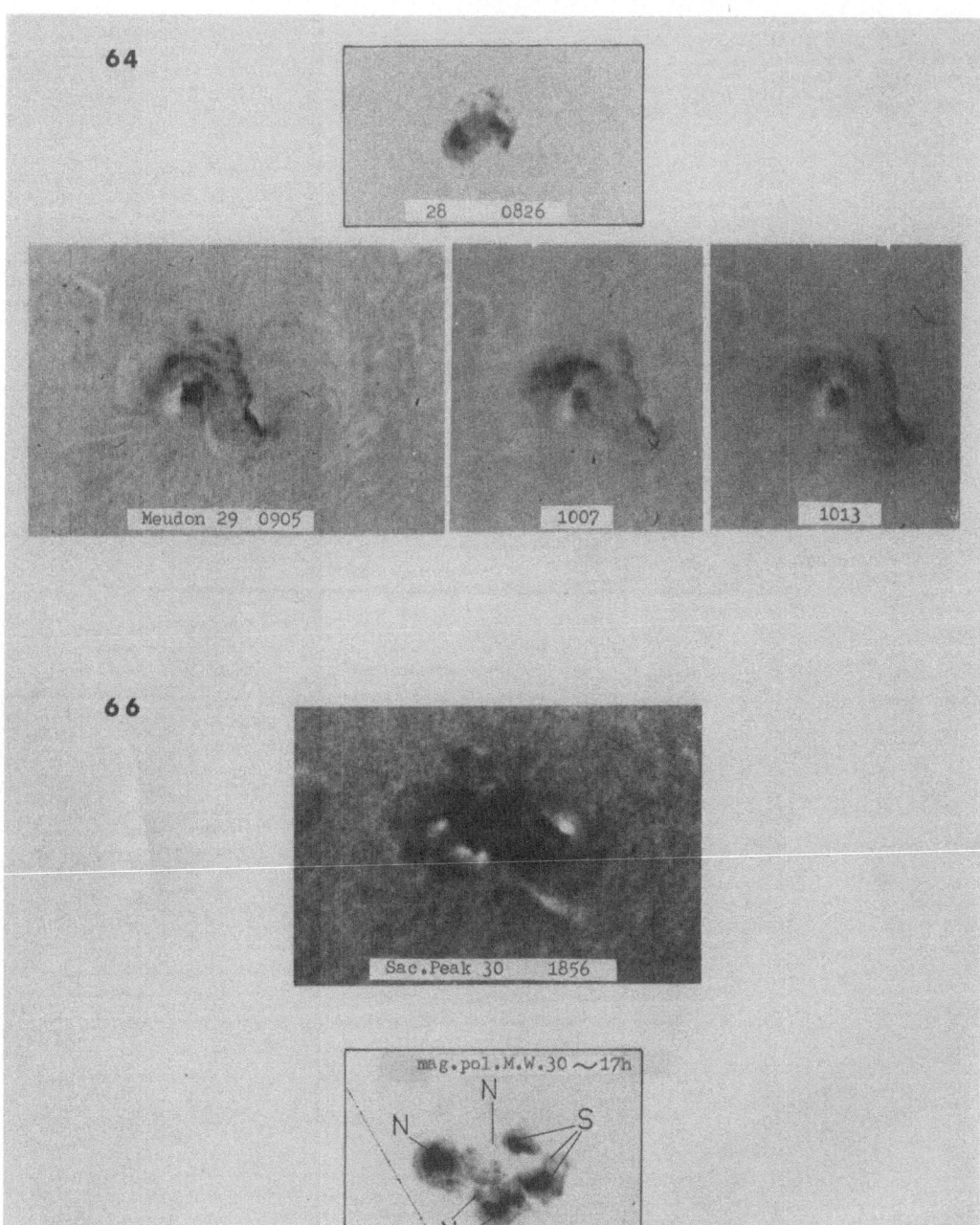

64

28   0826

Meudon 29 0905   1007   1013

66

Sac.Peak 30   1856

mag.pol.M.W.30 ∼17h
N
N   S
N
30   1149

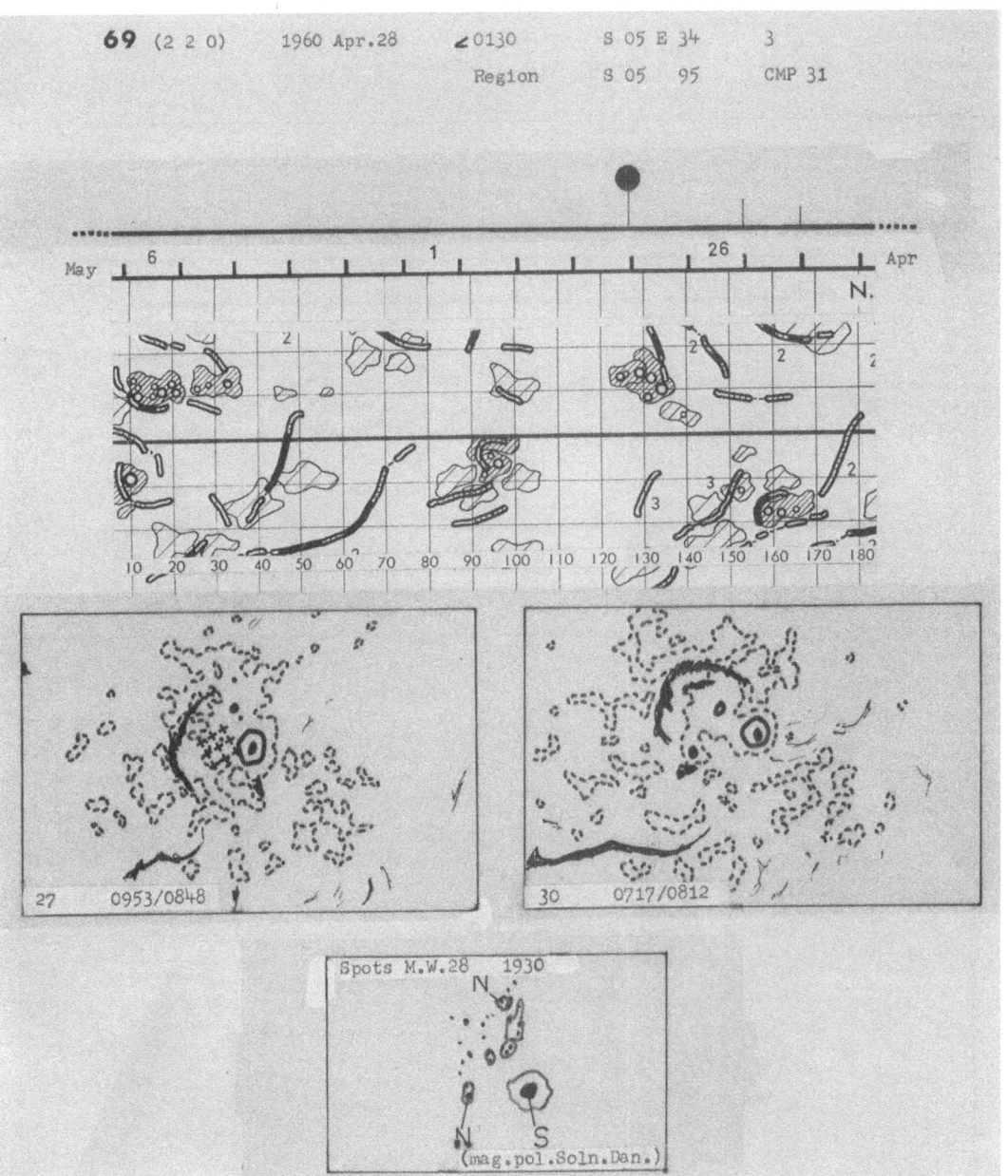

69 (2 2 0)   1960 Apr.28   ∠0130   S 05 E 34   3
Region   S 05   95   CMP 31

27   0953/0848

30   0717/0812

Spots M.W.28   1930
N
N   S
(mag.pol.Soln.Dan.)

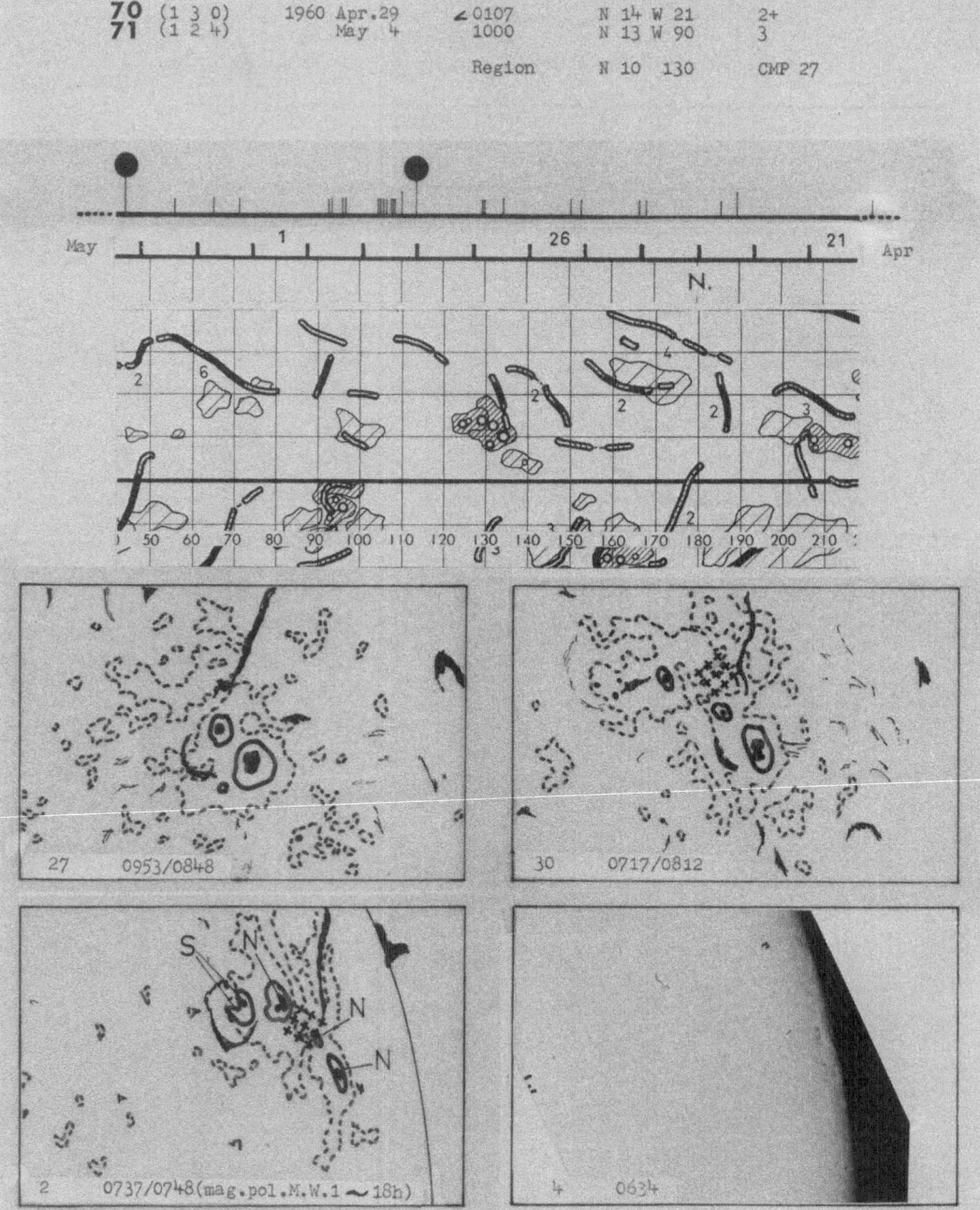

70 (1 3 0)   1960 Apr.29   ∠ 0107   N 14 W 21   2+
71 (1 2 4)   May 4   1000   N 13 W 90   3

                   Region   N 10 130   CMP 27

72 (1 3 0)   1960 May 6   1404   S 08 E 07   3+

Region   S 10   8   CMP 7

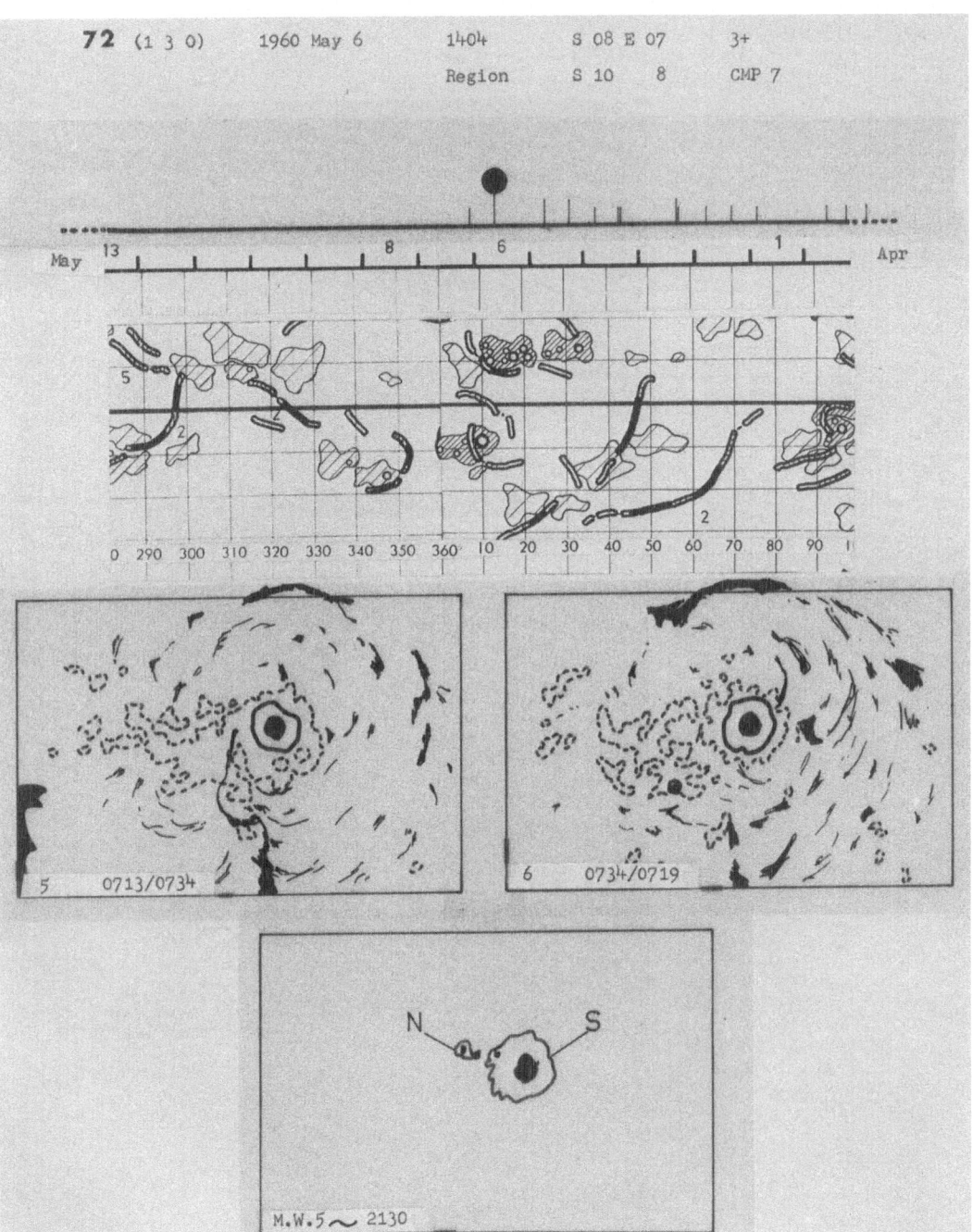

5   0713/0734

6   0734/0719

M.W.5 ∿ 2130

72

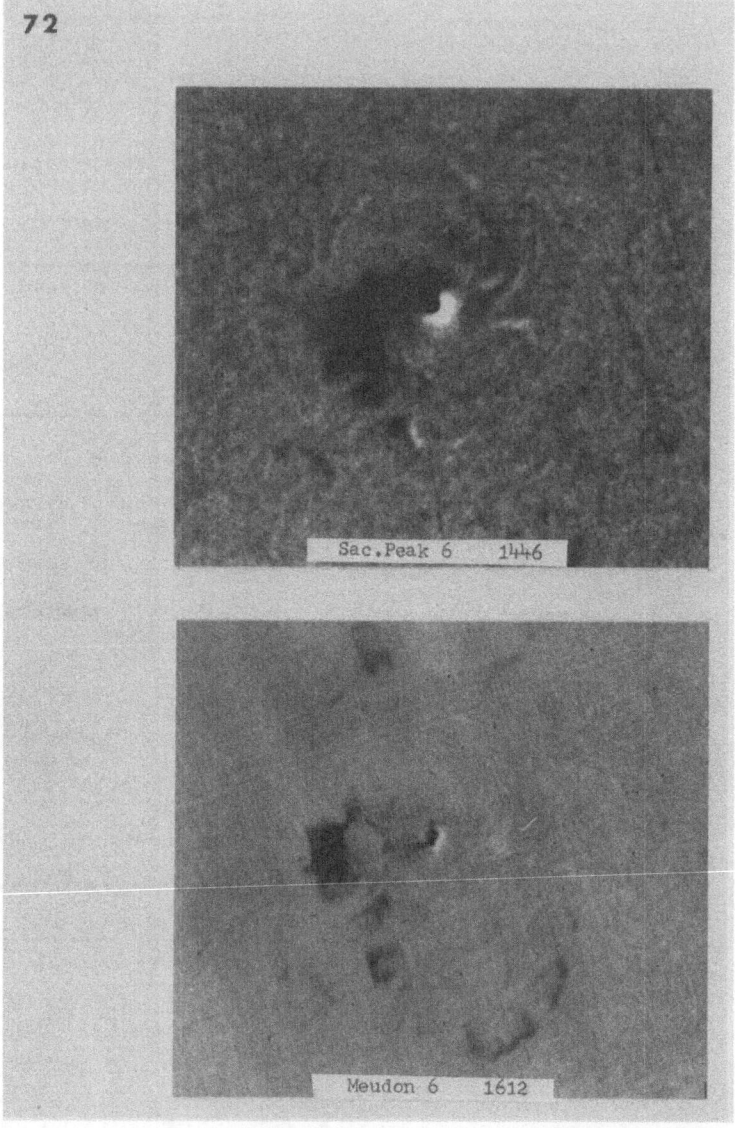

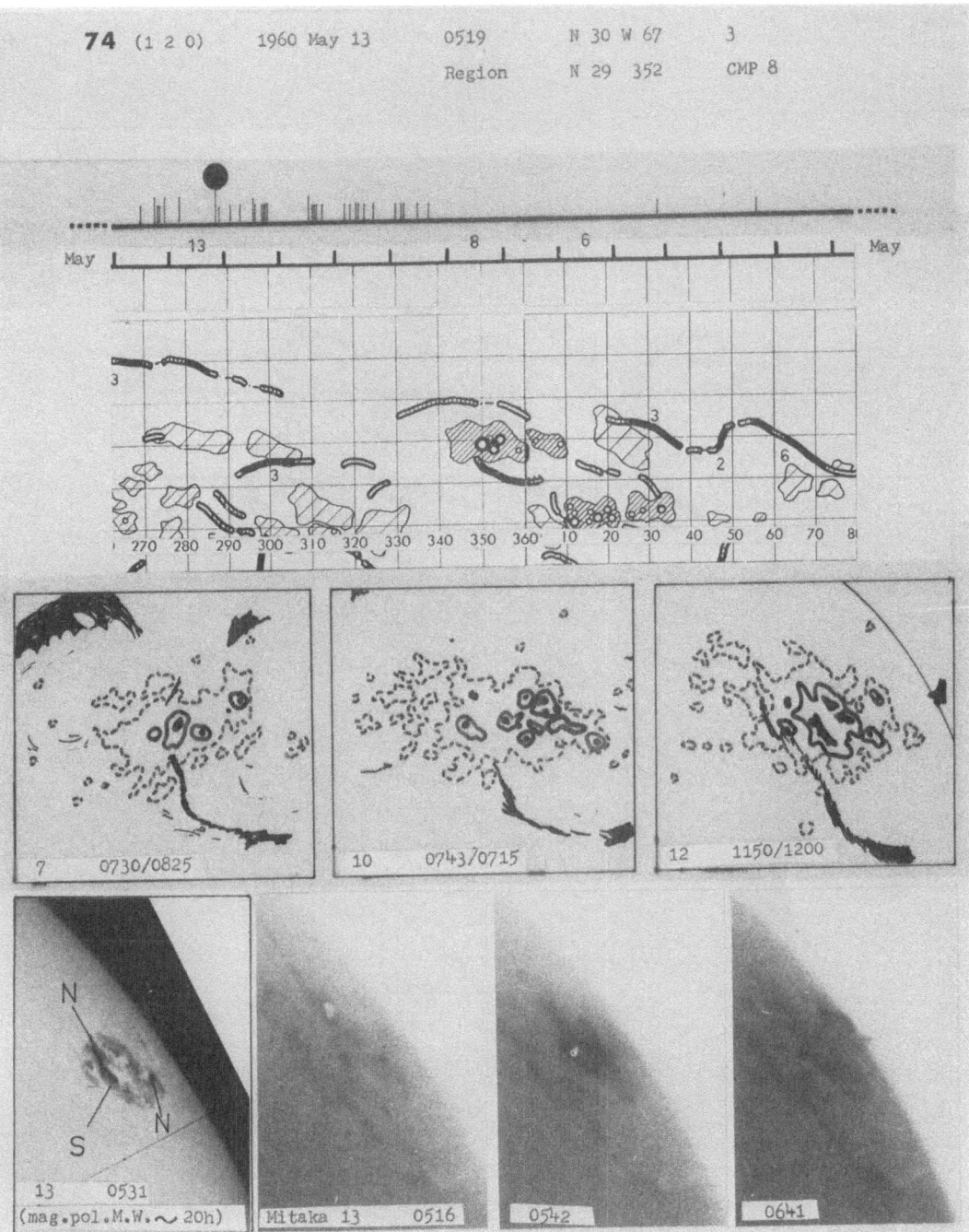

74 (1 2 0)   1960 May 13   0519   N 30 W 67   3
                             Region N 29 352   CMP 8

7   0730/0825

10   0743/0715

12   1150/1200

13   0531
(mag.pol.M.W. ~ 20h)

Mitaka 13   0516   0542   0641

76 (0 1 0)   1960 May 26   0850   N 14 W 15   2+

Region   N 12 128   CMP 25

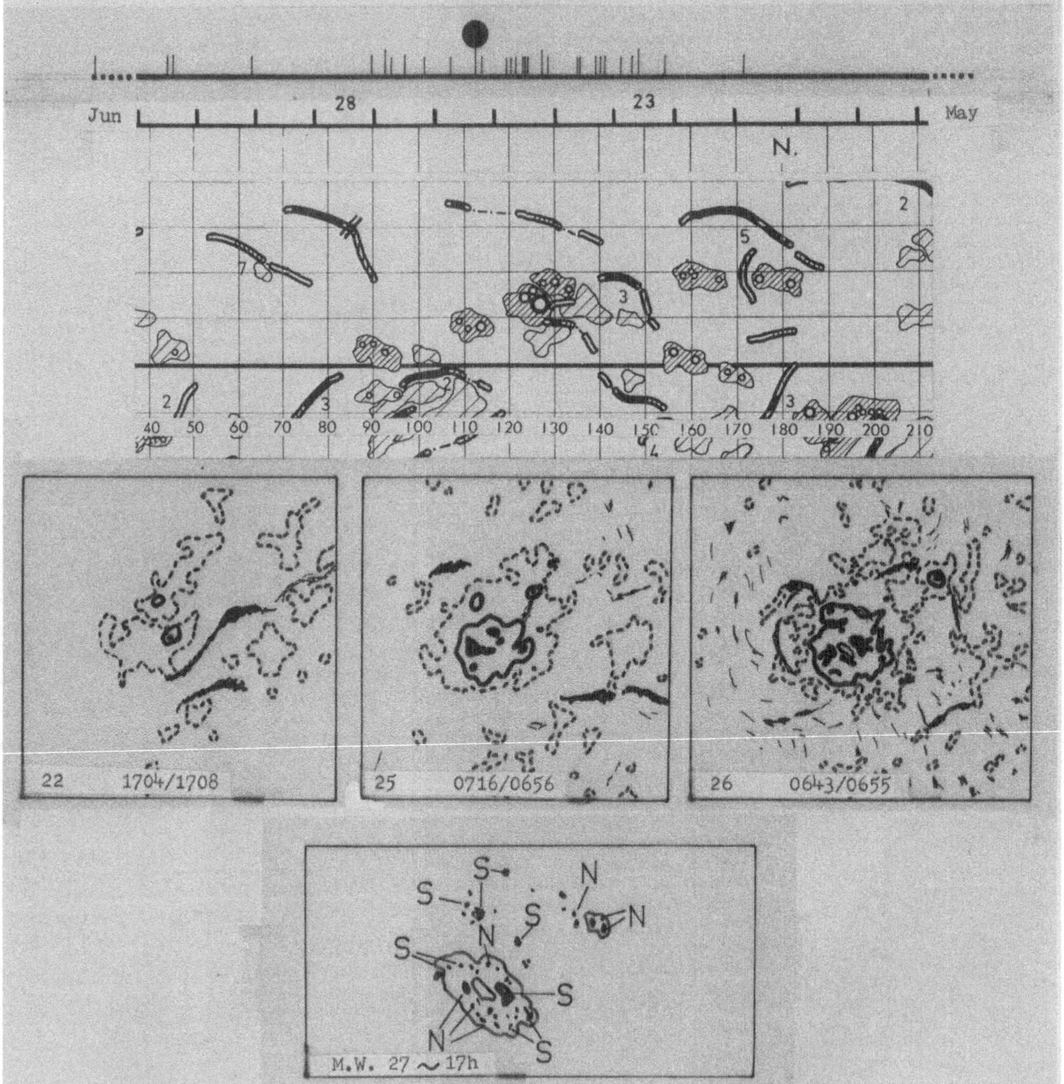

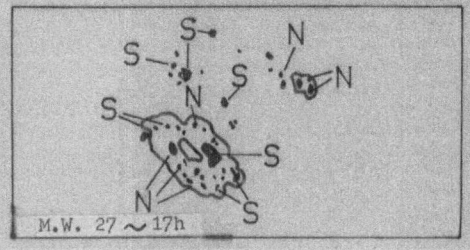

22   1704/1708   25   0716/0656   26   0643/0655

M.W. 27 ~ 17h

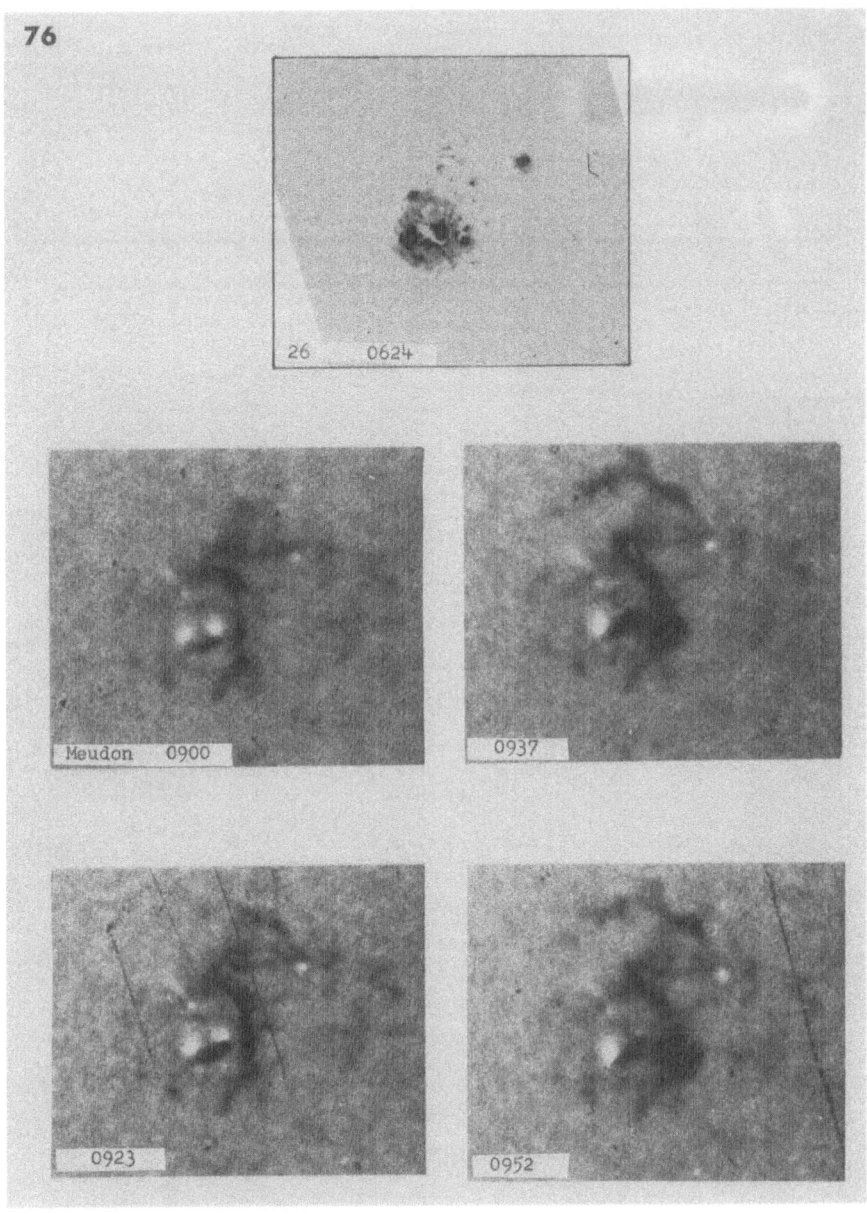

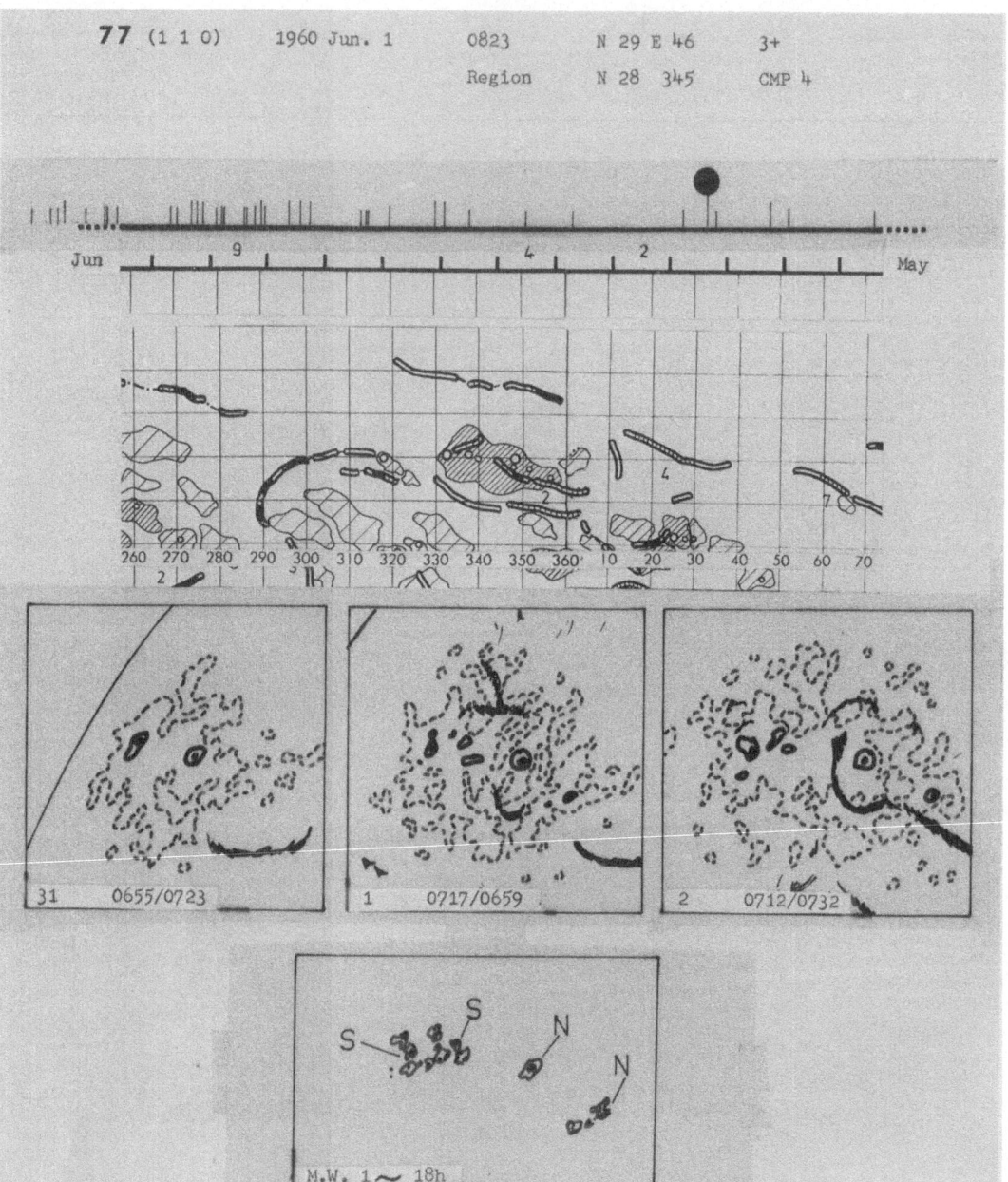

77 (1 1 0)   1960 Jun. 1   0823   N 29 E 46   3+
                  Region   N 28 345   CMP 4

31   0655/0723   1   0717/0659   2   0712/0732

M.W. 1 ~ 18h

77

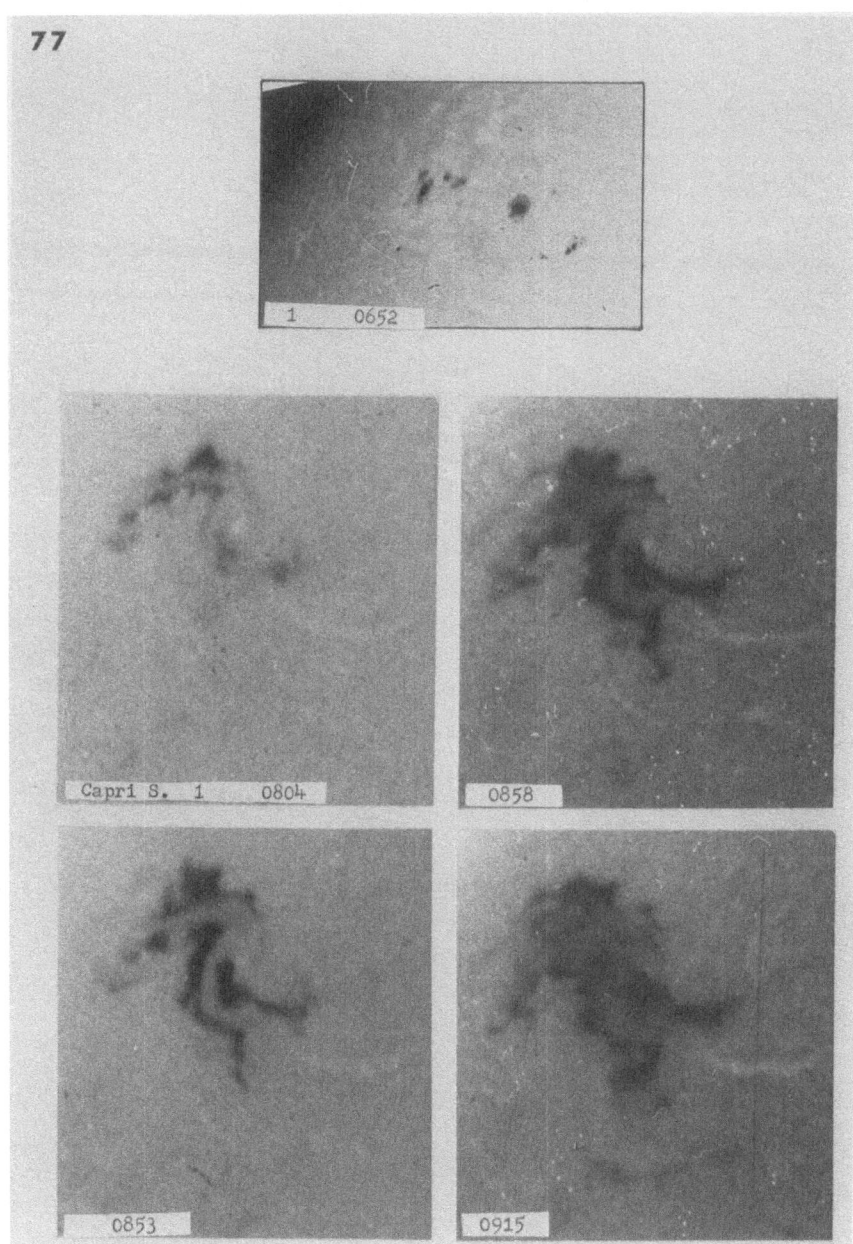

83 (2 2 1)   1960 Sep. 3   0037   N 18 E 88   2+

Region   N 18 148   CMP 9

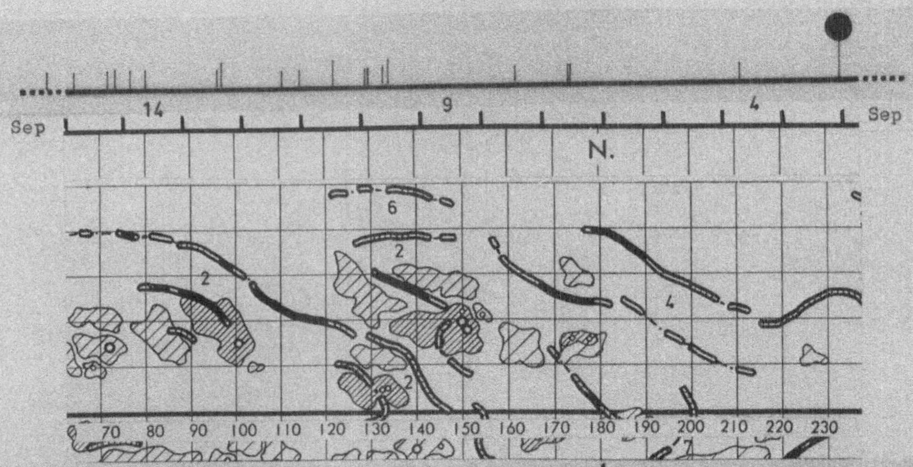

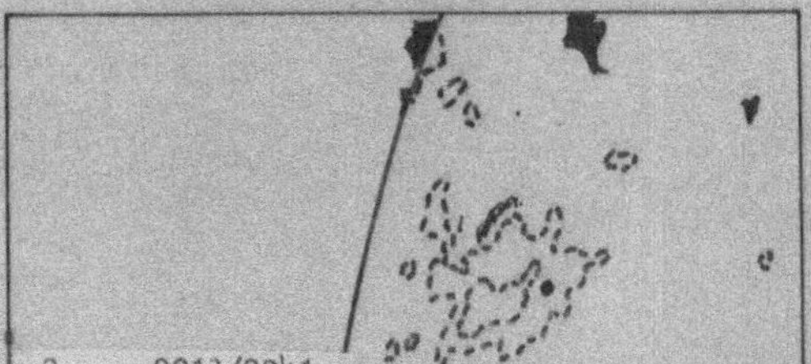

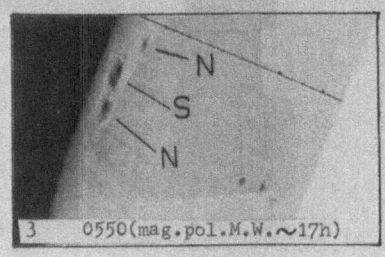

2   0913/0941

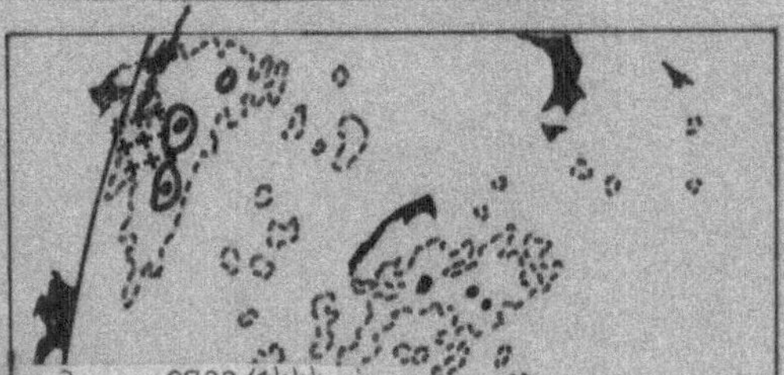

3   0720/1444

3   0550(mag.pol.M.W.~17h)

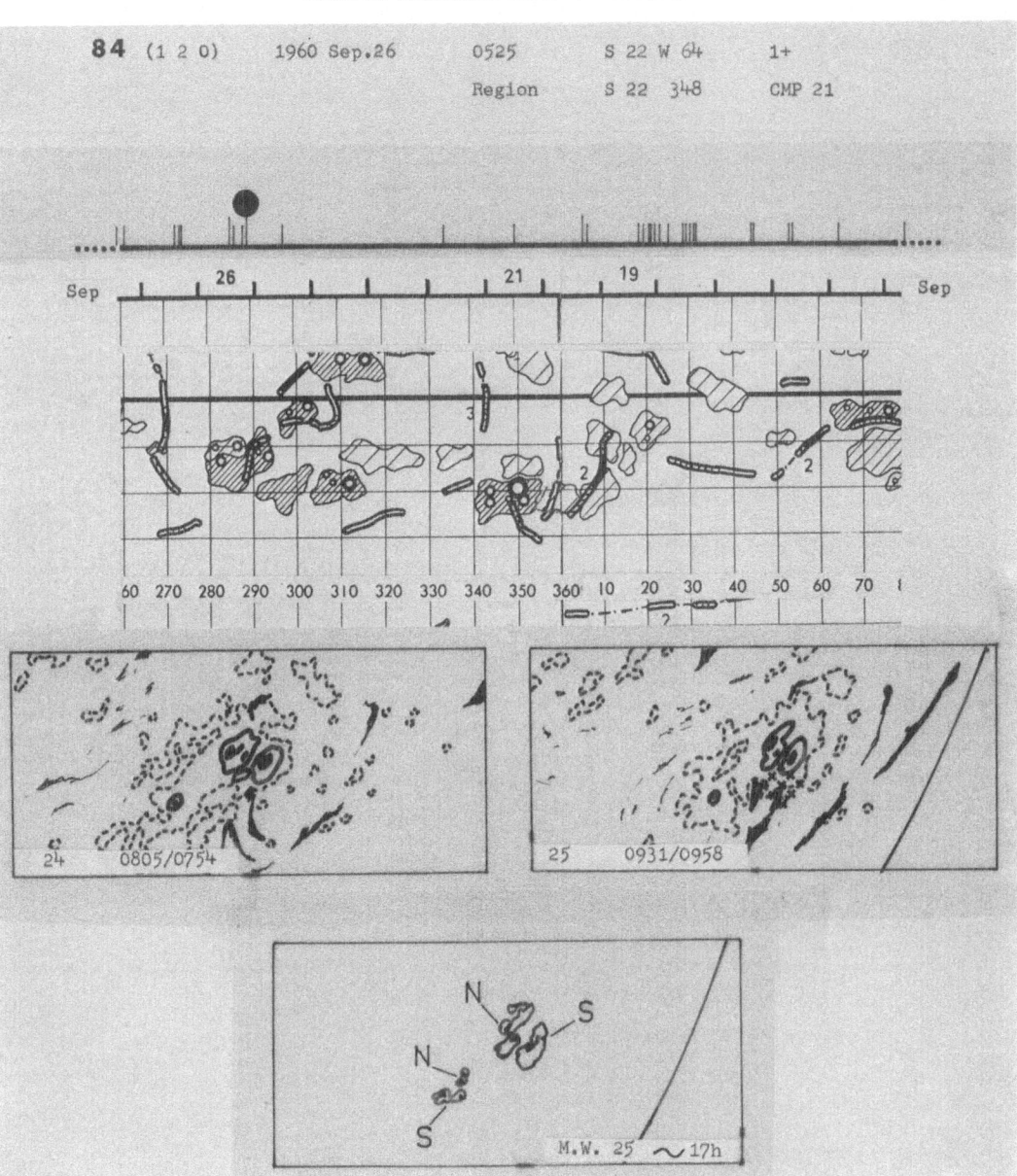

84 (1 2 0)   1960 Sep.26   0525   S 22 W 64   1+
                            Region   S 22 348   CMP 21

86 (X 1 0)   1960 Oct.29   1026   N 22 E 27   3
                                 Region   N 21 185   CMP 31

26   1218/1204   28   1041/1034   30   1219/1229

M.W. 30 ~ 13h

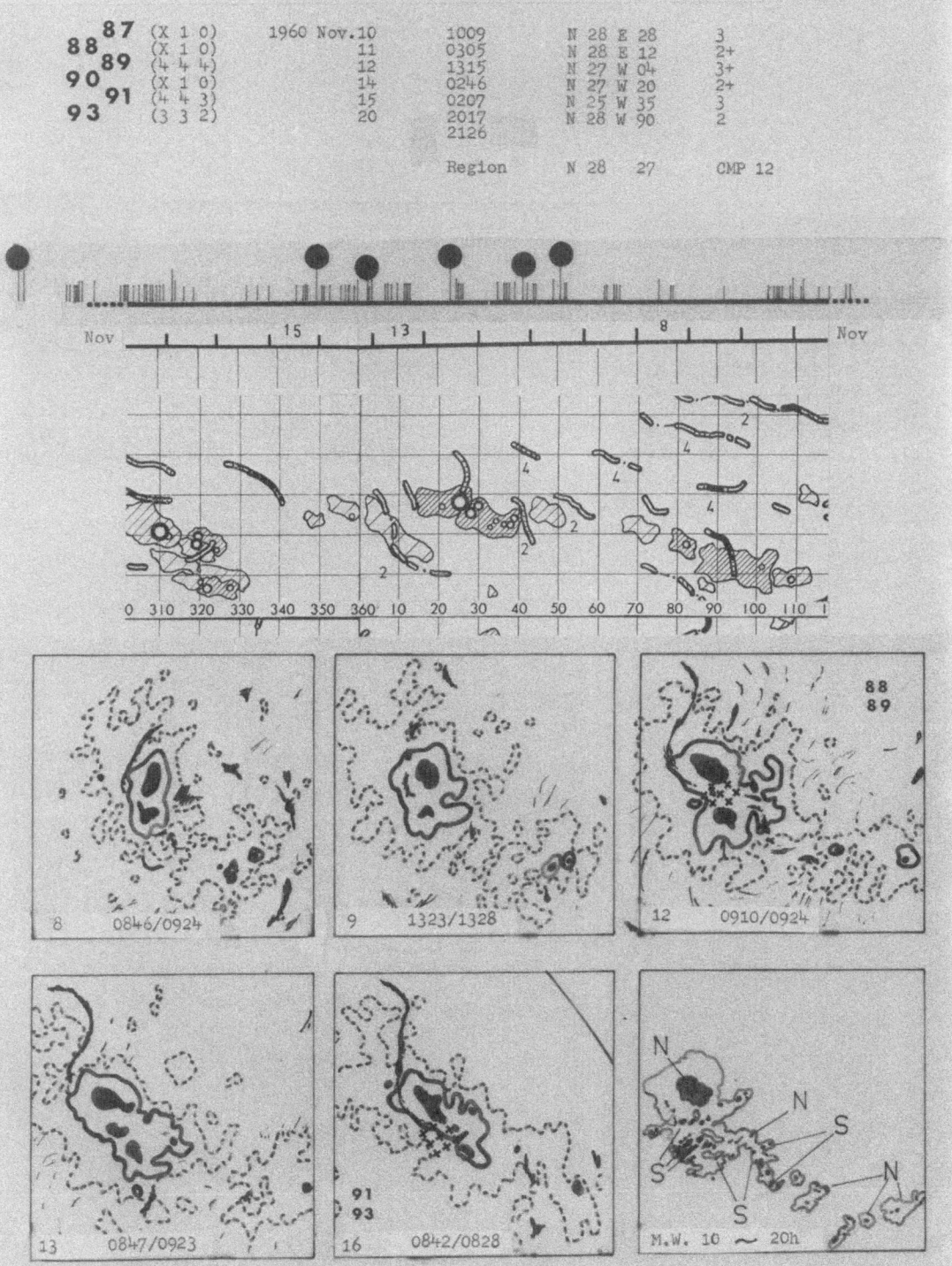

88	87	(X 1 0)	1960 Nov. 10	1009	N 28 E 28	3
	89	(X 1 0)	11	0305	N 28 E 12	2+
90		(4 4 4)	12	1315	N 27 W 04	3+
	91	(X 1 0)	14	0246	N 27 W 20	2+
93		(4 4 3)	15	0207	N 25 W 35	3
		(3 3 2)	20	2017	N 28 W 90	2
				2126		
			Region		N 28 27	CMP 12

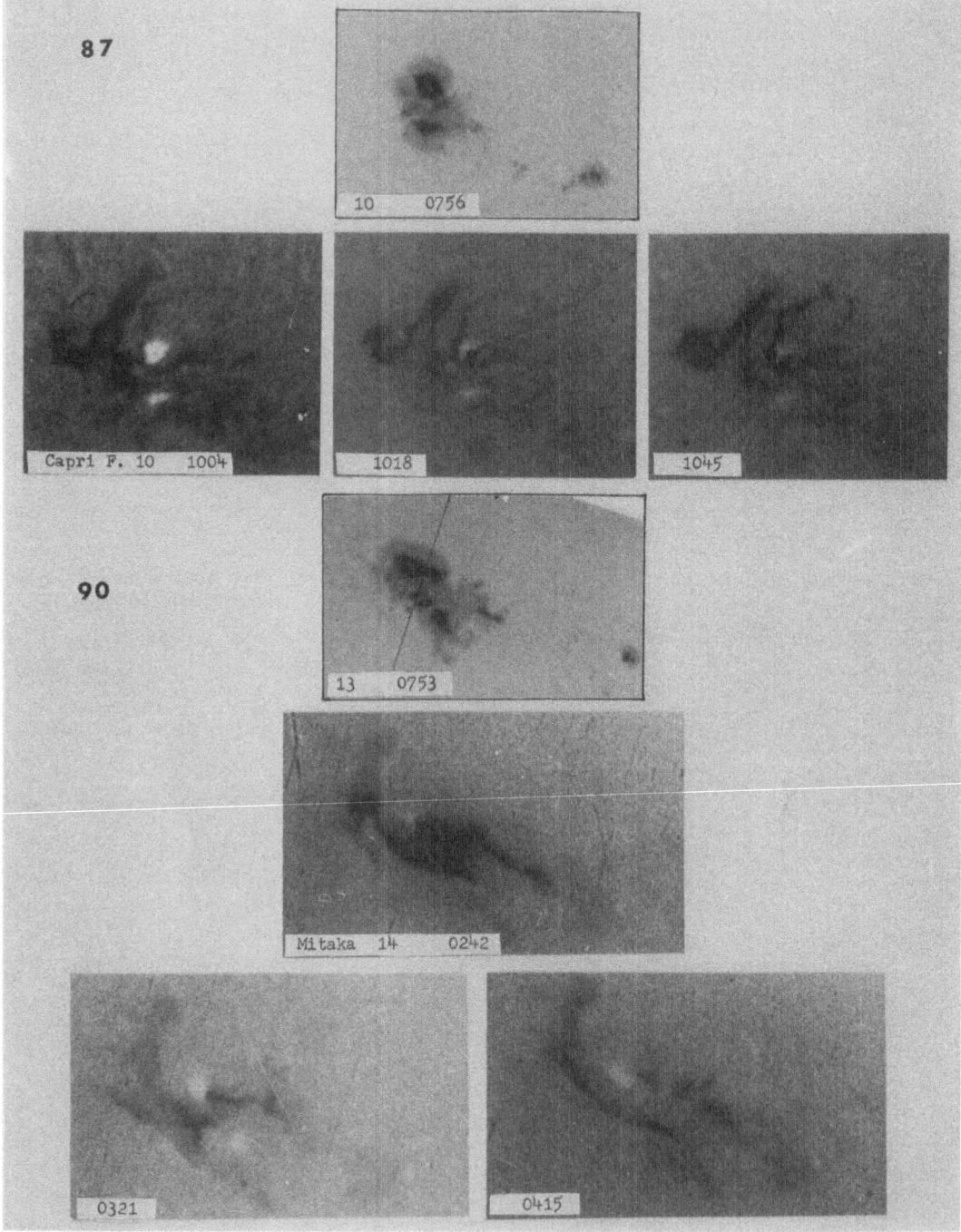

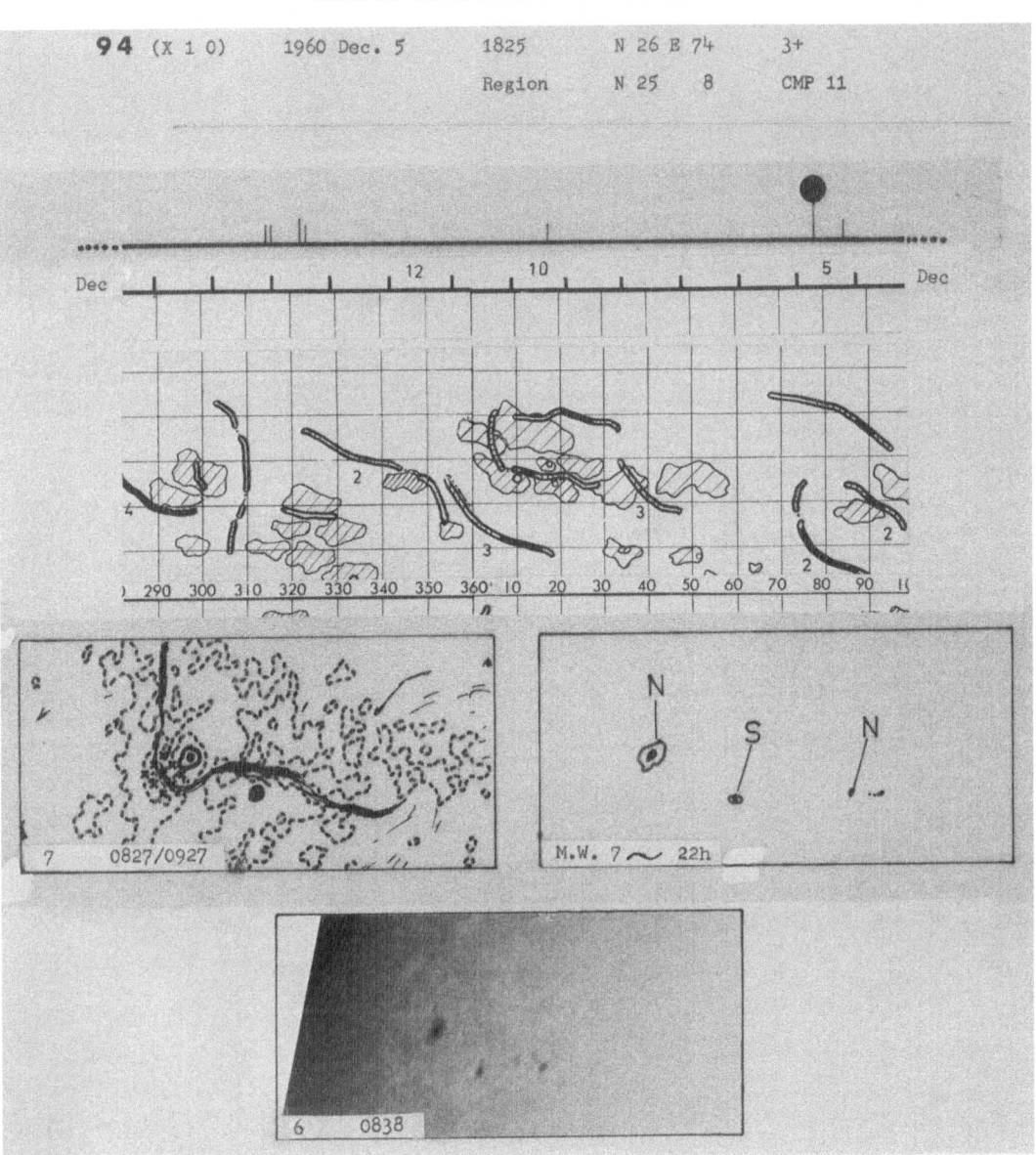

94 (X 1 0)   1960 Dec. 5   1825   N 26 E 74   3+
                    Region   N 25   8   CMP 11

97	96	(X 1 0)	1961 Jul.11	1615	S 07 E 32	3
	98	(2 4 0)	12	1000	S 07 E 23	3
99		(X 2 0)	15	1508	S 07 W 20	2
	100	(3 3 3)	18	0920	S 07 W 59	3+
		(1 2 2)	20	1553	S 06 W 90	3
				1823		
			Region		S 08 49	CMP 14

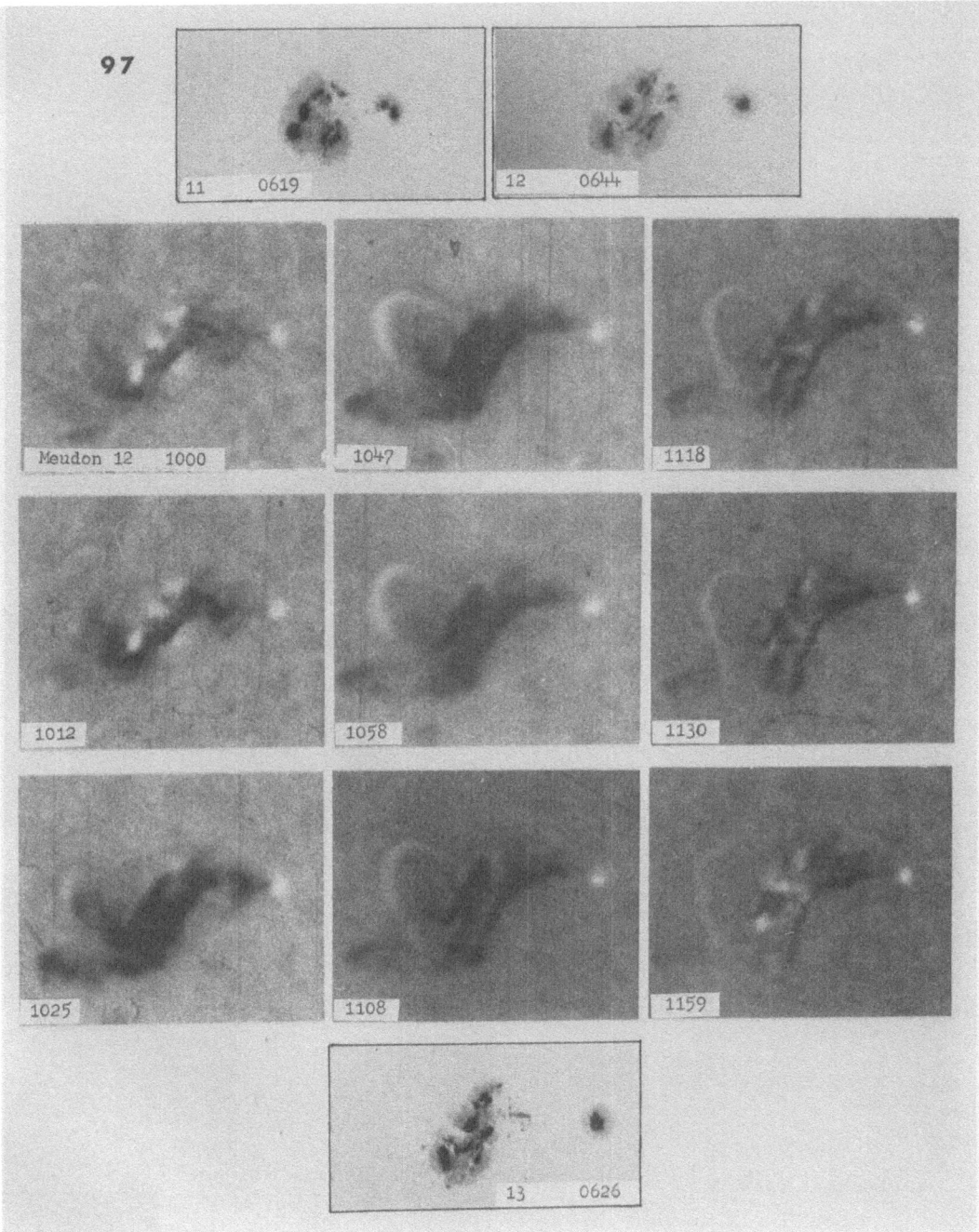

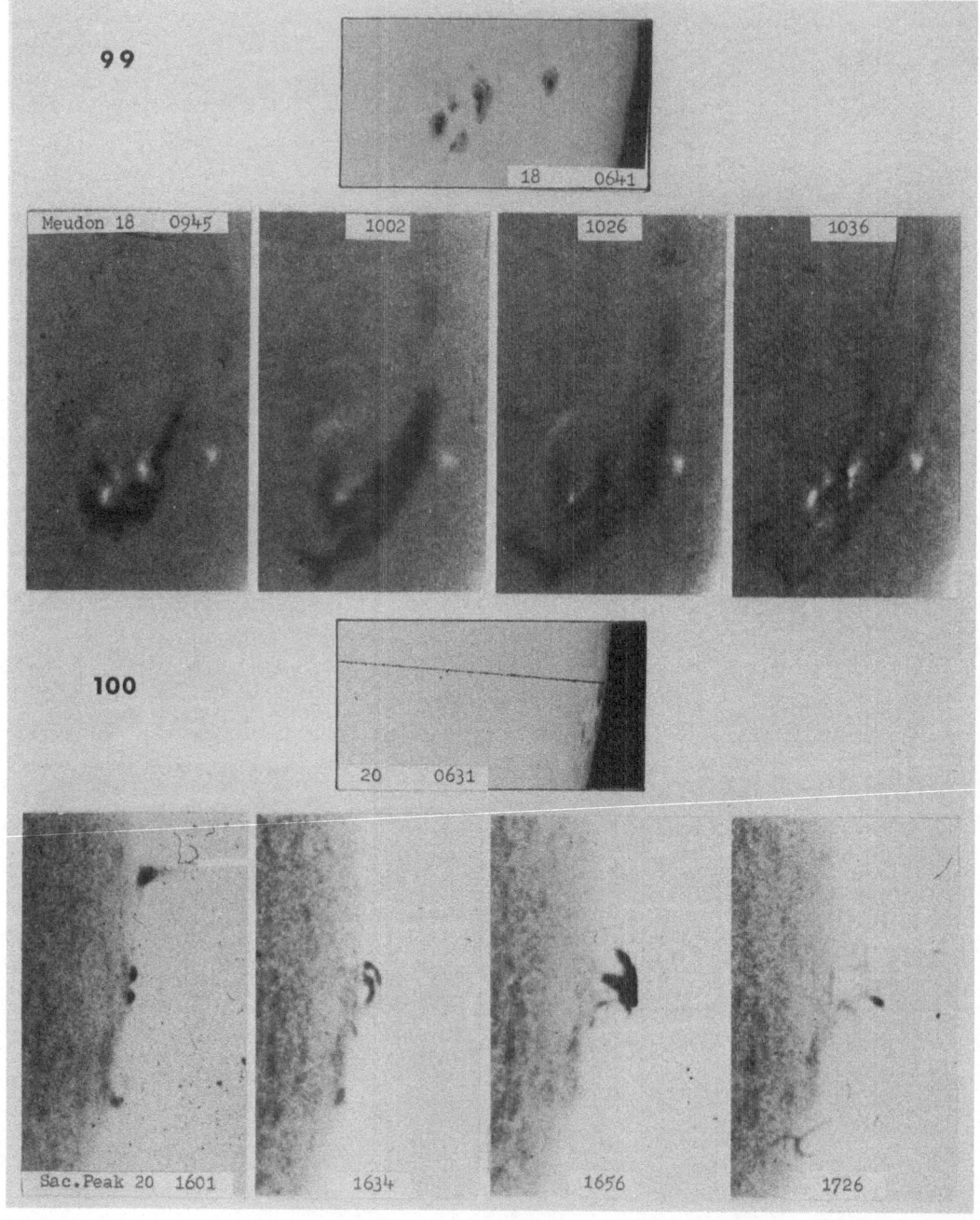

99

18   0641

Meudon 18   0945   1002   1026   1036

100

20   0631

Sac.Peak 20   1601   1634   1656   1726

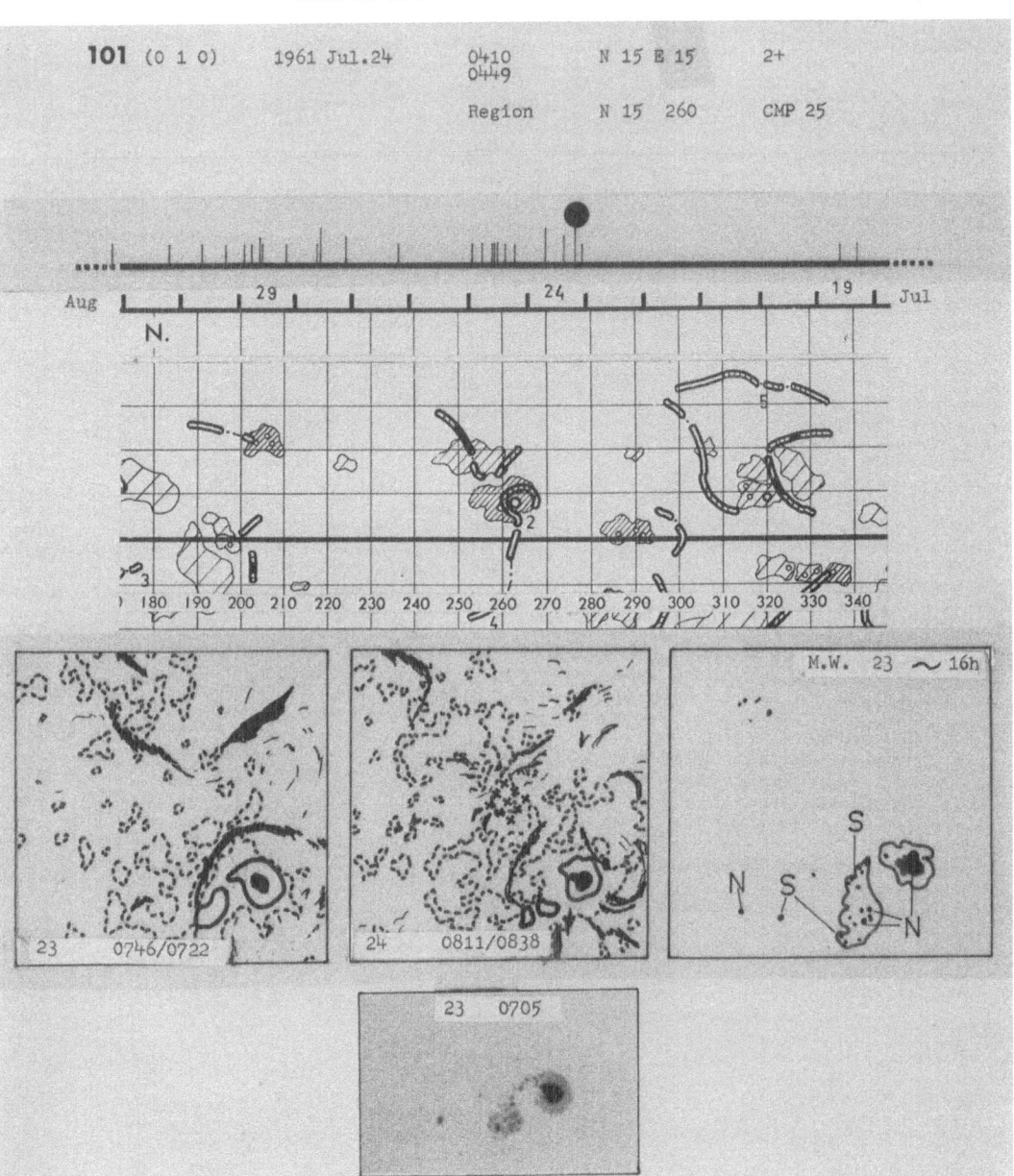

101 (0 1 0)   1961 Jul.24   0410   N 15 E 15   2+
                                  0449

                   Region   N 15 260   CMP 25

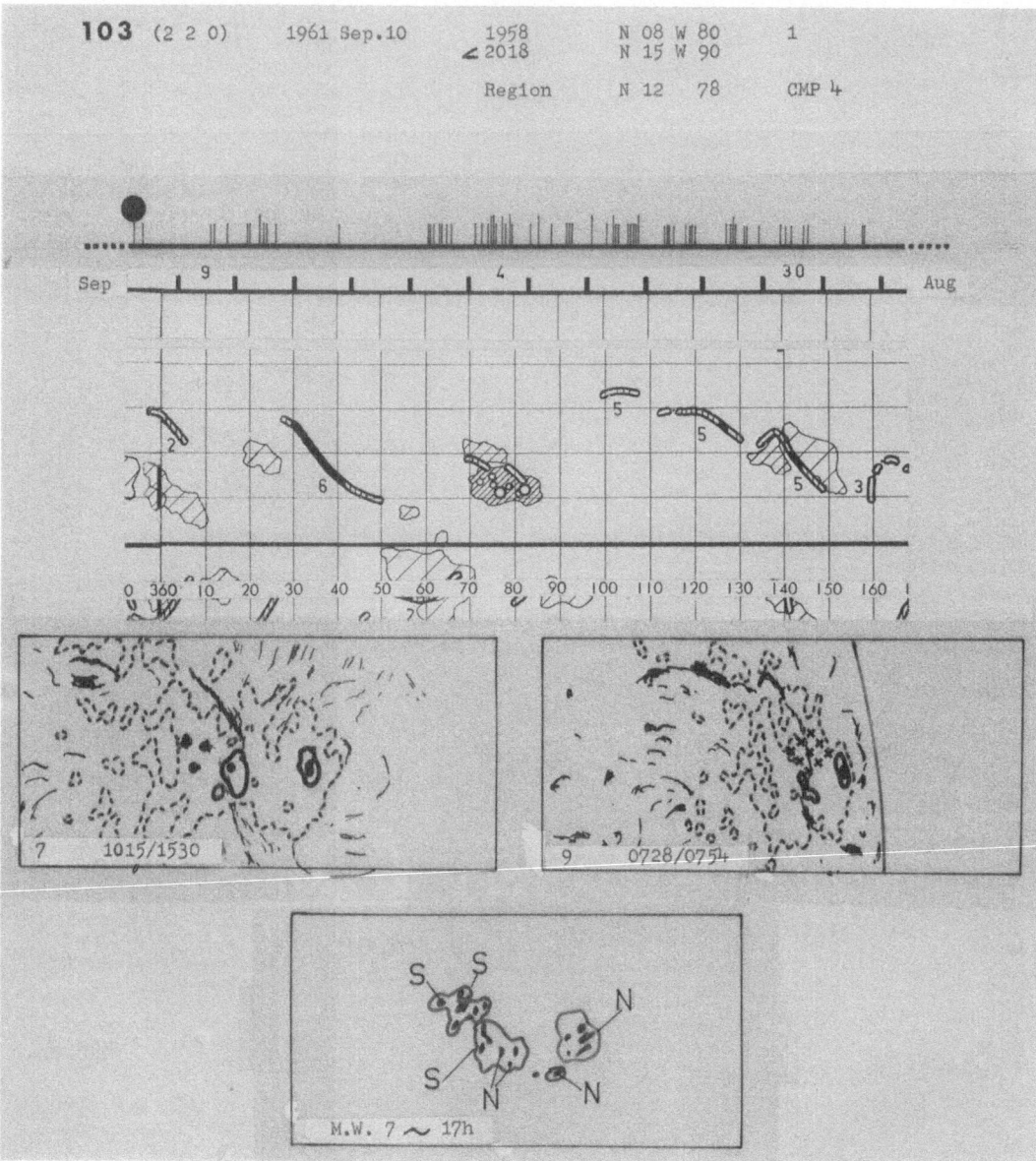

103 (2 2 0)   1961 Sep.10   1958   N 08 W 80   1
                                ∠ 2018   N 15 W 90

                                  Region   N 12   78   CMP 4

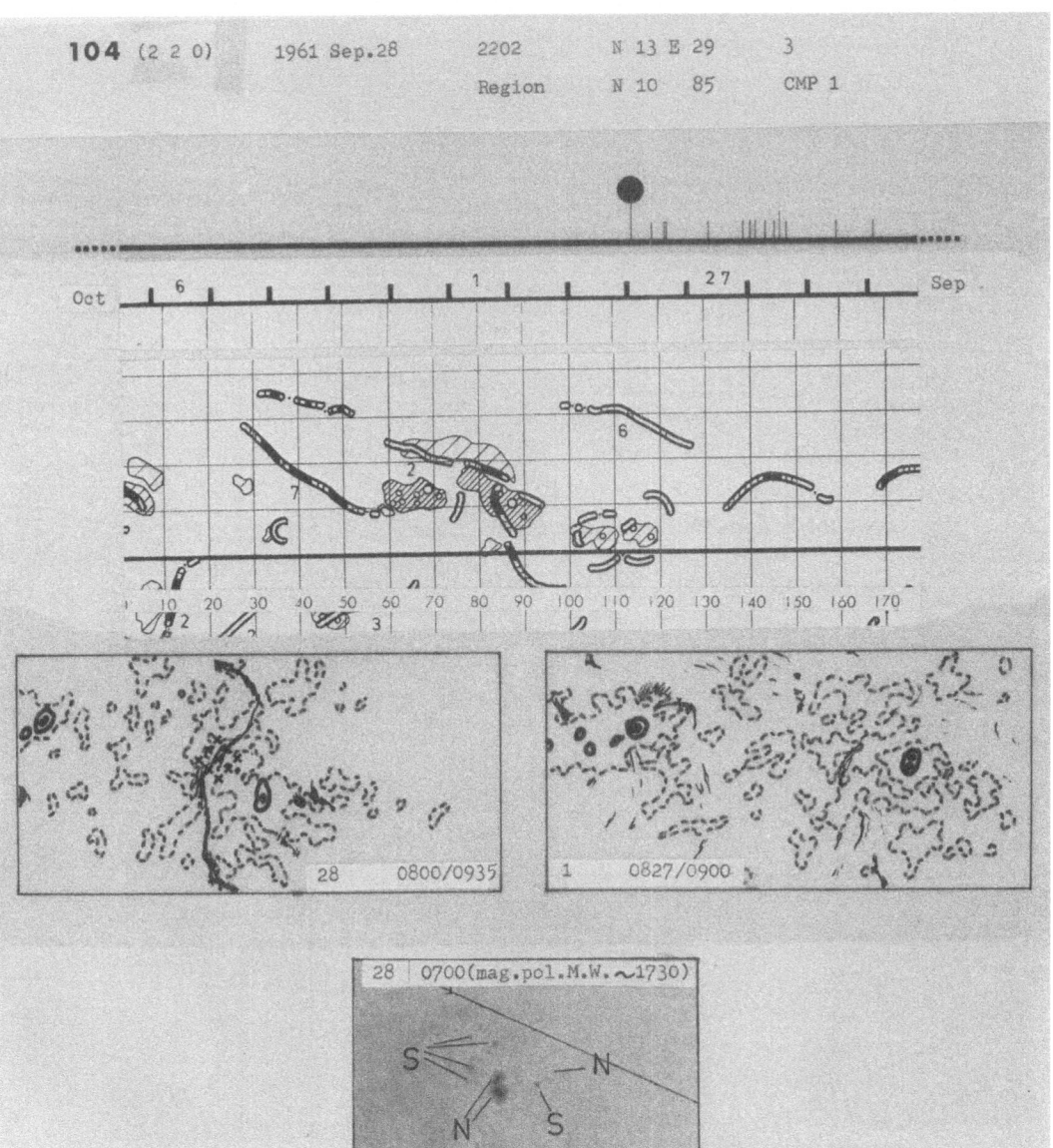

104 (2 2 0)   1961 Sep.28   2202   N 13 E 29   3
                                 Region   N 10   85   CMP 1

28   0800/0935

1   0827/0900

28   0700(mag.pol.M.W. ∼1730)

106 (1 1 0)   1961 Nov.10   1434   N 19 W 90   1+

                                      Region   N 10   5   CMP 3

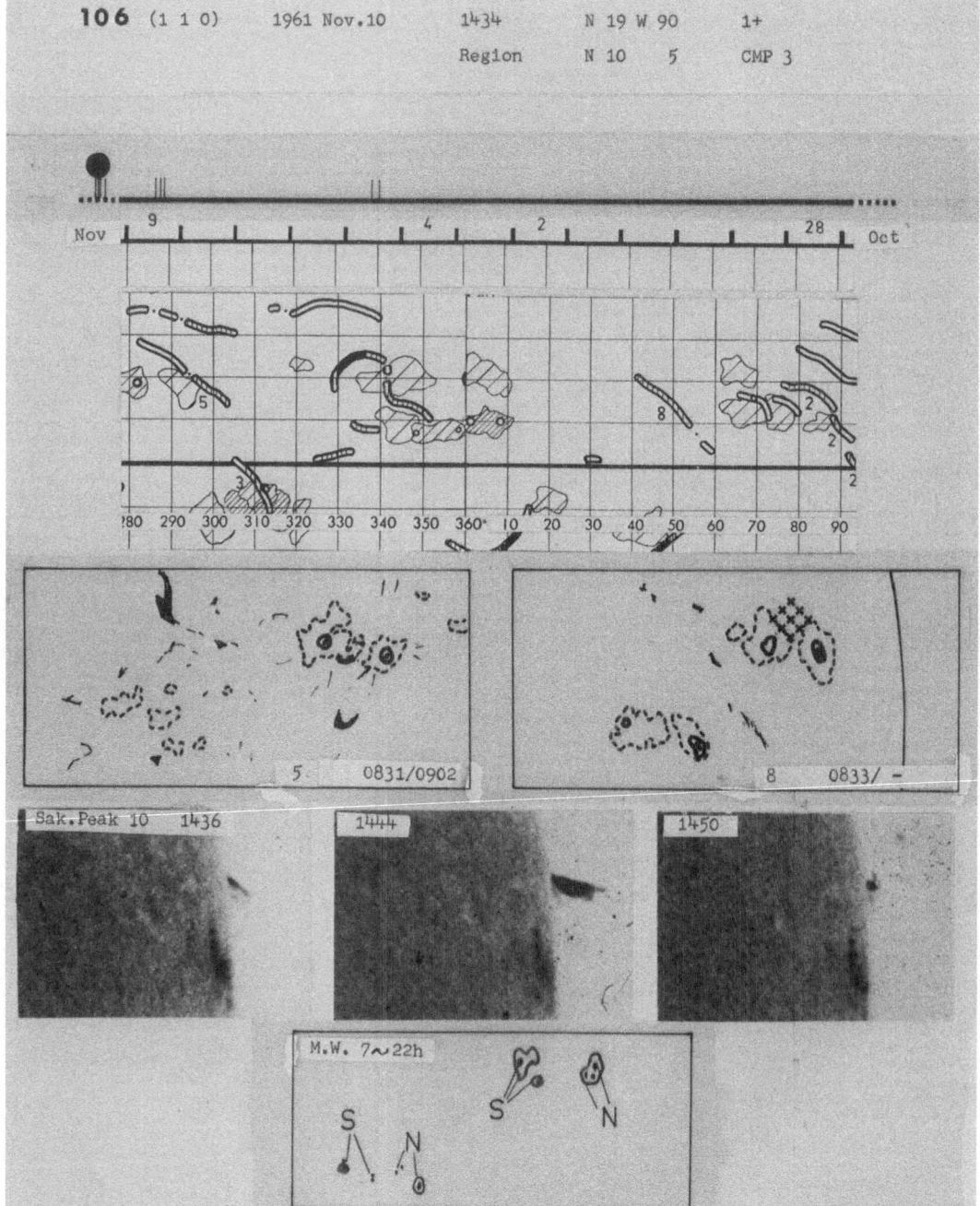

Nov 9   4   2   28 Oct

280 290 300 310 320 330 340 350 360° 10 20 30 40 50 60 70 80 90

5   0831/0902

8   0833/ -

Sak.Peak 10   1436   1444   1450

M.W. 7~22h

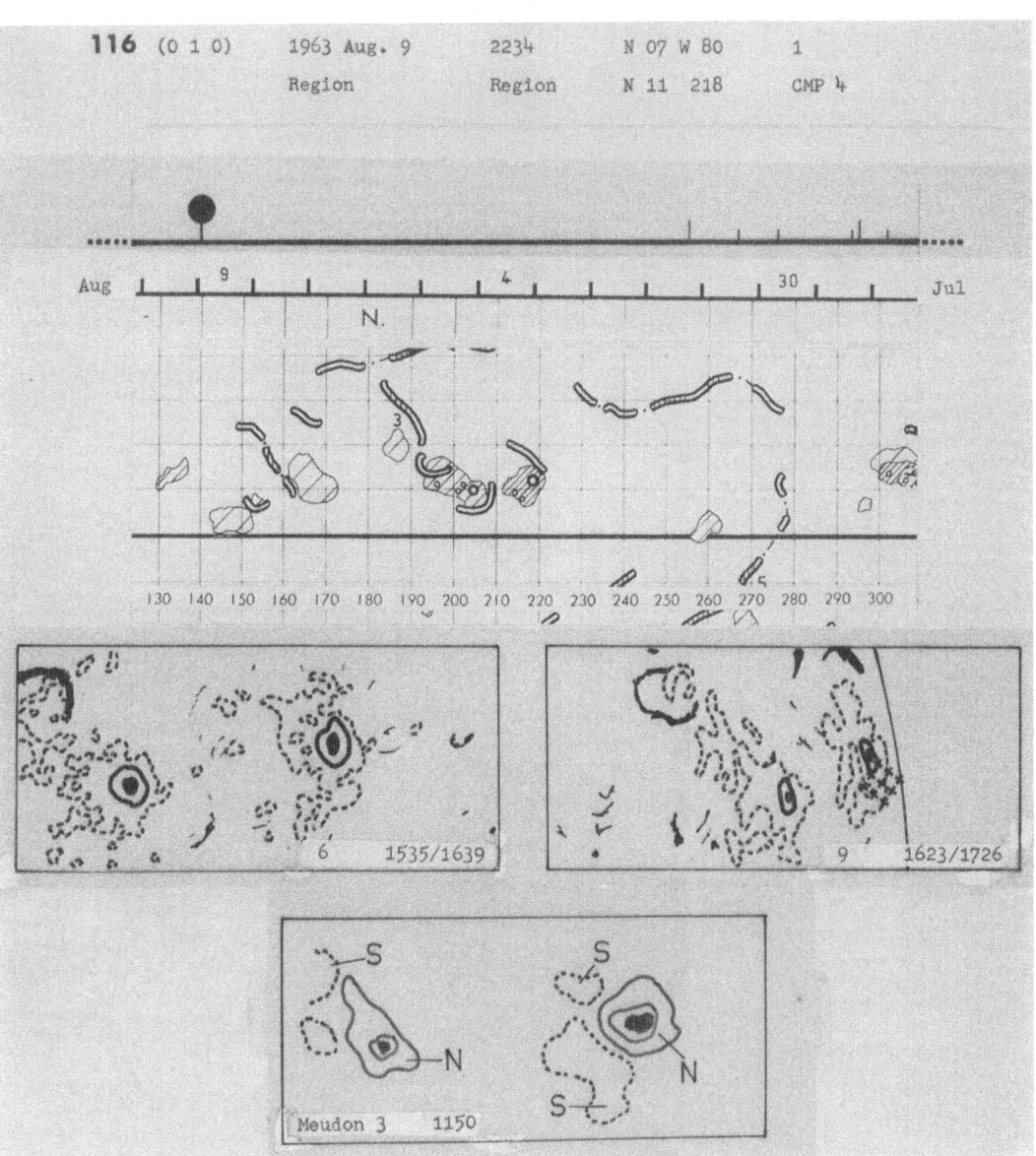

116 (0 1 0)   1963 Aug. 9   2234   N 07 W 80   1
                 Region   Region   N 11 218   CMP 4

6   1535/1639

9   1623/1726

Meudon 3   1150

117	(X 1 0)	1963 Sep.15	0015	N 15 E 75	2
118	(X 1 0)	16	1430	N 12 E 48	2
119	(1 2 0)	20	2314	N 10 W 09	2
120	(1 2 0)	26	2351 <0638	N 13 W 78	3
			Region	N 12 310	CMP 20

Sep
22
17
15
Sep
3
3
230 240 250 260 270 280 290 300 310 320 330 340 350 360 10 20 30

117
118
15 1634/0839
16 1023/0928
119
18 1040/1006

N
S
N
N
S
Meudon 15 1440
S
N
N
S
N
Meudon 17 1530

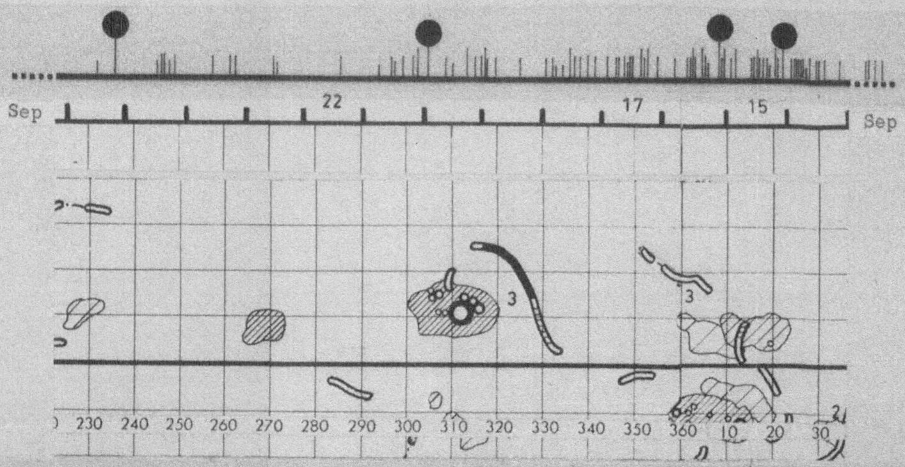

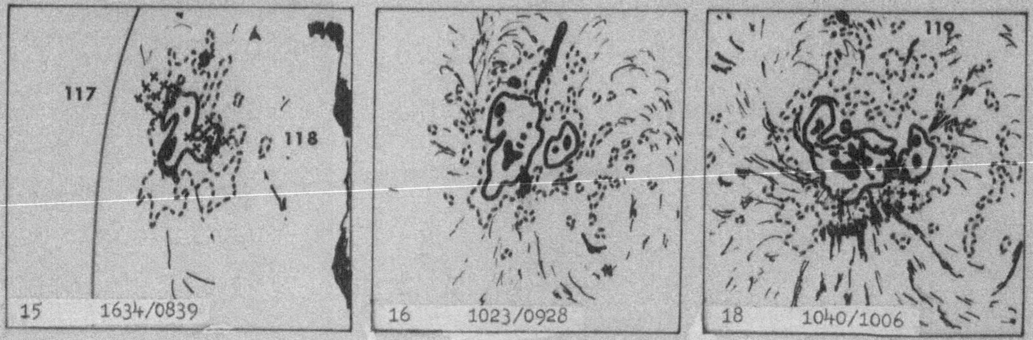

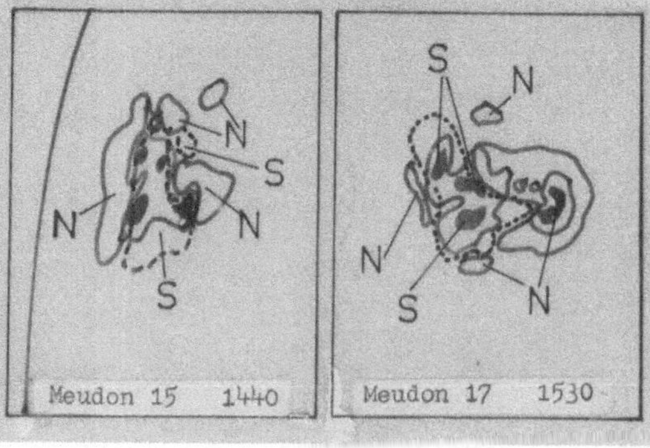

120

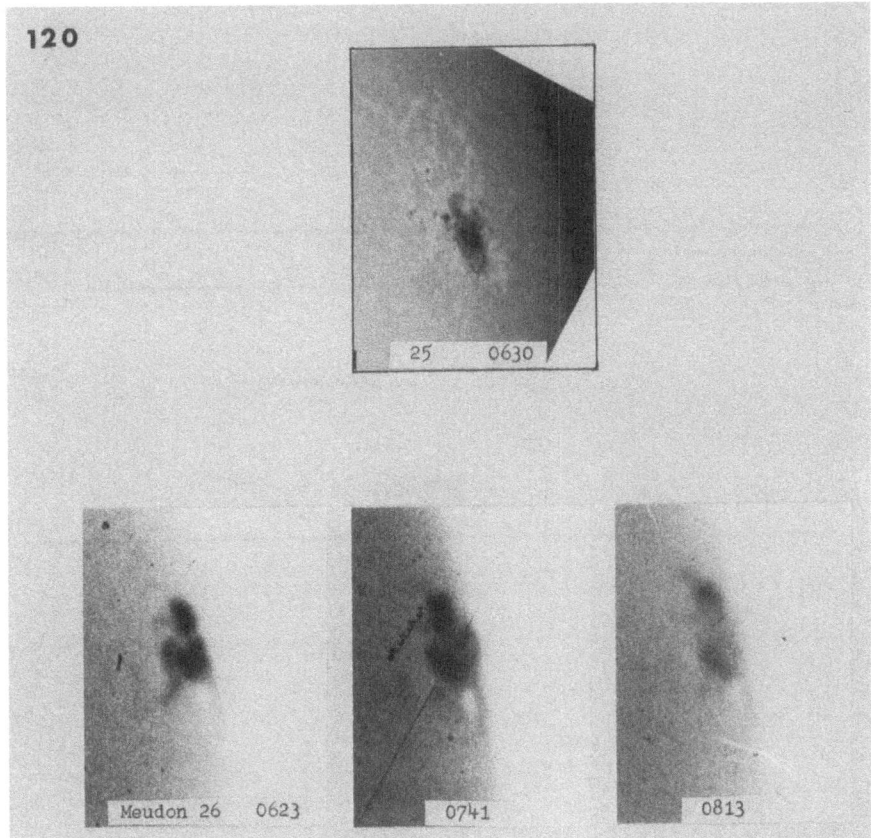

Meudon 26   0623   0741   0813   25   0630

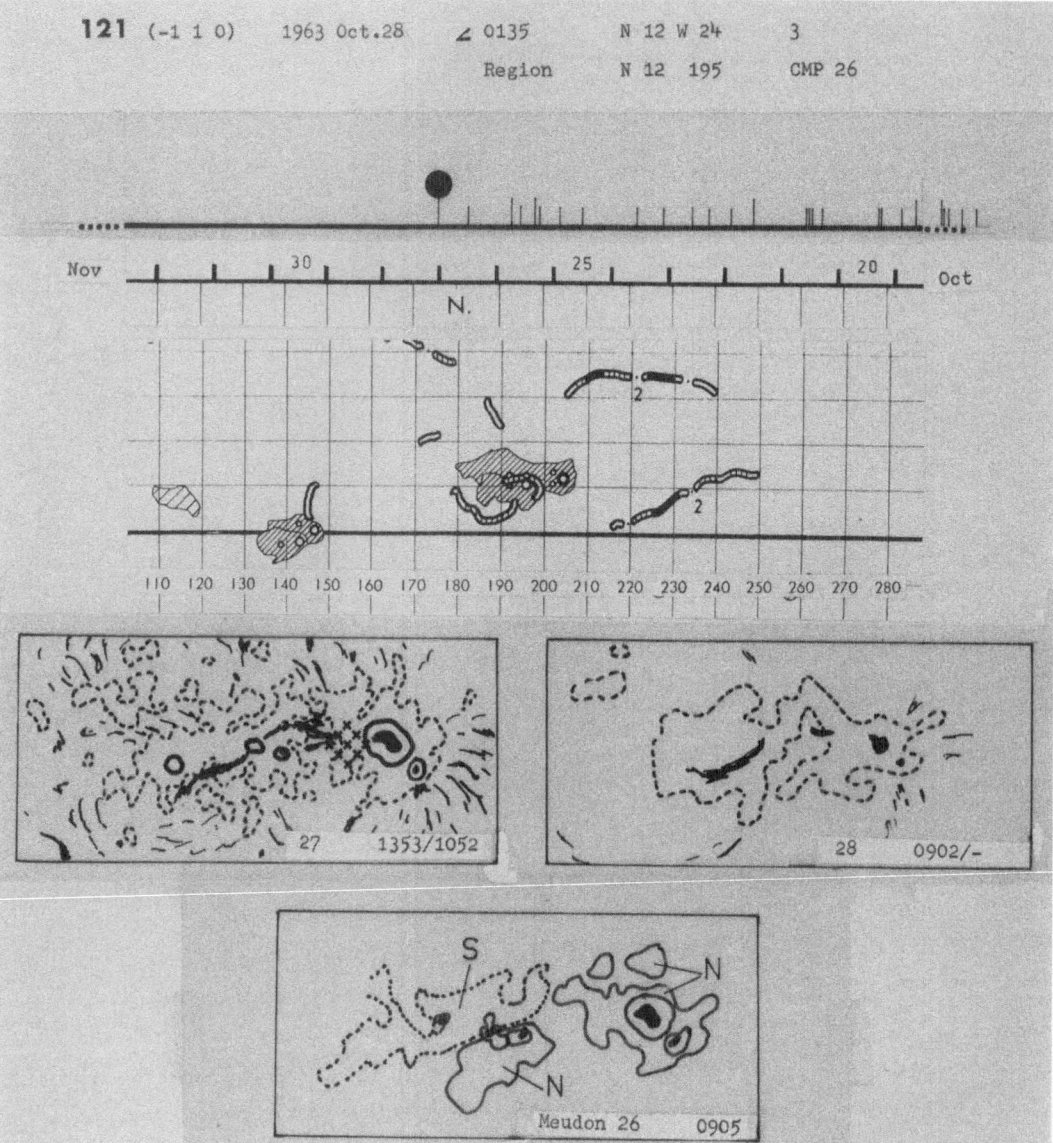

121 (-1 1 0)   1963 Oct.28   ∠ 0135   N 12 W 24   3
                               Region   N 12 195   CMP 26

27   1353/1052

28   0902/-

Meudon 26   0905

122 (0 1 0)   1964 Mar.16   1553   N 05 W 73   1+
                            Region   N 04 190   CMP 11

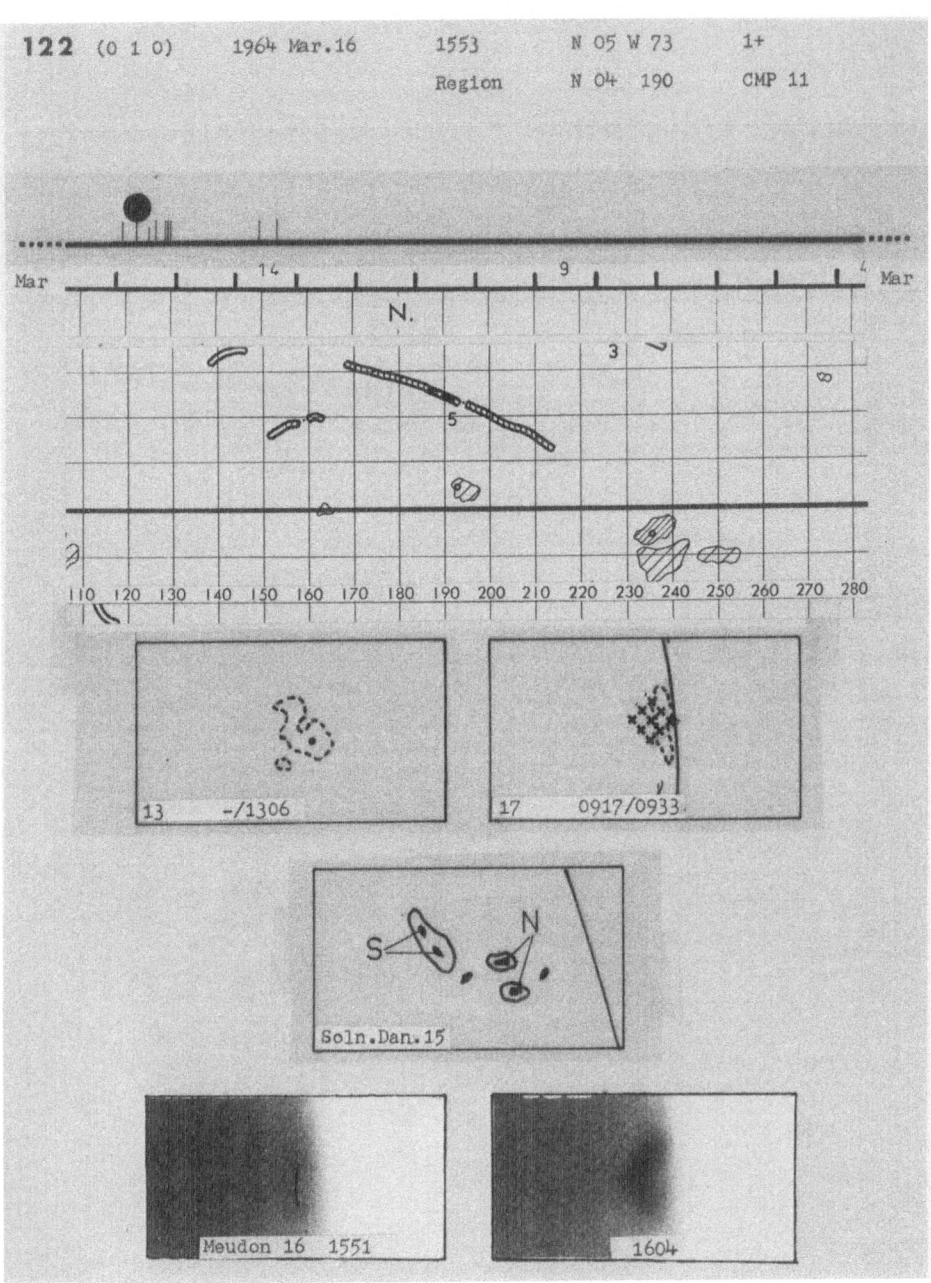

13   -/1306

17   0917/0933

Soln.Dan.15

Meudon 16 1551

1604

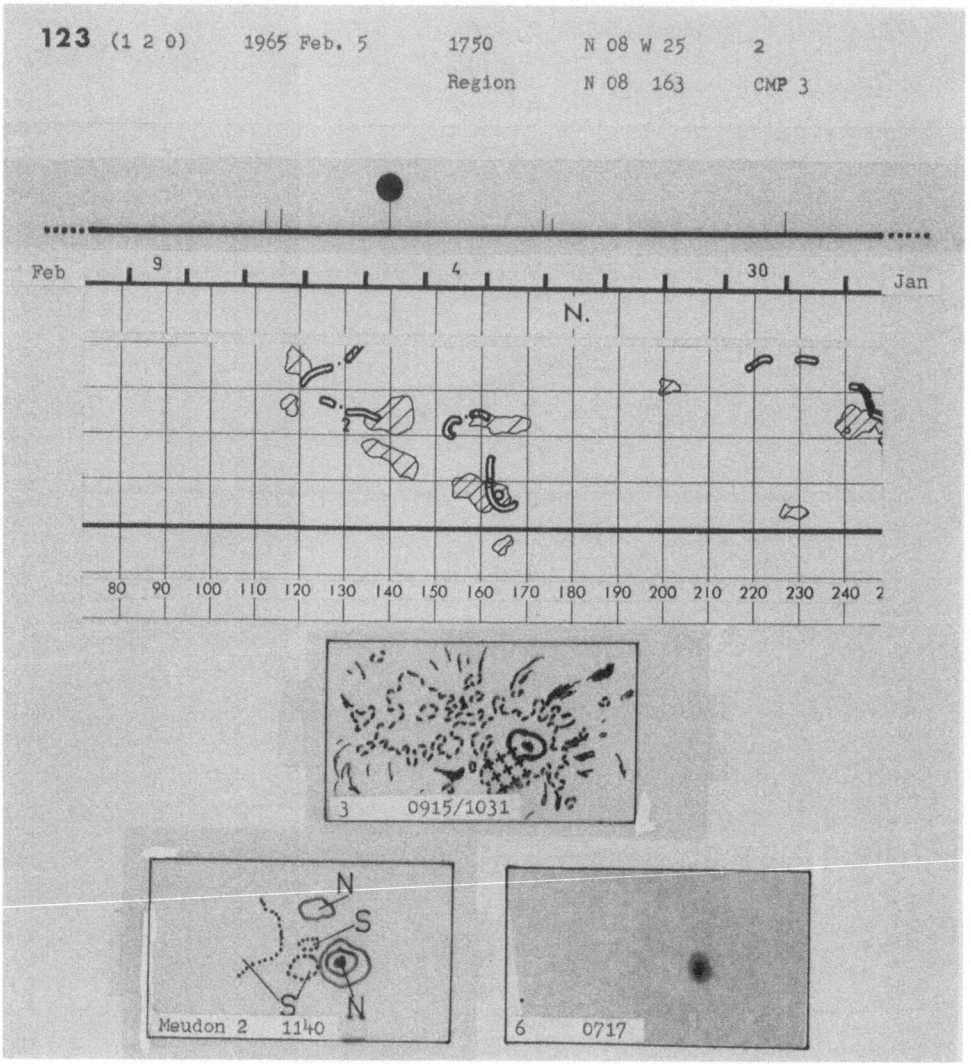

123 (1 2 0)   1965 Feb. 5   1750   N 08 W 25   2

Region   N 08   163   CMP 3

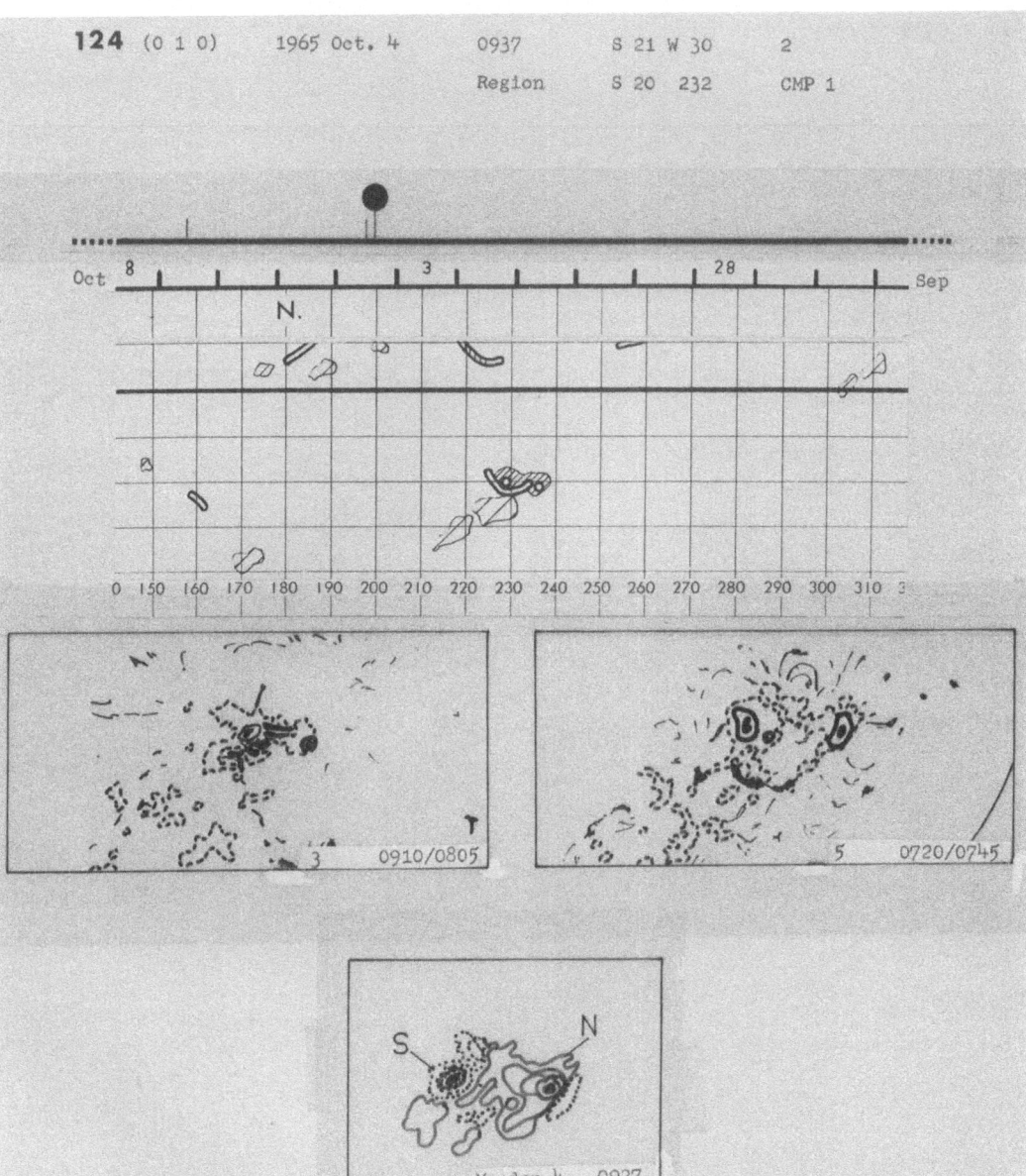

124 (0 1 0)   1965 Oct. 4   0937   S 21 W 30   2
                            Region   S 20 232   CMP 1

124

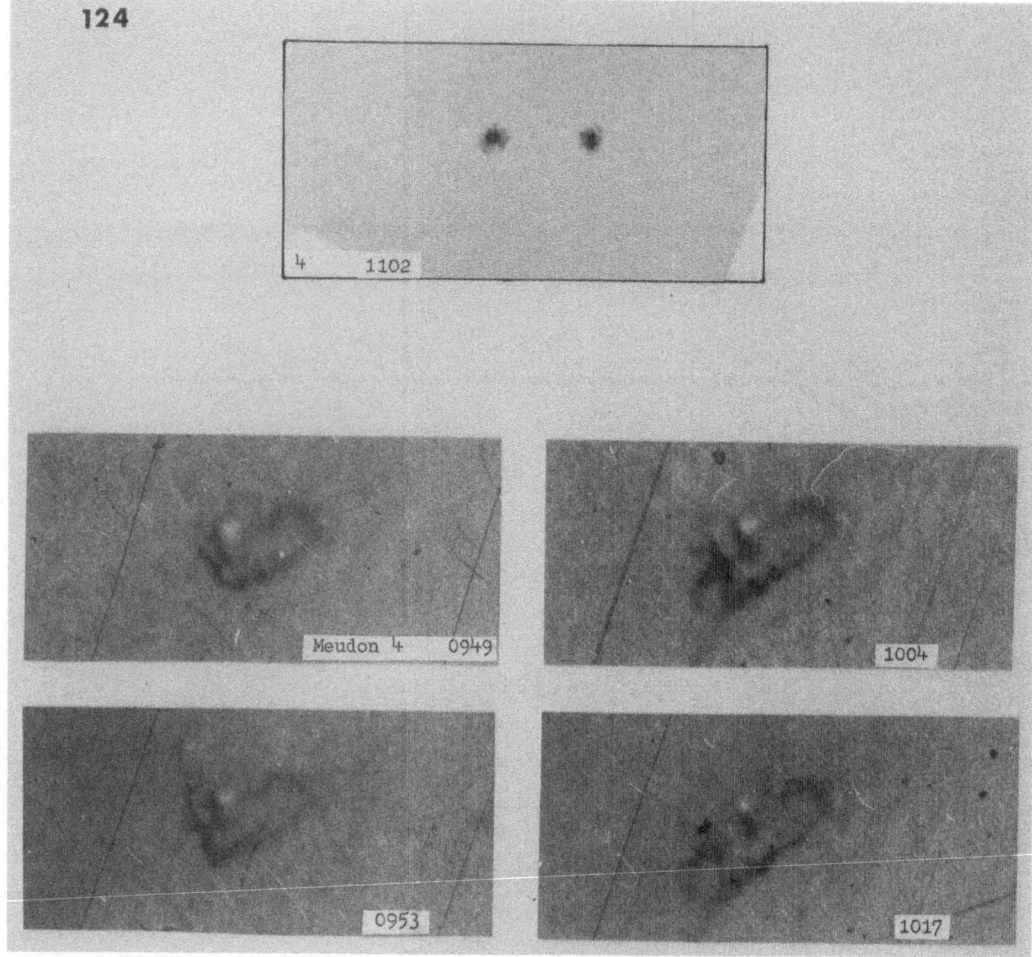

125 (-1 0 0)   1965 Dec.30   0006   N 09 W 70   1?
Region   N 10 205   CMP 24

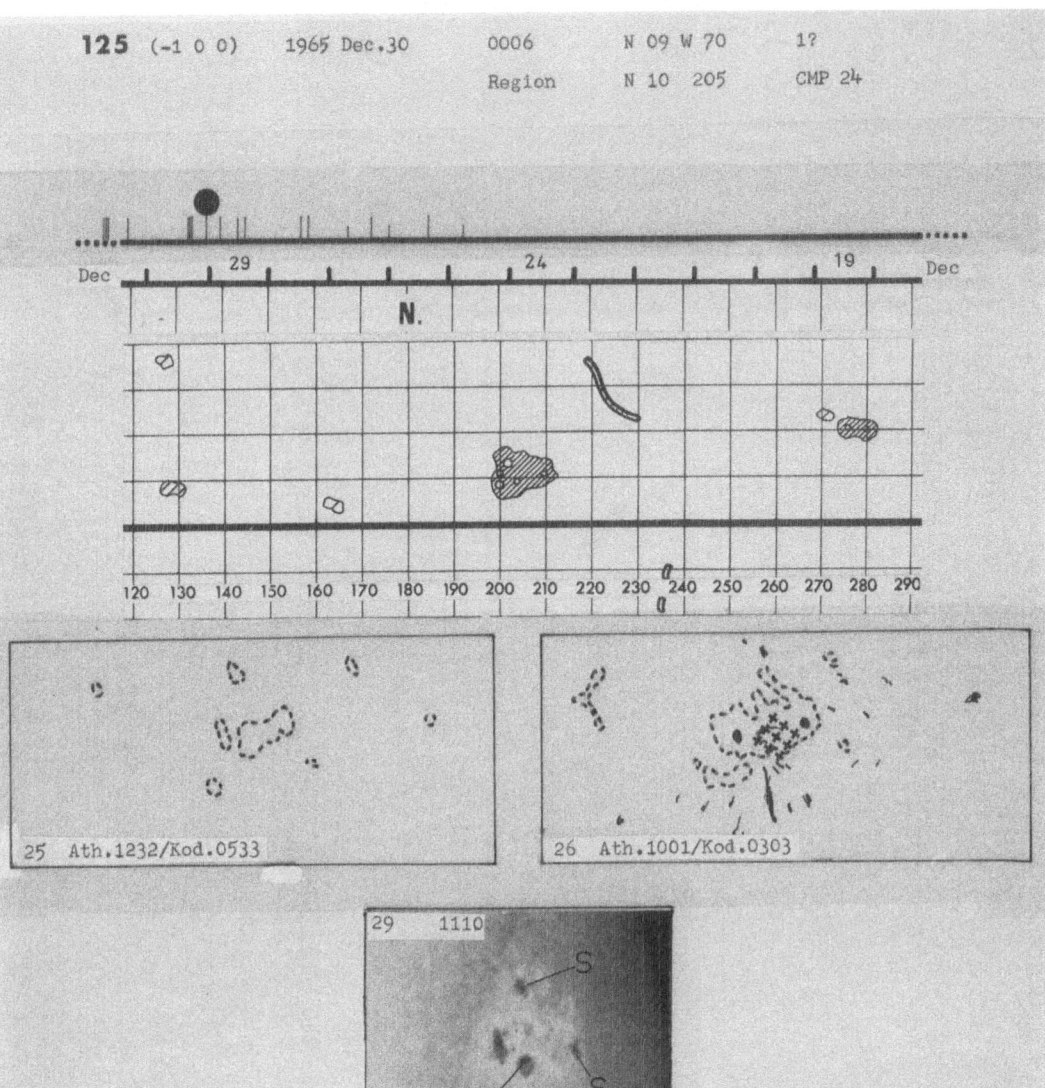

25 Ath.1232/Kod.0533

26 Ath.1001/Kod.0303

29 1110
(magn.pol. Soln.Dan.)

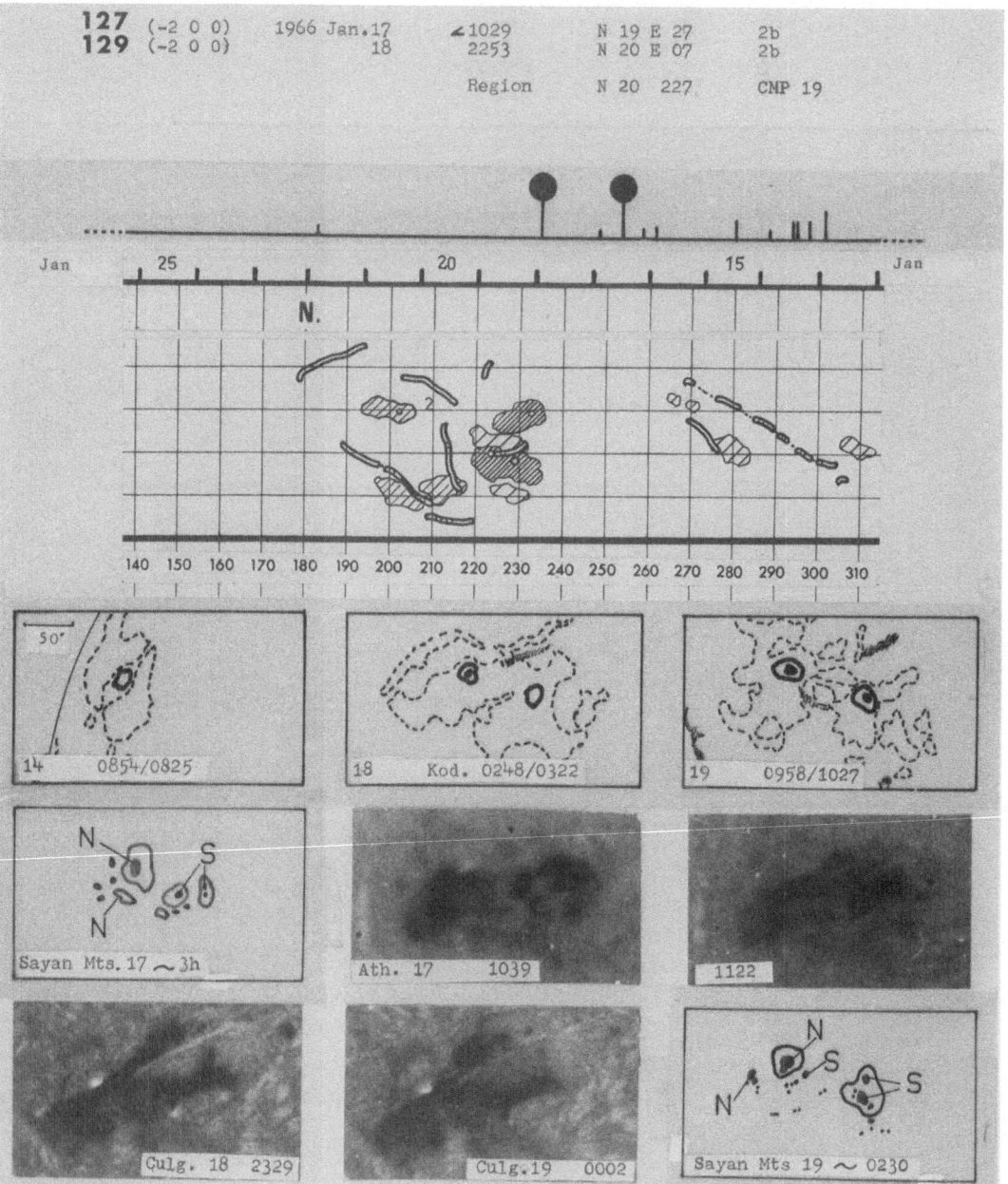

127 (-2 0 0) 1966 Jan.17 ≮1029 N 19 E 27 2b
129 (-2 0 0) 18 2253 N 20 E 07 2b

Region N 20 227 CMP 19

Jan 25 20 15 Jan

N.

140 150 160 170 180 190 200 210 220 230 240 250 260 270 280 290 300 310

50°

14 0854/0825

18 Kod. 0248/0322

19 0958/1027

N S N

Sayan Mts. 17 ~3h

Ath. 17 1039

1122

Culg. 18 2329

Culg.19 0002

N S S N

Sayan Mts 19 ~ 0230

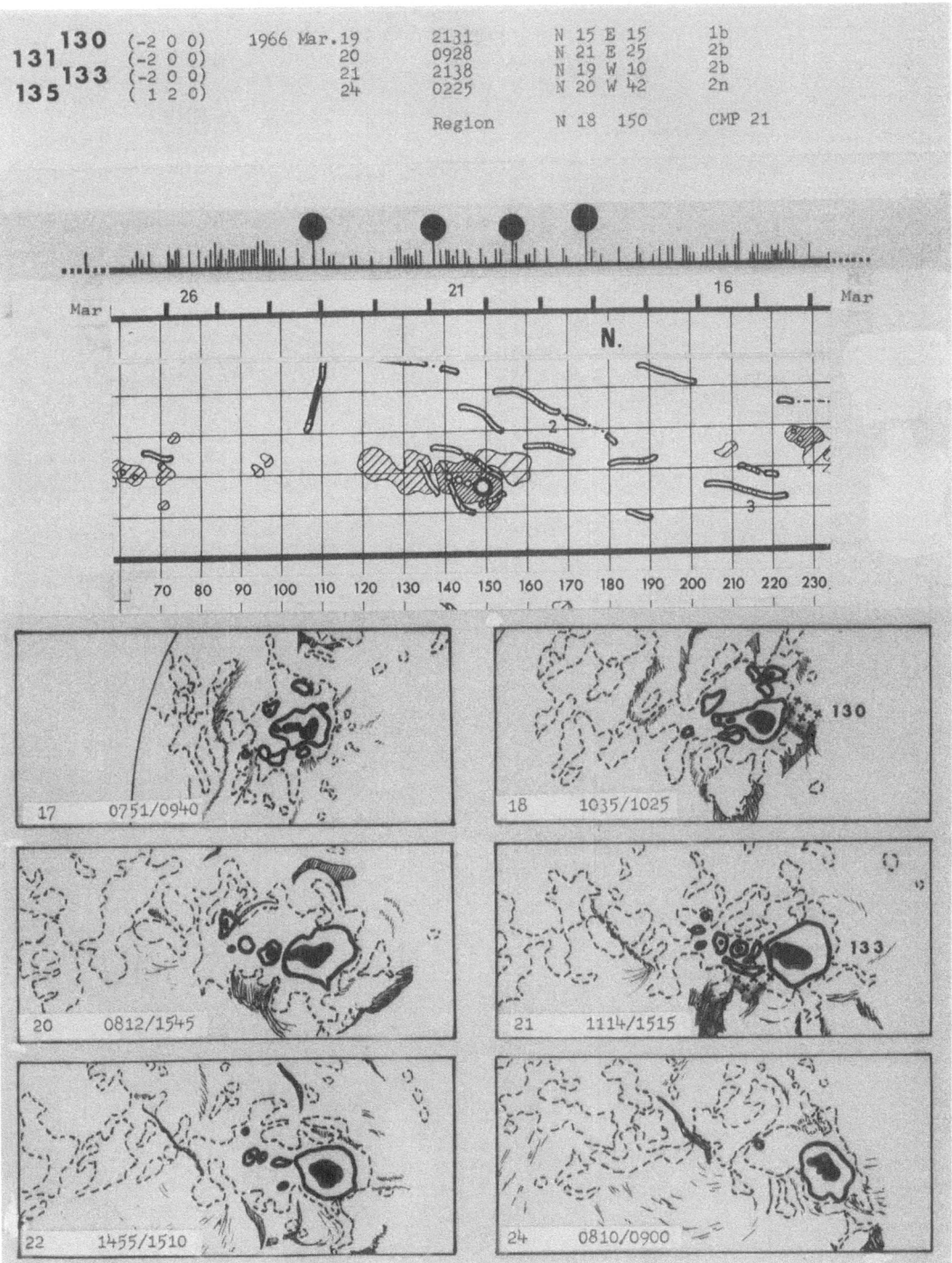

130	(-2 0 0)	1966 Mar.19	2131	N 15 E 15	1b
131	(-2 0 0)	20	0928	N 21 E 25	2b
133	(-2 0 0)	21	2138	N 19 W 10	2b
135	(1 2 0)	24	0225	N 20 W 42	2n
		Region		N 18 150	CMP 21

131

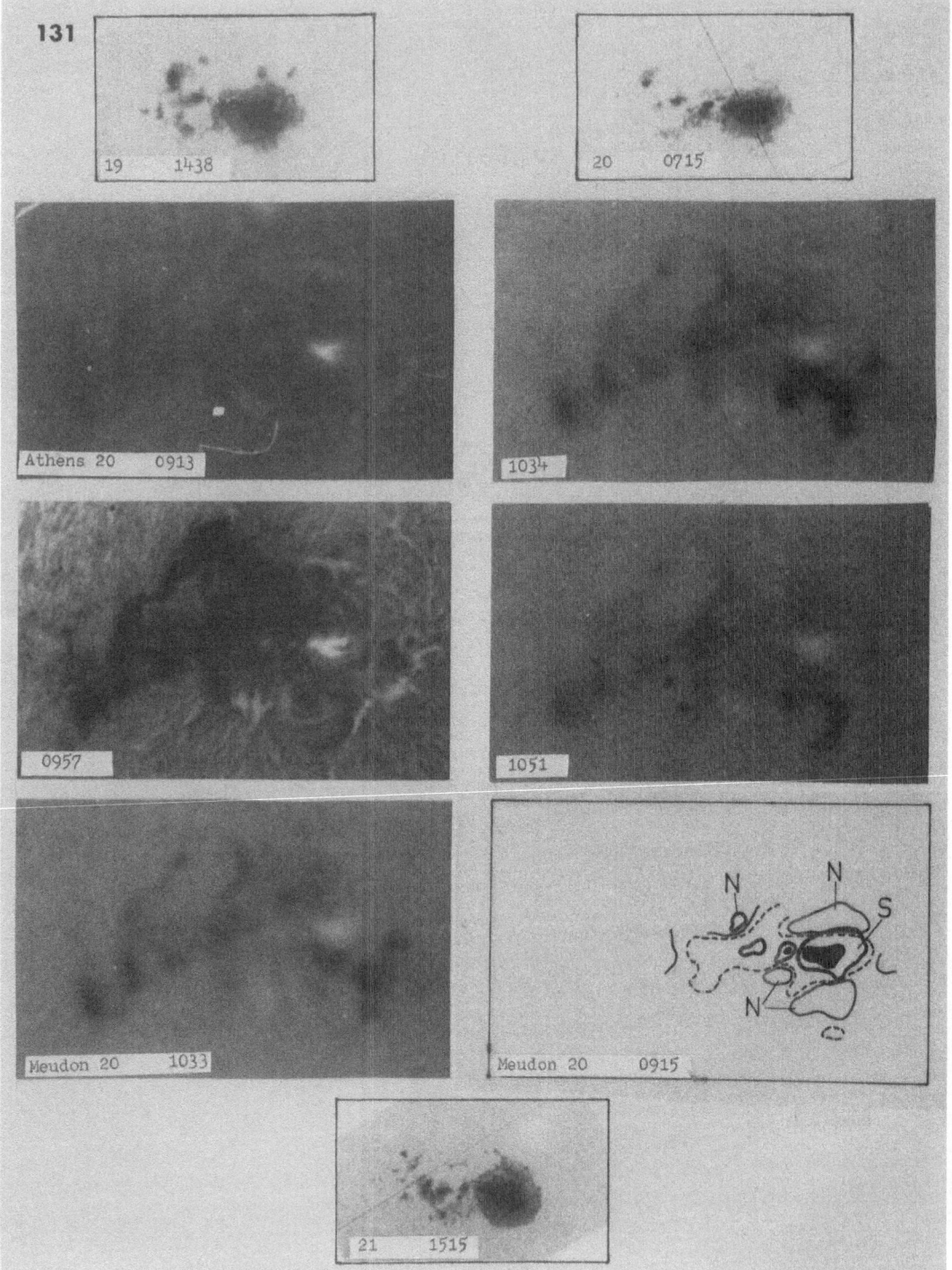

19   1438

20   0715

Athens 20   0913

1034

0957

1051

Meudon 20   1033

Meudon 20   0915

21   1515

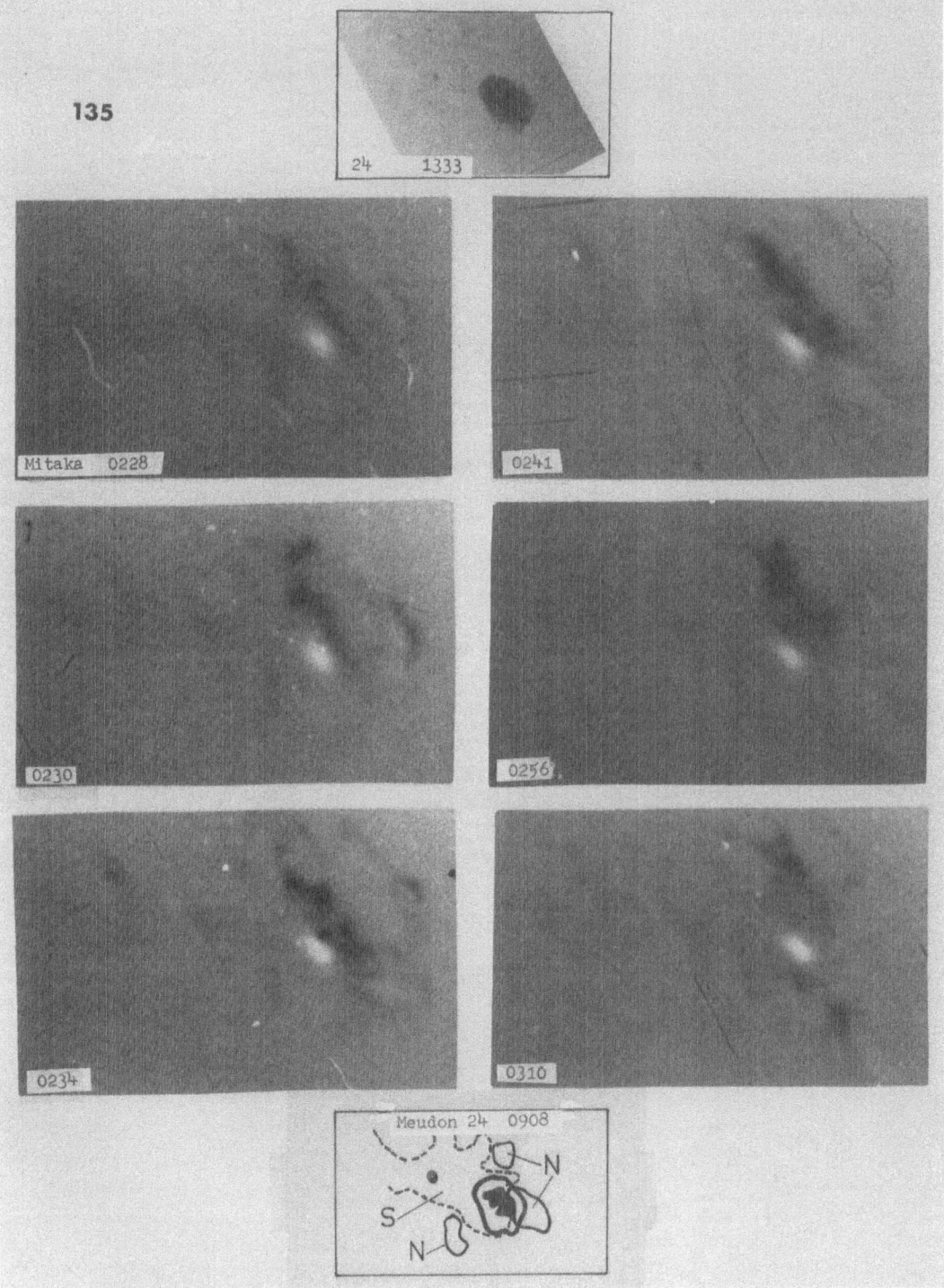

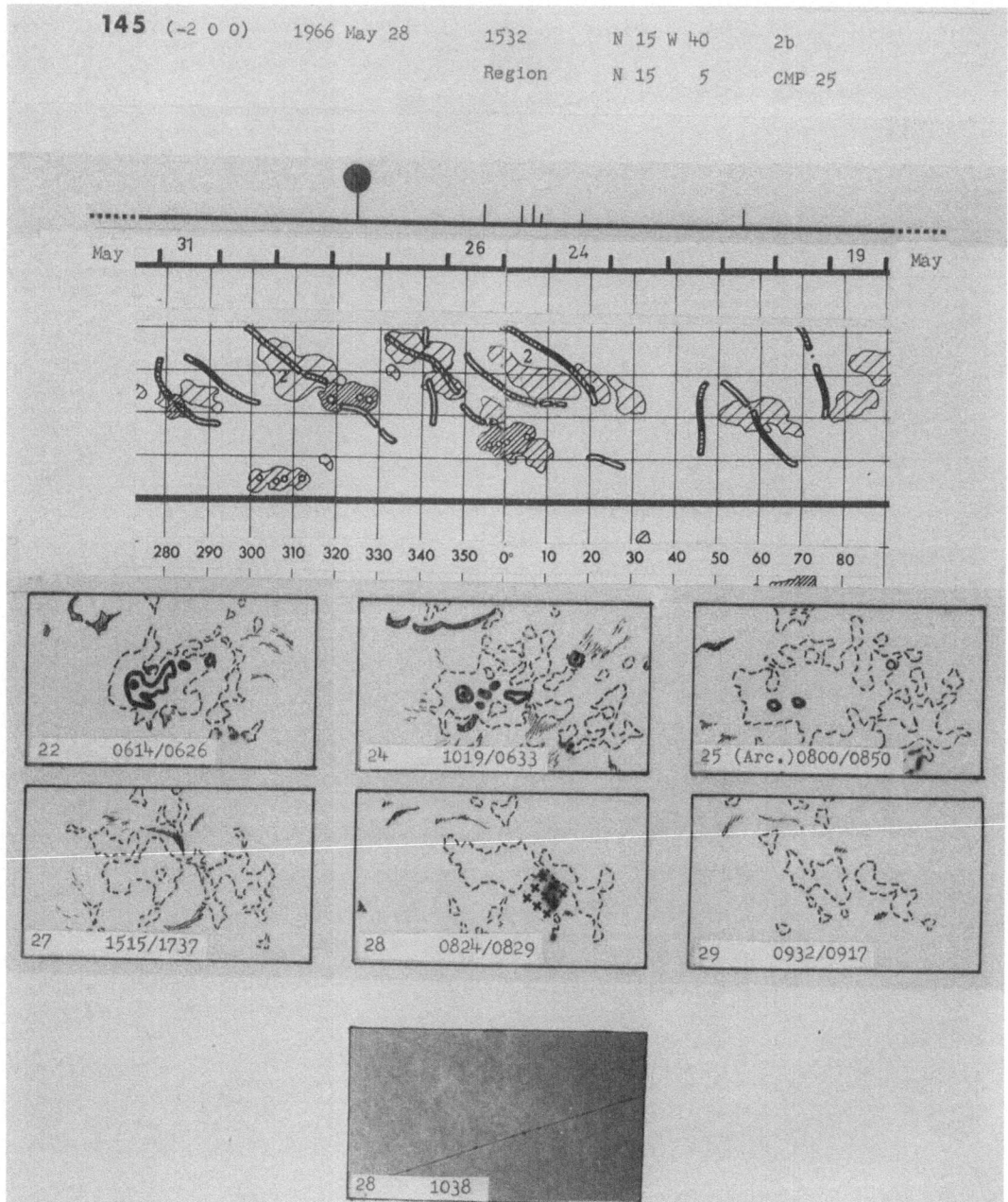

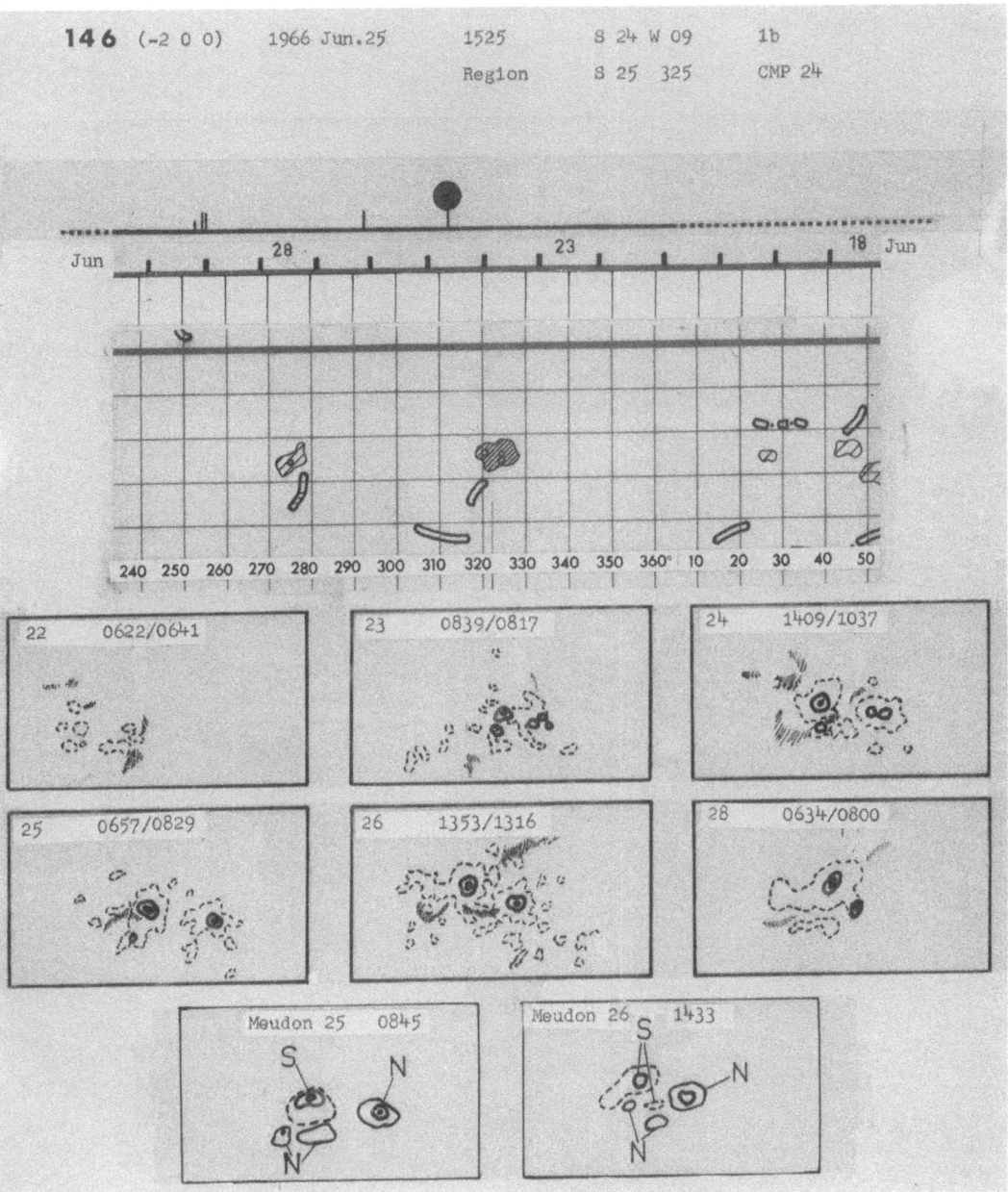

146 (-2 0 0) 1966 Jun.25 1525 S 24 W 09 1b
Region S 25 325 CMP 24

146

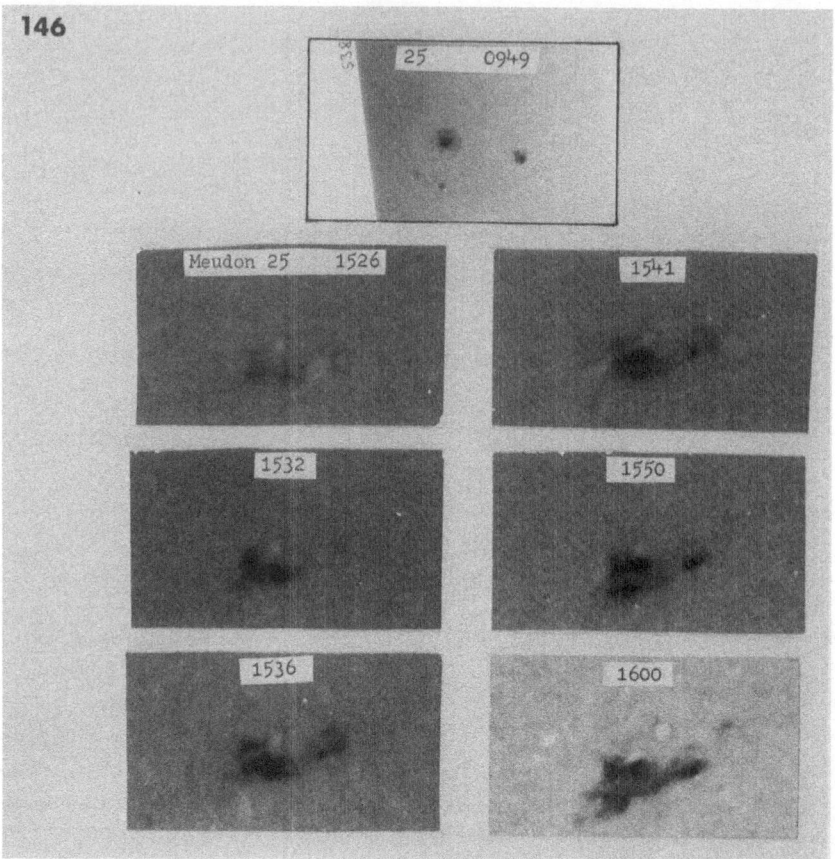

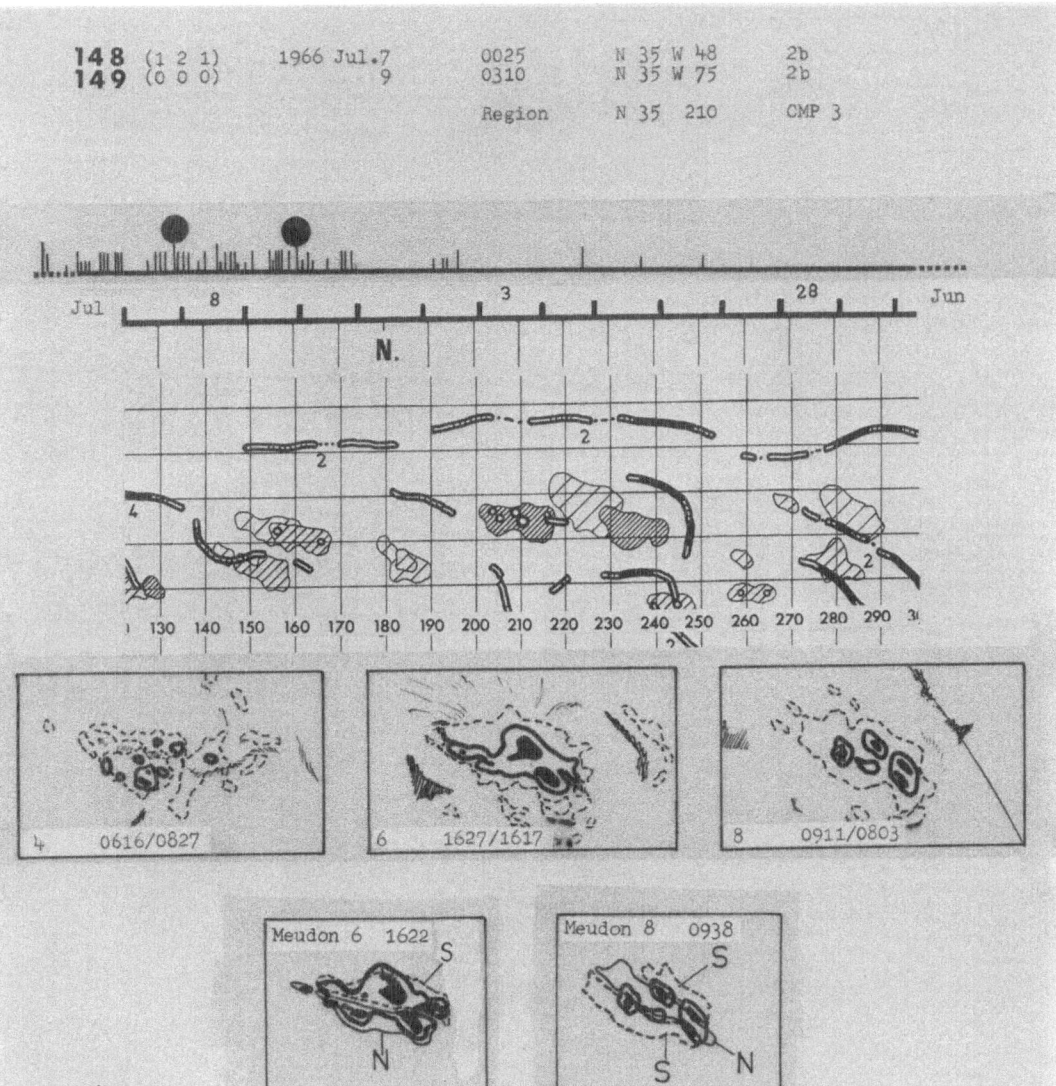

148 (1 2 1)   1966 Jul.7   0025   N 35 W 48   2b
149 (0 0 0)   9   0310   N 35 W 75   2b

              Region   N 35   210   CMP 3

4   0616/0827

6   1627/1617

8   0911/0803

Meudon 6   1622

Meudon 8   0938

148

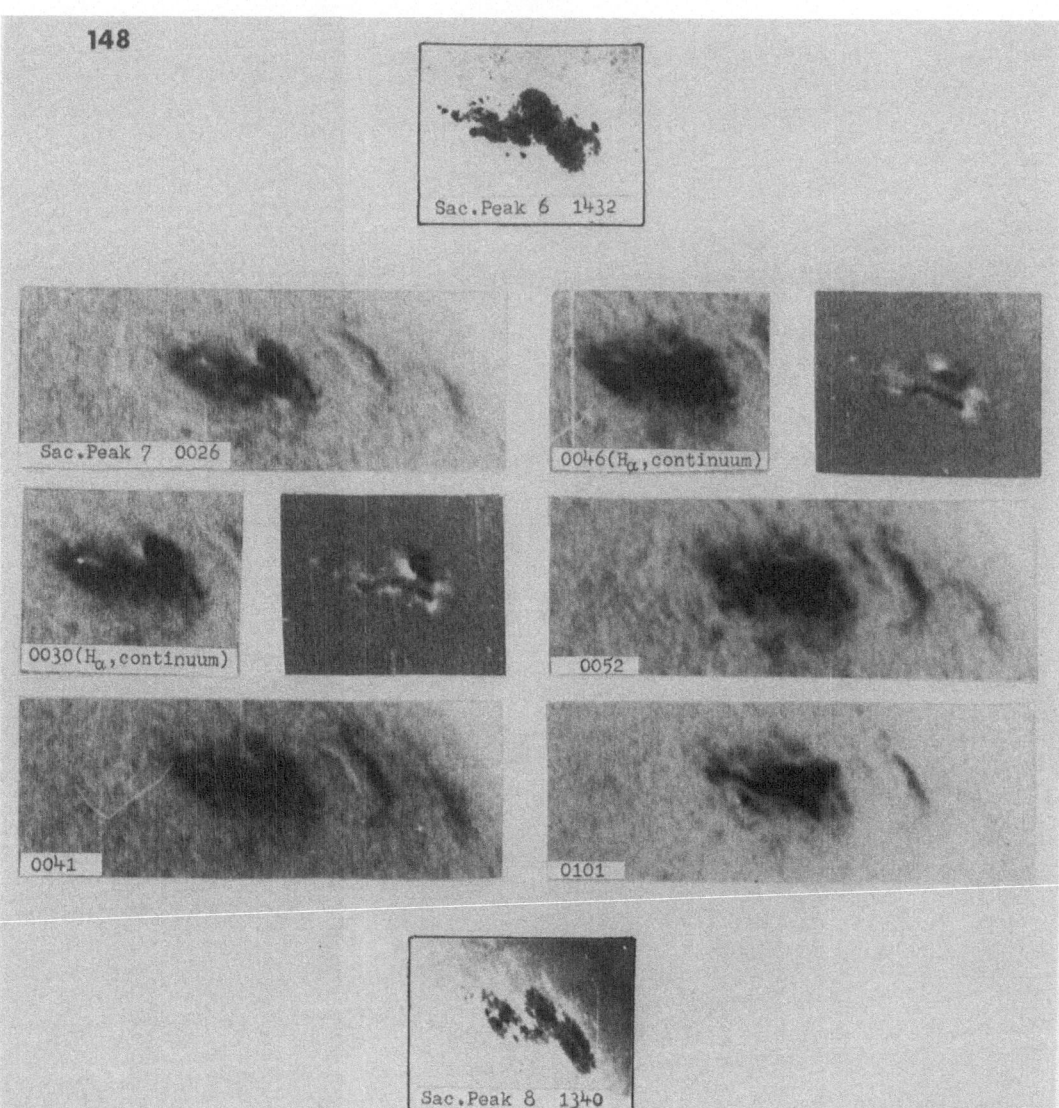

Sac.Peak 6   1432

Sac.Peak 7   0026

0046(H$_\alpha$,continuum)

0030(H$_\alpha$,continuum)

0052

0041

0101

Sac.Peak 8   1340

149

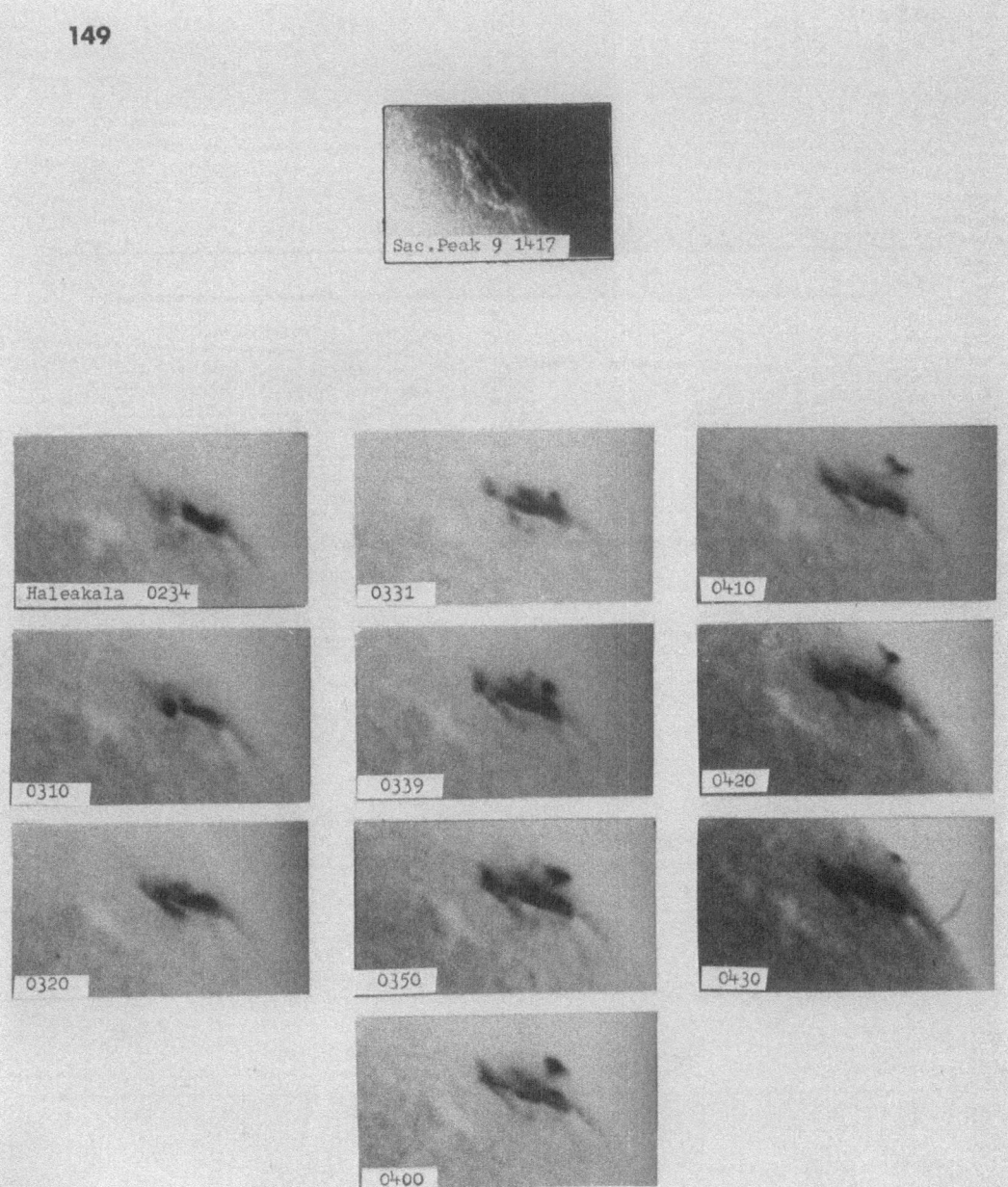

Sac.Peak 9 1417

Haleakala 0234

0331

0410

0310

0339

0420

0320

0350

0430

0400

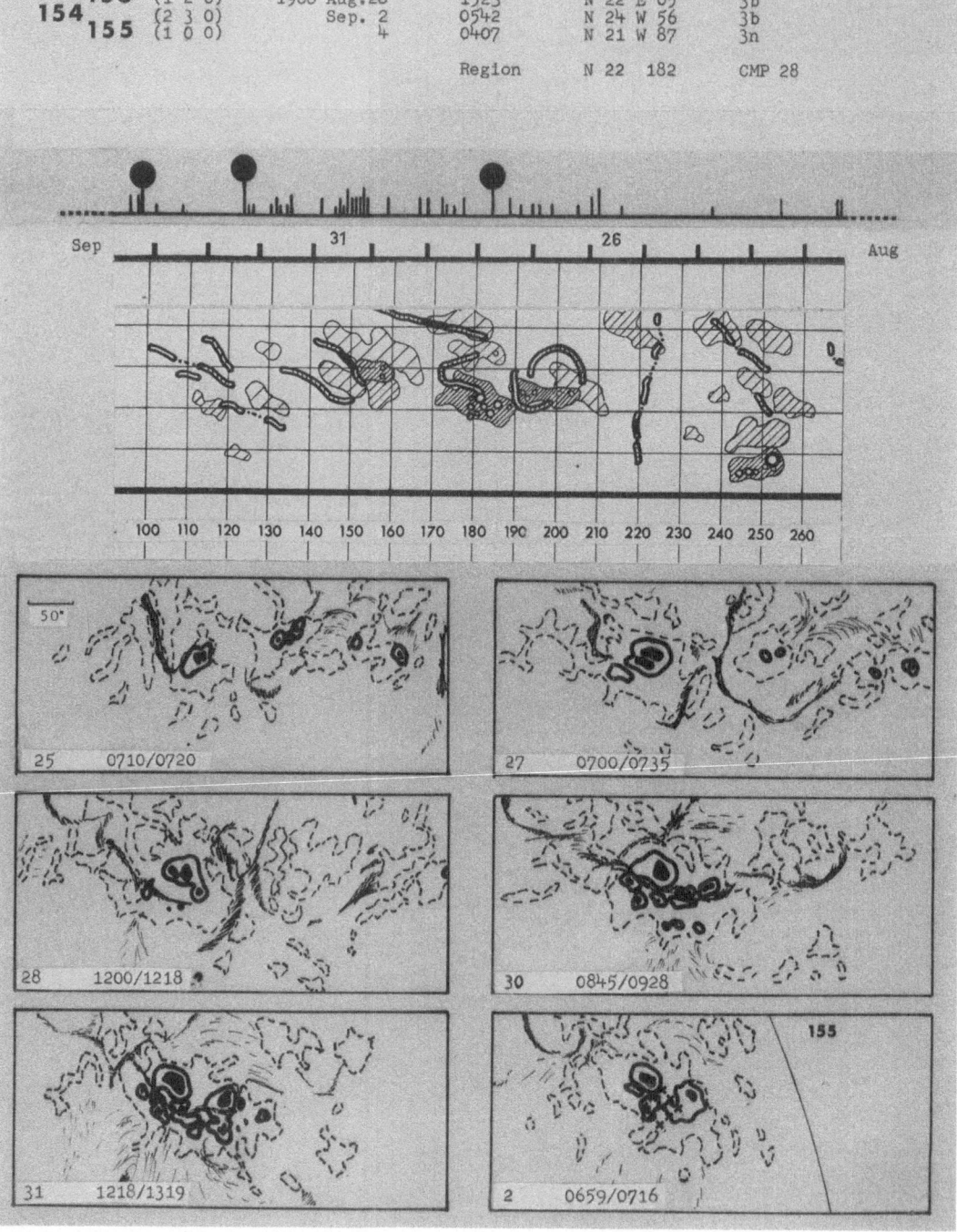

153	(1 2 0)	1966 Aug.28	1523	N 22 E 05	3b
154	(2 3 0)	Sep. 2	0542	N 24 W 56	3b
155	(1 0 0)	4	0407	N 21 W 87	3n
		Region		N 22 182	CMP 28

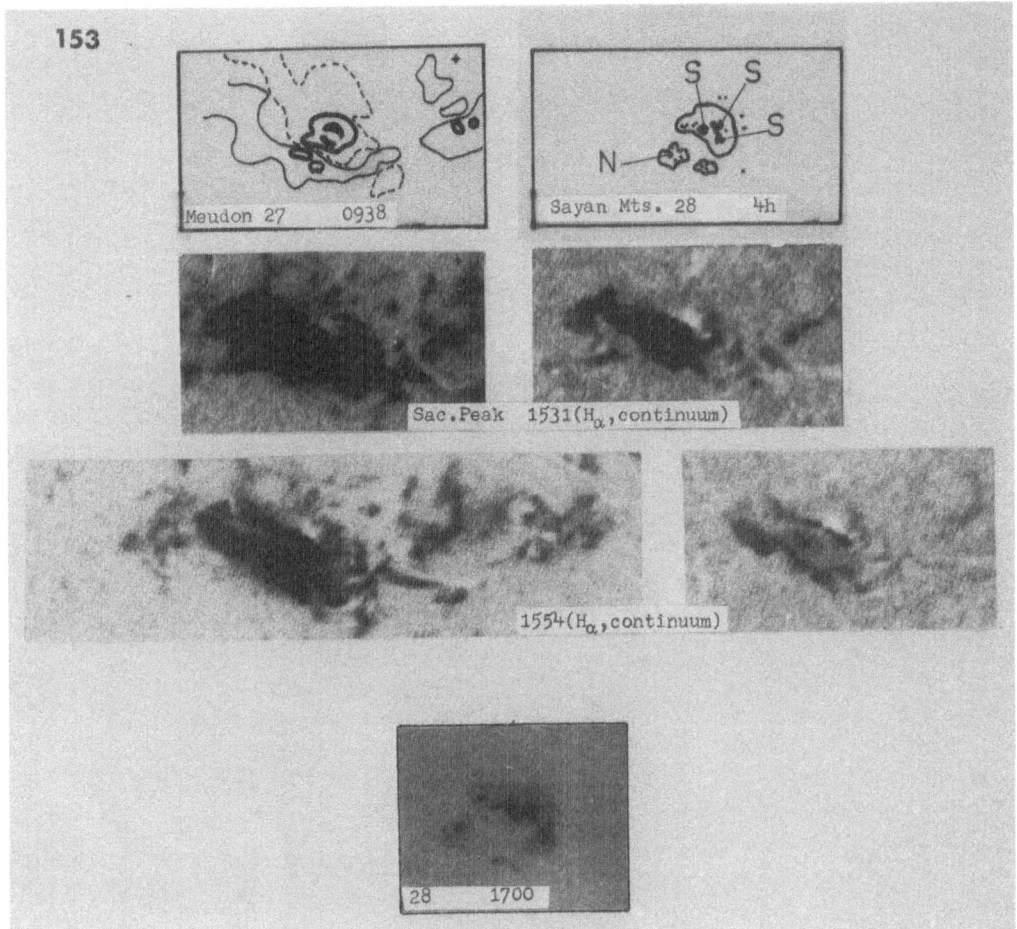

153

Meudon 27   0938

Sayan Mts. 28   4h

Sac.Peak   1531(H_α,continuum)

1554(H_α,continuum)

28   1700

154

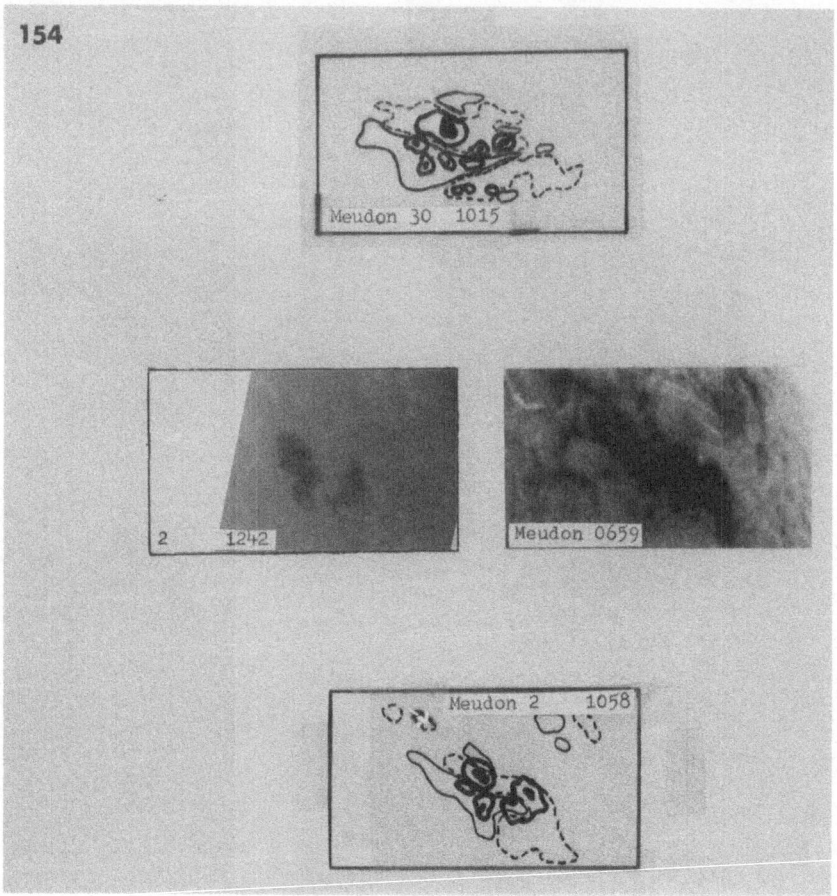

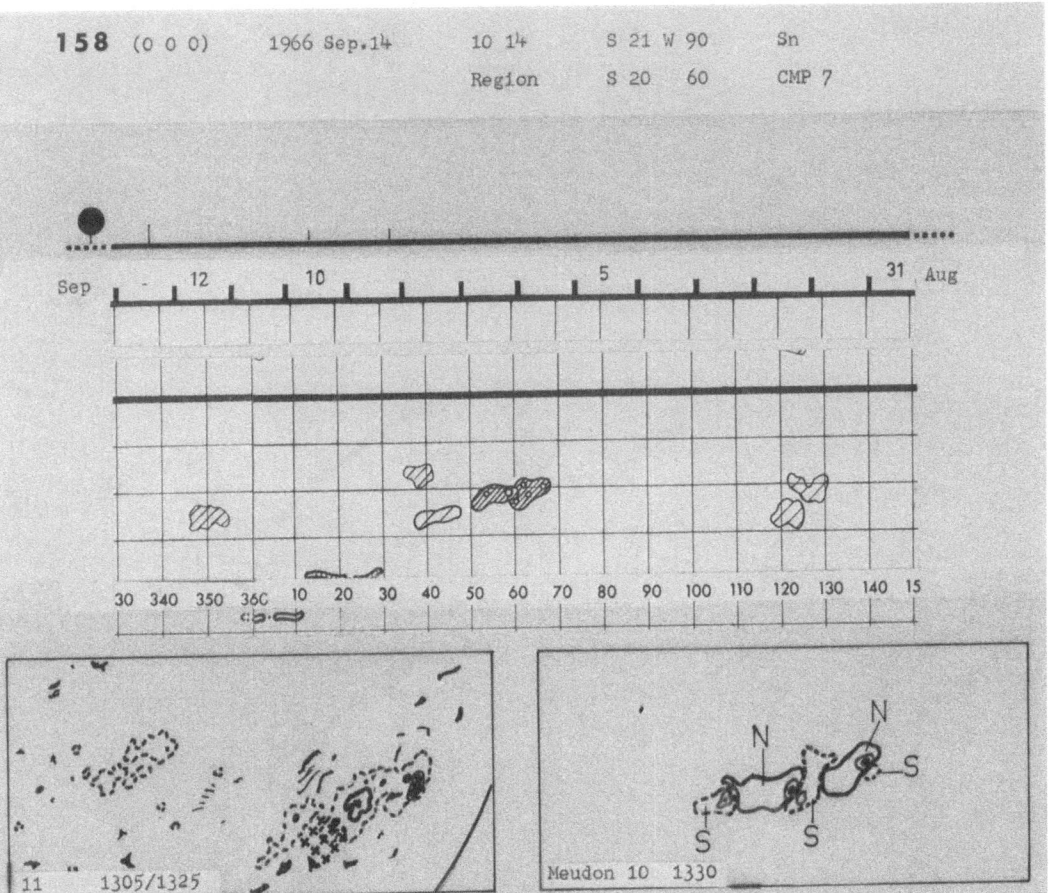

158 (0 0 0)   1966 Sep.14   10 14   S 21 W 90   Sn
                              Region   S 20   60   CMP 7

Sep   12   10   5   31 Aug

30 340 350 360 10 20 30 40 50 60 70 80 90 100 110 120 130 140 15

11   1305/1325

Meudon 10   1330

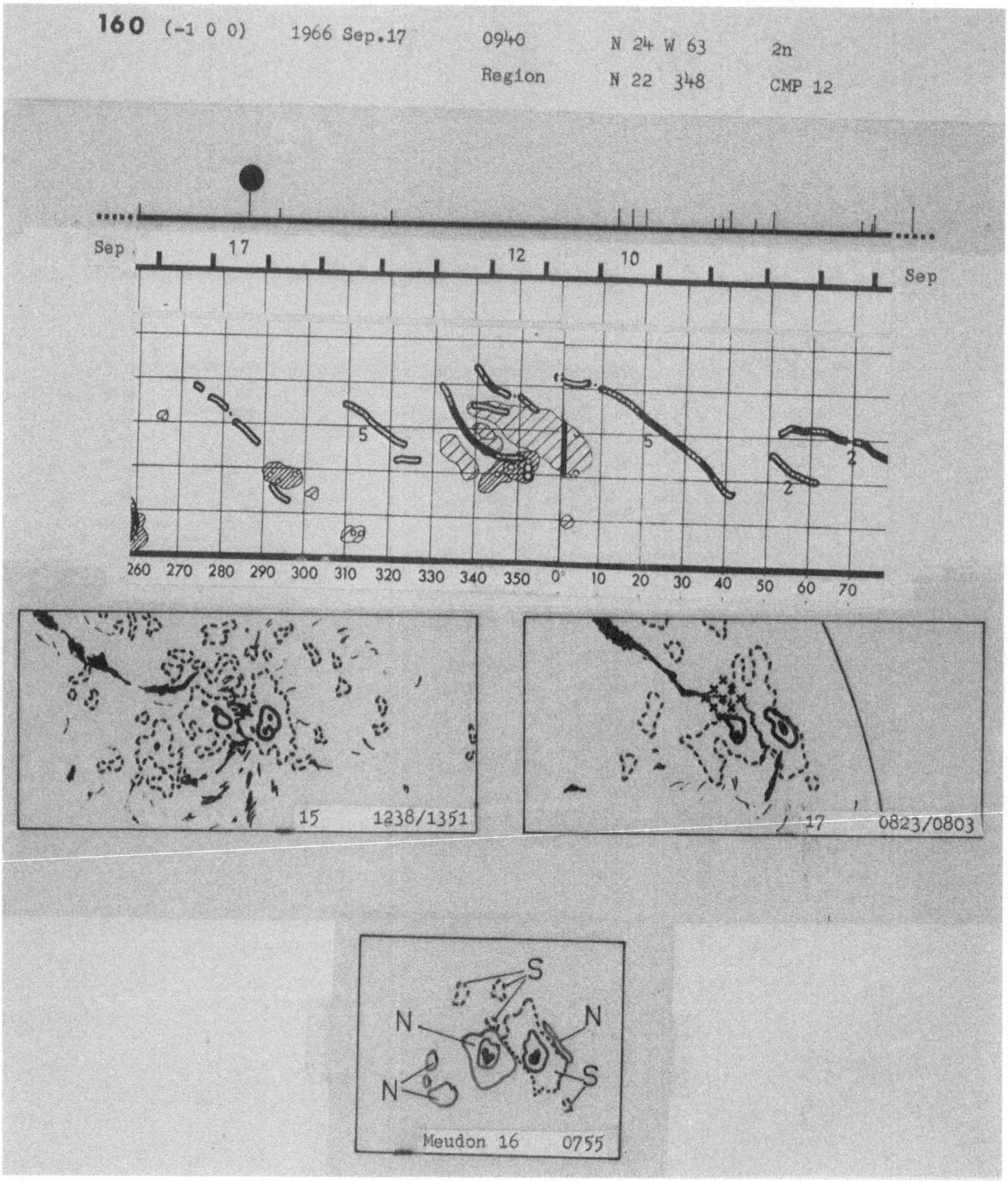

160 (-1 0 0) 1966 Sep.17 0940 N 24 W 63 2n
Region N 22 348 CMP 12
Sep 17 12 10 Sep
5 5 2 2
260 270 280 290 300 310 320 330 340 350 0 10 20 30 40 50 60 70
15 1238/1351
17 0823/0803
S N N S N
Meudon 16 0755

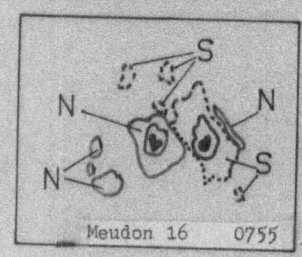

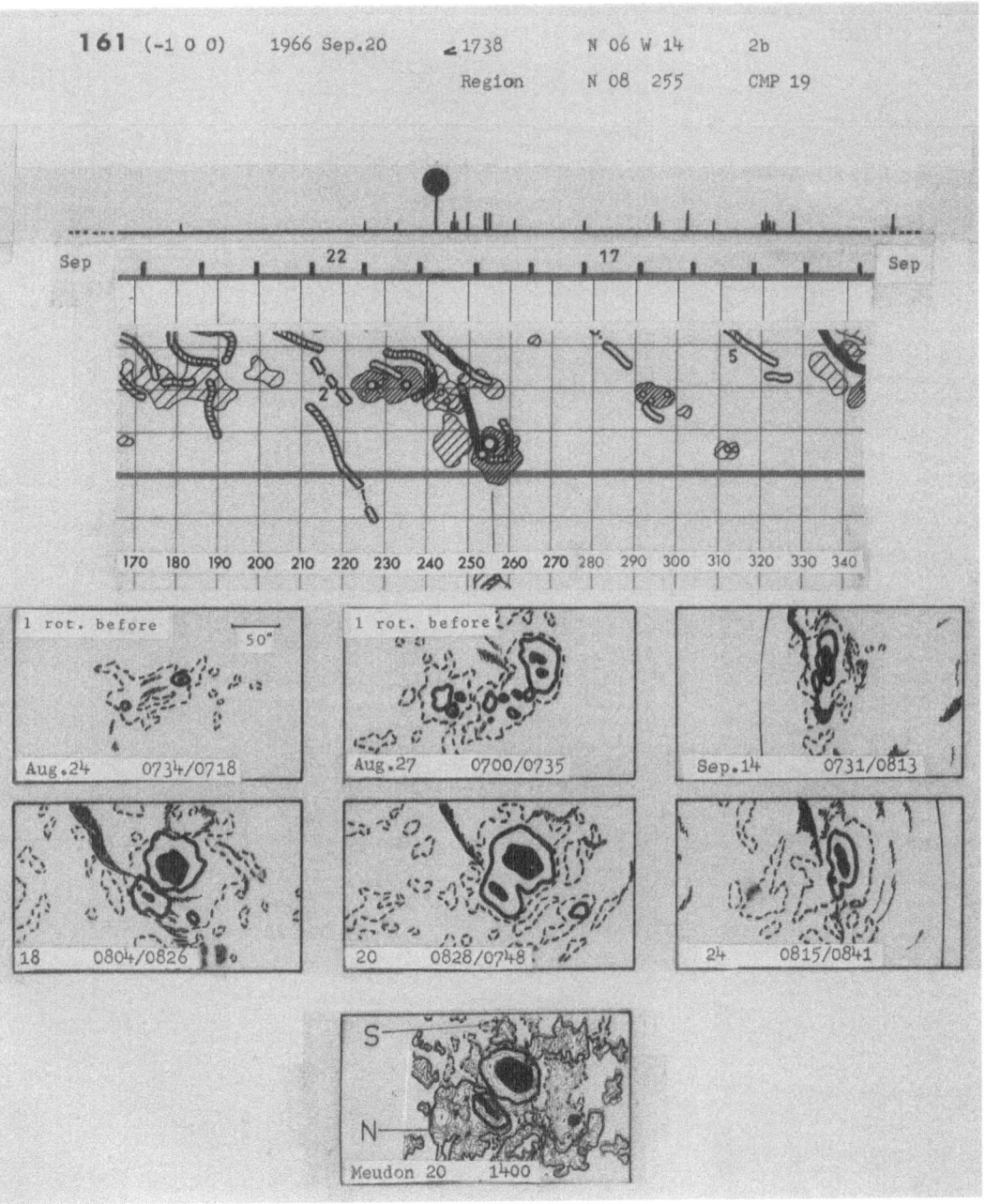

161 (-1 0 0) 1966 Sep.20 ∠1738 N 06 W 14 2b
Region N 08 255 CMP 19

161

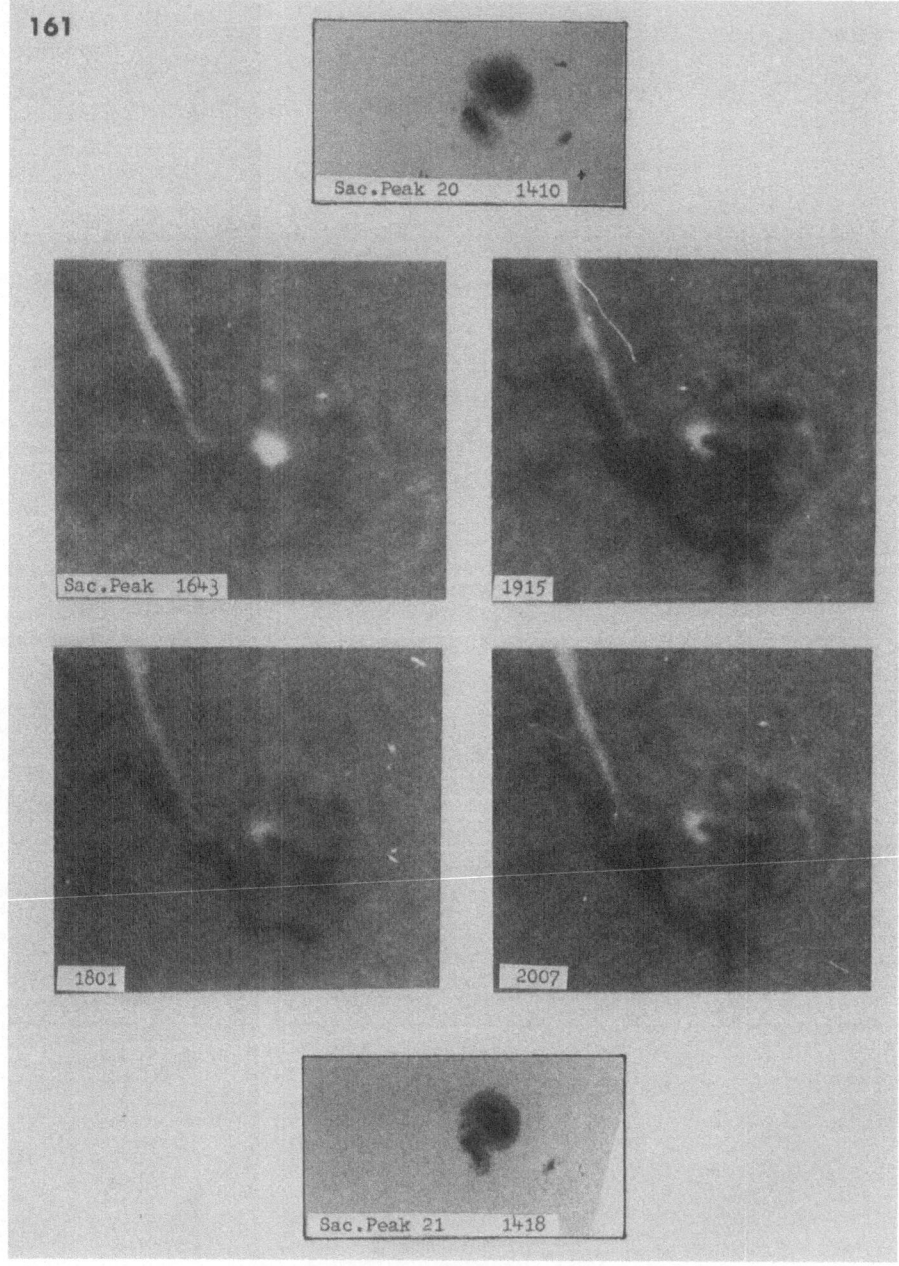

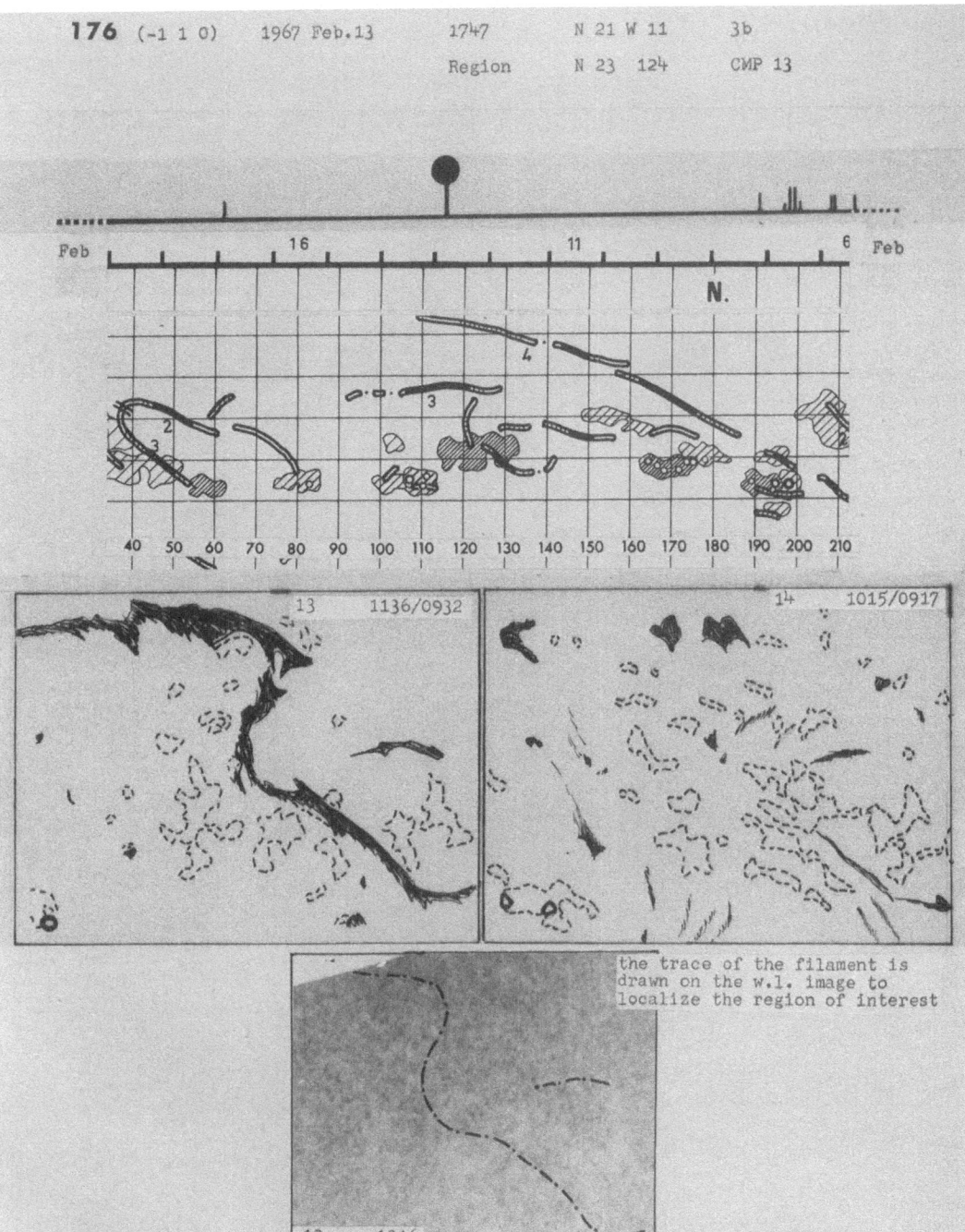

176 (-1 1 0)   1967 Feb.13   1747   N 21 W 11   3b
                      Region   N 23   124   CMP 13

the trace of the filament is
drawn on the w.l. image to
localize the region of interest

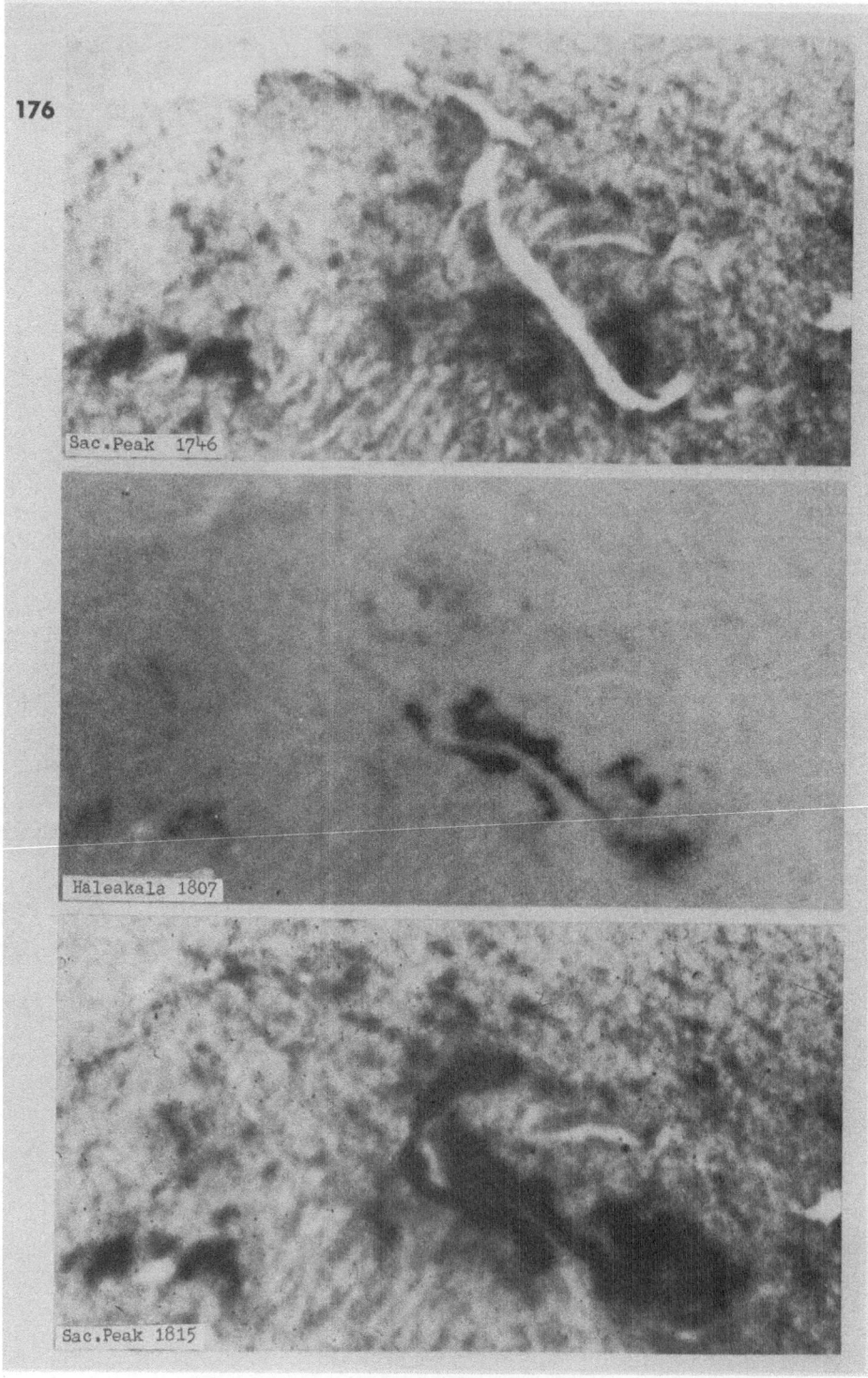

176

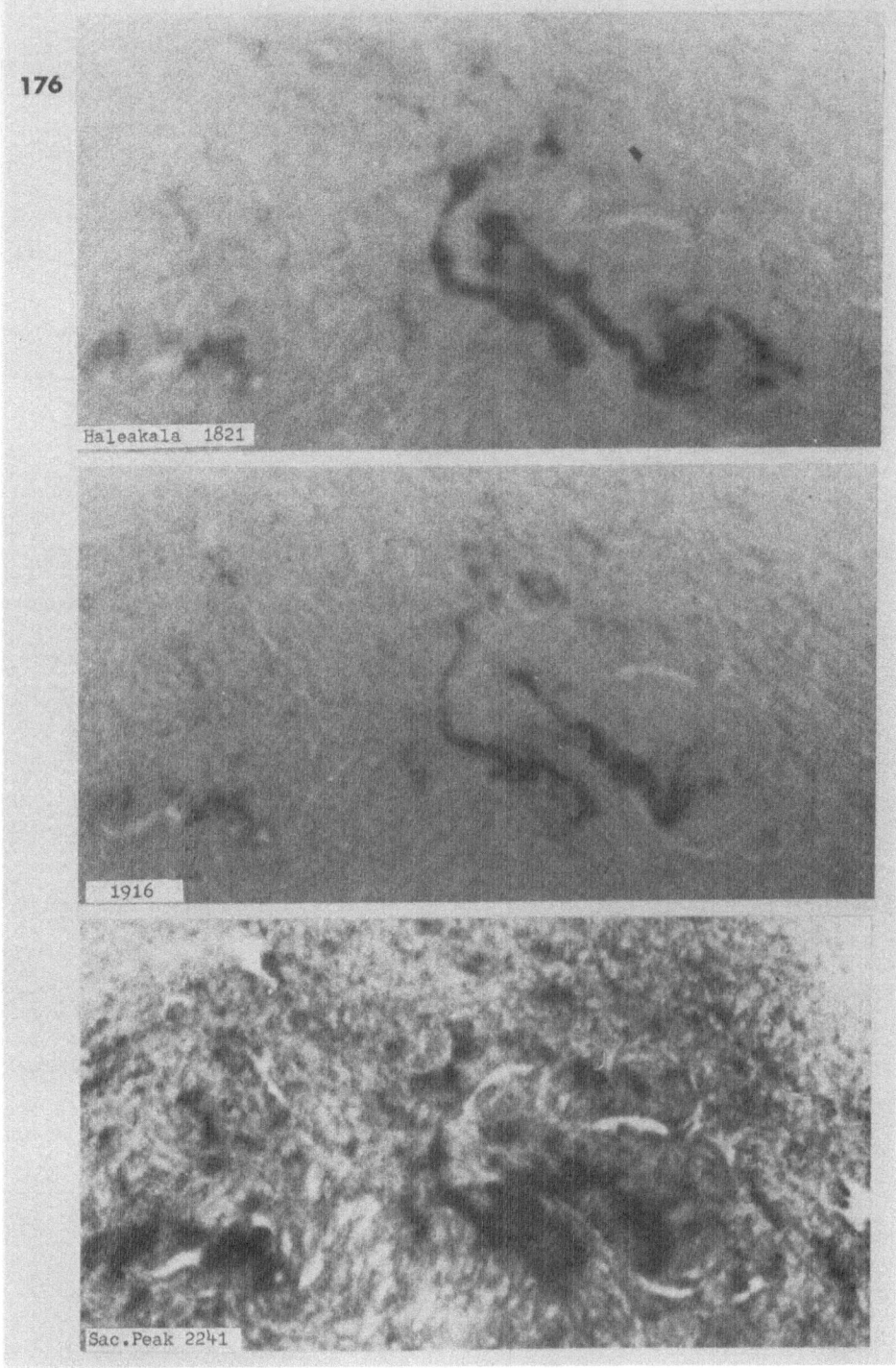

176

Haleakala 1821

1916

Sac.Peak 2241

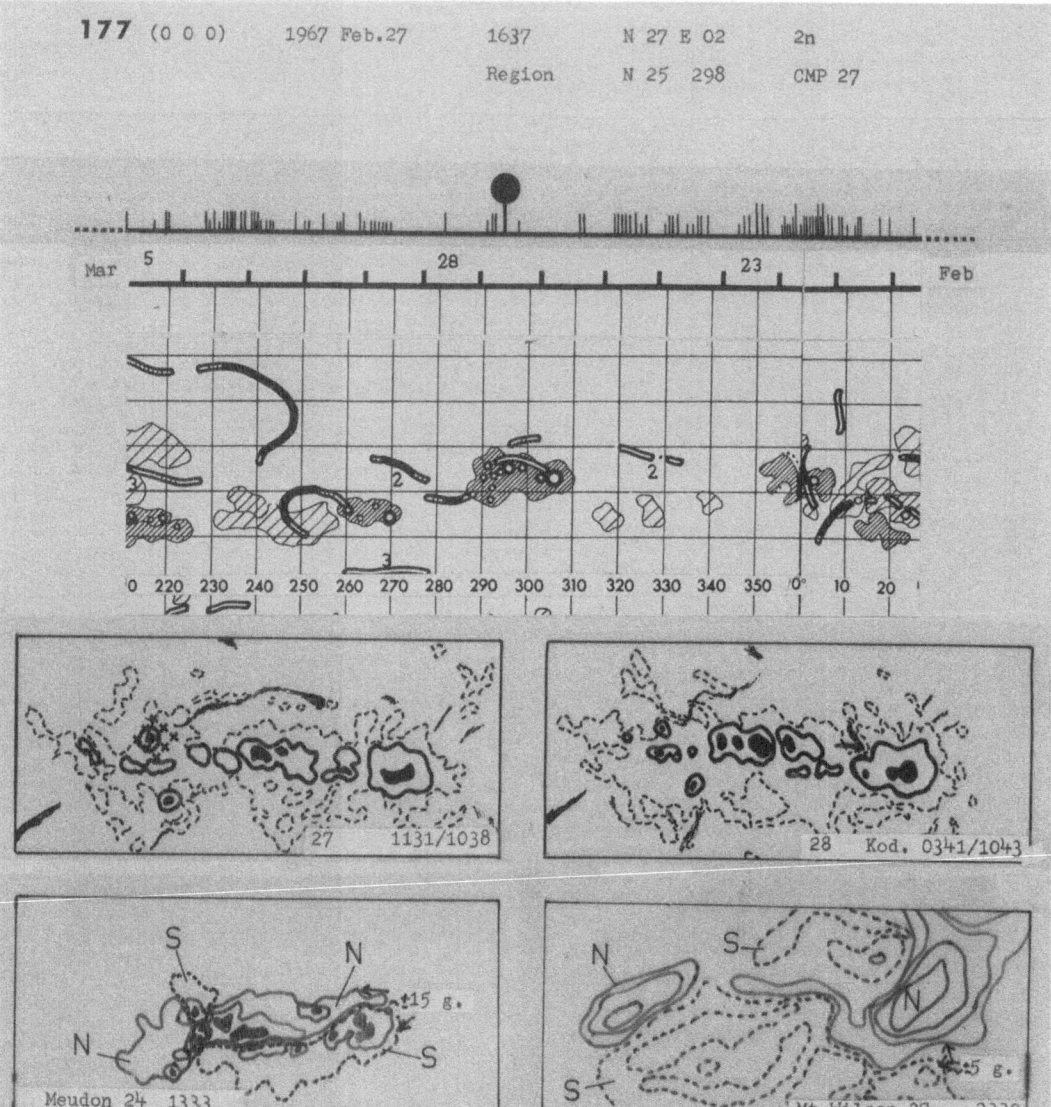

177 (0 0 0)   1967 Feb.27   1637   N 27 E 02   2n
Region   N 25   298   CMP 27

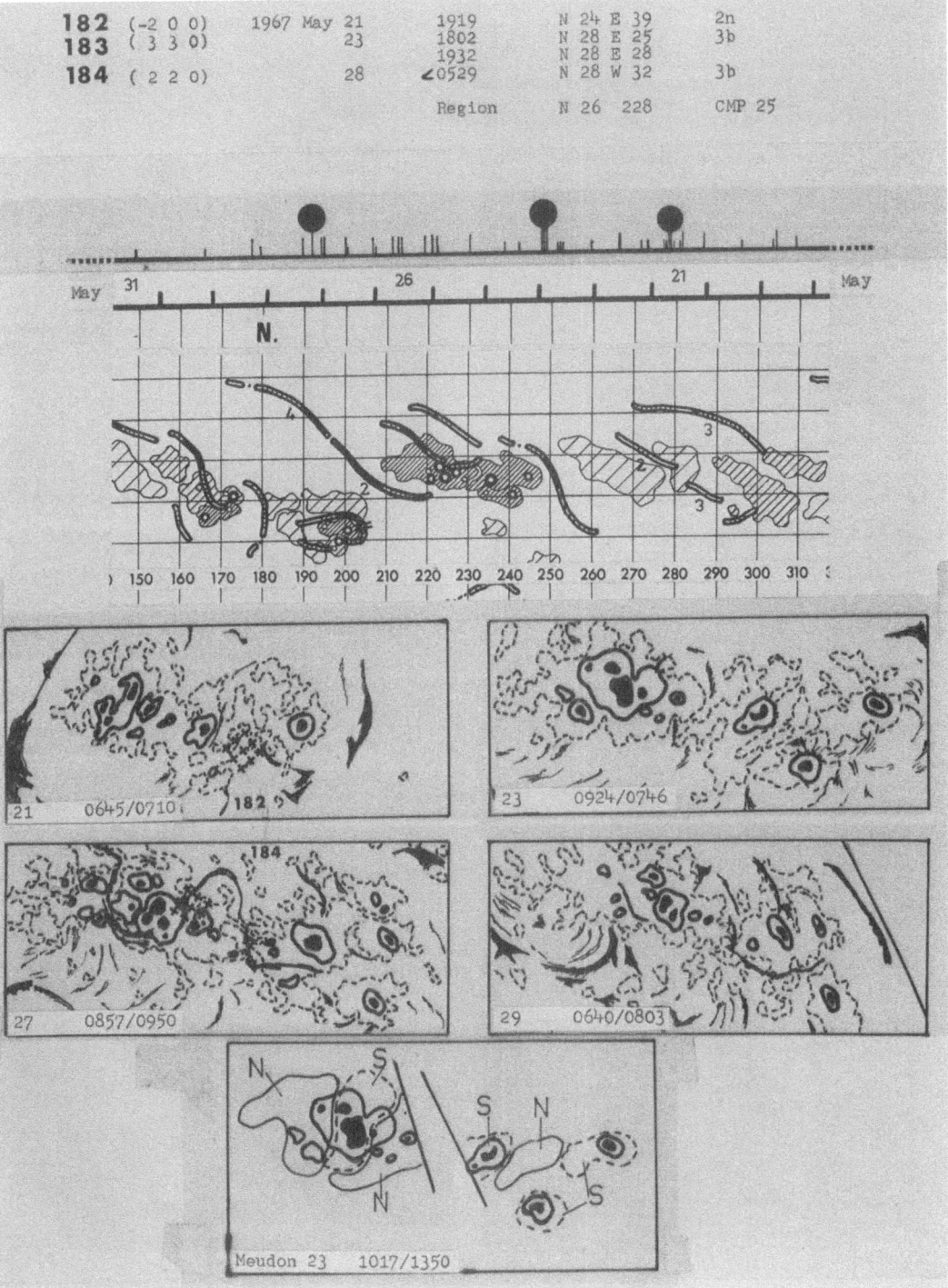

182	(-2 0 0)	1967 May 21	1919	N 24 E 39	2n
183	(3 3 0)	23	1802	N 28 E 25	3b
			1932	N 28 E 28	
184	(2 2 0)	28	<0529	N 28 W 32	3b
		Region		N 26 228	CMP 25

21   0645/0710   182

23   0924/0746

27   0857/0950   184

29   0640/0803

Meudon 23   1017/1350

183

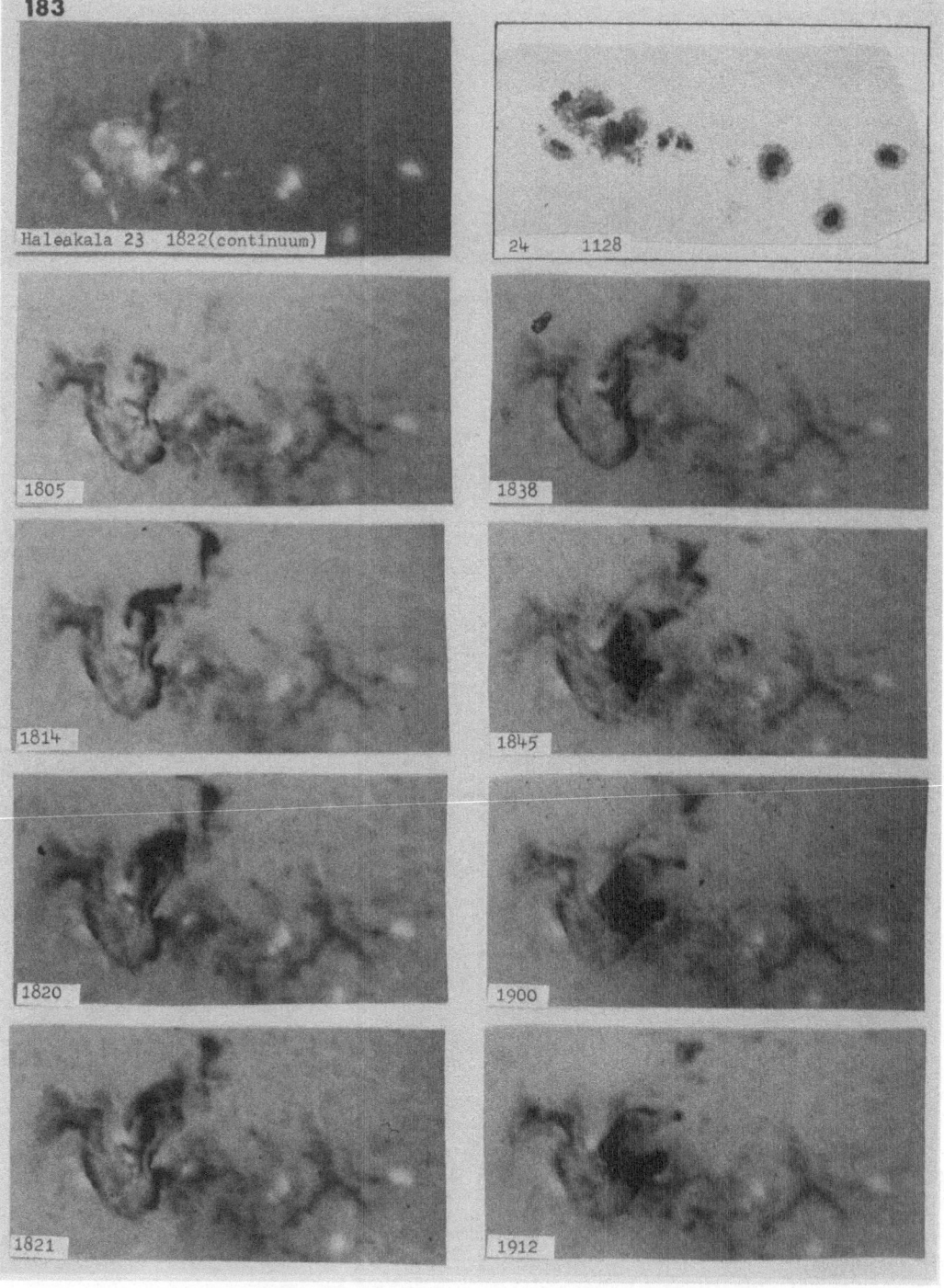

Haleakala 23 1822(continuum)

24   1128

1805

1838

1814

1845

1820

1900

1821

1912

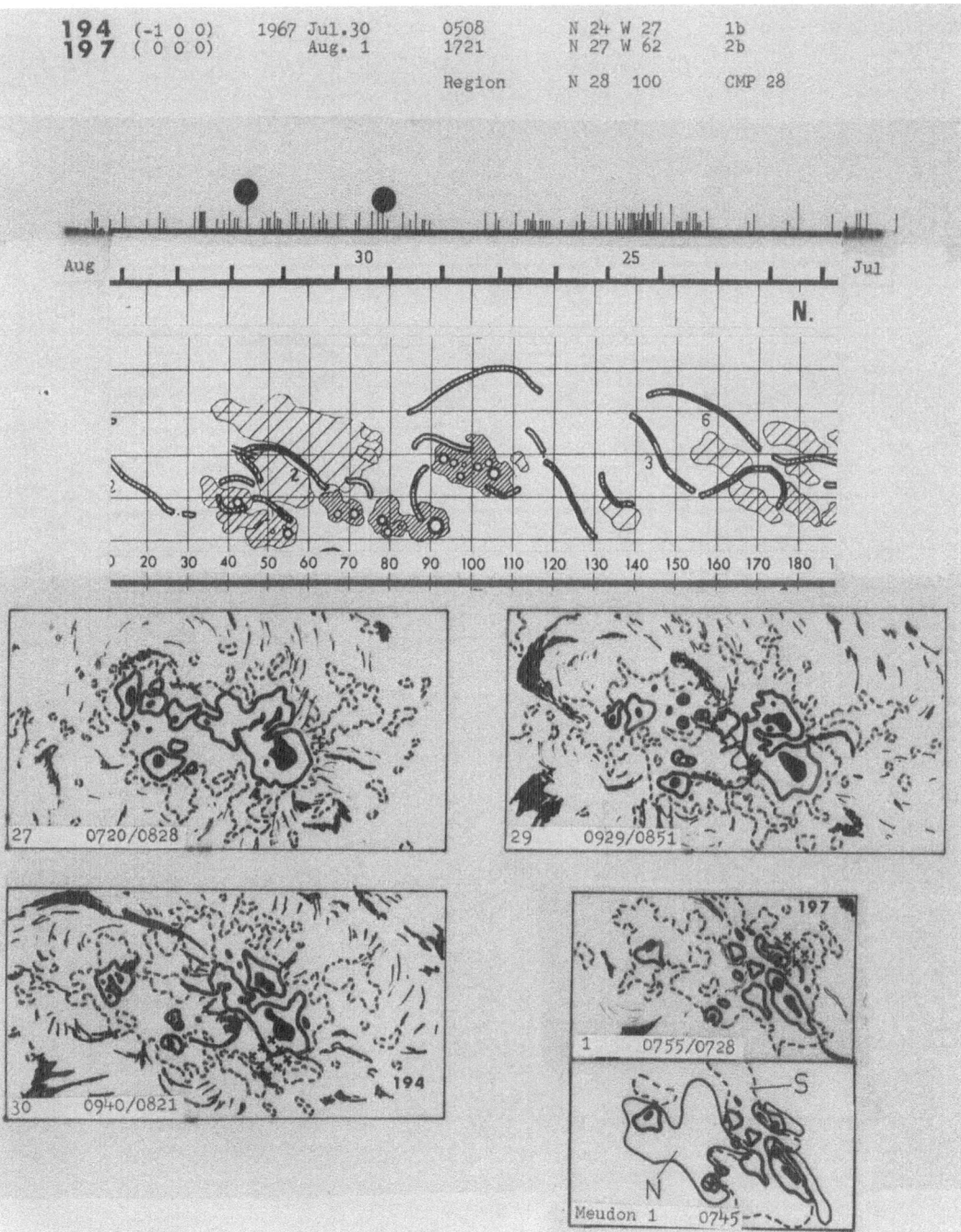

194 (-1 0 0) 1967 Jul.30 0508 N 24 W 27 1b
197 (0 0 0) Aug. 1 1721 N 27 W 62 2b

Region N 28 100 CMP 28

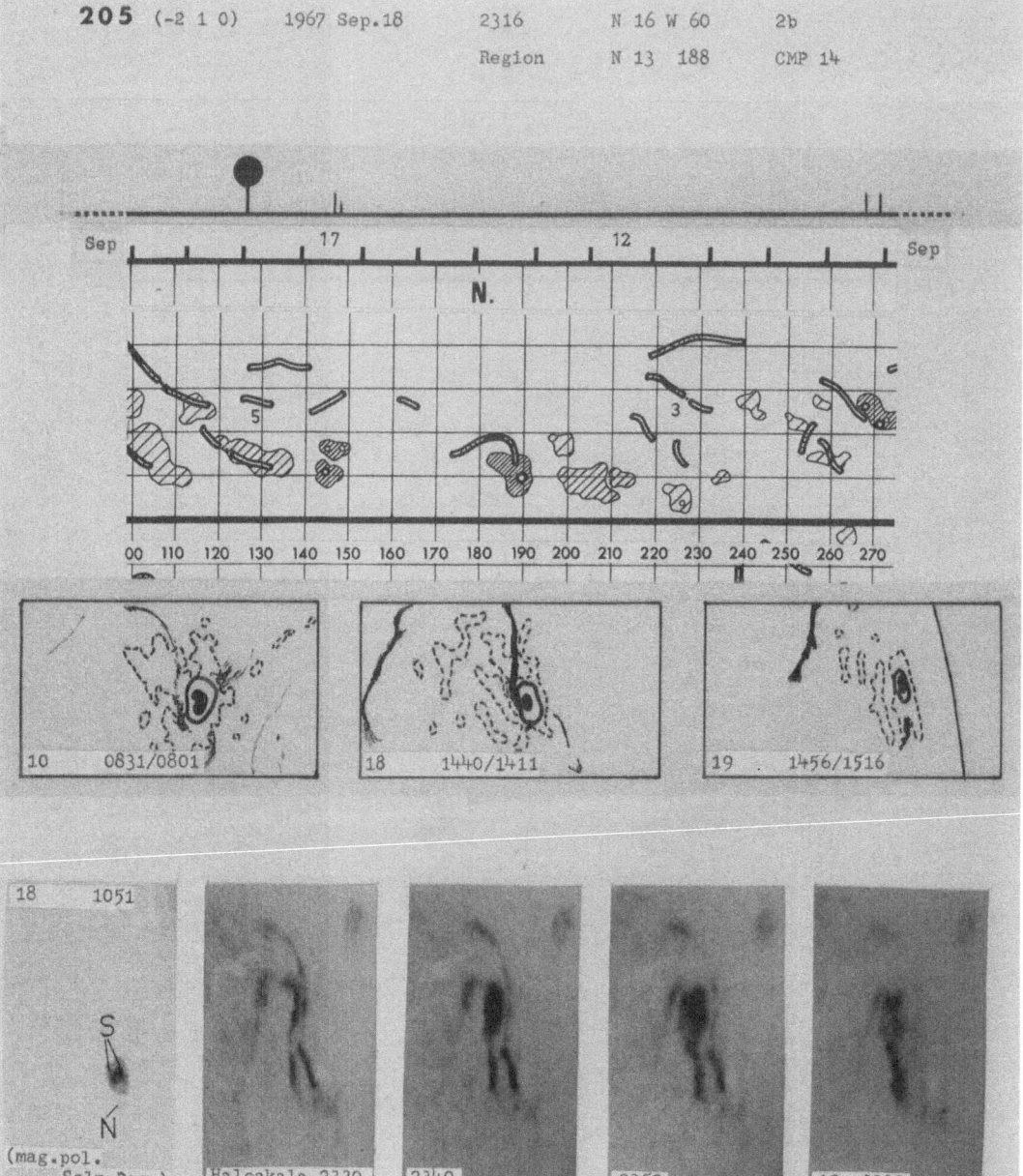

205 (-2 1 0)   1967 Sep.18   2316   N 16 W 60   2b
                                 Region   N 13 188   CMP 14

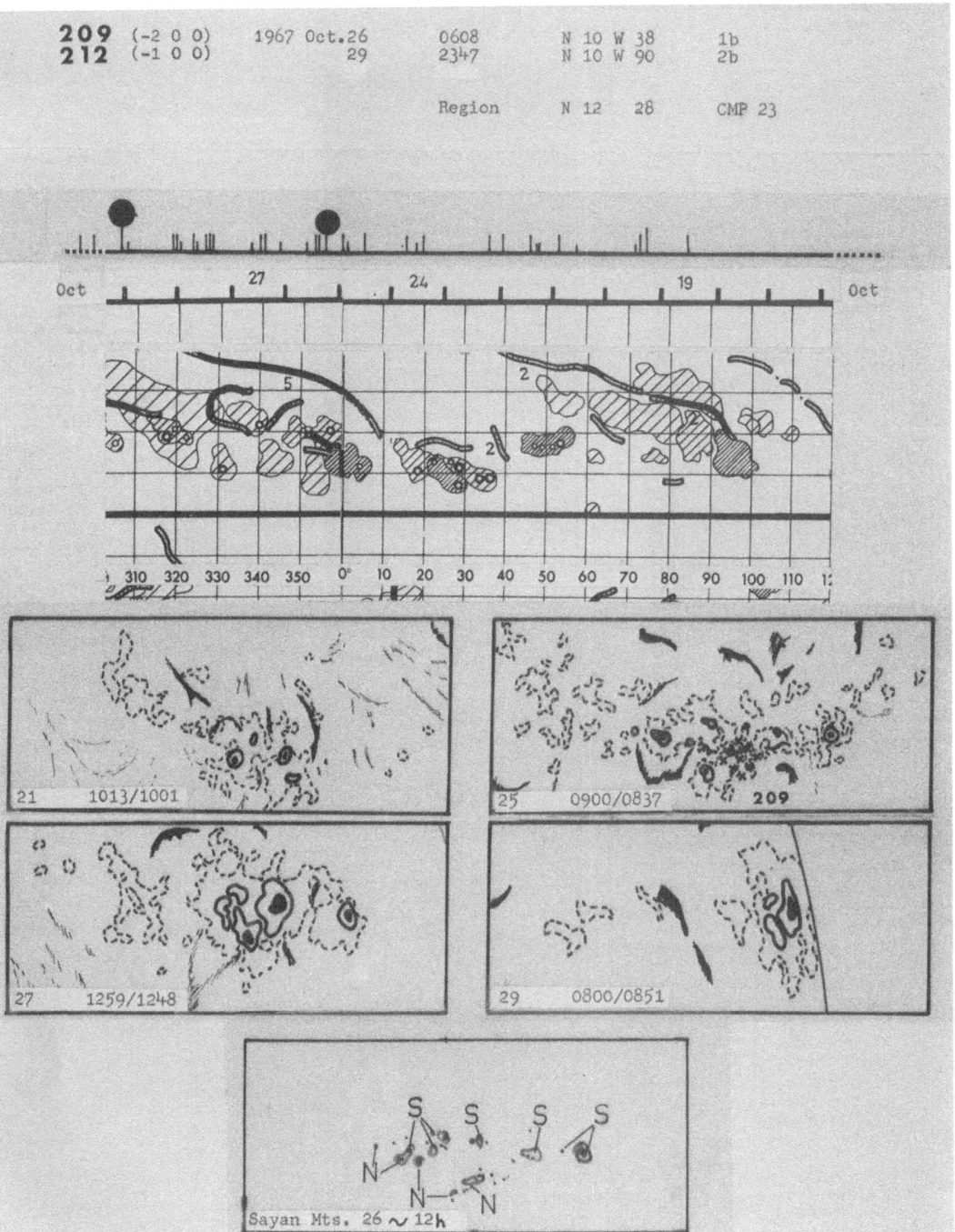

209 (-2 0 0)   1967 Oct.26   0608   N 10 W 38   1b
212 (-1 0 0)   29   2347   N 10 W 90   2b

Region   N 12   28   CMP 23

21   1013/1001

25   0900/0837   **209**

27   1259/1248

29   0800/0851

Sayan Mts. 26 ~ 12ʰ

212

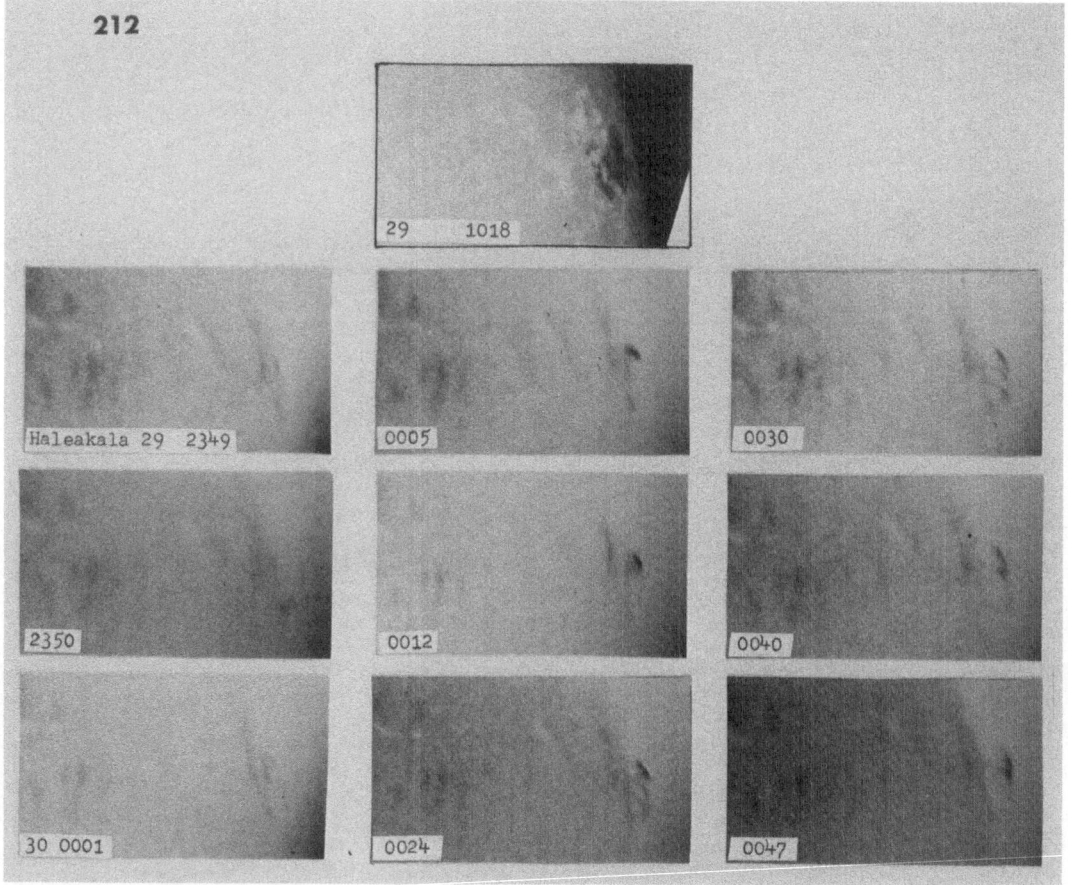

213 (0 1 0)   1967 Nov.2   0852   S 18 W 02   2b
214 (-1 0 0)   4   1151   S 18 W 33   1b

                              Region   S 20 250   CMP 3

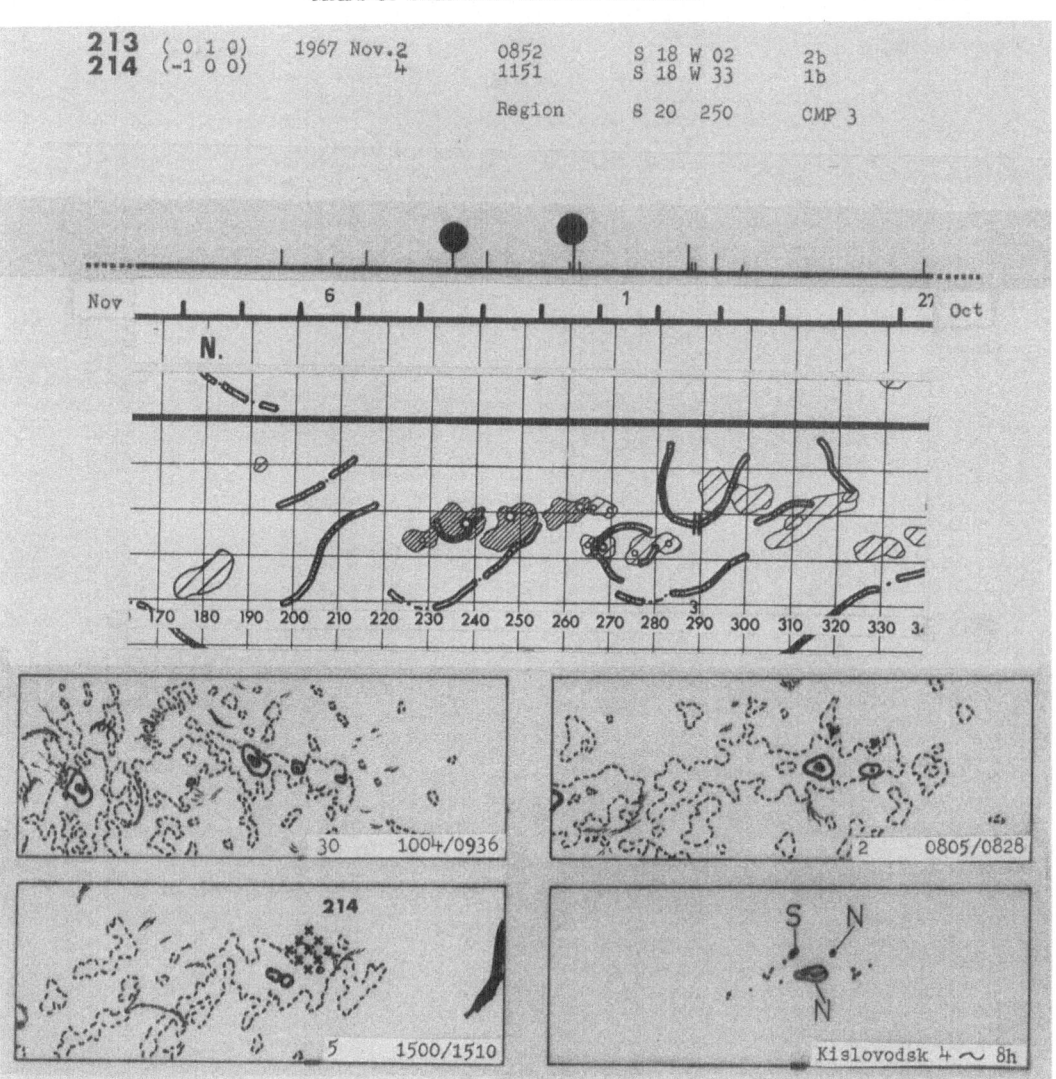

213

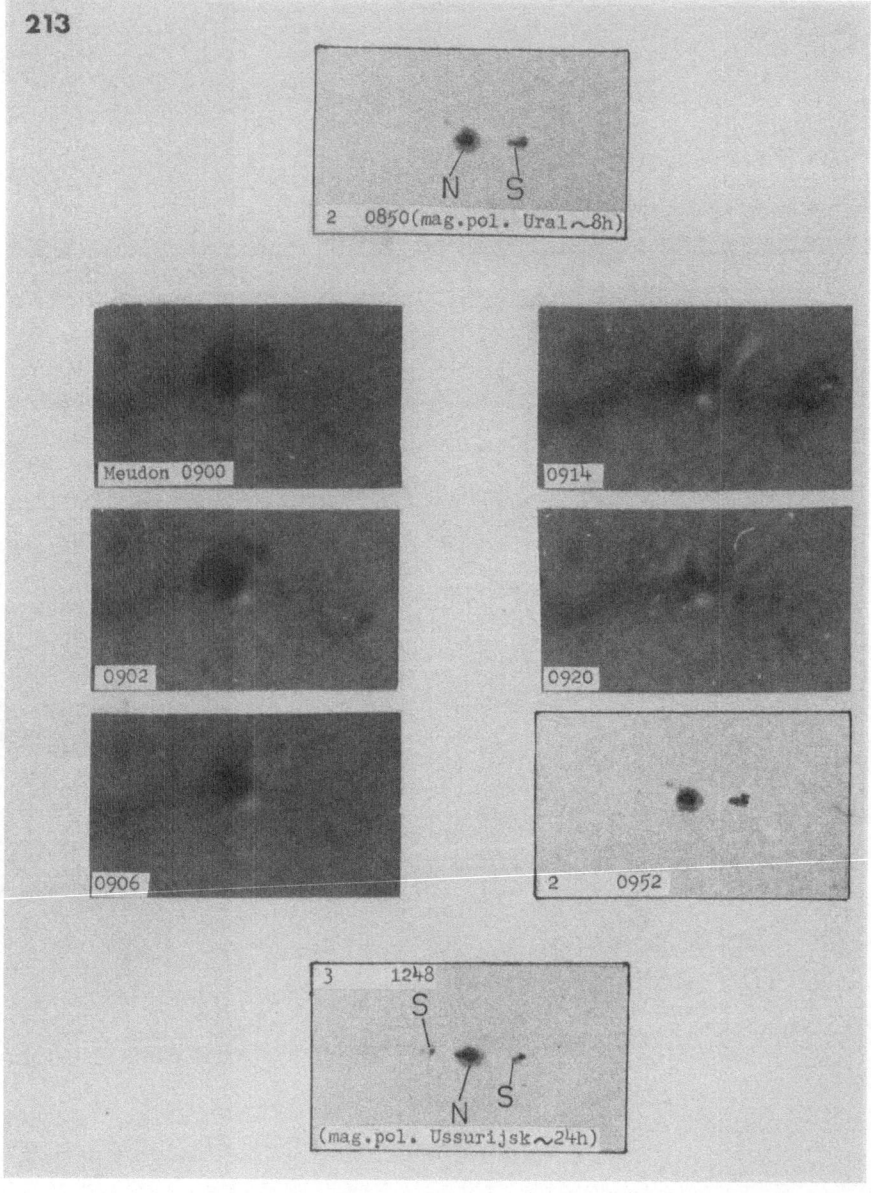

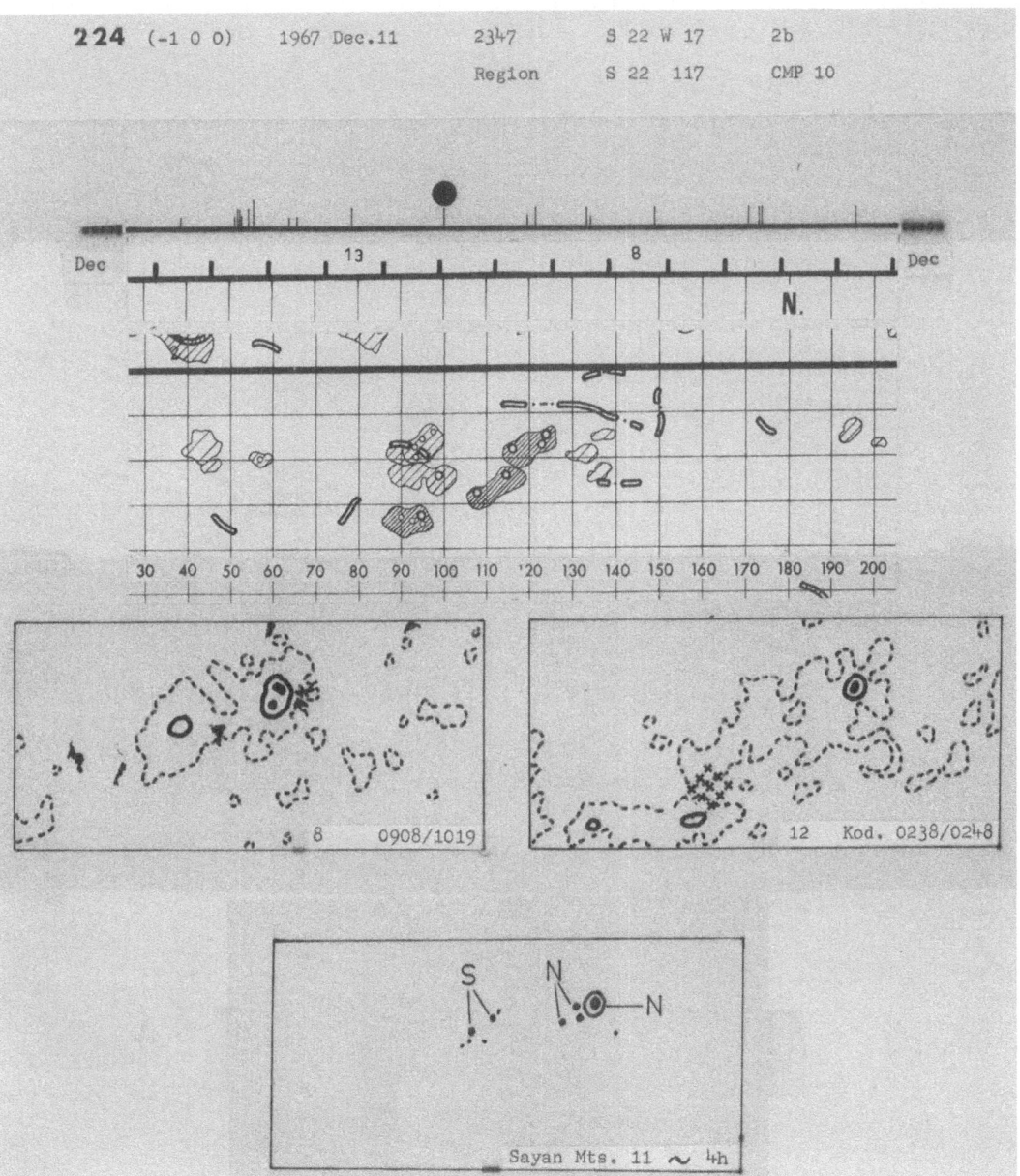

224 (-1 0 0)   1967 Dec.11   2347   S 22 W 17   2b
                 Region   S 22 117   CMP 10

8   0908/1019

12   Kod. 0238/0248

Sayan Mts. 11 ∼ 4h

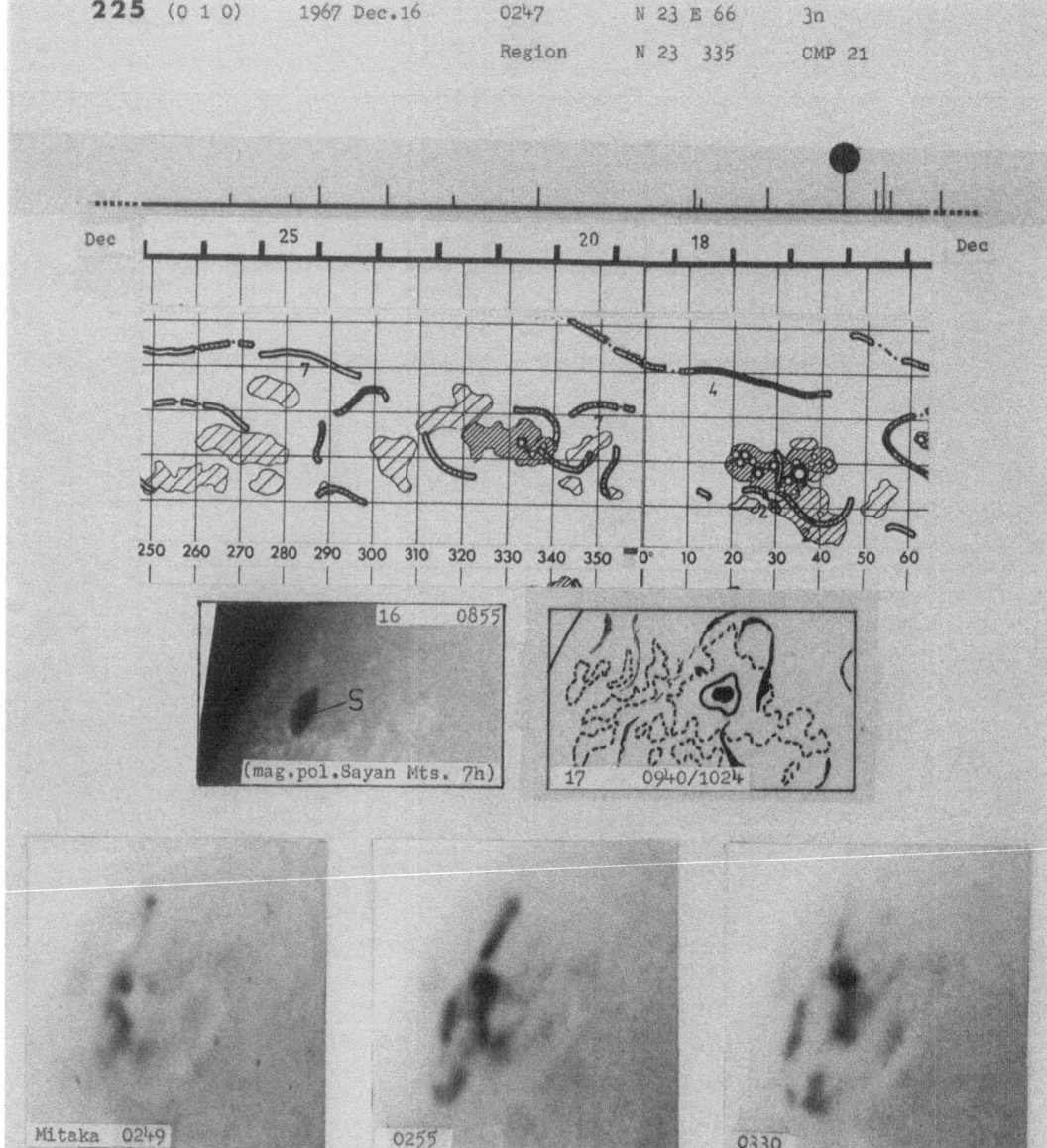

225 (0 1 0)   1967 Dec.16   0247   N 23 E 66   3n

Region   N 23   335   CMP 21

Dec   25   20   18   Dec

250 260 270 280 290 300 310 320 330 340 350 0° 10 20 30 40 50 60

16   0855

S

(mag.pol.Sayan Mts. 7h)

17   0940/1024

Mitaka 0249   0255   0330

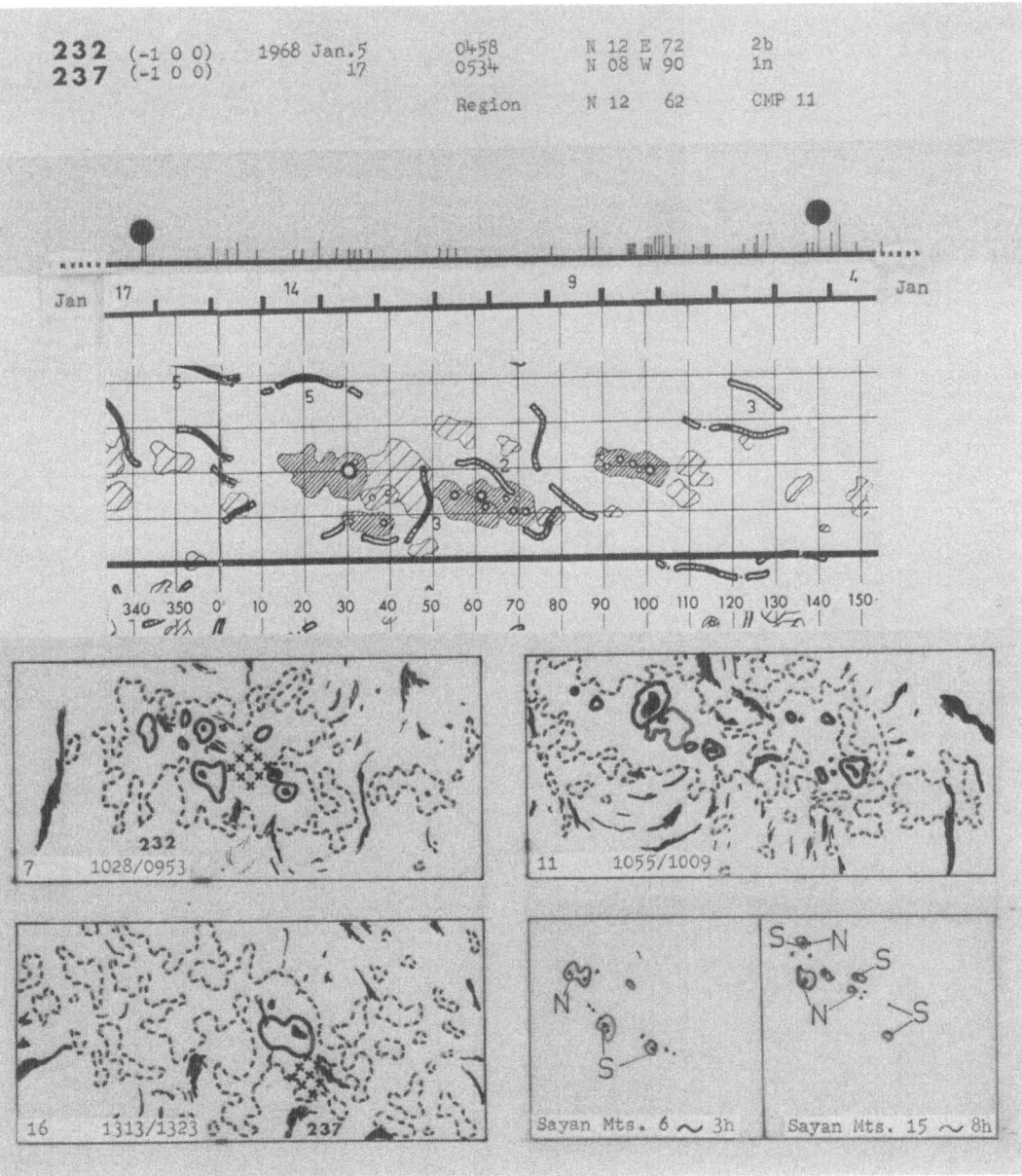

232 (-1 0 0)   1968 Jan.5   0458   N 12 E 72   2b
237 (-1 0 0)   17   0534   N 08 W 90   1n

                                   Region   N 12 62   CMP 11

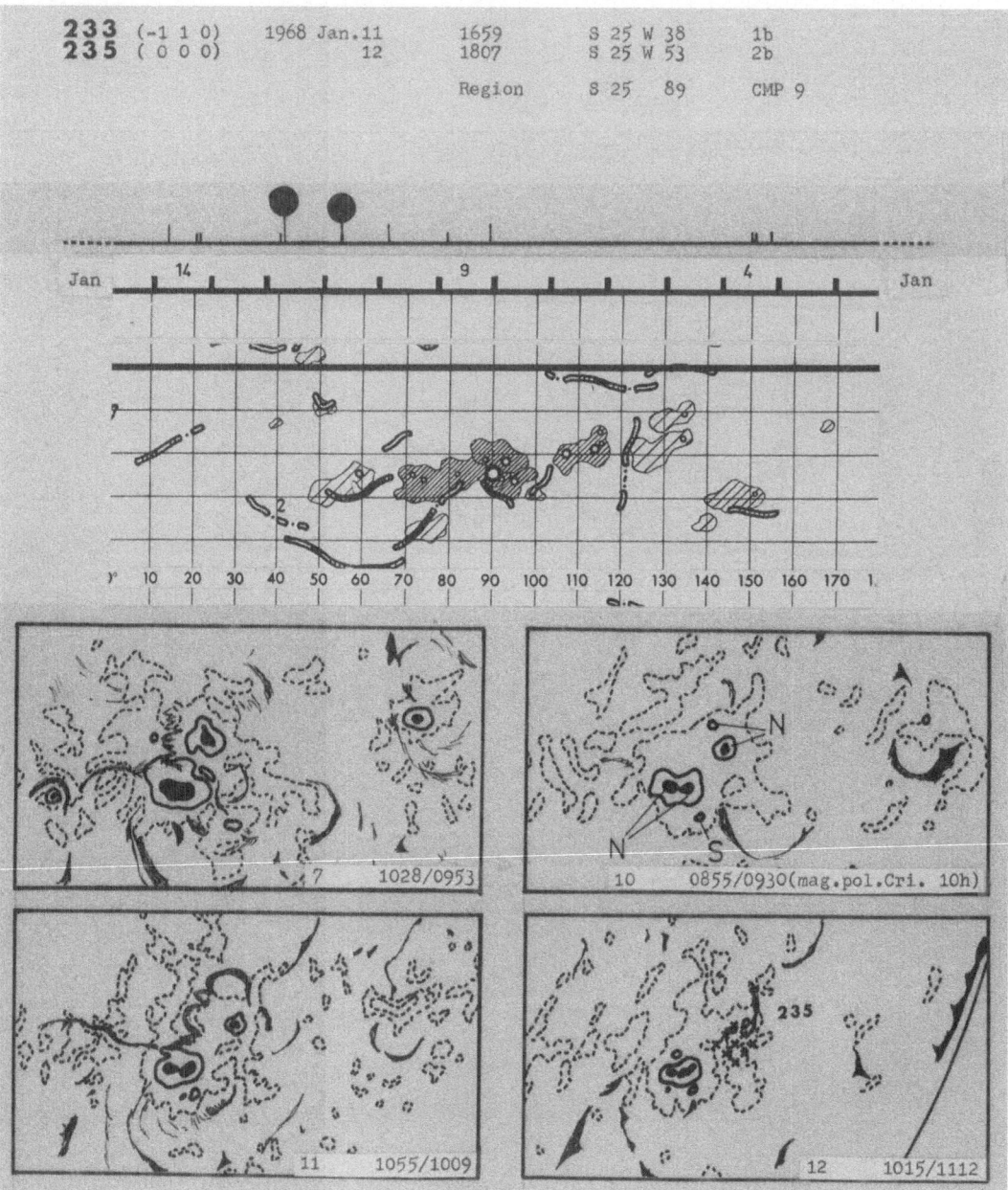

233 (−1 1 0)   1968 Jan.11   1659   S 25 W 38   1b
235 (0 0 0)   12   1807   S 25 W 53   2b

                              Region   S 25   89   CMP 9

233

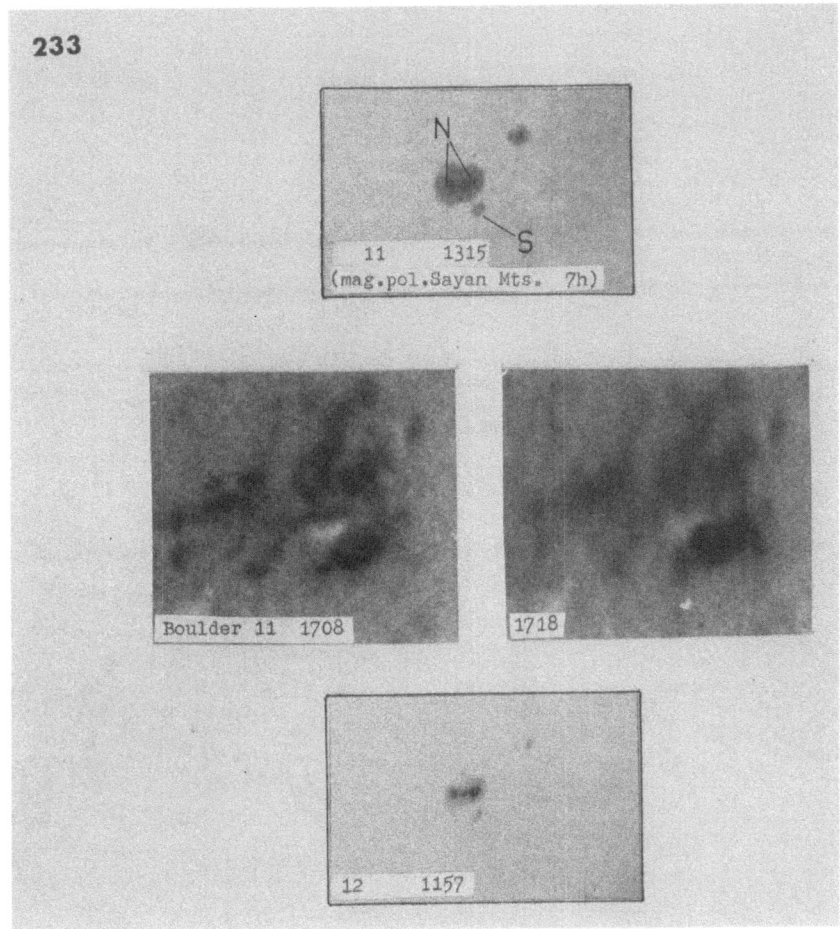

240 (-1 0 0)   1968 Feb.2   0541   N 15 W 24   1b
                               Region   N 13 163   CMP 30

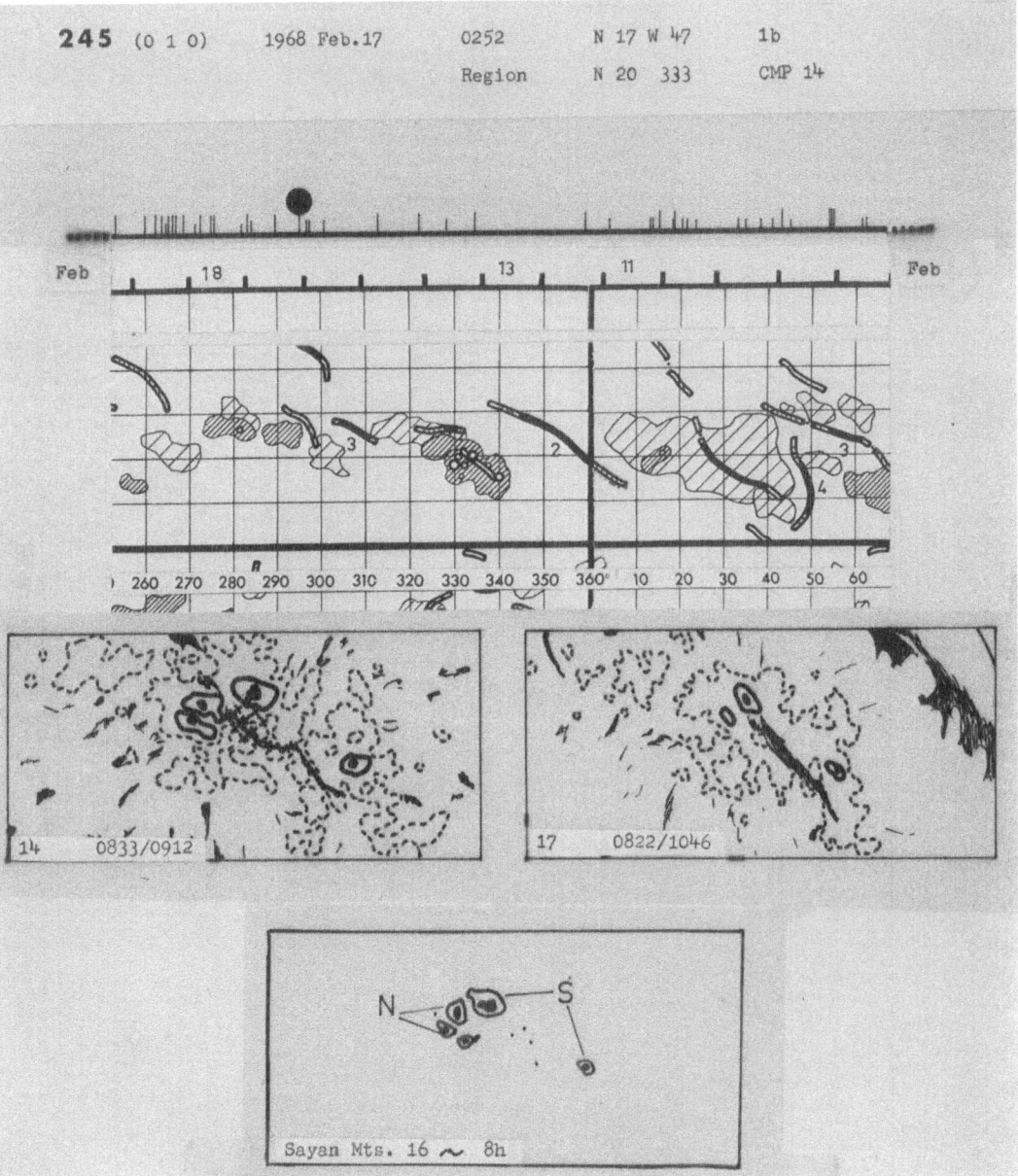

245 (0 1 0) 1968 Feb.17 0252 N 17 W 47 1b

Region N 20 333 CMP 14

Sayan Mts. 16 ∿ 8h

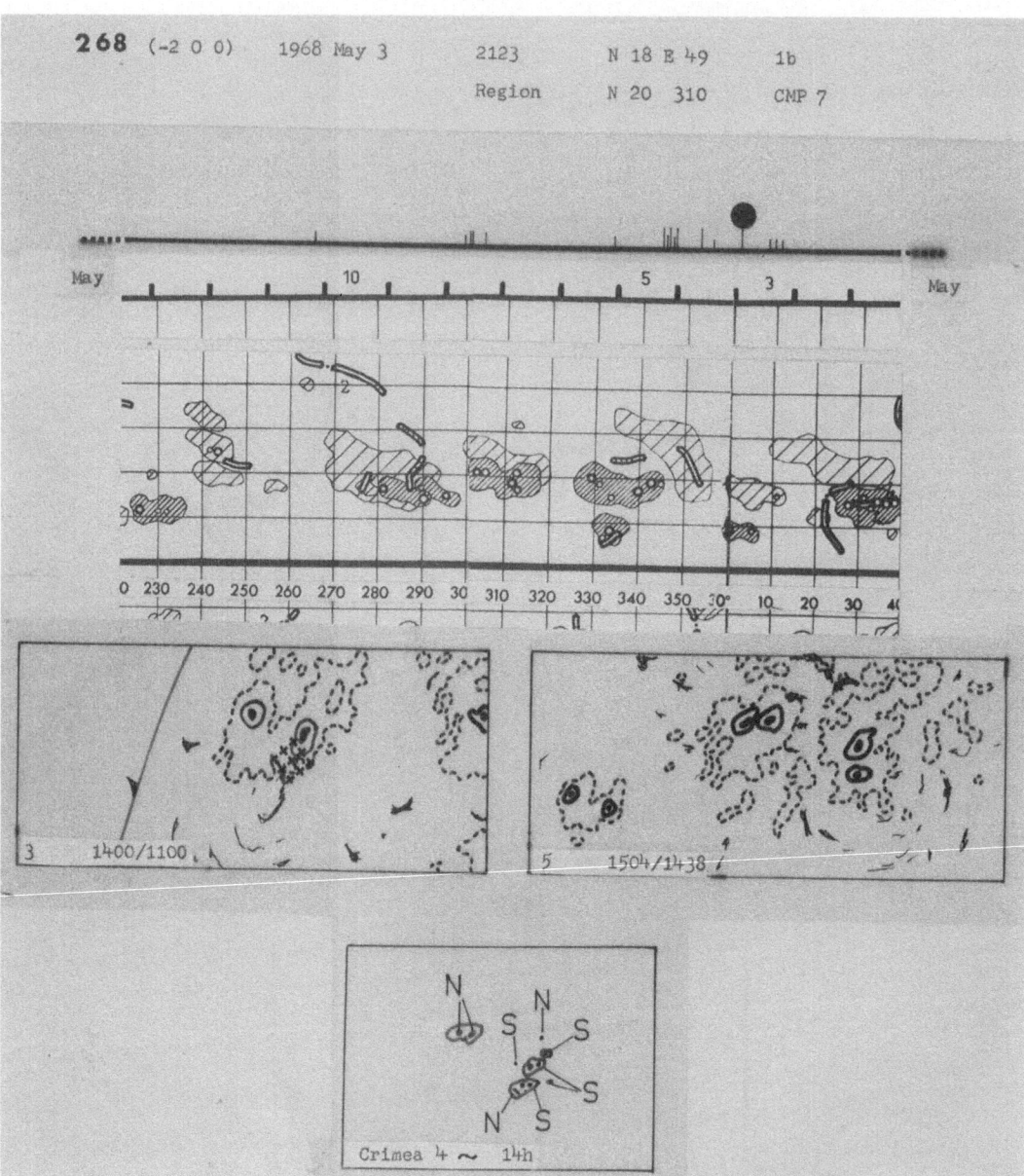

268 (-2 0 0)   1968 May 3   2123   N 18 E 49   1b

Region   N 20 310   CMP 7

3   1400/1100

5   1504/1438

Crimea 4 ~ 14h

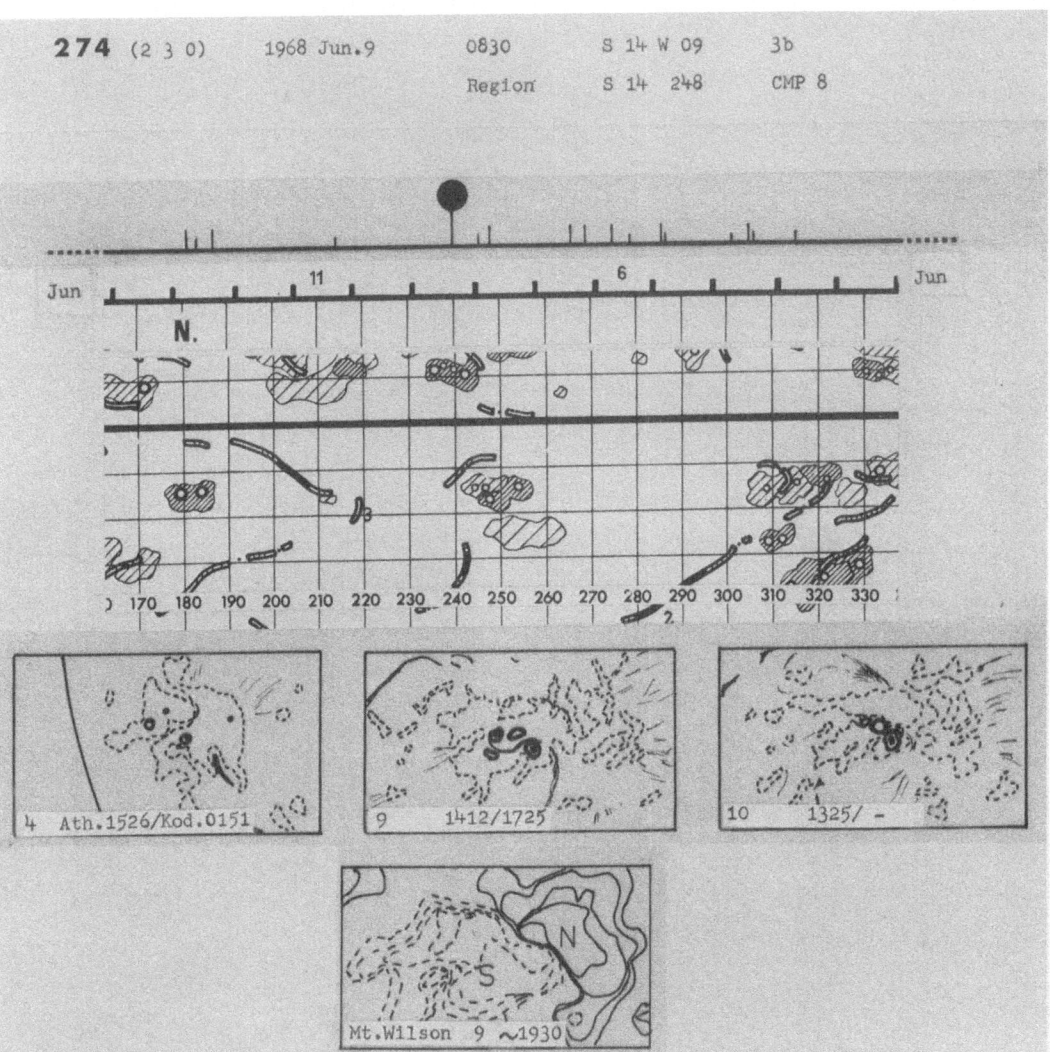

274 (2 3 0) 1968 Jun.9 0830 S 14 W 09 3b

Region S 14 248 CMP 8

4 Ath.1526/Kod.0151

9 1412/1725

10 1325/ –

Mt.Wilson 9 ~1930

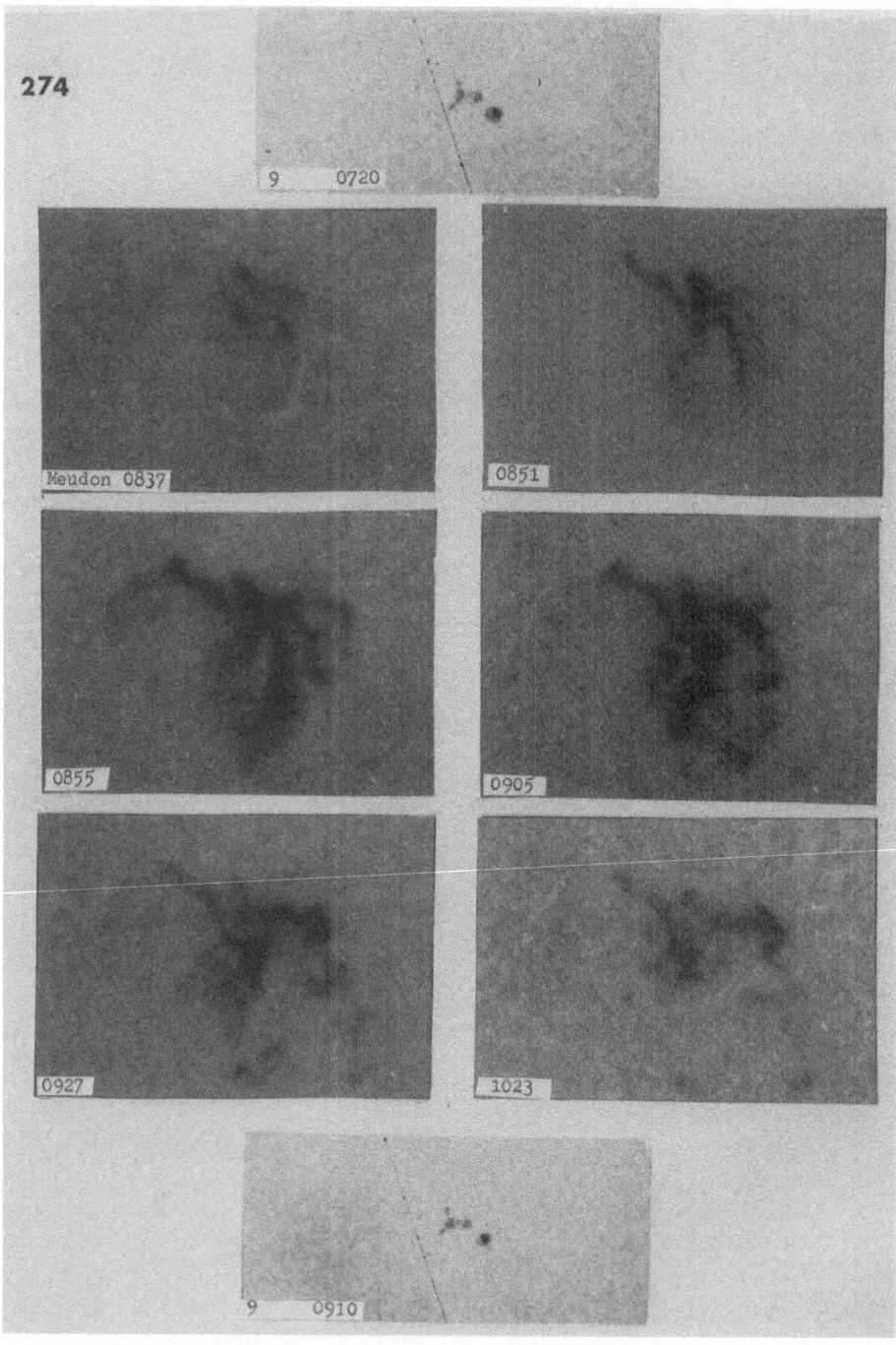

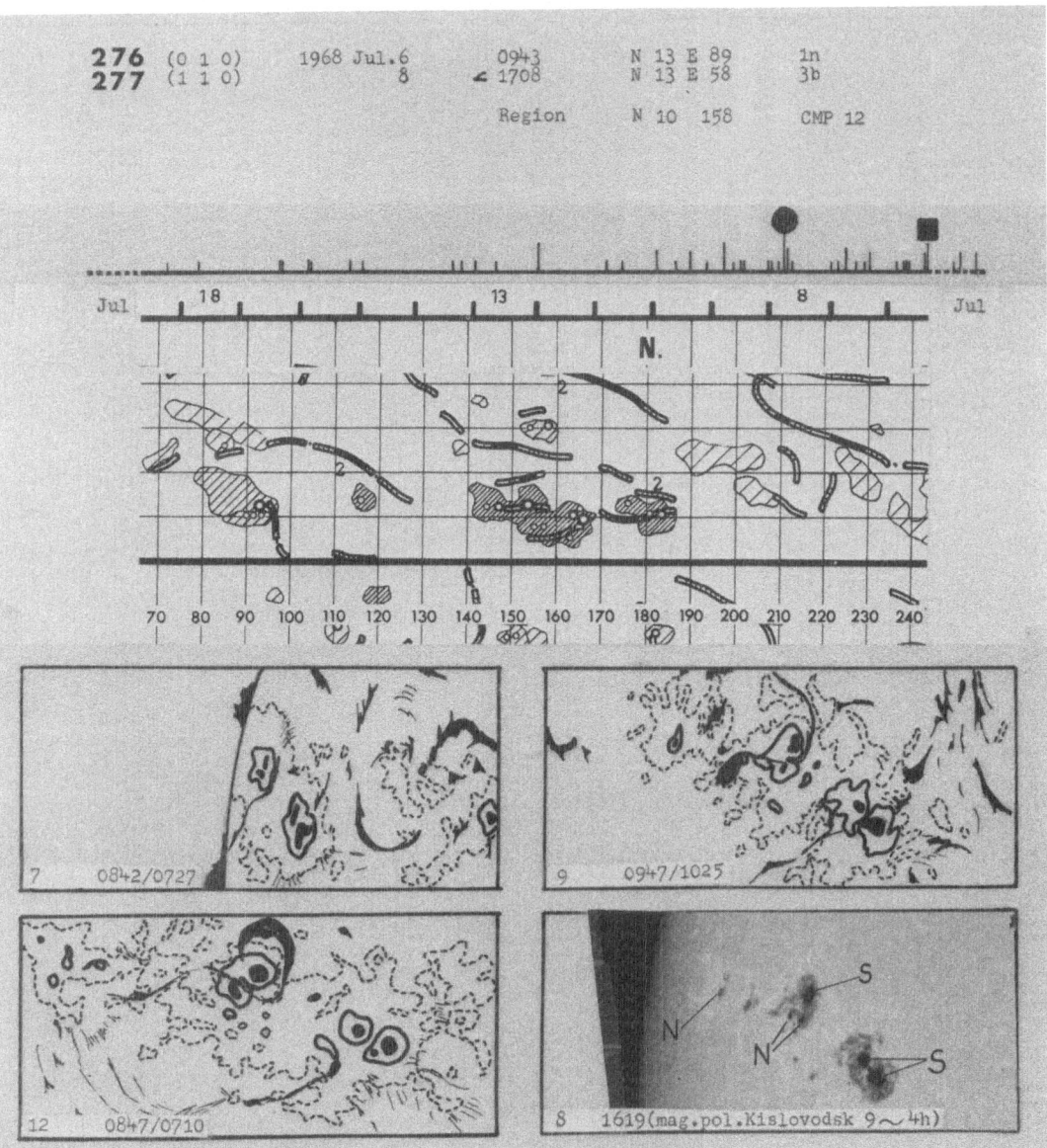

276 (0 1 0) 1968 Jul.6 0943 N 13 E 89 1n
277 (1 1 0) 8 ∠ 1708 N 13 E 58 3b

Region N 10 158 CMP 12

7 0842/0727
9 0947/1025
12 0847/0710
8 1619(mag.pol.Kislovodsk 9∼4h)

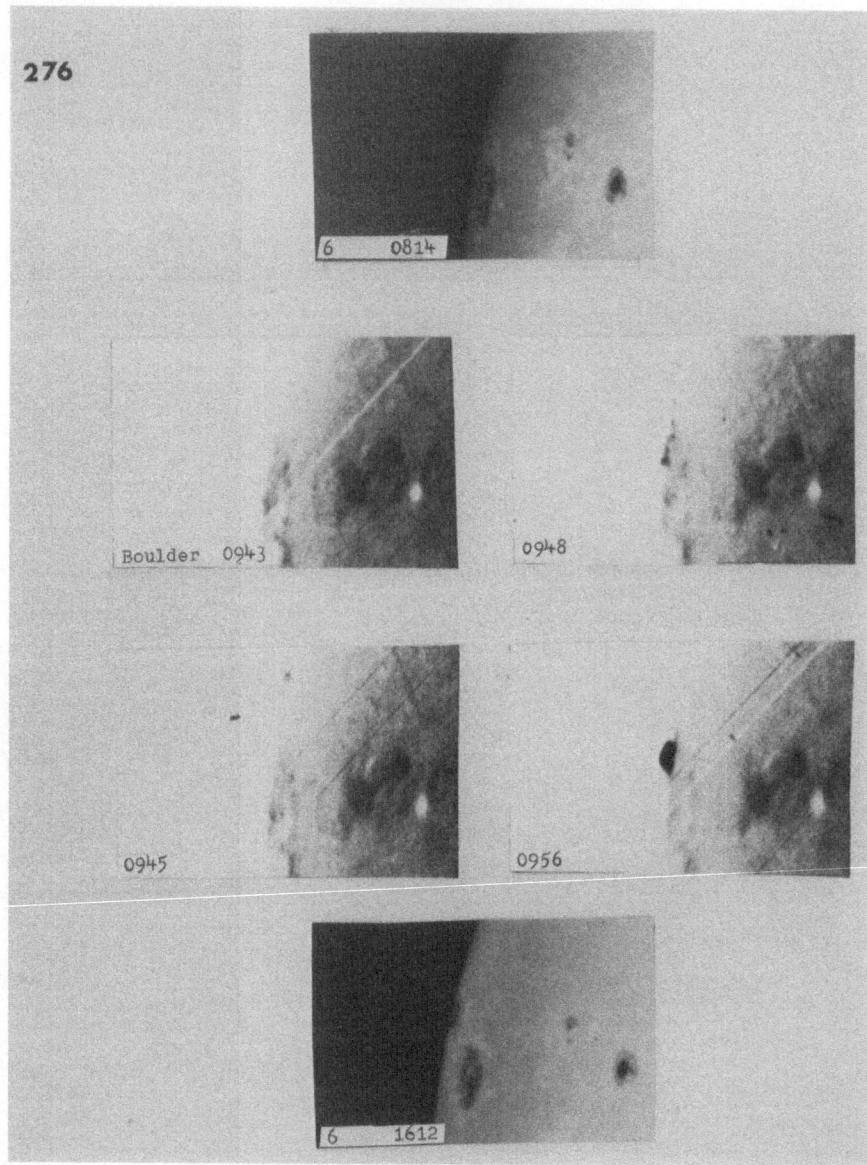

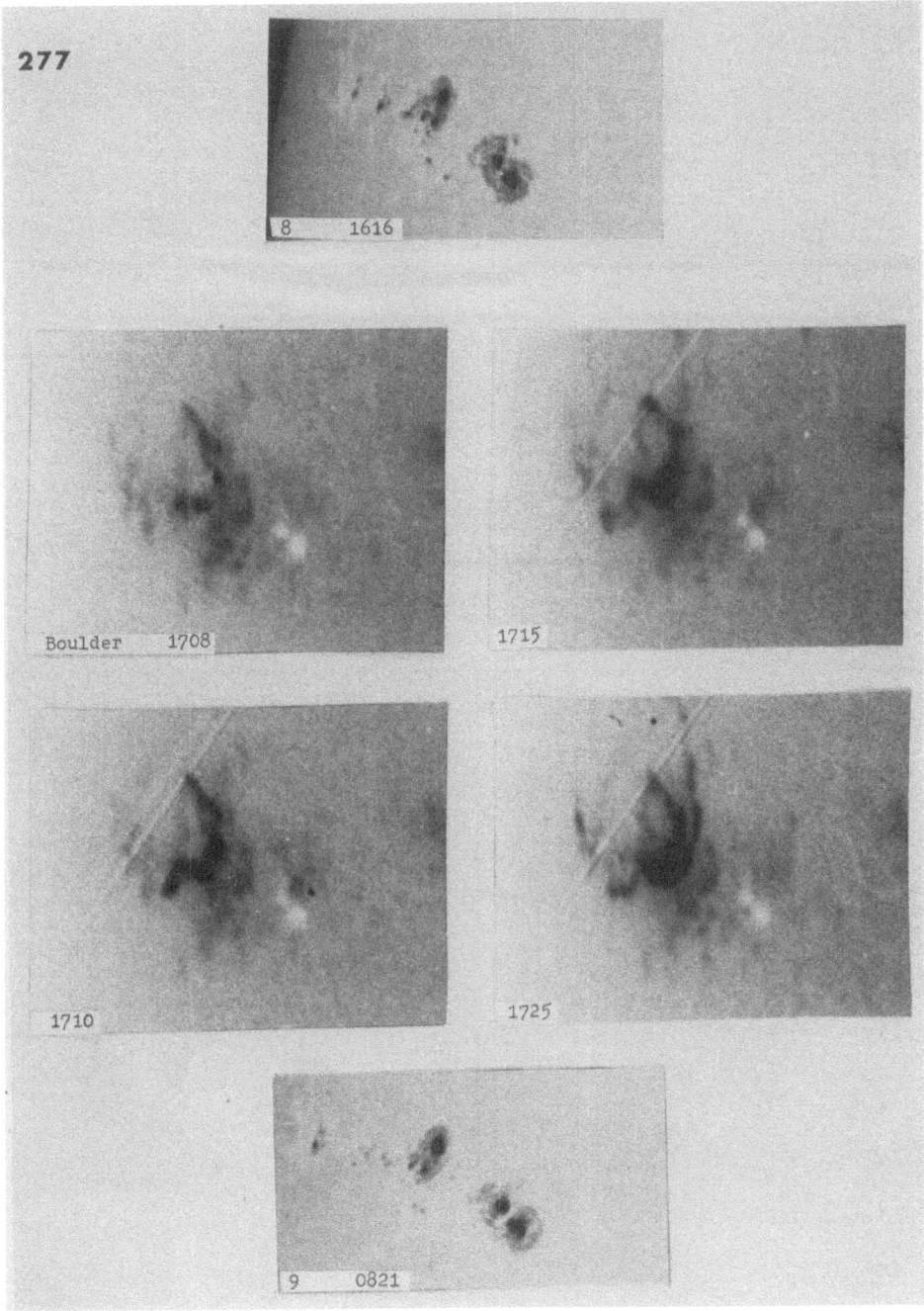

281 (-2 0 0)   1968 Aug.3   0313   N 10 E 75   2n
284 (-1 0 0)   14   ∠1327   N 13 W 80   1b
                               1400   N 13 W 82   1n

                               Region   N 12 160   CMP 8

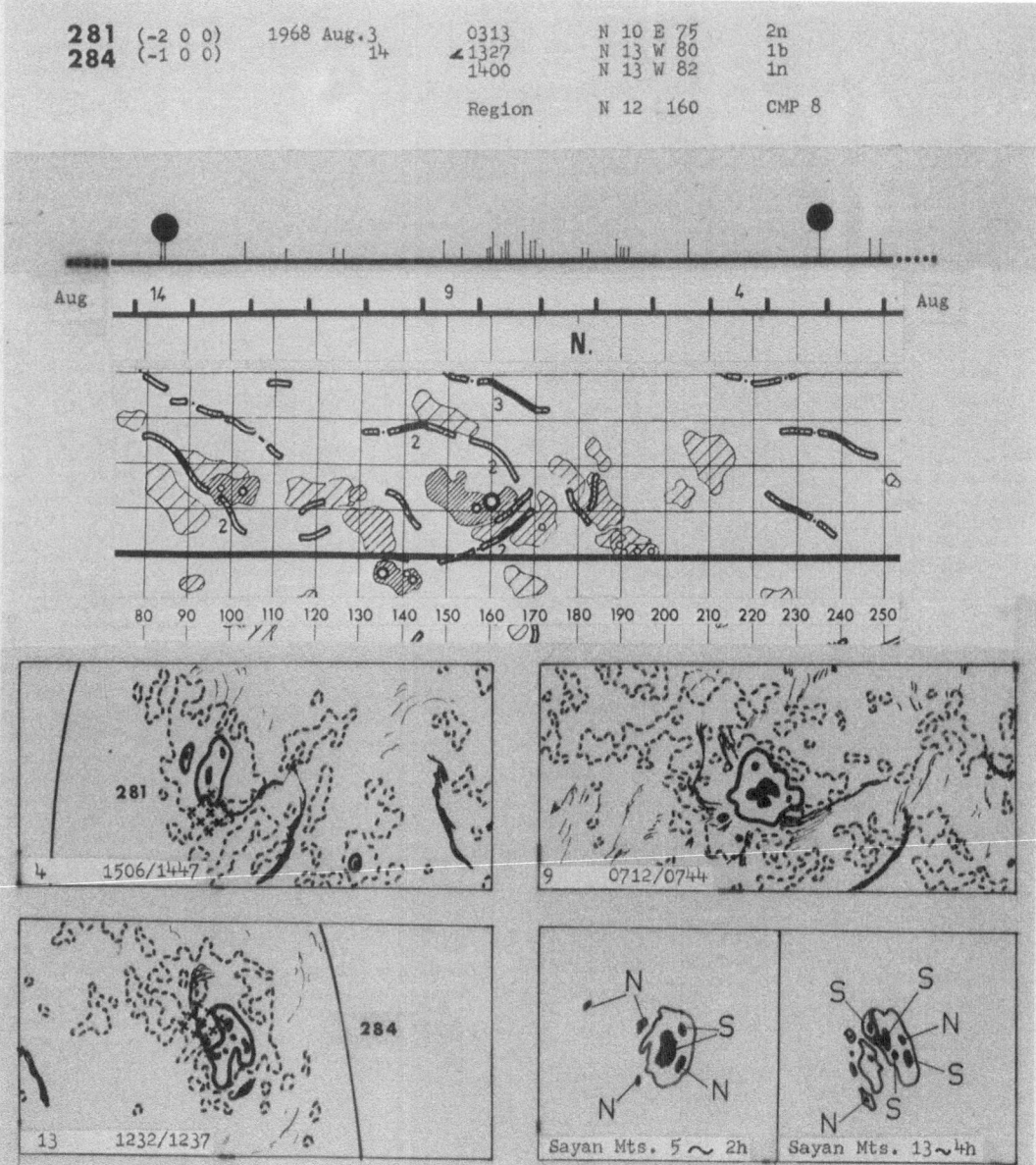

287 (1 1 0)   1968 Sep.26   0026   N 14 E 34   2b
Region   N 12 205   CMP 28

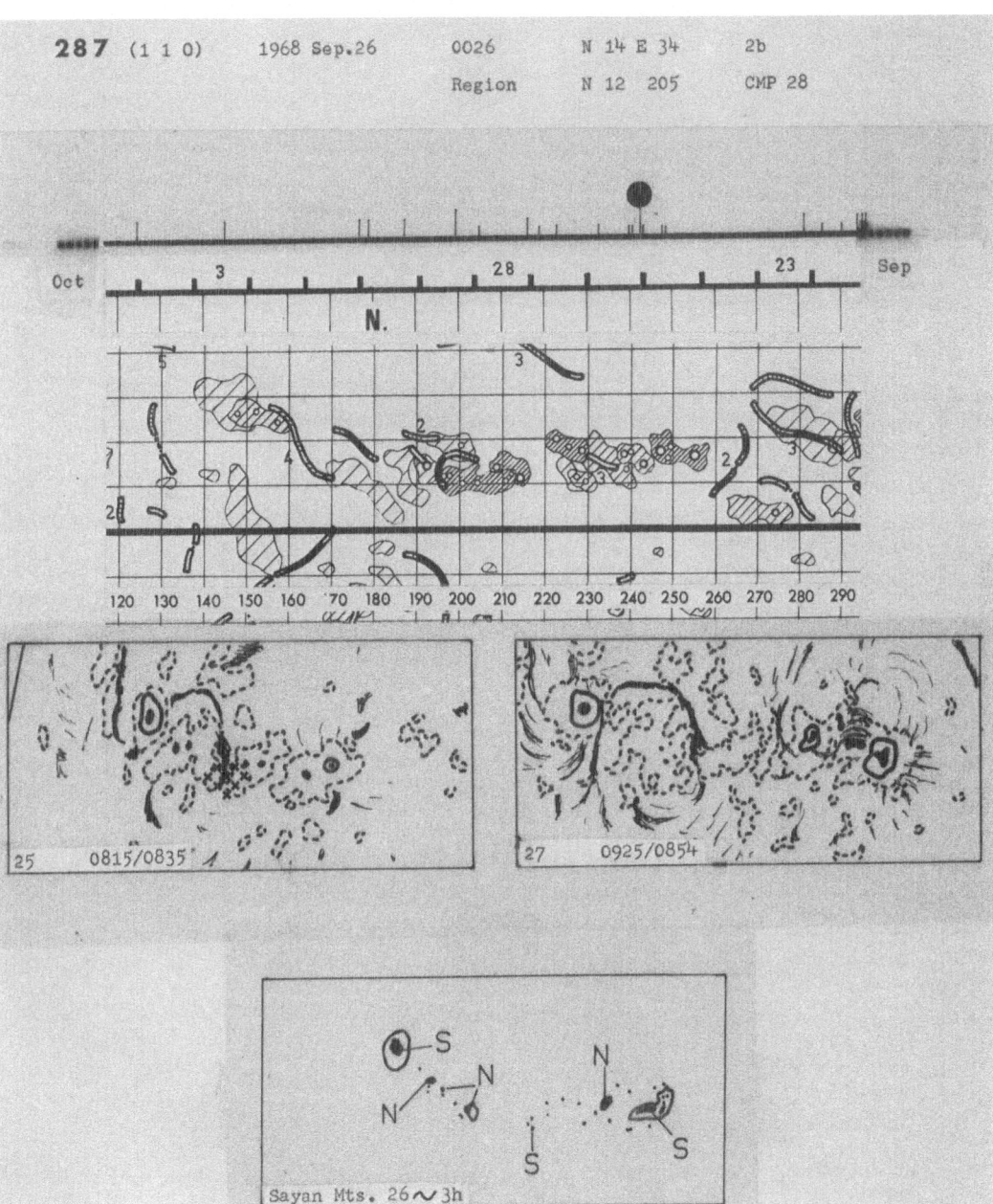

25   0815/0835

27   0925/0854

Sayan Mts. 26 ∿ 3h

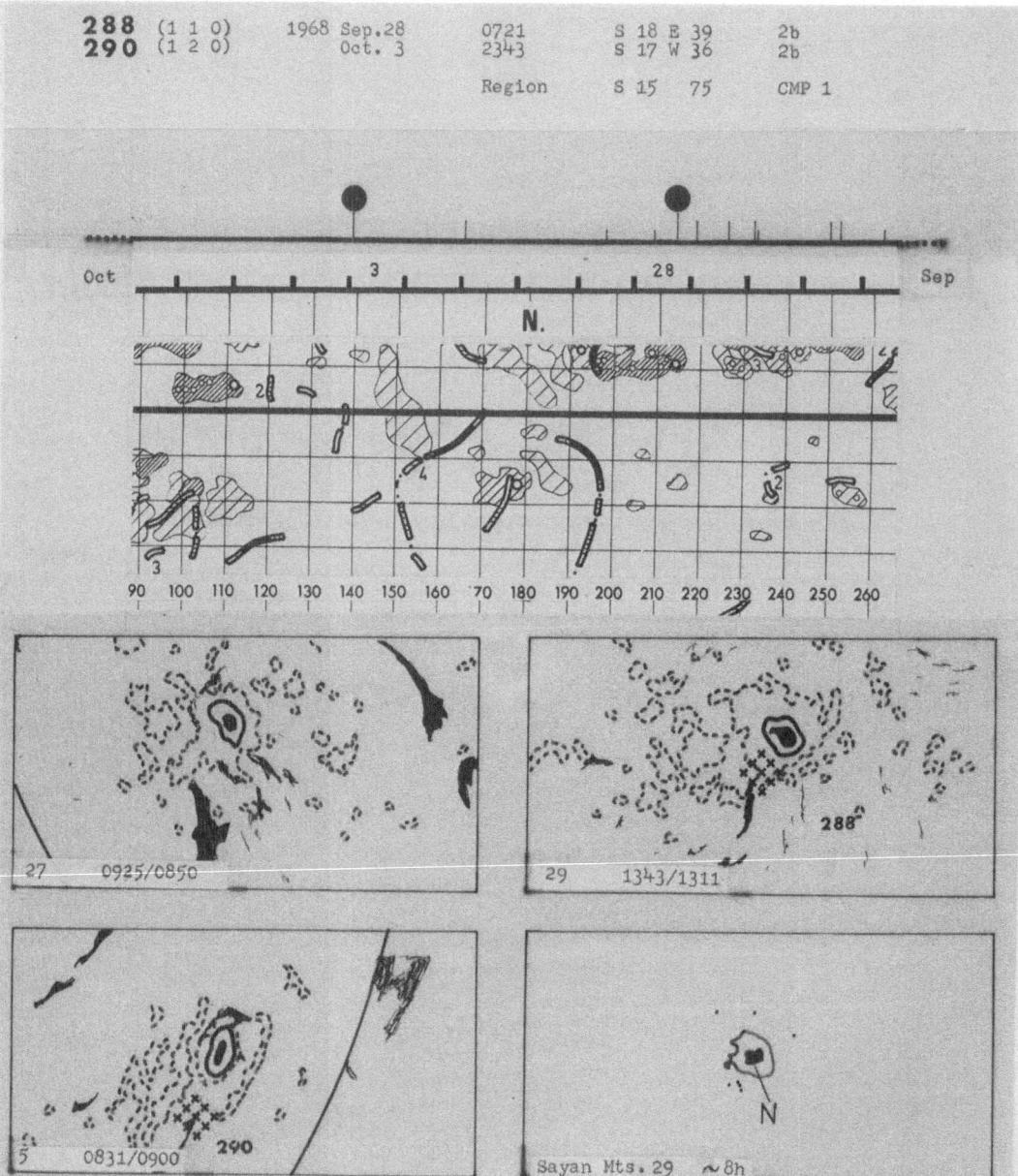

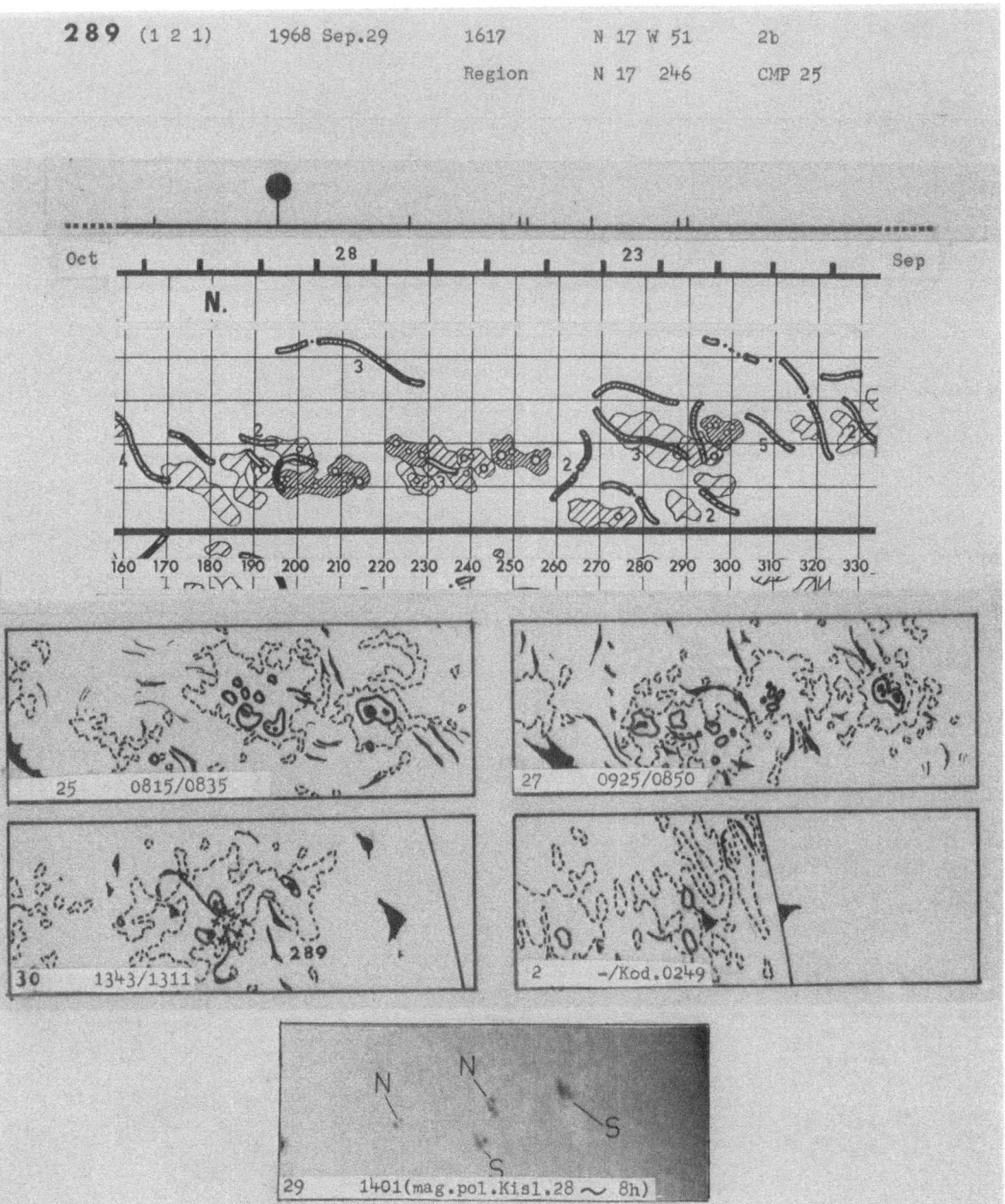

289 (1 2 1) 1968 Sep.29 1617 N 17 W 51 2b

Region N 17 246 CMP 25

293	291 295	(0 0 0) (-1 0 0) (0 0 0)	1968 Oct.23 27 30	2352 1232 < 1318 < 1235	S 13 E 59 S 17 E 17 S 18 W 26	2n 1b 2n 2b
296	297	(2 3 0) (2 3 0)	30 Nov. 1	2340 0814	S 14 W 37 S 16 W 47	3b 2n
298		(1 2 0)	4	< 0524	S 15 W 90	1b

Region   S 15 172   CMP 28

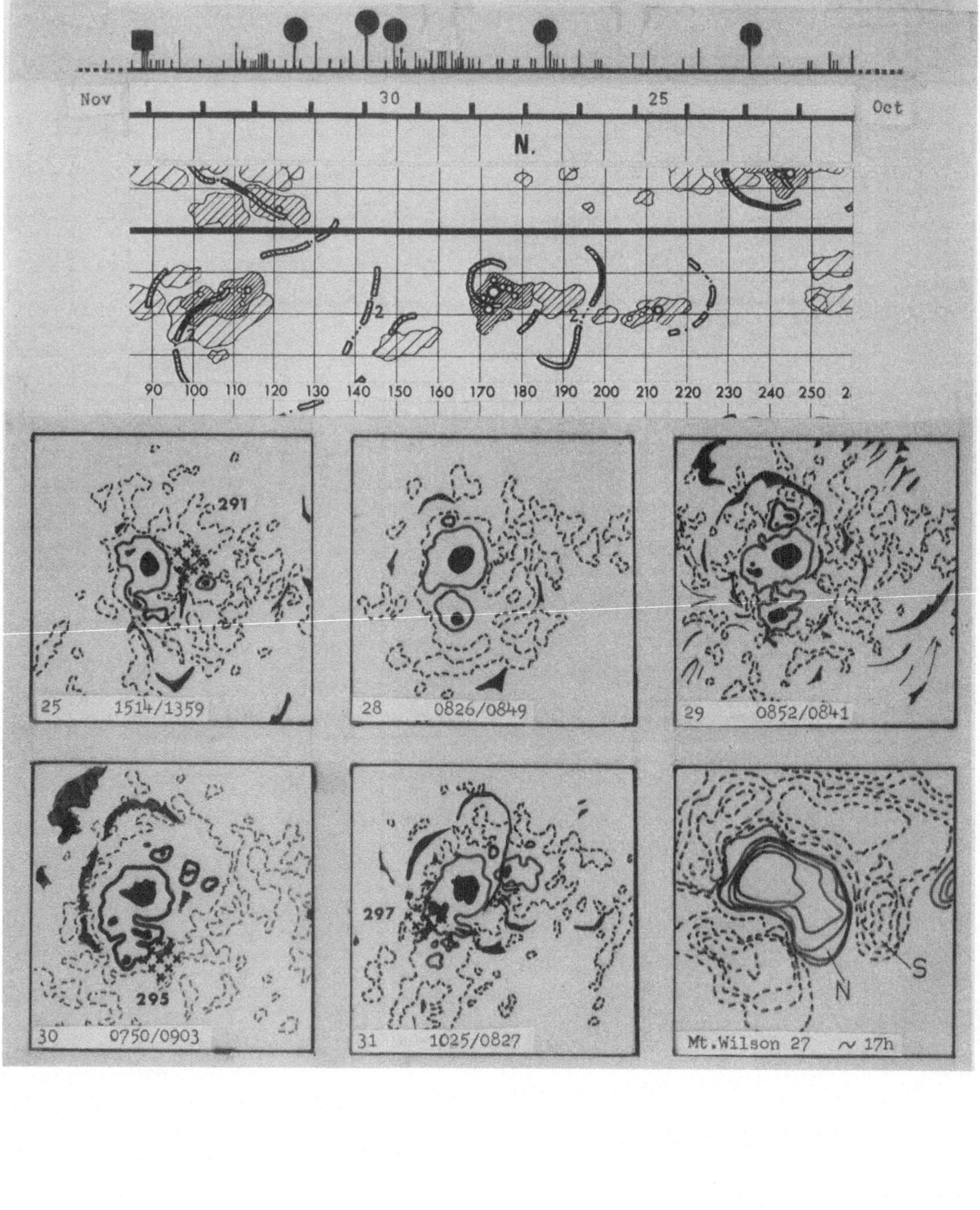

25   1514/1359

28   0826/0849

29   0852/0841

30   0750/0903

31   1025/0827

Mt.Wilson 27   ~ 17h

293

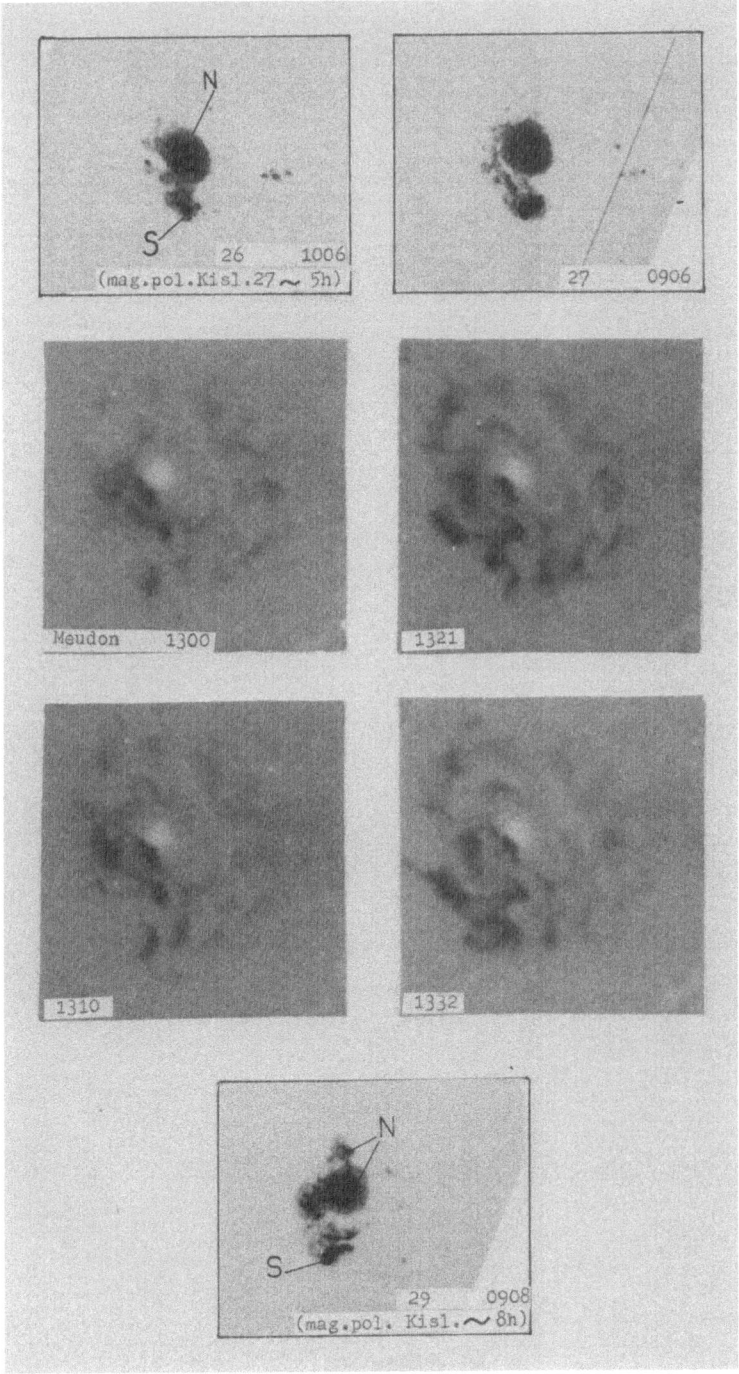

296

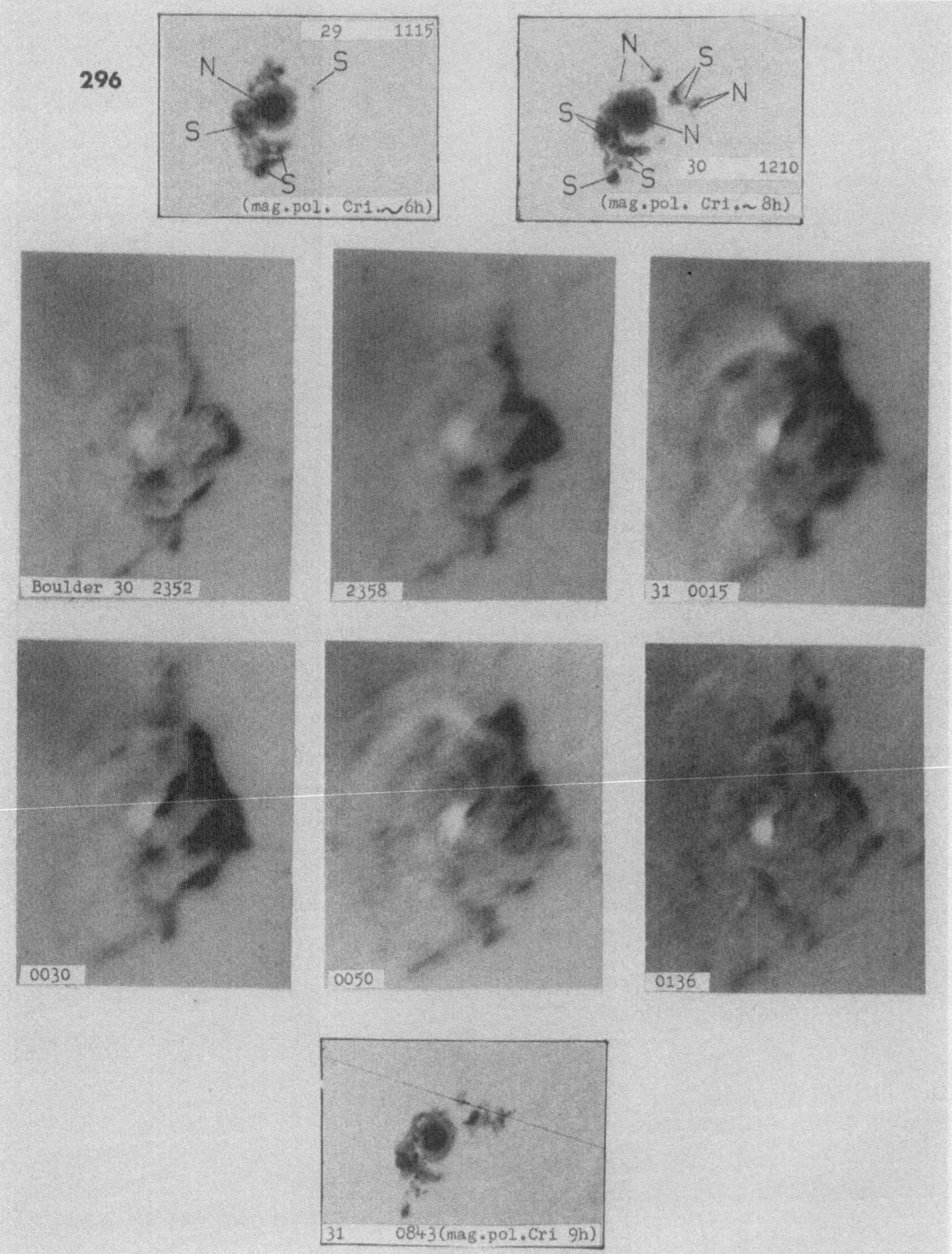

29   1115
N
S
S
S
(mag.pol. Cri. ~6h)

N
S
N
S
N
S
S
30   1210
S
(mag.pol. Cri. ~8h)

Boulder 30   2352

2358

31 0015

0030

0050

0136

31   0843(mag.pol.Cri 9h)

.298

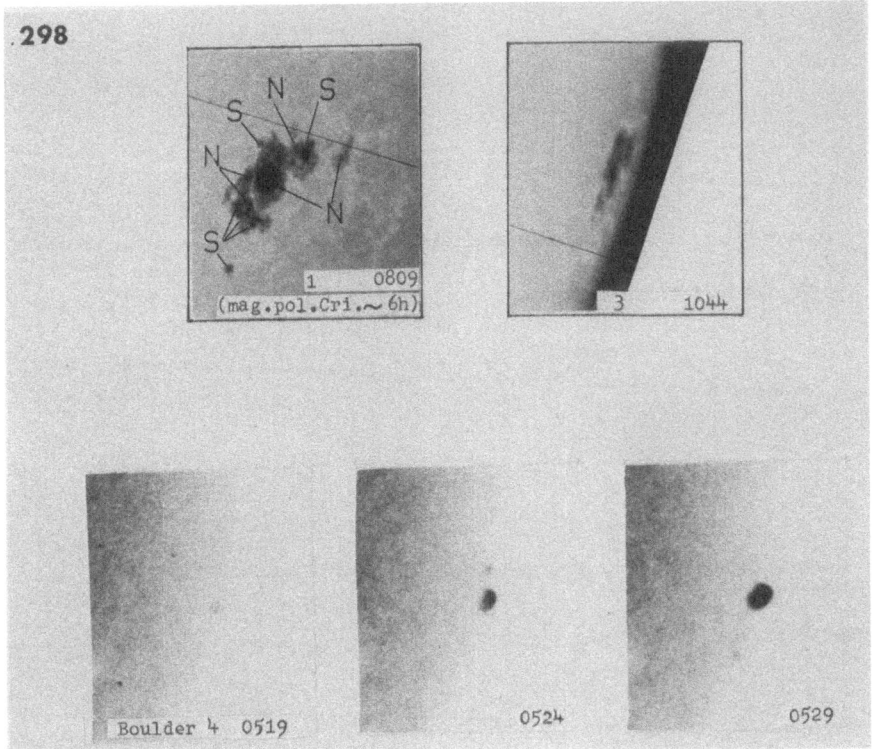

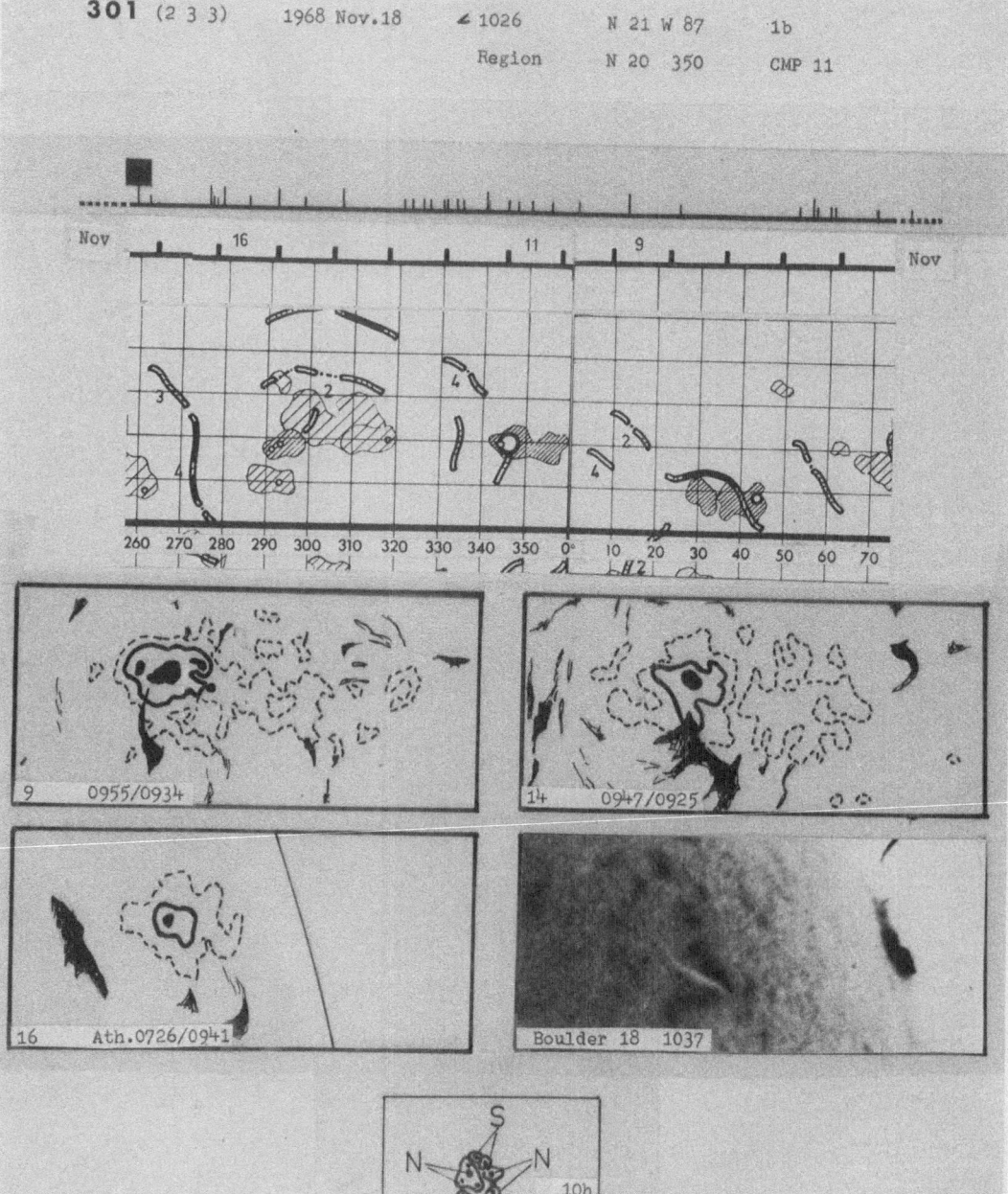

301 (2 3 3)   1968 Nov.18   ∠ 1026   N 21 W 87   1b
Region   N 20   350   CMP 11

9   0955/0934

14   0947/0925

16   Ath.0726/0941

Boulder 18   1037

Crimea 15   10h

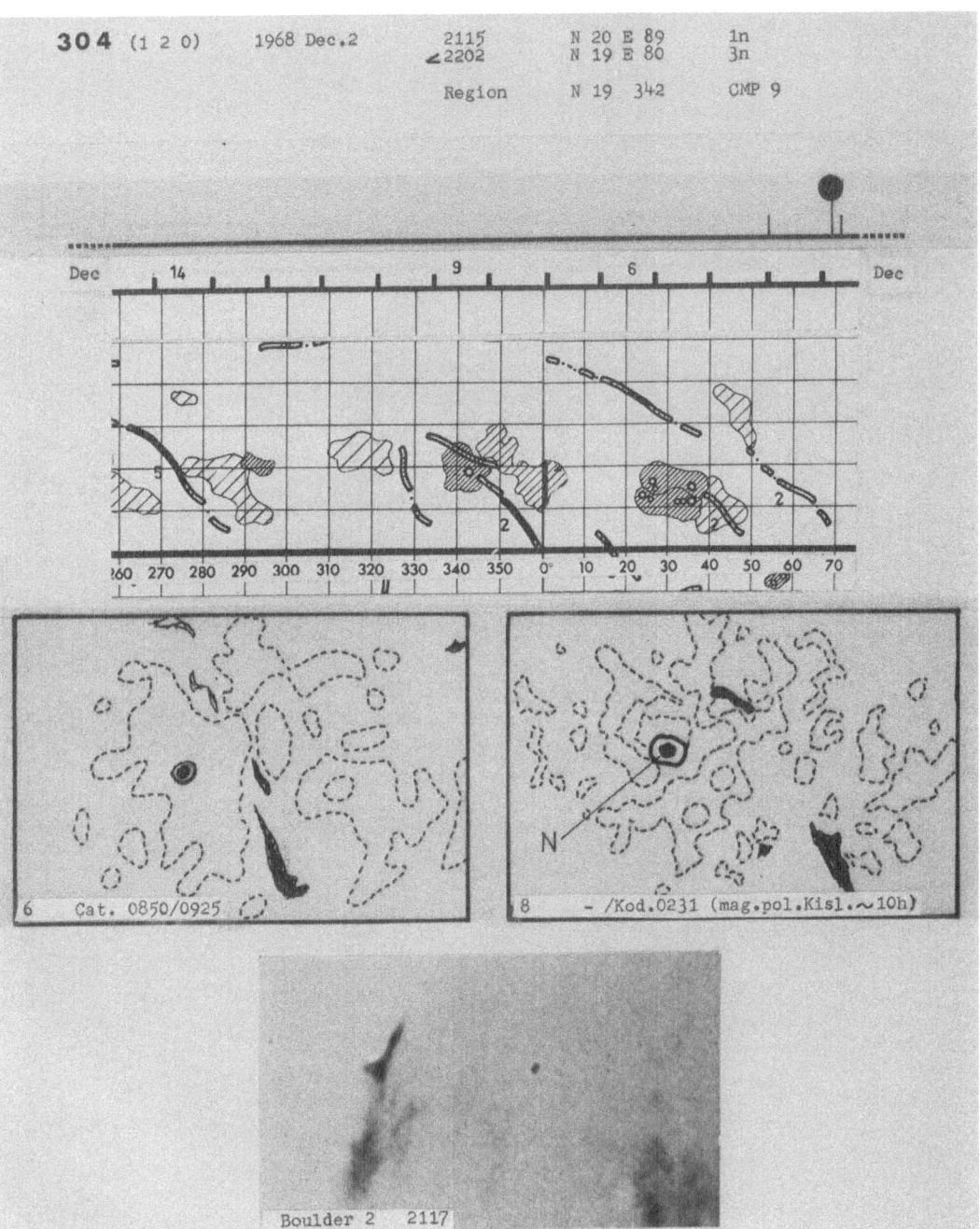

304 (1 2 0)   1968 Dec.2   2115   N 20 E 89   1n
                        2202   N 19 E 80   3n

                Region   N 19 342   CMP 9

6   Çat. 0850/0925

8   - /Kod.0231 (mag.pol.Kisl. ~10h)

Boulder 2   2117

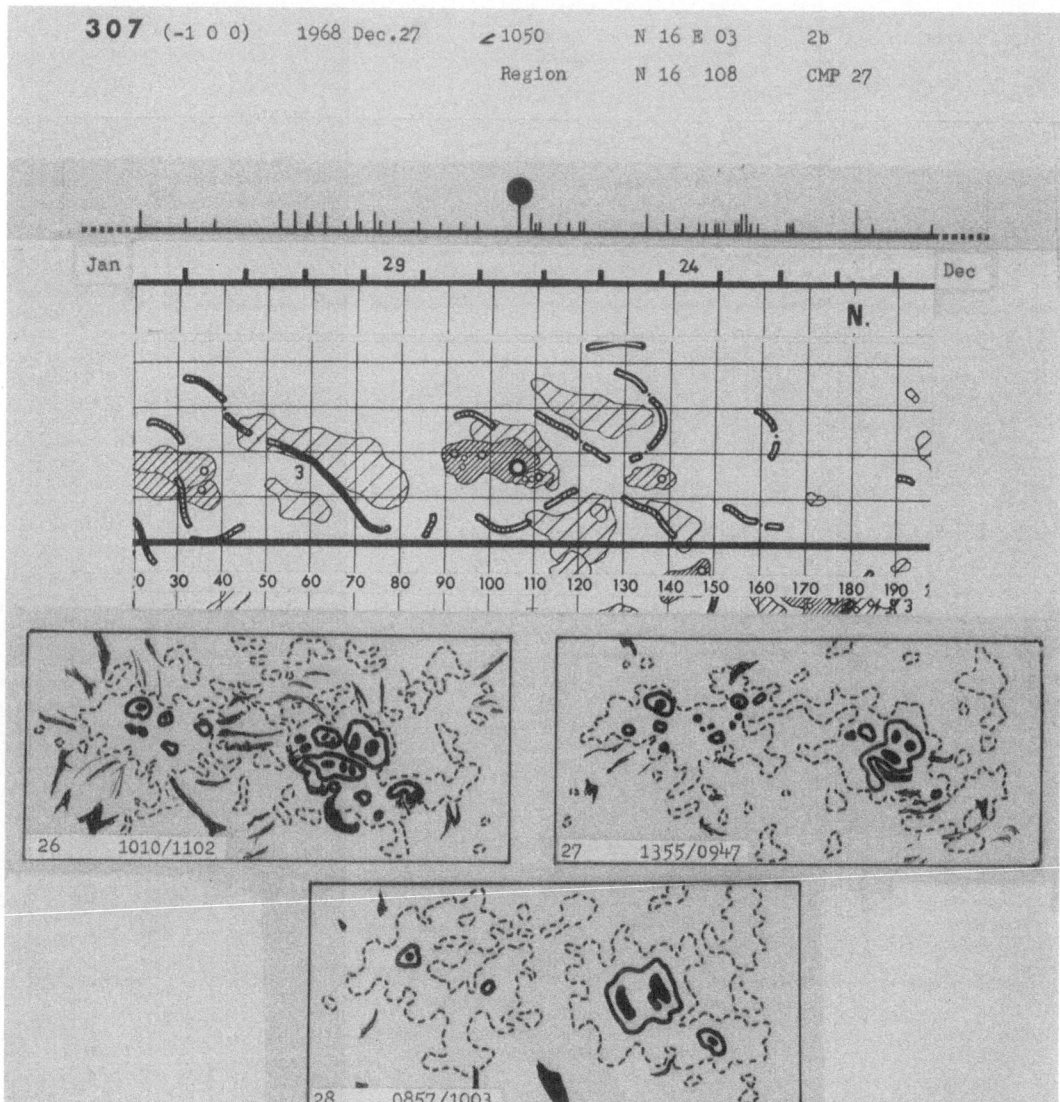

307 (-1 0 0) 1968 Dec.27 ∠1050 N 16 E 03 2b

Region N 16 108 CMP 27

307

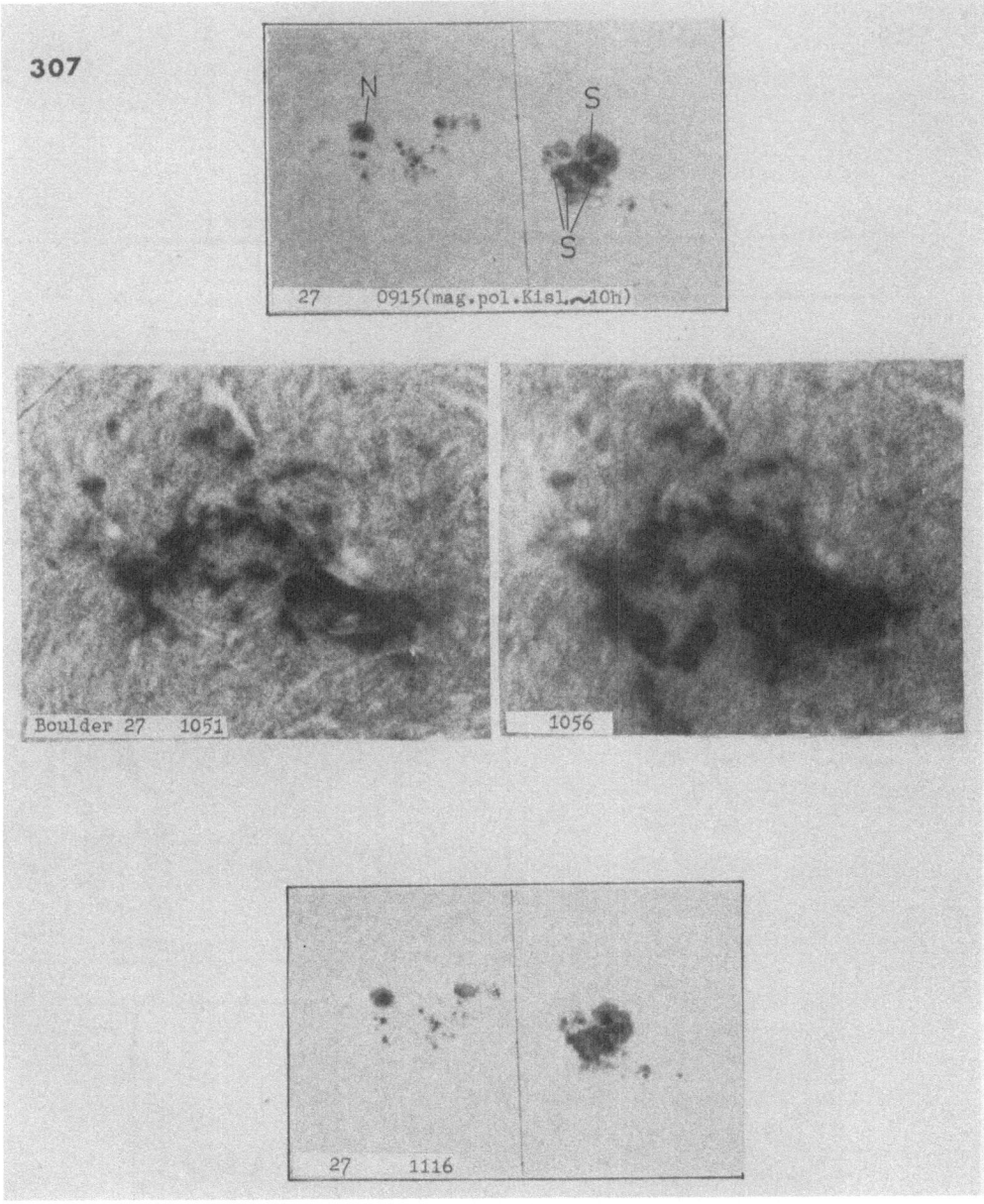

309 (0 1 0) 1969 Jan.24 ∠ 0706 N 20 W 08 3b

Region N 18 102 CMP 23

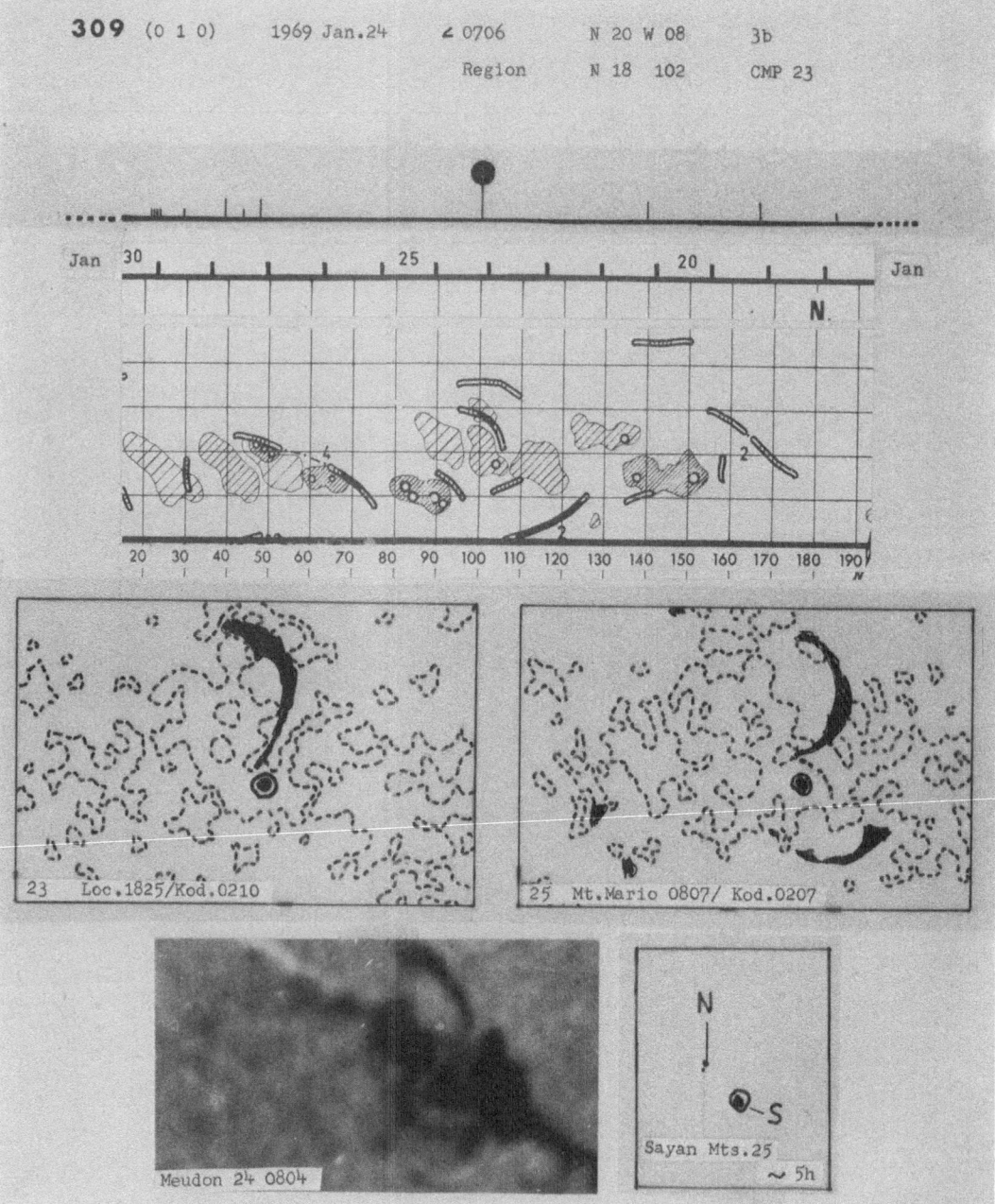

23 Loc.1825/Kod.0210

25 Mt.Mario 0807/ Kod.0207

Meudon 24 0804

Sayan Mts.25
~ 5h

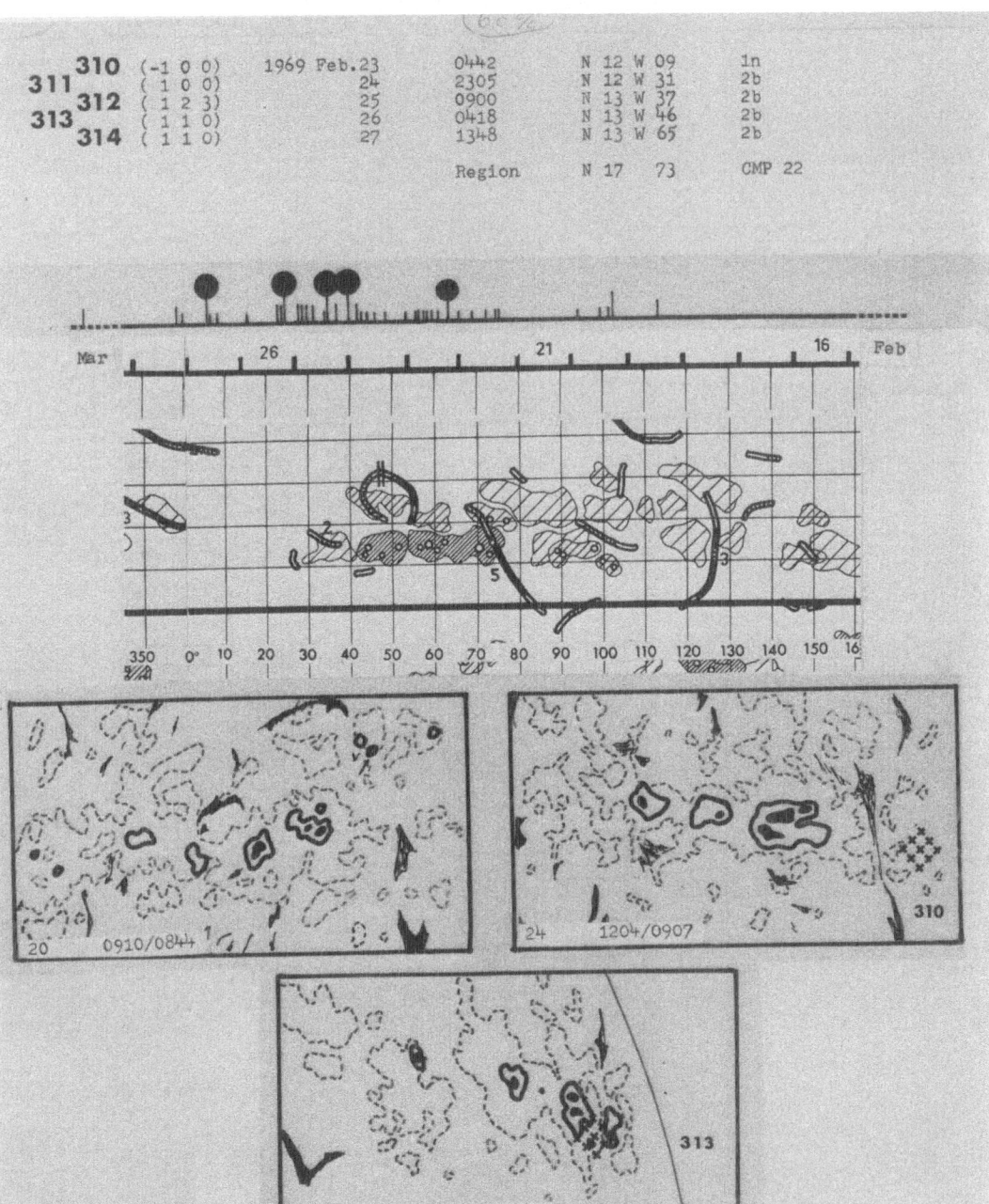

310	(-1 0 0)	1969 Feb.23	0442	N 12 W 09	1n
311	(1 0 0)	24	2305	N 12 W 31	2b
312	(1 2 3)	25	0900	N 13 W 37	2b
313	(1 1 0)	26	0418	N 13 W 46	2b
314	(1 1 0)	27	1348	N 13 W 65	2b

Region   N 17   73   CMP 22

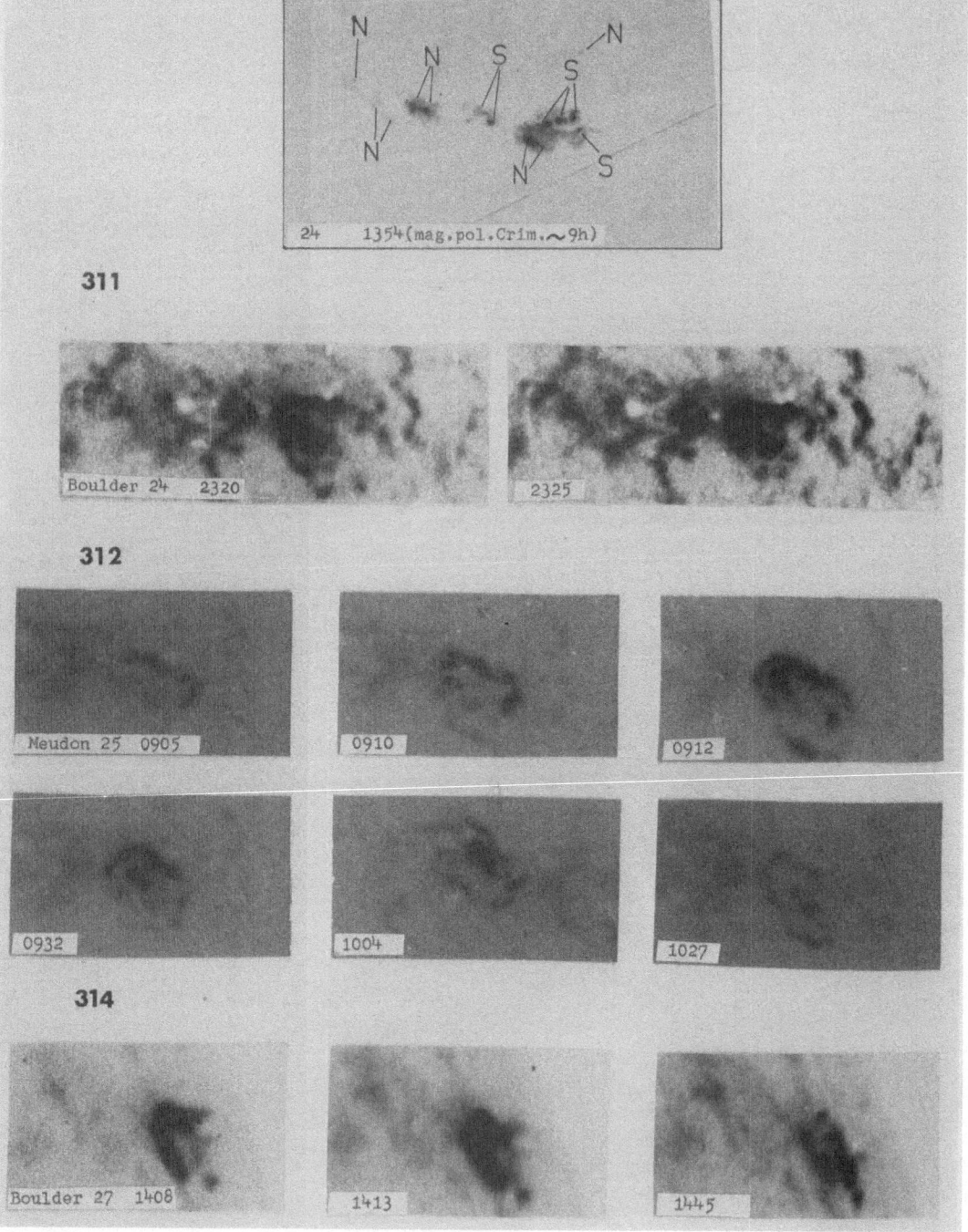

24   1354(mag.pol.Crim.～9h)

311

Boulder 24   2320

2325

312

Neudon 25   0905

0910

0912

0932

1004

1027

314

Boulder 27   1408

1413

1445

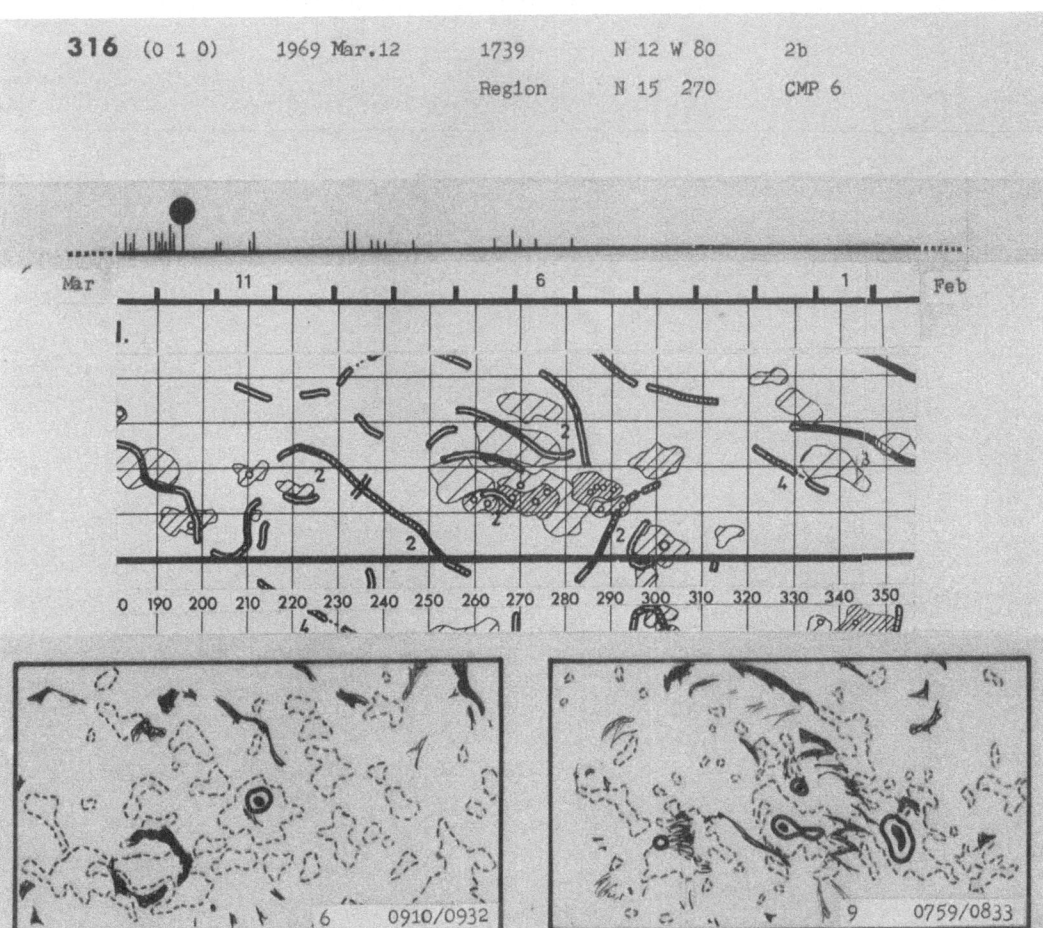

316 (0 1 0) 1969 Mar.12 1739 N 12 W 80 2b

Region N 15 270 CMP 6

316

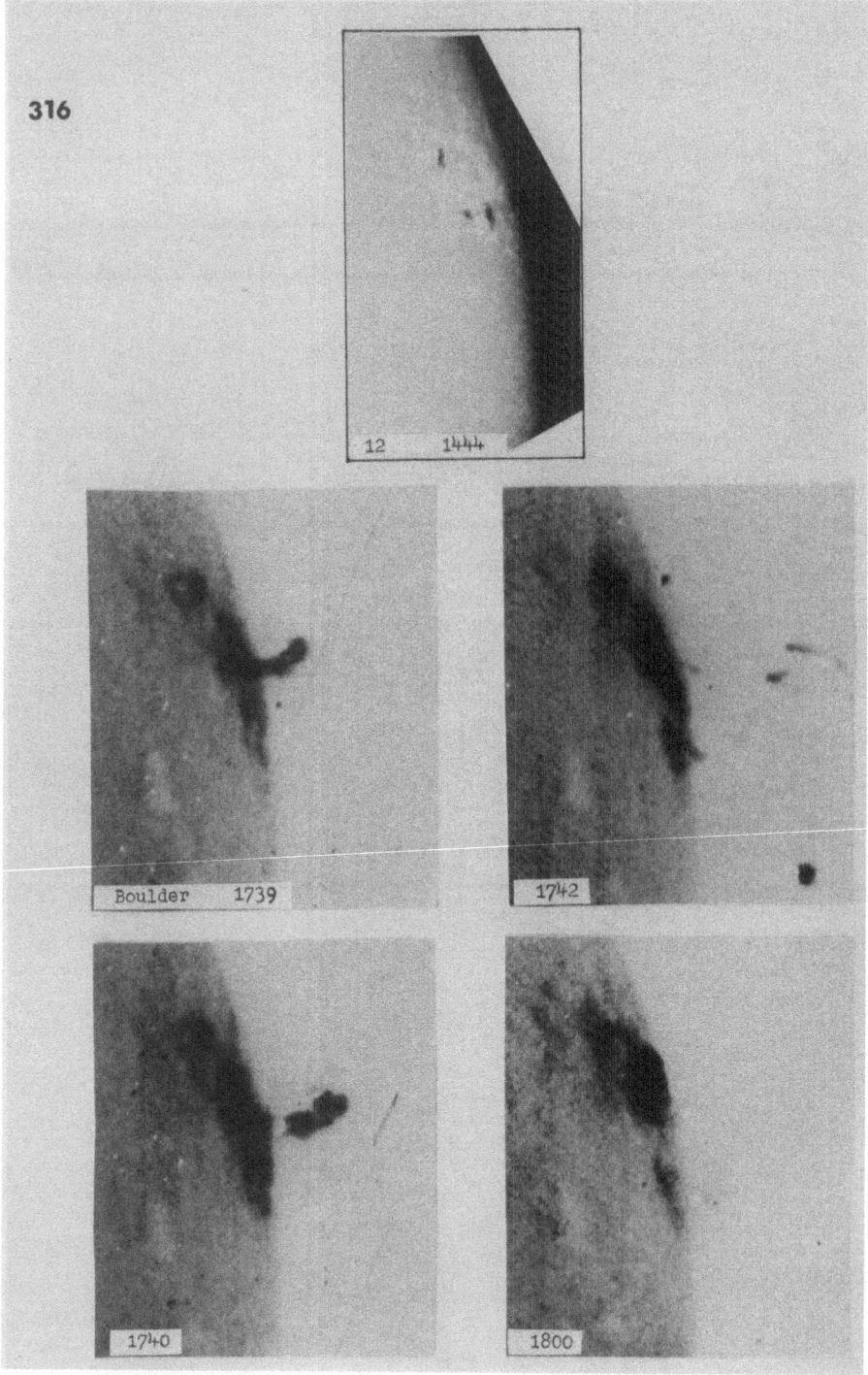

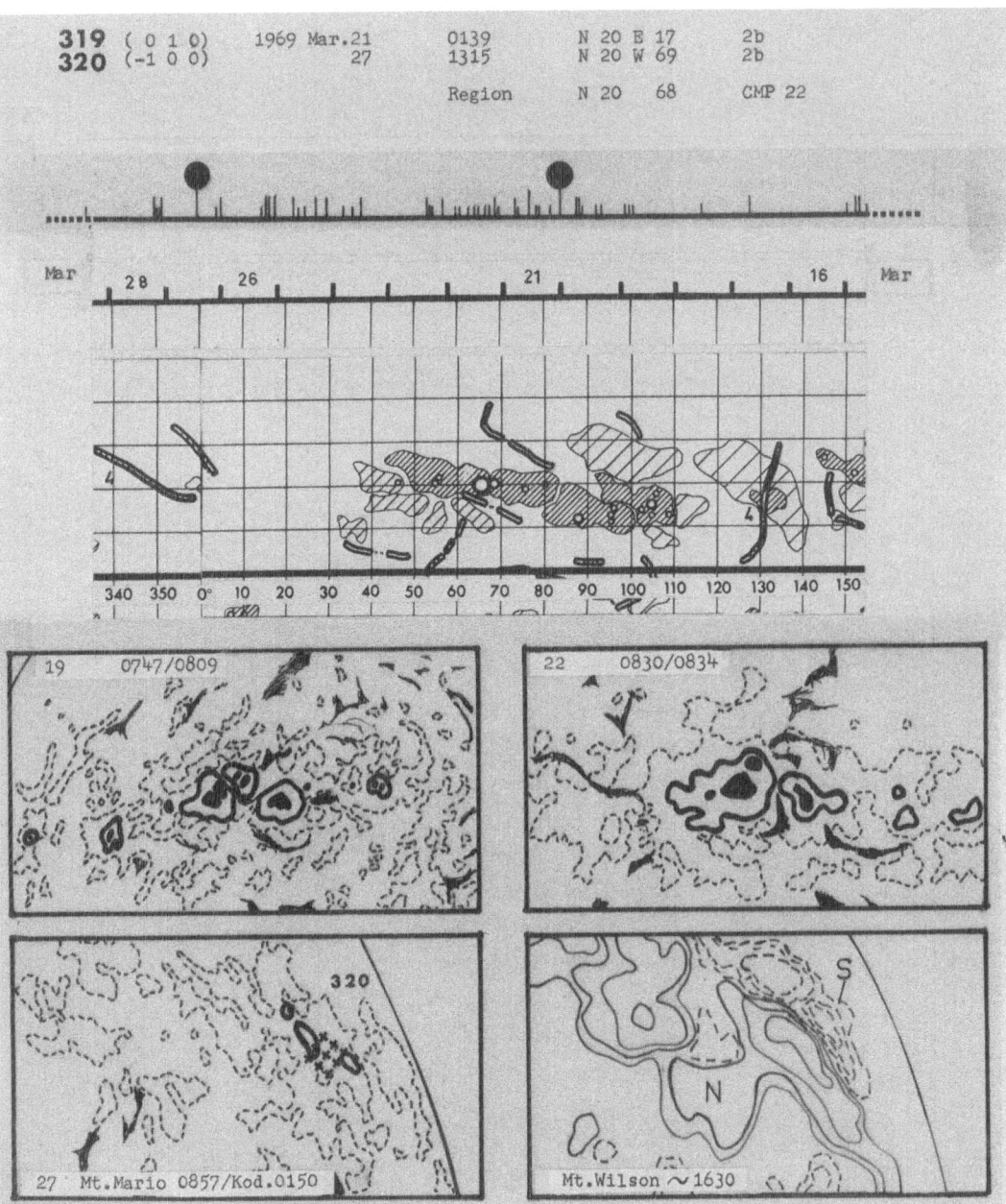

319 (0 1 0)   1969 Mar.21   0139   N 20 E 17   2b
320 (-1 0 0)   27   1315   N 20 W 69   2b

                      Region   N 20   68   CMP 22

19   0747/0809

22   0830/0834

27 Mt.Mario 0857/Kod.0150

320

Mt.Wilson ~1630

N

S

319

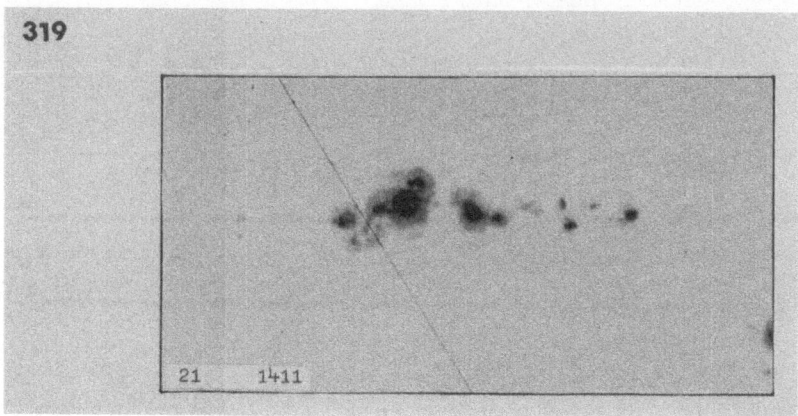

326 (–2 0 0) 1969 May 2   1745   N 09 W 40   1b
327 (–2 0 0)   5   1237   N 08 W 77   Sn

                                     Region   N 08 277   CMP 29

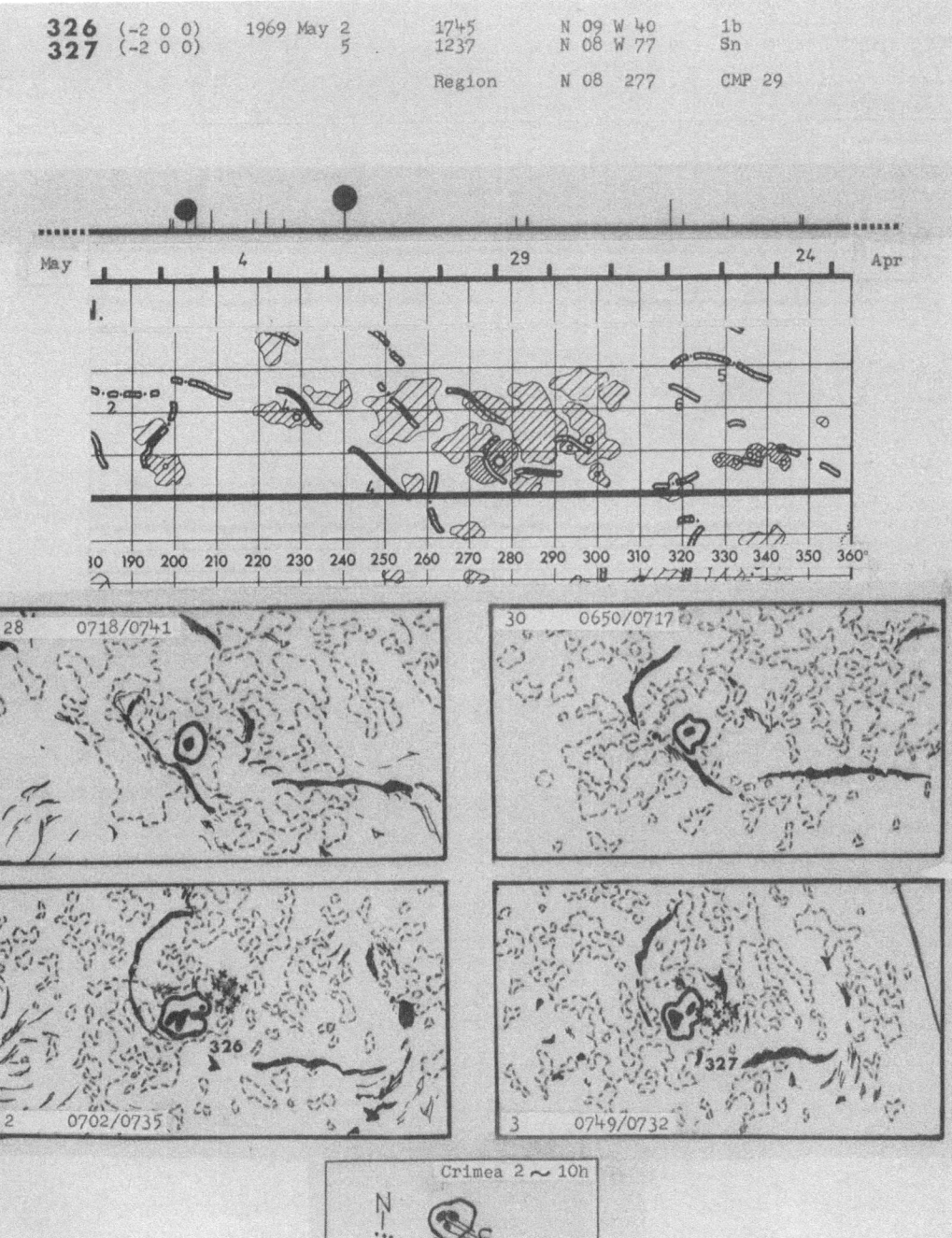

Crimea 2 ∼ 10h

329 (-2 0 0)   1969 May 28   1241   N 10 W 59   1b
                                    Region   N 10 310   CMP 24

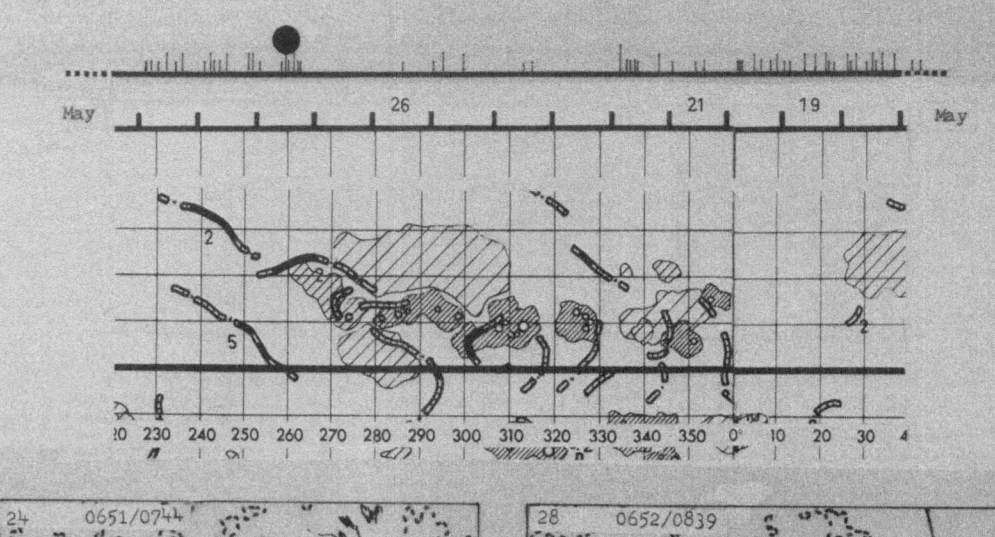

May   26   21   19   May

20 230 240 250 260 270 280 290 300 310 320 330 340 350 0° 10 20 30 4

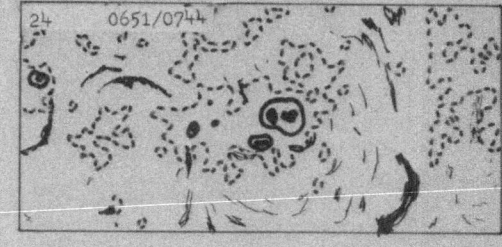

24   0651/0744

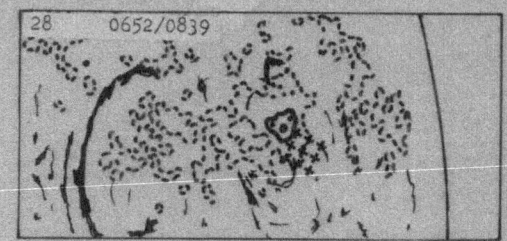

28   0652/0839

Sayan Mts. 23 ~ 3h

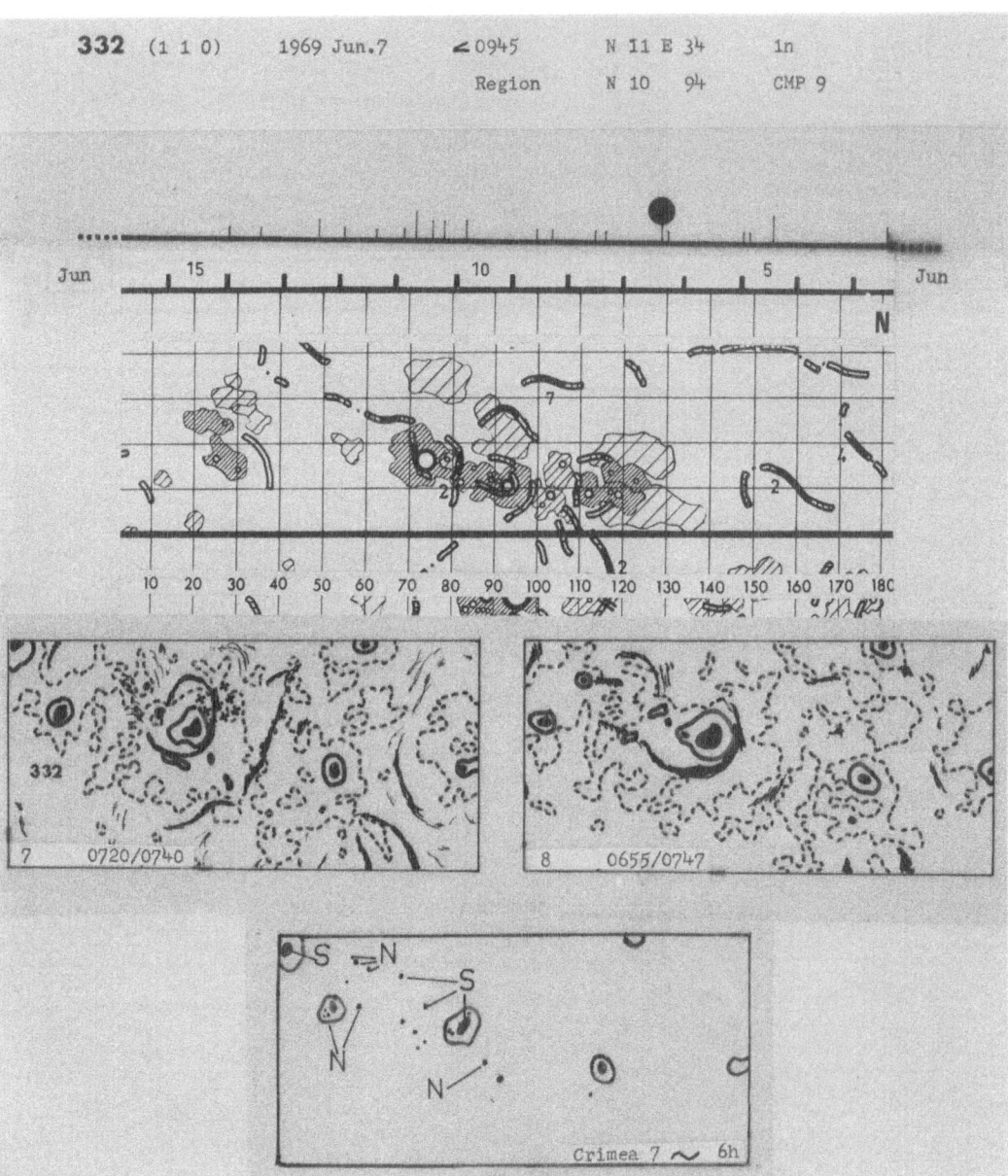

332 (1 1 0)   1969 Jun.7   ∠0945   N 11 E 34   1n

Region   N 10   94   CMP 9

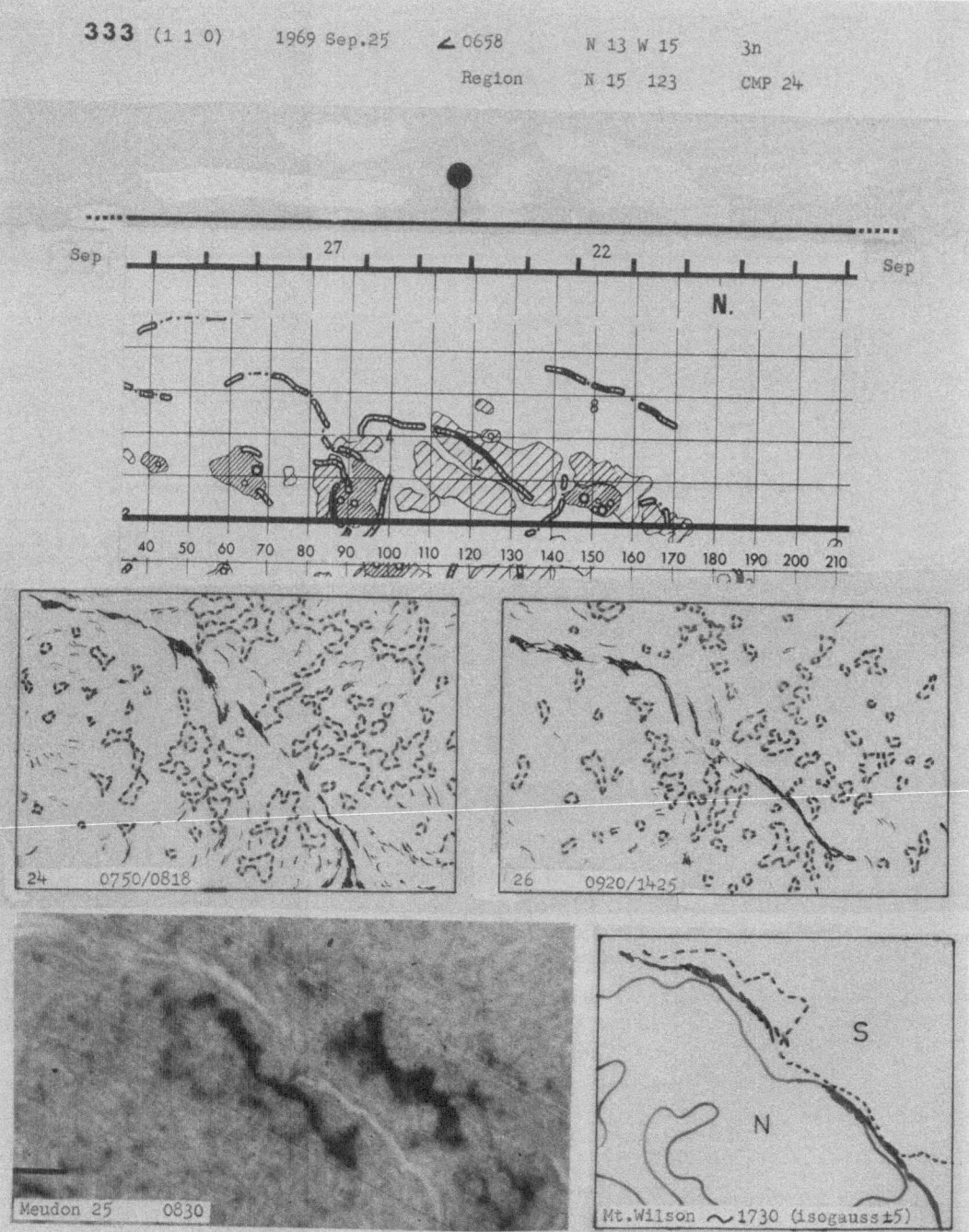

333 (1 1 0)   1969 Sep.25   ∠ 0658   N 13 W 15   3n

Region   N 15 123   CMP 24

Sep   27   22   Sep

N.

40 50 60 70 80 90 100 110 120 130 140 150 160 170 180 190 200 210

24   0750/0818

26   0920/1425

Meudon 25   0830

Mt.Wilson ∼1730 (isogauss ±5)

S

N

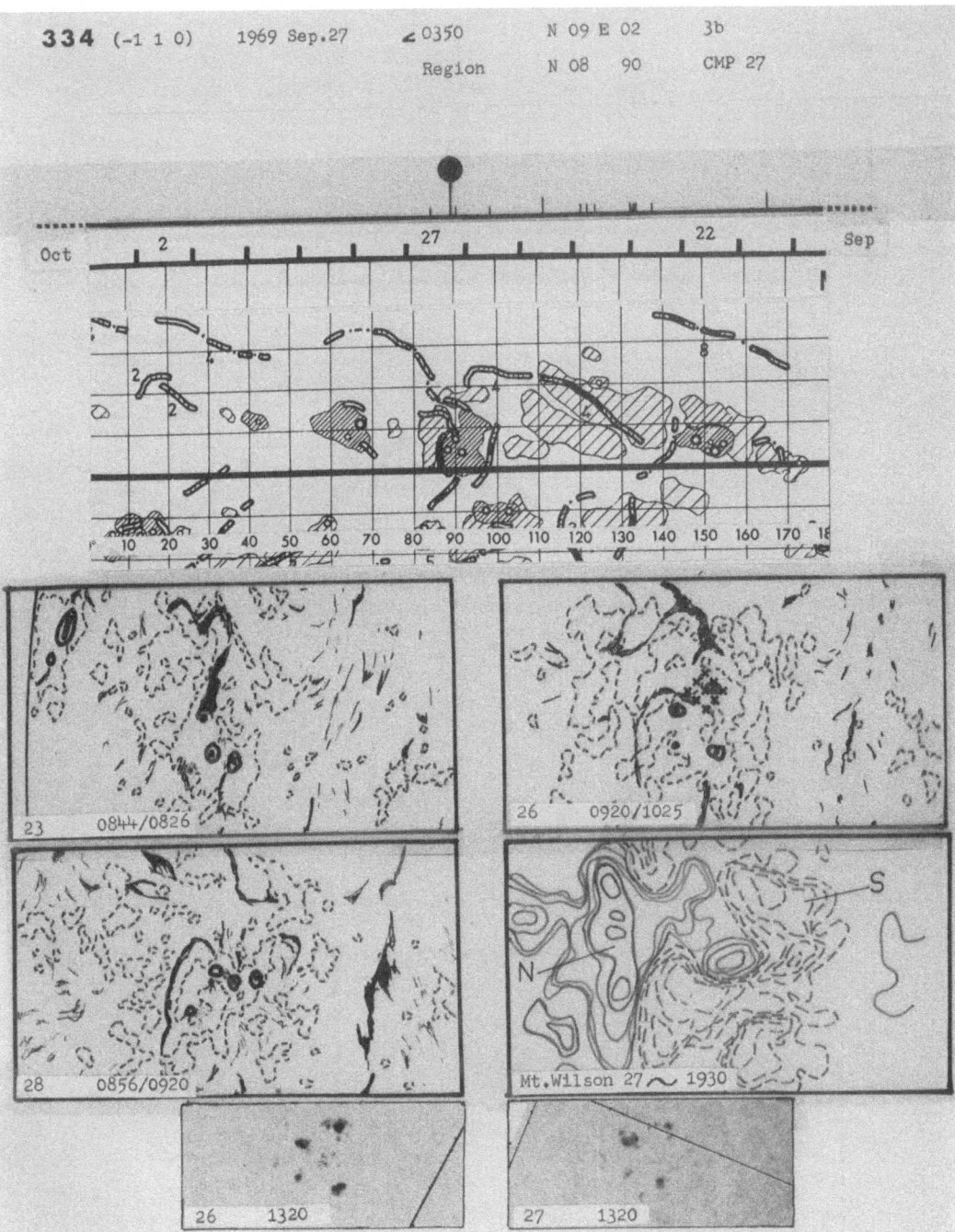

334 (-1 1 0)   1969 Sep.27   ∠0350   N 09 E 02   3b
Region   N 08   90   CMP 27

Oct   2   27   22   Sep

10 20 30 40 50 60 70 80 90 100 110 120 130 140 150 160 170 18

23   0844/0826

26   0920/1025

28   0856/0920

Mt.Wilson 27 ∼ 1930   N   S

26   1320

27   1320

337 (-1 0 0)   1969 Oct.14   0539   N 25 W 71   2n

Region   N 25 298   CMP 8

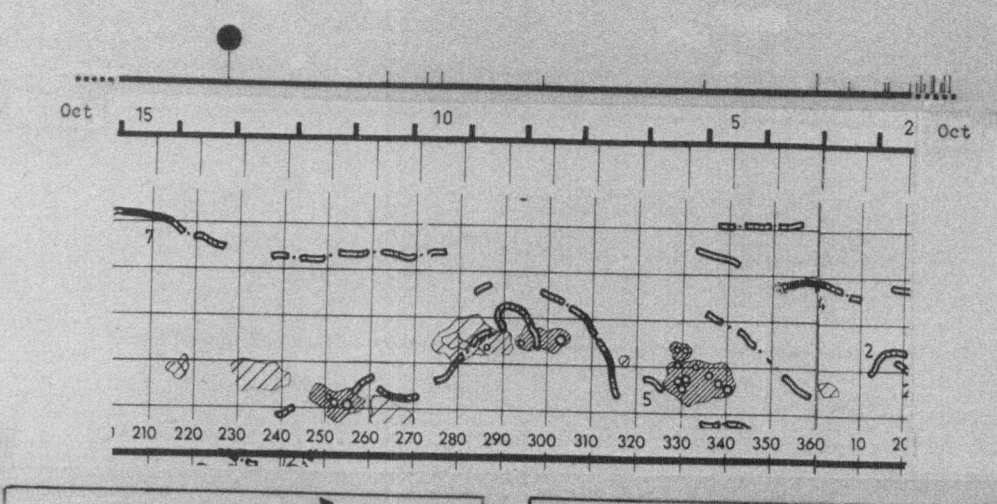

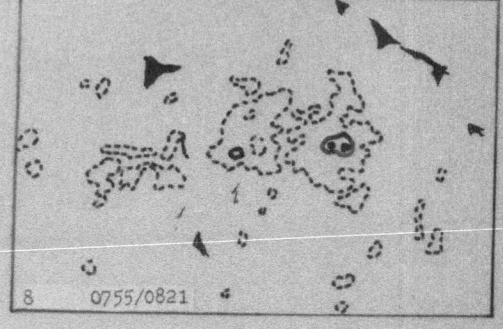

8   0755/0821

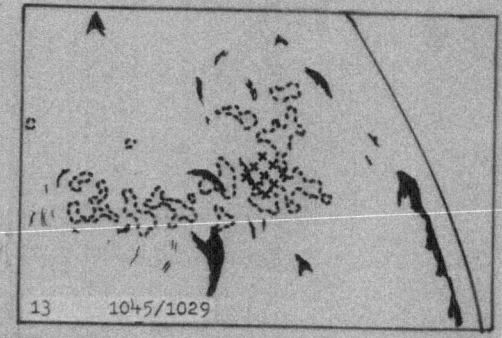

13   1045/1029

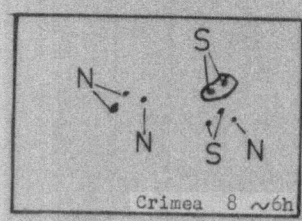

Crimea 8 ∼6h

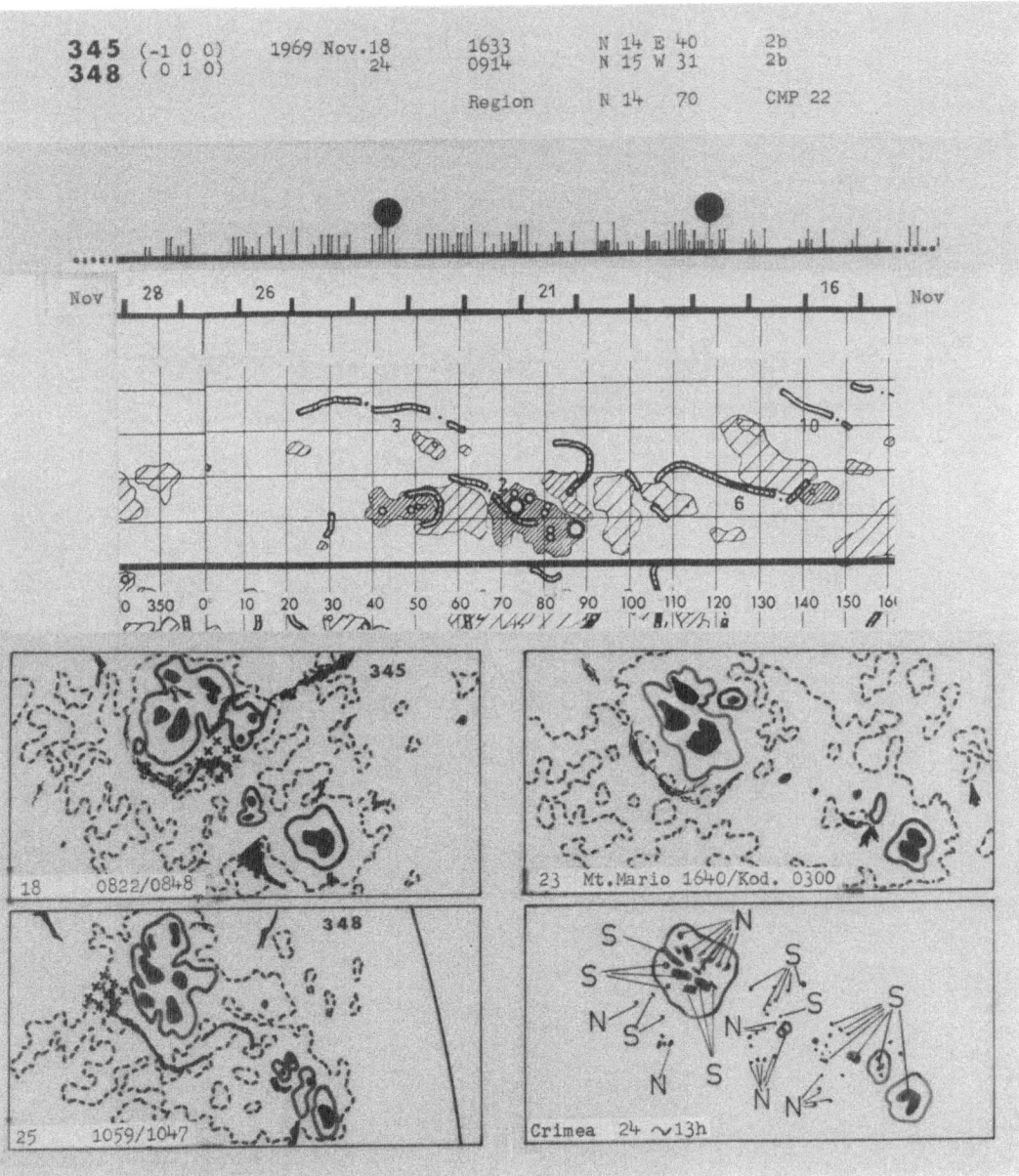

345 (-1 0 0) 1969 Nov.18 1633 N 14 E 40 2b
348 (0 1 0)   24 0914 N 15 W 31 2b

                      Region N 14 70 CMP 22

352 (0 1 0)   1969 Dec.30   1927   S 14 W 85   1n
                                      Region   S 13   7   CMP 24

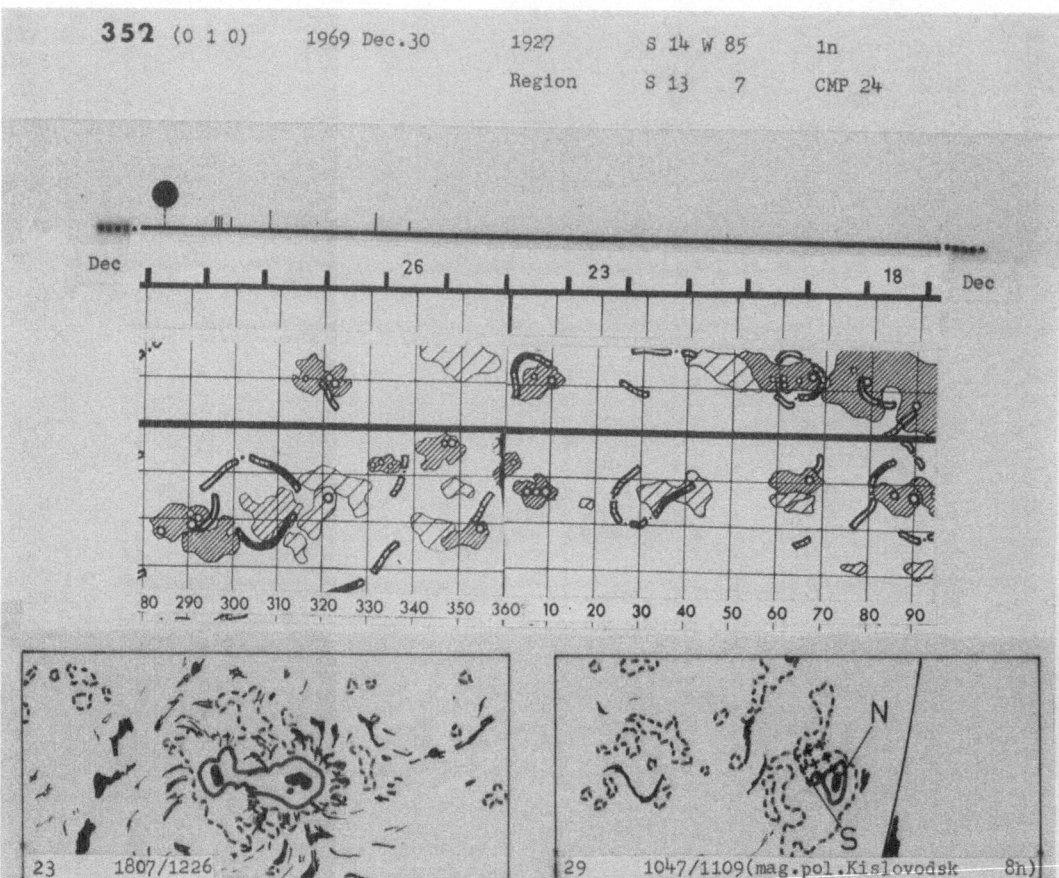

23   1807/1226

29   1047/1109(mag.pol.Kislovodsk   8h)

APPENDIX B

LIST OF SELECTED ACTIVE REGIONS WHICH WERE SOURCES OF PARTICLE EVENTS

McM 3065	N36	13.3 Jan. 1955

16 Jan. < 2130

Age 2	large spot, 7–9 and 17–19 Jan. – H; 10–16 Jan. – E.
CMP:	Ca no data; spots 758/13; E; γ
Ref.:	45(6), 140(2)

McM 3400	N22	17.8 Feb. 1956

23 Feb. ≤ 0334

Age 2	large elongated group, developing to 22 Feb.; after 22 Feb. – fast decrease in the area. 12–15 and 20–22 Feb. – E, 16–19 Feb. – F
CMP:	Ca no data; spots 2670/63; F; γ
Ref.:	32(6), 33(3,6), 36(3,6), 39(1,3,4,6), 53(3,4,6), 100(6), 114(3,4), 135(3,4,6), 136(3,4,6), 147(3), 178(6) 179(5,6), 197(1,3)

McM 3432	N22	17.0 Mar. 1956

10 Mar. < 0515

Age 2	multipolar group, 2 maxima of development – 14 and 18 Mar.
CMP:	Ca no data; spots 667/33; E; βγ
No ref.	

McM 3474	N16	25.3 Apr. 1956

27 Apr. 2050

Age 2	unstable group of small spots and pores
CMP:	Ca no data; spots 124/10; C; αp
No ref.	

McM 3643	N18	01.9 Sept. 1956

28 Aug. 1520, < 2220; 31 Aug. 1226

New	gradual decrease in the area, before 4 Sept. – E, 4–6 Sept. – G, 7 Sept. – H
CMP:	Ca no data; spots 851/37; E; γ
Ref.:	82(3,4,6), 94(3), 106(6), 112(3), 136(3,4,6), 167(1,2,3,4,6), 169(3), 190(1,3,4), 203(3)

McM 3751	S17	09.7 Nov. 1956

7 Nov. 1109; 14 Nov. 1037

New	large spot, surrounded by small spots, 5–7 and 12 Nov. – D; 8–11 Nov. – E
CMP:	Ca no data ; spots 938/12; E; βp
Ref.:	21(3,4,5,6), 55(4), 153(1,3,6)

McM 3753	N18	13.0 Nov. 1956

13 Nov. < 1430

Age 2   after 12 Nov. − gradual decrease in the area, before 12 Nov. − E, after − G
CMP: Ca no data;   spots 494/26;   G; β
No ref.

McM 3820	S28	19.0 Jan. 1957

20 Jan. < 1100

Age 3+6   return of McM 3794 and 3797; large spot surrounded by pores
CMP: Ca 8000/3.0;   spots 600/3;   H; αp
Ref.:   13(4), 21(3,4,5,6), 87(3,4,5), 129(6), 137(4,5,6), 142(4,5), 202(3)

McM 3856	N14	19.1 Feb. 1957

21 Feb. 1605

Age 4   return of McM 3823
CMP: Ca 1300/2.0;   spots 130/5;   A; αp
No ref.

McM 3907	S19	30.2 Mar. 1957

3 Apr. 0825

Age 6   return of McM 3872; fast development from pore to spot group
CMP: Ca 3700/3;   spots 40/6;   (B); β
Ref.:   11(5), 12(5), 87(3,4,5), 142(4,5), 150(4,5)

McM 3916	S23	08.0 Apr. 1957

12 Apr. 1850

Age 3   return of McM 3881; disintegrating
CMP: Ca 3800/2.0;   spots 300/4;   C; βp
No ref.

McM 3923	S23	12.3 Apr. 1957

11 Apr. 1722

Age 7   return of McM 3888
CMP: Ca 6100/3.5;   spots 640/19;   E; β
Ref.:   185(5)

McM 3941	N27	23.0 Apr. 1957

17 Apr. 2000

Age 2	return of McM 3908; fast decrease in the area after 21 Apr. (from E to J, A)
CMP:	Ca 6000/3.0; spots 480/10; D; $\beta\gamma$
Ref.:	12(5), 185(5)

McM 3944	S16	22.6 Apr. 1957

18 Apr. < 1310

Age 2	return of McM 3899; two large spots, accompanied by unstable small spots and pores
CMP:	Ca 7500/3.0; spots 780/12; C; αp
Ref.:	95(3), 105(4), 124(4)

McM 4024	N19	22.4 June 1957

19 June 1609

Age 1–2	group of large spots, mostly new, in position of old McM 3989 and part of McM 3991
CMP:	Ca 8000/3.5; spots 2460/21; F; $\beta\gamma$
Ref.:	11(5), 12(5), 73(4.5.6), 87(3,4,5), 105(4), 122(5), 124(4), 137(4,5,6), 150(4,5)

McM 4039	N12	30.3 June 1957

30 June 0924; 3 July 0712

Age 2	return of McM 4001; large spot surrounded by unstable small spots and pores
CMP:	Ca 4000/4.0; spots 560/16; G; αp
Ref.:	11(5), 12(5), 20(3,4), 21(3,4,5,6), 28(6), 56(4), 58(4,5,6), 72(1), 73(4,5,6), 75(4,5,6), 84(3,5,6), 85(3,5), 91(3), 95(3), 97(6), 111(3), 118(4,5,6), 122(5), 124(4), 127(6), 133(3), 137(4,5,6), 142(4,5), 150(4,5), 155(6), 187(4), 188(3), 190(1,3,4), 202(3), 207(3)

McM 4070	S22	22.0 July 1957

24 July 1712

Age 3	return of McM 4030; regular spot surrounded by very unstable small spots and pores
CMP:	Ca 5500/3.0; spots 680/16; (B); βp
Ref.:	12(5), 71(5), 73(4,5,6), 75(4,5,6), 86(3,5) 97(6), 105(4), 118(4,5,6), 122(5), 124(4), 128(5), 137(4,5,6), 142(4,5), 150(4,5), 174(4,5), 188(3)

McM 4082	S29	01.6 Aug. 1957

9 Aug. 1330

Age 3	return of McM 4044; regular spot, accompanied by unstable pores; since 5 Aug. – bipolar group
CMP:	Ca 6500/3.5; spots 1290/19; H; αp
No ref.	

McM 4099   S11   15.4 Aug. 1957

---

9 Aug. 0619

Age 2	return of McM 4066 and 4072; group of pores
CMP:	Ca 3000/3.0; spots 80/4; A; α
Ref.:	11(5), 12(5), 56(4), 71(5), 72(1), 73(4,5,6), 91(3), 97(6), 122(5), 174(4,5), 206(4)

McM 4124   N22   31.5 Aug. 1957

---

31 Aug. 0544, 1257; 2 Sept. 1257; 3 Sept. 1412

Age 3	return of McM 4083 and 4084; fast changing group, very large irregular unstable spots and small spots and pores. Before 31 Aug. and after 4 Sept. – E, from 31 Aug. to 4 Sept. – F
CMP:	Ca 18000/2.0; spots 2590/51; F; βγ
Ref.:	7(4,5,6), 11(5), 12(5), 28(6), 56(4), 58(4,5,6), 72(1), 73(4,5,6), 75(4,5,6), 84(3,5,6), 85(3,5), 91(3), 97(6), 105(4), 108(3), 111(3), 118(4,5,6), 122(5), 124(4), 126(4,6), 128(5), 137(4,5,6), 142(4,5), 150(4,5), 170(3), 175(3,4,5), 181(3,5), 185(5), 187(4), 190(1,3,4), 206(4), 207(3)

McM 4125   S29   29.8 Aug. 1957

---

28 Aug. ≤ 0913, 2010; 31 Aug. 0521

New	large irregular spot preceeded by unstable small spots
CMP:	Ca (6400)/(3.0); spots 760/11; G; no data
Ref.:	4(6), 7(4,5,6), 11(5), 12(5), 56(4), 72(1), 73(4,5,6), 75(4,5,6), 85(3,5), 87(3,4,5), 81(3), 97(6), 99(6), 105(4), 118(4,5,6), 122(5), 124(4), 126(4,6), 128(5), 137(4,5,6), 142(4,5), 150(4,5), 174(4,5), 176(3,4,5), 181(3,5), 185(5), 187(4), 188(3), 190(1,3,4), 204(5,6), 206(4), 207(3)

McM 4134   N12   10.2–11.5 Sept. 1957

---

11 Sept. 0236; 12 Sept. 1512

Age 2	return of McM 4100 and 4098, three close large groups of spots
CMP:	Ca 7600/3.5; spots 1540/54; H + E + J; βγ
Ref.:	12(5), 17(4), 28(6), 56(4), 58(4,5,6), 72(1), 73(4,5,6), 75(4,5,6), 78(3,5), 84(3,5,6), 85(3,5), 97(6), 99(6), 111(3), 117(3,4,6), 118(4,5,6), 122(5), 126(4,6), 128(5), 134(3,4), 135(3,4,6), 137(4,5,6), 142(4,5), 150(4,5), 155(6), 170(3), 185(5), 187(4), 202(3), 207(3)

McM 4151   N20   19.1 Sept. 1957

---

18 Sept. < 1303, < 1722; 19 Sept. < 0350

Age 5	return of McM 4112; developing group of very large spots and unstable small spots and pores
CMP:	Ca 8000/4.0; spots 2260/34; F; βγ
Ref.:	12(5), 17(4), 31(3,4,6), 72(1), 84(3,5,6), 97(6), 105(4), 122(5), 124(4), 142(4,5), 185(5)

McM 4152	N11	19.8–20.8 Sept. 1957

21 Sept. 1330

Age 2	return of McM 4114; very large irregular spot accompanied by very unstable small spots. Before 20 Sept. – E, after – G, H
CMP:	Ca 3500/3.0; spots (1050)/(15); E; βp
Ref.:	11(5), 28(6), 72(1), 73(4,5,6), 78(3,5), 91(3), 97(6), 122(5), 124(4), 128(5), 137(4,5,6), 142(4,5), 150(4,5), 155(6), 170(3)

McM 4159	N22	27.8 Sept. 1957

26 Sept. 1907

Age 4	return of McM 4124; regular spot and unstable small spots
CMP.	Ca 22000/3.0; spots 1660/51; C; βp
Ref.:	1(6), 12(5), 28(6), 72(1), 73(4,5,6), 75(4,5,6), 84(3,5,6), 97(6), 105(4), 118(4,5,6), 122(5), 128(5), 137(4,5,6), 142(4,5), 150(4,5)

McM 4189	S25	17.5 Oct. 1957

20 Oct. 1637

Age 2	return of McM 4155; very large spot of complicated structure and large number of surrounding large and small unstable spots
CMP:	Ca 13100/4.0; spots 3560/35; F; βf
Ref.:	11(5), 12(5), 28(6), 58(4,5,6), 72(4,5,6), 72(4,5,6), 75(4,5,6), 90(4), 97(6), 99(6), 105(4), 116(5,6), 118(4,5,6), 122(5), 124(4), 127(5), 128(5), 137(4,5,6), 142(4,5), 150(4,5), 170(3), 185(5)

McM 4207	S18	31.9 Oct. 1957

5 Nov. < 1205

Age 3+4	return of McM 4167 and 4175; chain of unstable small spots and disintegrating spot
CMP:	Ca 15000/3.0; spots 1890/25; G; βf
Ref.:	27(4), 28(6), 71(5), 72(1), 97(6), 116(5,6)

McM 4263	S16	27.1 Nov. 1957

24 Nov. 0848

Age 4+5	return of McM 4207; rather large spots accompanied by unstable small spots and pores, 21 Nov. – H, 22–26 Nov. – G, 27–29 Nov. – E, 30 Nov. – D, 1–2 Dec. – C
CMP:	Ca 8800/4.0; spots 610/12; E; βp
Ref.:	28(6), 58(4,5,6), 75(4,5,6), 91(3), 97(6), 122(5), 124(4), 137(4,5,6), 142(4,5), 206(4)

McM 4314	N18	20.0 Dec. 1957

16 Dec. 1125; 17 Dec. < 0734

New	scattered group of unstable spots
CMP:	Ca 10000/3.5; spots 1360/37; E; βγ
Ref.:	56(4), 58(4,5,6), 75(4,5,6), 87(3,4,5), 91(3), 97(6), 99(6), 124(4), 142(4,5), 171(3), 185(5), 187(4)

| McM 4321 | N21 | 24.2 Dec. 1957 |

28 Dec. 2229

New	very elongated group of very large stable spots, before 23 Dec. and after 27 Dec. − E, 23−26 Dec. − F
CMP:	Ca 4000/3.0; spots 1160/13; F; βp
Ref.:	12(5), 28(6), 71(5), 97(6), 105(4), 122(5), 124(4), 174(4,5), 185(5)

| McM 4372 | S24 | 20.6 Jan. 1958 |

25 Jan. 0915

Age 6	return of McM 4319; fast developing from pores, 19 Jan. − A, 20 Jan. − B, 21 Jan. − D, 22−25 Jan. − E
CMP:	Ca 1500/1.5; spots 170/3; B; β
Ref.:	12(5), 13(4), 71(5), 72(1), 84(3,5,6), 91(3), 97(6), 111(3), 185(5), 187(4), 188(3), 206(4)

| McM 4382 | S17 | 26.0 Jan. 1958 |

25 Jan. 1205

Age 6	return of McM 4333; very unstable spot group
CMP:	Ca 4000/3.0; spots 870/18; D; βp
Ref.:	47(3,4,6), 73(4,5,6), 97(6), 124(4), 142(4,5), 174(4,5), 187(4)

| McM 4400 | S14 | 07.9 Feb. 1958 |

9 Feb. 2108

Age 4	return of McM 4355; regions 4400, 4402 and 4404 merged and were called 4400 before CMP; very large irregular spot surrounded by pores
CMP:	Ca 25000/3.0; spots 2730/108; H; αp
Ref.:	1(6), 2(6), 11(5), 12(5), 14(3,5,6), 28(6), 47(3,4,6), 58(4,5,6), 60(4,5,6), 72(1), 75(4,5,6), 97(6), 99(6), 105(4), 113(5,6), 116(5,6), 118(4,5,6), 122(5), 128(5), 137(4,5,6), 142(4,5), 150(4,5), 155(6)

| McM 4445 | S15 | 07.7 Mar. 1958 |

3 Mar. < 1005; 14 Mar. 1454

Age 5	return of McM 4400; group of one stable regular spot and several irregular spots, after 9 Mar. − E
CMP:	Ca 8000/2.5; spots 2210/24; F; βp
Ref.:	12(5), 13(4), 47(3,4,6), 58(4,5,6), 72(1), 91(3), 97(6), 122(5), 124(4), 137(4,5,6), 185(5), 187(4)

| McM 4449 | N14 | 11.9 Mar. 1958 |

11 Mar. < 0030

Age 3	return of McM 4410; elongated group of unstable small spots and a stable spot
CMP:	Ca 8300/3.0; spots 1060/12; D; β
No ref.	

McM 4476	S12	28.5 Mar. 1958

23 Mar. 0947

New	very large spot accompanied by disintegrating spots
CMP:	Ca 12000/3.5; spots 2090/37; E; βγ
Ref.:	11(5), 12(5), 21(3,4,5,6), 28(6), 47(3,4,6), 48(4,5) 58(4,5,6), 71(5), 72(1), 73(4,5,6), 75(4,5,6), 84(3,5,6), 87(3,4,5), 93(5), 97(6), 99(6), 118(4,5,6), 122(5), 128(5), 137(4,5,6), 142(4,5), 150(4,5), 155(6), 164(3,4,6), 171(3), 174(4,5), 175(3,4,5), 185(5), 187(4), 206(4)

McM 4578	N24	30.3 May 1958

4 June < 2147; 6 June 0436

Age 2	return of McM 4529; very small spots degenerating to pores
CMP:	Ca 6500/3.0; spots 160/6; B; β
Ref.:	12(5), 28(6), 48(4,5), 58(4,5,6), 75(4,5,6), 97(6), 105(4), 122(5), 124(4), 137(4,5,6), 150(4,5), 170(3), 185(5), 187(4), 188(3)

McM 4634	N28	07.2 July 1958

7 July 0020

Age 2	return of McM 4596; two irregular spots developed from pores, 5–9 July – E
CMP:	Ca 4500/3.5; spots 860/11; E; no data
Ref.:	11(5), 12(5), 14(3,5,6), 17(4), 24(4,5,6), 28(6), 48(4,5), 58(4,5,6), 61(4,5,6), 73(4,5,6), 75(4,5,6), 93(5), 97(6), 99(6), 118(4,5,6), 123(5,6), 128(5), 137(4,5,6), 150(4,5), 170(3), 174(4,5), 190(1,3,4), 202(3)

McM 4659	S20	26.3 July 1958

29 July < 0259

Age 2	return of McM 4622; developing; stable spot accompanied by very unstable spots, 25–26 July – E, after 26 July – F
CMP:	Ca 14000/3.5; spots 1790/67; E; βγ
Ref.:	12(5), 17(4), 47(3,4,6), 48(4,5), 58(4,5,6), 61(5,6), 72(1), 73(4,5,6), 75(4,5,6), 97(6), 115(4,6), 118(4,5,6), 122(5), 128(5), 137(4,5,6), 150(4,5), 170(3), 199(3,4)

McM 4686	S13	12.4 Aug. 1958

16 Aug. 0433

Age 2	return of McM 4653; very large spot and unstable small spots, 7–10 Aug. – G
CMP:	Ca 6000/3.0; spots 2140/30; G; βγ
Ref.:	11(5), 12(5), 14(3,5,6), 28(6), 48(4,5), 58(4,5,6), 72(1), 73(4,5,6), 75(4,5,6), 85(3,5), 93(5), 97(6), 99(6), 118(4,5,6), 121(4,5,6), 124(4), 137(4,5,6), 150(4,5), 155(6), 162(3), 170(3), 174(4,5), 176(3,4,5), 181(3,5), 188(3), 190(1,3,4), 199(3,4), 202(3), 206(4)

McM 4708   N18   21.6 Aug. 1958

---

19 Aug. 2118; 20 Aug. 0042; 22 Aug. 1428; 26 Aug. 0005

New   close group of unstable irregular spots
CMP: Ca 7000/3.5;   spots 1540/13;   E; $\beta\gamma$ Feb. – E
Ref.:   5(5,6), 11(5), 12(5), 14(3,5,6), 16(4), 17(4), 28(6), 43(4), 47(3,4,6), 48(4,5), 58(4,5,6), 72(1),
            73(4,5,6), 75(4,5,6), 84(3,5,6), 85(3,5), 87(3,4,5), 91(3), 93(5), 97(6), 99(6), 105(4), 118(4,5,6),
            122(5), 123(5,6), 124(4), 128(5), 137(4,5,6), 142(4,5), 143(4), 146(3), 150(4,5), 155(6), 162(3),
            164(3,4,6), 168(3,4,5,6), 170(3), 174(4,5), 177(3,4,5,6), 187(4), 196(3), 199(3,4), 200(3)

McM 4741   S08   09.0 Sept. 1958

---

14 Sept. 0822

New   developing close group of unstable irregular spots, after 10 Sept. – E
CMP: Ca 5000/2.5;   spots 530/20;   F; γ
Ref.:   13(4), 28(6), 58(4,5,6), 72(1), 75(4,5,6), 87(3,4,5), 91(3), 97(6), 108(3), 122(5), 124(4), 137(4,5,6),
            142(4,5), 162(3), 164(3,4,6), 174(4,5), 185(5), 188(3), 206(4)

McM 4765   S16   20.1 Sept. 1958

---

22 Sept. 0738

Age 4   return of McM 4712; large spot divided into two parts after 17 Sept., 14–17 Sept. – H, after – E
CMP: Ca 14000/3.5;   spots 1220/40;   E; βp
Ref.:   11(5), 13(4), 47(3,4,6), 48(4,5), 56(4), 72(1), 73(4,5,6), 97(6), 118(4,5,6), 128(5), 137(4,5,6),
            142(4,5), 150(4,5), 181(3,5), 187(4), 202(3), 206(4)

McM 4969   N17   20.9 Jan. 1959

---

26 Jan. 0842, 1027

Age 3   return of McM 4932; spreading group of unstable spots
CMP: Ca 13000/3.0;   spots 1140/60;   B + E;   no data
Ref.:   13(4), 37(3,6), 48(4,5), 71(5), 72(1), 81(3,6), 92(3), 97(6), 122(5), 124(4), 142(4,5), 185(5),
            187(4), 206(4)

McM 5009   N12   16.2 Feb. 1959

---

12 Feb. < 2301

Age 4   return of McM 4969; large spot accompanied by small spots and pores; 10–11 and 21 Feb. – H,
            12–16 Feb. – E, 17–20 Feb. – G
CMP: Ca 4500/3.0;   spots 950/23;   E;   no data
Ref.:   11(5), 12(5), 28(6), 48(4,5), 71(5), 72(1), 75(4,5,6), 87(3,4,5), 97(6), 105(4), 122(5), 124(4),
            137(4,5,6), 142(4,5), 174(4,5), 185(5), 195(3)

McM 5148	N14	14.8 May 1959

10 May 2102; 11 May 2006; 13 May 0509

Age 6 + 2	return of McM 5095 and 5100; fast changing group of unstable spots
CMP:	Ca 13000/3.5; spots 2600/32; E + E; no data
Ref.:	11(5), 12(5), 17(4), 21(3,4,5,6), 28(6), 43(4), 46(6), 47(3,4,6), 48(4,5), 62(4), 72(1), 73(4,5,6), 75(4,5,6), 84(3,5,6), 92(3), 97(6), 99(6), 102(3), 105(4), 118(4,5,6), 123(5,6), 124(4), 128(5), 137(4,5,6), 142(4,5), 150(4,5), 156(3,4,5,6), 174(4,5), 199(3,4), 200(3)

McM 5204	N19	17.1 June 1959

9 June 1707; 12 June 0735

Age 7 + 3	return of McM 5157; bipolar group with complex components; 12–14 June – E; 15–20 June – G, 21–22 June – H
CMP:	Ca 9000/3.5; spots 1130/15; G; no data
Ref.:	28(6), 43(4), 57(4), 70(3,4,6), 71(5), 72(1), 73(4,5,6), 87(3,4,5), 97(6), 105(4), 122(5), 124(4), 137(4,5,6), 142(4,5), 151(4), 174(4,5), 185(5), 187(4,5), 188(3), 193(3)

McM 5265	N16	13.7 July 1959

9 July 1930; 10 July 0206; 14 July < 0325; 16 July 2114

Age 4 + 2	return of McM 5204 and 5218; at first – unstable pores, then – fast developed regular spot; 9–16 July – H; after – E
CMP:	Ca 12000/3.5; spots 1430/13; H; no data
Ref.:	1(6), 3(6), 11(5), 12(5), 14(3,5,6), 17(4), 21(3,4,5,6), 28(6), 30(3), 39(1,3,4,6), 46(6), 47(3,4,6), 48(4,5), 59(4,5,6), 72(1), 73(4,5,6), 75(4,5,6), 84(3,5,6), 85(3,5), 92(3), 93(5), 97(6), 99(6), 105(4), 107(3), 118(4,5,6), 123(5,6), 124(4), 128(5), 134(3,4), 137(4,5,6), 142(4,5), 143(4), 150(4,5), 157(5,6), 164(3,4,6), 170(3), 174(4,5), 176(3,4,5), 177(3,4,5,6), 181(3,5), 185(5), 188(3), 190(1,3,4), 196(3), 198(3), 199(3,4), 200(3), 202(3)

McM 5323	N14	16.0 Aug. 1959

18 Aug. 1014

Age 4	return of McM 5280; bipolar group with stable main spots and fast changing large spot and pores between them
CMP:	Ca 6000/3.0 spots 530/8; E; no data
Ref.:	28(6), 37(3,6), 48(4,5), 71(5), 73(4,5,6), 87(3,4,5), 92(3), 97(6), 122(5), 124(4), 137(4,5,6), 142(4,5), 170(3), 176(3,4,5), 181(3,5), 185(5), 187(4), 196(3), 206(4)

McM 5329	N09	20.0 Aug. 1959

18 Aug. 1654

New	bipolar group with regular eastern spot, disintegrating western spot and unstable small spots between them, before 18 Aug. – E, after – F
CMP:	Ca 6500/3.0; spots 880/33; F; no data
Ref.:	48(4,5), 90(4), 97(6)

McM 5355	N18	06.7 Sept. 1959

1 Sept. 1923

Age 6 + 4   return of McM 5315; decreasing spot
CMP: Ca 3000/1.5;   spots 70/1;   C;   no data
Ref.:   12(5), 48(4,5), 69(3), 97(6), 137(4,5,6), 142(4,5), 185(5)

McM 5476	N07	01.5 Dec. 1959

30 Nov. 0247, 1720

Age 8   return of McM 5443; very large irregular spot, accompanied by unstable small spots and pores
CMP: Ca 8500/4.0;   spots 2500/24;   F;   no data
No ref.

McM 5494	S06	17.5 Dec. 1959

21 Dec. 0043

New   regular spot, before 20 Dec. – H, 21 Dec. – J, 22 Dec. – C
CMP: Ca 2300/3.0;   spots 410/1;   H;   no data
No ref.

McM 5525	S16	10.2 Jan. 1960

15 Jan. 1334

New   bipolar group of stable spots, between them a stormy changing, considerably greater spot, 4–13 Jan. – E, 14 Jan. – C, 15 Jan. – J
CMP: Ca 6300/3.0;   spots 1020/6;   E;   no data
Ref.:   12(5), 47(3,4,6), 48(4,5), 57(4), 64(3,6), 124(4), 142(4,5), 185(5)

McM 5527	N20	11.6 Jan. 1960

11 Jan. < 2040

Age 2   return of McM 5491; regular rather stable spot accompanied by pores
CMP: Ca 3500/2.5;   spots 730/5;   H;   no data
Ref.:   12(5), 48(4,5), 75(4,5,6), 87(3,4,5), 105(4), 122(5), 124(4), 137(4,5,6), 142(4,5), 185(5)

McM 5592	N23	11.0 Mar. 1960

10 Mar. 1716

Age 2   return of McM 5566; bipolar group, small spot – on W, and pores – on E
CMP: Ca 3800/3.0;   spots 310/3;   C;   no data
No ref.

McM 5615   N11   31.6 Mar 1960

28 Mar. 2042; 29 Mar. 0650; 30 Mar. 0216, 1455; 1 Apr. 0843; 5 Apr. 0215

New   unstable fast developing from A (→ J, H, E) to F group of large spots and pores, F – since 31 Mar.
CMP: 3500/3.0;   spots 1730/48;   F;   no data
Ref.:   11(5), 12(5), 13(4), 14(3,5,6), 21(3,4,5,6), 46(6), 47(3,4,6), 48(4,5), 54(5,6), 63(5,6), 64(3), 72(1),
          73(4,5,6), 75(4,5,6), 84(3,5,6), 87(3,4,5), 90(4), 93(5), 105(4), 116(5,6), 120(5,6), 122(5), 124(4),
          131(4,6), 137(4,5,6), 142(4,5), 143(4), 150(4,5), 164(3,4,6), 170(3), 177(3,4,5,6), 185(5),
          190(1,3,4), 192(3), 199(3,4), 202(3), 208(4)

McM 5642   N11   27.4 Apr. 1960

29 Apr. 0107; 4 May 1000

Age 2   return of McM 5615; two close spots and pores
CMP: Ca 3700/3.0;   spots 690/7;   H;   no data
Ref.:   11(5), 12(5), 14(3,5,6), 17(4), 39(1,3,4,6), 44(4), 46(6), 47(3,4,6), 48(4,5), 54(5,6), 72(1),
          73(4,5,6), 75(4,5,6), 76(4,6), 84(3,5,6), 85(3,5), 87(3,4,5), 93(5), 101(6), 122(5), 123(5,6),
          131(4,6), 137(4,5,6), 142(4,5), 150(4,5), 175(3,4,5,6), 185(5), 195(3), 199(3,4), 202(3)

McM 5645   S08   30.6 Apr. 1960

28 Apr. < 0130

Age 2   return of McM 5618; decreasing regular spot and pores, 24–27 Apr. and 4–5 May – J,
          28–30 Apr. – D, 1–3 May – C
CMP: Ca 4000/3.0;   spots 570/7;   D;   no data
Ref.:   11(5), 12(5), 46(6), 48(4,5), 54(5,6), 73(4,5,6), 105(4), 122(5), 137(4,5,6), 142(4,5), 150(4,5)

McM 5653   S11   07.6 May 1960

6 May 1404

Age 2   return of McM 5625; regular stable spot; 1–3 May and 12 May – H, 4–11 May – G
CMP: Ca 4000/3.0;   spots 440/1;   G;   no data
Ref.:   11(5), 12(5), 14(3,5,6), 21(3.4.5.6), 64(3), 72(1), 73(4,5,6), 75(4,5,6), 86(3,5,6), 87(3,4,5), 93(5),
          105(4), 137(4,5,6), 142(4,5), 150(4,5), 199(3,4)

McM 5654   N29   08.2 May 1960

12 May < 1342; 13 May 0519

New   developing group of irregular spots, 2–3 May – J, 4–6 May – H, 7–8 May – D, 9–10 May – E,
          11–14 May – F
CMP: Ca 2000/3.0;   spots 440/6;   D;   no data
Ref.:   12(5), 14(3,5,6), 17(4), 21(3,4,5,6), 44(4), 47(3,4,6), 64(3), 72(1), 73(4,5,6), 79(3,4,6), 84(3,5,6),
          85(3,5), 87(3,4,5), 93(5), 137(4,5,6), 141(3), 142(4), 143(4), 150(4,5), 152(4), 176(3.4.5),
          177(3,4,5,6), 185(5), 199(3,4)

McM 5657	S10	13.0 May 1960

9 May < 0704

Age 3	return of McM 5630; decreasing small spot and pores
CMP:	Ca 4500/2.5; spots 180/2; C; no data
Ref.:	12(5), 48(4,5), 122(5), 141(3), 185(5)

McM 5669	N12	24.7 May 1960

26 May 0850

Age 3	return of McM 5642; fast changing group, the most part of spots surrounded by common penumbra
CMP:	Ca 4000/3.0; spots 720/30; H + B; no data
Ref.:	12(5), 44(4), 48(4,5), 85(3.5), 87(3,4,5), 103(3,4), 122(5), 137(4,5,6), 142(4,5), 185(5), 208(4)

McM 5680	N29	04.6 June 1960

1 June 0823

Age 2	return of McM 5654; elongated bipolar group of unstable spots and pores
CMP:	Ca 6800/3.0; spots 240/11; C + C; no data
Ref.:	12(5), 38(3,4,6), 44(4), 48(4,5), 64(3), 72(1), 84(3,5,6), 85(3,5), 87(3,4,5), 122(5), 137(4,5,6), 142(4,5), 164(3,4,6), 185(5), 190(1,3,4), 199(3,4), 208(4)

McM 5713	N18	26.4 June 1960

25 June 1136, 1659, 2039; 27 June 0418, 2140; 29 June 0125

New	regular stable spot and pores; before 27 June – C, after – H
CMP:	Ca 2400/3.5; spots 730/5; C; no data
Ref.:	12(5), 19(6), 21(3.4.5.6), 30(3), 48(4,5), 64(3), 72(1), 75(4,5,6), 85(3,5), 87(3,4,5), 102(3), 105(4), 122(5), 137(4,5,6), 141(3), 142(4,5), 152(4), 164(3,4,6), 185(5), 199(3,4), 200(3), 208(4)

McM 5719	S08	29.6 June 1960

26 June 2358

Age 4	return of McM 5679; regular stable spot and sometimes pores
CMP:	Ca 3000/3.0; spots 220/1; C; no data
Ref.:	17(4), 105(4), 142(4,5)

McM 5794	N20	13.4 Aug. 1960

11 Aug. 0233, 1916; 14 Aug. 0511

Age 2	return of McM 5749; large group of one stable spot on W and large breading spot and small unstable spots and pores on E; 8–13 and 15–16 Aug. – E, 14 and 19 Aug. – C, 17 and 18 Aug. – D
CMP:	Ca 11000/3.5; spots 790/57; E; no data
Ref.:	12(5), 17(4), 48(4,5), 64(3), 75(4,5,6), 84(3,5,6), 85(3,5), 105(4), 122(5), 137(4,5,6), 142(4,5), 145(4), 152(4), 177(3,4,5,6), 185(5), 194(3), 195(3), 199(3,4)

McM 5837	N24	10.0 Sept. 1960

3 Sept. 0037

Age 3	return of McM 5794; pores
CMP:	Ca 9000/3.0; spots 150/5; A; no data
Ref.:	11(5), 12(5), 21(3,4,5,6), 27(4), 48(4,5), 72(1), 74(4), 75(4,5,6), 84(3,5,6), 85(3,5), 86(3,5), 105(4), 137(4,5,6), 142(4,5), 150(4,5), 159(3,4,6), 174(4,5), 185(5)

McM 5858	S19	20.9 Sept. 1960

26 Sept. 0525

Age 2	return of McM 5828; close group of large unstable spots
CMP:	Ca 5500/3.5; spots 1210/10; E; no data
No ref.	

McM 5880	S14	07.5 Oct. 1960

11 Oct. 0517

Age 3	return of McM 5839; regular stable spot surrounded by unstable small spots and pores, before 7 Oct. – C, after – D
CMP:	Ca 5000/3.5; spots 520/11; D; no data
Ref.:	47(3,4,6), 142(4,5), 145(4), 177(3,4,5,6), 181(3,5), 199(3,4), 202(3)

McM 5909	N24	31.3 Oct. 1960

29 Oct. 1026

New	fast decreasing regular spot, accompanied by chain of pores
CMP:	Ca 2800/3.0; spots 500/22; D; no data
Ref.:	13(4,6), 47(3,4,6), 48(4,5), 72(1), 85(3,5), 87(3,4,5), 142(4,5), 172(3), 181(3,5), 190(1,3,4)

McM 5925	N24	11.9 Nov. 1960

10 Nov. 1009; 11 Nov. 0305; 12 Nov. 1315; 14 Nov. 0246; 15 Nov. 0207; 19 Nov. 1522; 20 Nov. 2017, 2126

New	developed bipolar group with small spot on W and large spot on E
CMP:	Ca 9100/3.5; spots 1740/20; F; no data
Ref.:	11(5), 12(5), 15(5), 21(3,4,5,6), 26(4), 39(1,3,4,6), 40(3,4,6), 44(4), 46(6), 47(3,4,6), 48(4,5), 64(3), 72(1), 73(4,5,6), 75(4,5,6), 84(3,5,6), 85(3,5), 86(3,5), 87(3,4,5), 104(3,6), 105(4), 119(4,5,6), 122(5), 137(4,5,6), 139(6), 142(4,5), 144(4,6), 145(4), 164(3,4,6), 170(3), 173(5,6), 174(4,5), 175(3,4,5), 177(3,4,5,6), 185(5), 189(3), 190(1,3,4), 195(3), 196(3), 199(3,4), 208(4)

McM 5959	N30	10.5 Dec. 1960

5 Dec. 1825

Age 2	return of McM 5925; small spot and pores
CMP:	Ca 6800/3.0; spots 120/3; J; no data
Ref.:	48(4,5), 72(1), 75(4,5,6), 89(3,5), 105(4), 122(5), 142(4,5), 185(5)

McM 6171   S10   14.4 July 1961

11 July 1615; 12 July 1000; 15 July 1508; 17 July 0710; 18 July 0920; 20 July 1553

Age 2   return of McM 6144; bipolar group with regular western spot and irregular eastern spot,
               before 11 July – H, after – E
CMP: Ca 5100/3.5;   spots 1420/14;   E;   no data
Ref.:   9(4,5), 11(5), 12(5), 13(4,6), 14(3,5,6), 21(3,4,5,6), 22(3,4,5,6), 41(3,4,6), 44(4), 46(6), 48(4,5),
               63(5,6), 65(3), 72(1), 84(3,5,6), 86(3,5), 87(3,4,5), 103(3,4), 105(4), 122(5), 124(4), 125(5,6),
               138(4), 142(4,5), 150(4,5), 158(6), 161(3), 163(3,4,6), 173(5,6), 174(4,5), 176(3,4,5), 177(3,4,5,6),
               185(5), 191(3), 199(3,4), 202(3)

McM 6172   N13   17.2 July 1961

15 July 1433

Age 2   return of McM 6151; regular spot and pores
CMP: Ca 4000/3.0;   spots 220/1;   A;   no data
Ref.:   48(4,5), 72(1), 122(5), 125(5,6), 138(4), 142(4,5), 150(4,5), 158(6), 173(5,6), 185(5)

McM 6178   N08   25.1 July 1961

24 July 0410

Age 2   return of McM 6155; regular stable spot, surrounded by pores
CMP: Ca 3900/3.5;   spots 550/5;   H;   no data
Ref.:   89(5), 122(5), 125(5,6), 142(4,5), 173(4,6), 184(4,5,6), 185(5), 199(3,4)

McM 6212   N16   04.2 Sept. 1961

10 Sept. 1958

Age 2   return of McM 6197; close chain of large irregular spots, 30–31 Sept. – D, 1 and 8–9 Oct. – E,
               2–7 Oct. – F
CMP: Ca 5300/2.0;   spots 990/25;   F;   no data
No ref.

McM 6227   N16   22.0 Sept. 1961

16 Sept. 1057

Age 4   return of McM 6206; regular stable spot and sometimes pores
CMP: Ca 1800/3.0;   spots 170/1;   J;   no data
Ref.:   13(4,6), 47(3,4,6), 65(3), 72(1), 141(3), 148(3,4,6), 184(3,5,6), 185(5)

McM 6235   N12   01.2 Oct. 1961

28 Sept. 2202

Age 3   return of McM 6212; disintegrating spot and pores; 25 Oct. – J, 26 Oct.–2 Nov. – C, 3–5 Nov. – Λ
CMP: Ca 3600/3.5;   spot 70/3;   C;   no data
Ref.:   6(4), 9(4,5), 14(3,5,6), 47(3,4,6), 48(4,5), 63(5,6), 89(5), 93(5), 102(3), 105(4), 122(5), 142(4,5),
               150(4,5), 158(6), 170(3), 184(4,5,6)

McM 6264	N14	03.6 Nov. 1961
10 Nov. 1434		

Age 4	return of McM 6240; fast developing bipolar group of irregular spots; 3–4 Nov. – A, 5–8 Nov. – D, 9 Nov. – C
CMP:	Ca 1500/3.0; spots 50/4; A; no data
No ref.	

McM 6581	N03	18.4 Oct. 1962
23 Oct. 1642		

Age 2	return of McM 6563 A, disintegrating bipolar group
CMP:	Ca 2800/3.5; spots 160/4; C; β
Ref.:	12(5), 23(3,4,6), 48(4,5), 50(5,6), 142(4,5), 185(5)

McM 6766	S12	14.8 Apr. 1963
10 Apr. 1034		

New	unstable bipolar group of large spot on E and pores on W, before 18 Apr. – D, after – H
CMP:	Ca 2000/3.5; spots 480/6; D; no data
Ref.:	12(5), 13(4,6), 72(1), 95(3), 122(5), 138(4), 142(4,5), 185(5), 186(3,4,5)

McM 6790	N17	04.7 May 1963
1 May 0525		

Age 2	return of part of McM 6759 that experienced a rejuvenation on the disk after Apr. 6; unstable spot and pores, 29 May – 4 June – H, 5–7 June – C, 8–10 June – J
CMP:	Ca 5000/3.5; spots 270/4; H; no data
Ref.:	49(4), 72(1), 87(3,4,5), 122(5), 184(3), 185(5), 186(3,4,5), 202(3)

McM 6832	N12	12.0 June 1963
14 June 022		

Age 2	return of McM 6803; unstable group of large irregular spot and chain of pores, 8–9 June – D, 10–14 June – E, 15 June – G, 16 June – C, 17 June – H
CMP:	Ca 3000/3.5; spots 1140/19; E; no data
No ref.	

McM 6908	N12	04.2 Aug. 1963
9 Aug. 2234		

New	regular stable spot and pores, after 4 Aug. – H
CMP:	Ca 2000/3.5; spots 490/1; H; βp
No ref.	

McM 6909   N11   05.2 Aug. 1963

---

6 Aug. 0855

Age 4   return of McM 6870; bipolar group of regular spots
CMP: Ca 2400/3.5;   spots 330/11;   C;   βp
Ref.:   13(**4,6**), 48(4,5), 67(3,6), 72(1), 142(4,5), 185(4,5), 186(3,4,5)

McM 6964   N12   20.3 Sept. 1963

---

15 Sept. 0015; 16 Sept. 1430; 20 Sept. 2314; 26 Sept. 0638

Age 3   return of McM 6931; large developed group of irregular fast changing spots, small spots and pores,
               before 24 Sept. − E, after − H
CMP: Ca 3900/3.5;   spots 1770/15;   E;   γ
Ref.:   12(5), 13(**4,6**), 21(3,4,5,6), 29(6), 34(**3,4,6**), 35(3,6), 48(4,5), 67(**3,6**), 72(1), 83(**4,6**), 84(3,5,6),
               85(3,5), 86(3,5), 87(3,4,5), 88(**3,4,6**), 109(2), 110(3), 122(5), 138(4), 142(4,5), 160(3), 165(3),
               185(4,5), 186(3,4,5), 199(3,4), 200(3), 202(3), 205(**3**)

McM 7003   N12   26.4 Oct. 1963

---

28 Oct. < 0135

Age 2   return of McM 6980, large spot accompanied by small spots and chain of pores, 21−28 Oct. − E,
               29 Oct. − C, 30−31 Oct. − J
CMP: Ca 5500/4.0;   spots 1370/11;   E;   βγ
Ref.:   72(1), 122(4,5,6)

McM 7182   M 1478−08   N04   11.3 Mar. 1964

---

16 Mar. 1553

New   fast developed group of irregular spots
CMP: Ca 200/2.0;   spots (530)/(7);   (B);   (β)
No ref.

McM 7661   M 1490−09   N09   04.0 Feb. 1965

---

2 Feb. 2043; 5 Feb. 1750

New   regular stable spot, before 6 Feb. − C, after − J
CMP: Ca 2000/3.0;   spots 315/1;   C;   (αp)
Ref.:   18(**4,5,6**), 52(**4,5,6**), 89(3,5,6), 98(3,4,5,6), 154(5)

McM 7809   M 1494−06   N23   21.2 May 1965

---

25 May 2239

New   fast changing group of disintegrating spots
CMP: Ca 4000/3.5;   spots 290/55;   D;   no data
No ref.

McM 7842	M 1494–14	S11	02.0 June 1965

5 June 1807

New	close constellation of unstable small spots and pores, 3 June – A, 4–5 June – C, 6 June – J, 7 June – A
CMP:	Ca(400)/(3.5); spots (10)/(6); (A); (βp)
Ref.:	8(4,5), 10(5)

McM 7847	M 1495–02	N22	13.4 June 1965

13 June 0257

Age 2	return of McM 7813; very small spots, degenerating to pores
CMP:	Ca 3200/3.0; spots (10)/(1); (A); (αp)
No ref.	

McM 8005	M 1499–04	N23	02.8 Oct. 1965

8 Oct. 1603

New	scattered group of unstable spots and pores, 27–28 Aug. and 7 Oct. – C, 29–30 Aug. – D, 1–6 Oct. – E, 8 Oct. – A
CMP:	Ca 3900/3.5; spots 270/87; E; γ
No ref.	

McM 8012	M 1499–01	S19	02.0 Oct. 1965

4 Oct. 0932

New	bipolar group with regular main spots
CMP:	Ca (1100)/(3.5); spots (100)/(16); (A); (βf)
Ref.:	9(4,5), 42(1,3), 52(5), 68(3,4,6), 87(3,4,5), 89(3,5,6), 206(4)

McM 8105	M 1502–07	N11	24.9 Dec. 1965

27 Dec. 0620; 29 Dec. 1123; 30 Dec. 0006; 31 Dec. 0805

New	close group of unstable spots developed from pores, after 25 Dec. – D
CMP:	Ca (1400)/(3.5); spots 30/15; J; ($\beta\gamma$)
Ref.:	183(5,6)

McM 8131	M 1503–10	N17	19.7 Jan. 1966

17 Jan. < 1029; 18 Jan. 2253

New	development from H through E to D
CMP:	Ca (3500)/(3.0); spots (290)/(53); E; ($\beta\gamma$)
Ref.:	212(6), 231(1,6), 240(4), 272(6), 44(4), 435(4)

McM 8207	M 1505−05	N19	22.1 Mar. 1966

18 Mar. 2345; 19 Mar. 0338, 2131; 20 Mar. 0928; 21 Mar. 2138; 24 Mar. 0225

Age 1+2	following portion is the return of McM 8174, leading portion − new; development from H through D, C to H
CMP:	Ca 9500/3.0; spots 900/82; D; (γ)
Ref.:	213(4,5), 9(4,5), 10(5), 231(1,6), 234(4,6), 235(1,4), 247(4,6), 42(1−5), 272(6), 274(1,2,6), 52(6), 289(6), 296(6), 318(1,3), 319(1,2,4), 326(6), 237(4,5), 328(3), 353(3,4), 357(6), 443(6)

McM 8310	M 1507−35	N16	25.7 May 1966

28 May 1532

New	fast decrease in the area after 25 May
CMP:	Ca 3900/3.0; spots 110/55; D; βf
Ref.:	9(4,5), 262(1,3), 291(1), 327(4,5), 409(1−4,6)

McM 8348	M 1509−05	S24	25.1 June 1966

25 June 1525

New	
CMP:	Ca 1000/3.0; spots 90/20; D; (βf)
No ref.	

McM 8362	M 1509−15	N35	03.4 July 1966

7 July 0025; 9 July 0310

New	has developed near the position of weak remnants of old McM 8331, remarkable growth in area and brightness on and after June 30th simultaneously with the appearance and growth of its spot group (spots area ~ 3400 near 7 July); development from A through C, D to E (after 5 July)
CMP:	Ca 1400/3.0; spots 60/17; D; (βγ)
Ref.:	212(6), 213(4,5), 9(4,5), 214(4,5), 218(6), 221(6), 228(1,2), 231(1,6), 233(4), 234(4,6), 238(4,6), 241(4,5), 242(6), 247(4,6), 259(1), 260(1,3,6), 263(1,3), 272(6), 274(1,2,6), 276(5), 281(6), 286(3,5), 289(6), 292(6), 300(6), 301(6), 308(6), 309(6), 313(6), 314(6), 323(6), 327(4,5), 330(6), 331(3), 338(6), 339(1,3), 347(4−6), 348(1,6), 349(2,3), 350(1,2), 356(3,5), 358(4), 363(1,2), 365(3), 367(6), 368(1), 369(1−6), 370(6), 371(6), 372(1−6), 380(6), 381(1−6), 386(5), 403(6), 407(6), 410(3), 412(6), 416(6), 418(6), 423(6), 428(2,3), 429(6), 431(6), 434(6), 438(6), 439(6), 443(6)

McM 8461	M 1511−19	N23	29.1 Aug. 1966

28 Aug. 1523; 29 Aug. 1324; 2 Sept. 0542; 4 Sept. 0407, 1430

Age 1−3	return of McM 8414; McM 8459 and McM 8461 are very close to each other in position on the solar disk. Many of the flares spread throughout both plages. Fast development and growth of the spots area from H through D to E (30, 31 Aug., 1 Sept.) and then D; δ-configuration after 30 Aug.
CMP:	Ca 3700/5.0; spots 810/27; D; (βγ+δ)
Ref.:	212(6), 213(4,5), 9(4,5), 10(5), 214(4,5), 225(4), 231(1,6), 232(4), 233(4), 235(1,4), 239(1,4,6), 247(4,6), 255(4), 257(1,3,4), 258(1,3), 263(1,3), 264(4,6), 269(6), 270(6), 272(6), 273(6), 274(1,2,6), 285(1,3), 289(6), 290(1,3), 291(1), 297(4,6), 298(6), 302(1,6), 309(6), 394(3−5), 324(3−5), 327, 332(6), 344(6), 348(1,6), 350(1,2), 359(4), 372(1−6), 374(1−6), 375(3−6), 377(5), 384(4), 385(5), 386(5), 397(4,6), 406(1,3−5), 407(6), 408(4), 412(6), 418(6), 419(6), 422(6), 424(6), 428(2,3), 433(4,6), 436(3,6), 437(4), 438(6), 439(6), 440(3), 442(6), 443(6), 444(3)

McM 8484	M 1511−28/29	S21	07.6 Sept. 1966

14 Sept. 1014

New
CMP: Ca 800/3.0;   spots 10/8;   B;   βp
No ref.

McM 8496	M 1512−02/04	N20	12.6 Sept. 1966

17 Sept. 0940

Age 4   return of McM 8438
CMP: Ca 2100/3.5;   spots 120/20;   D;   βp
No ref.

McM 8505	M 1512−11	N07	20.0 Sept. 1966

12 Sept. 0925; 20 Sept. < 1738

Age 2   return of McM 8454; δ-configuration after 30 Sept.
CMP: Ca 4000/3.0;   spots 650/51;   H;   (γ+δ)
Ref.:   223(1,4), 231(1,6), 240(4), 289(6), 345(1,2), 346(1,2), 360(4), 378(1−3), 393(1,3,4), 430(6)

McM 8687	M 1517−32	N22	12.9 Feb. 1967

13 Feb. 1747

New   has formed among the remnants of McM 8641; decrease in the area and disintegration
CMP: Ca 4300/3.0;   spots 10/5;   B;   αp
Ref.:   220(6), 226(6), 229(6), 261(1,3,4), 262(1,3), 273(6), 289(6), 317(6), 322(6), 333(6), 336(6),
          386(5), 396(3,5), 407(6), 421(6), 431(6), 432(6)

McM 8704	M 1518−07	N23	27.4 Feb. 1967

27 Feb. 1637

Age 2   return of McM 8670; δ-configuration after 23 Feb.
CMP: Ca (8500)/(3.5);   spots 1600/132;   F;   (γ+δ)
Ref.:   247(4,6), 274(1,2,6)

McM 8740	M 1519−10	N22	27.4 Mar. 1967

1 Apr. 1410

Age 3   return of McM 8704; southern group is stable, northern disappeared after 26 Mar.
CMP: Ca (9800)/3.5;   spots   500/2;   B;   (βf)
                                   340/39;   E;   (βγ)
Ref.:   265(6), 288(5), 289(6), 337(5)

| McM 8818 | M 1521–16 | N25 | 25.4 May 1967 |

21 May 1919; 23 May 1802; 28 May < 0529

Age 2	return of McM 8785 and McM 8793; δ-configuration after 21 May
	150/15; C
CMP:	Ca 10000/4.0; spots 1010/4; F; $(\gamma+\delta)$
	540/54; C
Ref.:	209(6), 236(3–6), 237(4), 240(4), 247(4,6), 254(3,6), 256(1,3), 274(1,2,6), 278(6), 280(6), 283(6), 284(4), 287(6), 289(6), 295(5,6), 303(1,3), 309(6), 310(6), 311(6), 312(6), 316(1,3,4–6), 320(3), 321(6), 325(2), 329(3), 336(8,6), 343(6), 361(4), 362(3), 373(6), 387(5), 399(6), 401(1,6), 402(6), 404(6), 405(6), 408(4), 411(6), 414(6), 425(6), 437(4)

| McM 8905 | M 1523–37 | N27 | 28.3 July 1967 |

30 July 0508; 31 July 0808; 1 Aug. < 0535, 1721; 3 Aug. 0918

Age 3	return of McM 8871; δ-configuration after 28 July
CMP:	Ca 7200/3.5; spots 1500/35; F; $(\gamma+\delta)$
Ref.:	266(4), 315(5), 342(6), 343(6), 394(3,6), 415(6)

| McM 8949 | M 1524–32 | S21 | 26.1 Aug. 1967 |

26 Aug. 0014

New	born on the disk on 23 Aug.
CMP:	Ca 1400/3.5; sunspot 90/33; B; (βp)
No ref.	

| McM 8973 | M 1525–24 | N17 | 14.8 Sept. 1967 |

18 Sept. 2316

New	
CMP:	Ca 2400/3.5; spots 260/12; C; (βp)
Ref.:	240(4), 343(6), 394(3,6), 415(6)

| McM 9034 | M 1526–42/43 | N12 | 24.1 Oct. 1967 |

25 Oct. < 2312; 26 Oct. 0608; 27 Oct. 1105; 29 Oct. 0258, 2347

Age 2	return of McM 8988, fast developing from J through C, D to E (after 23 Oct.), max. spots area
	28 Oct. ~ 1400
CMP:	Ca 3700/3.5; spots 130/45; E; αp
Ref.:	251(5), 334(5), 343(6)

McM 9047	M 1527−14/17	S21	03.5 Nov. 1967

2 Nov. 0852; 4 Nov. 1151

Age 1−3	return of McM 9006 and McM 9014
	120/20; C; (αp)
CMP:	Ca (6200)/(3.5); spots 20/2 ; B
	290/19; H; (βp)
Ref.:	250(5), 287(6), 343(6)

McM 9108	M 1528−30/32	S18	10.5 Dec. 1967

11 Dec. 2347

Age 2 + 1	return of McM 9064
CMP:	Ca (3700/3.5); spots 110/14; C; βp
No ref.	

McM 9118	M 1529−05	N21	22.0 Dec. 1967

16 Dec. 0247

Age 4 + 2	return of McM 9082 and McM 9093 development from H through D, C to J, max. spots area
	18 Dec. ~ 640
CMP:	Ca (9100)/(3.0); spots 250/14; C; βf
Ref.:	250(5), 287(6), 343(6), 354(4), 405(6)

McM 9145	M 1529−38/40	S21	9.4 Jan. 1968

11 Jan. 1659; 12 Jan. 1807

Age 3 + 2	return of McM 9110 and McM 9108
CMP:	Ca (9000)/(3.0); spots 720/44; D; ($\beta\gamma$) (6.8/0.50− for 10 Jan.)
	30/26; J; (βf)
Ref.:	250(5), 343(6)

McM 9146	M 1529−43	N12	11.0 Jan. 1968

5 Jan. 0458; 12 Jan. 0019; 17 Jan. 0534

Age 2 + 1	return of McM 9114
CMP:	Ca 8500/2; spots 840/120; E; $\beta\gamma + \delta$
No ref.	

McM 9184	M 1530−30/32	N14	31.0 Jan. 1968

29 Jan. 1537; 1 Feb. 1802; 2 Feb. 0541

New	
CMP:	Ca 7900/3.5; spots 1970/95; F; ($\beta\gamma + \delta$)
No ref.	

McM 9204	M 1531−02	N20	14.8 Feb. 1968

17 Feb. 0252

New
has developed in the midst of old McM 9165
CMP:
Ca 6100/3.5; spots 570/38; E; (βf); (6.4/0.35 − for 16 Feb.)
No ref.

McM 9372	M 1534−07	N24	07.9 May 1968

3 May 2123

Age 2
return of McM 9324
CMP:
Ca 4300/2.5; spots 260/18; E; (βγ)
No ref.

McM 9429	M 1535−16/17	S12	09.0 June 1968

7 June 1226; 9 June 0830

New
disappeared 14 June, max. phase of development 6 and 7 June (D)
CMP:
Ca 3200/3.0; spots 40/26; C; (γ); 2.5/0.78
Ref.:
216(1−6), 224(4,6), 246(4,6), 247(4,6), 253(6), 293(5), 305(1,6), 307(2), 351(4), 391(5), 425(6), 441(6)

McM 9503	M 1536−20/24	N12	13.0 July 1968

6 July 0943; 8 July < 1708; 9 July 1809; 12 July < 0000

New
appeared near the location of old McM 9445
CMP:
Ca 8000/3.5; spots 1000/102; E; (βp); (26.3/0.72 − for 9 July)
Ref.:
243(1,4), 245(4), 246(4,6), 247(4,6), 248(4), 253(6), 275(1−6), 388(4,5), 400(1,3,4), 427(4)

McM 9567	M 1537−23/24	N13	08.9 Aug. 1968

3 Aug. 0313; 6 Aug. 1315, < 1344; 14 Aug. < 1327, 1400

Age 2
return of McM 9503; max. phase of development 6 and 7 Aug. (G)
CMP:
Ca 7600/4.0; spots $\frac{600/46;}{10/5}$ $\begin{array}{l}H;\\J;\end{array}$ (γ+δ); (13.9/0.95 − for 4 Aug.)
Ref.:
246(4,6)

McM 9593	M 1537−39	S17	18.0 Aug. 1968

21 Aug. 0146

Age 3
return of McM 9528; 14−19 Aug. − F, 19−22 Aug. − C, 23 Aug. − H
CMP:
Ca 6100/3.5; spots 660/48; F;
No ref.

McM 9630	M 1538−21	N15	02.8 Sept. 1968

4 Sept. < 0031

New	plage developed in the position of old region 9564 of the previous rotation
CMP:	Ca 3500/2.5; sunspots 250/14; D; (βγ)
No ref.	

McM 9678	M 1539−11	N16	25.7 Sept. 1968

29 Sept. 1617

Age 2 + 1	return of McM 9638; gradual decrease in the area, after 26 Sept. − D
CMP:	Ca 3800/3.0; spots $\begin{array}{l} 310/62; \\ 20/37; \end{array}$ $\begin{array}{l} E; \\ C; \end{array}$ $\begin{array}{l} (βγ) \\ (βp) \end{array}$ (8.6/0.53 − for 28 Sept.)
Ref.:	253(6)

McM 9687	M 1539−17/18	N16	29.4 Sept. 1968

26 Sept. 0026; 28 Sept. < 0155; 3 Oct. 1238

Age − 2 r.	return of McM 9630 and portion of McM 9623; after 30 Sept. − J
CMP:	Ca 5300/3.0; spots $\begin{array}{l} 300/31; \\ 190/27; \end{array}$ $\begin{array}{l} C; \\ C; \end{array}$ $\begin{array}{l} (βp) \\ (αp) \end{array}$
Ref.:	210(4), 247(4,6)

McM 9692	M 1539−21	S14	01.1 Oct. 1968

28 Sept. 0721; 3 Oct. 2343

Age 2	return of McM 9634; weak disintegration, max. phase of the development ~ 28, 29, 30 Oct. (H)
CMP:	Ca 3000/3.0; spots 260/10; J; (αp); 11.5/0.74
Ref.:	253(6), 299(6)

McM 9740	M 1540−20	S15	28.4 Oct. 1968

23 Oct, 2352; 26 Oct, 0046; 27 Oct. 1232, < 1318; 29 Oct, < 1222, 1515; 30 Oct. < 1235, 2340; 1 Nov. 0814;
4 Nov. < 0524

Age 3	return of McM 9692; after 27 Oct. − (γ+δ)
CMP:	Ca (7800)/(3.5); spots 910/45; H; (γ+δ); 24.0/0.85
Ref.:	211(1,3−5), 244(4), 246(4,6), 247(4,6), 248(4), 253(6), 294(6), 352(4,5), 392(6), 445(3)

McM 9760	M 1541−01	N18	11.5 Nov. 1968

18 Nov. < 1026

Age 2	return of McM 9736; after 7 Nov. − H, after 8 Nov. − (γ+δ)
CMP:	Ca 3000/4.0; spots $\begin{array}{l} 800/28; \\ -/-; \end{array}$ $\begin{array}{l} H; \\ A; \end{array}$ $\begin{array}{l} (γ+δ) \\ (βp) \end{array}$ (6.9/0.91 − for 17 Nov.)
Ref.:	246(4,6), 247(4,6), 248(4), 251(1,6), 252(1,6), 267(6), 304(2,3), 305(1,6), 343(6), 382(6), 383(6), 398(4), 417(6), 441(6)

McM 9802   M 1542−02   N21   08.7 Dec. 1968

2 Dec. 2115, < 2202

Age 3   return of McM 9760
                                                      200/13; J;   (αf)
CMP: Ca 4100/2.5;   spots   10/4 ;   A;   (αp);   (9.0/0.47 − for 3 Dec.)
                                                     (10)/(4); A;   (β)
Ref.:   247(4,6), 253(6), 343(6)

McM 9842   M 1542/29   N18   27.6 Dec. 1968

24 Dec. 2227; 27 Dec. < 1050

New   appeared before 24 Dec. in the position of old McM 9792, after 22 Dec. − (γ+δ), after 31 Dec. −
                   desintegration (F−D−C)
CMP: Ca (5900)/(3.5);   spots 680/52;   F;   (γ+δ);   6.5/0.47
Ref.:   245(4), 246(4,6), 248(4), 413(6)

McM 9879   M 1543−18/19   N22   24.3 Jan. 1969

24 Jan. < 0706

Age 2   return of McM 9842; weak desintegration
CMP: Ca (6300)/(3.5);   spots 60/4;   D;   (αp);   1.9/0.61
Ref.:   219(1,3,5,6), 247(4,6), 306(6), 389(6)

McM 9946   M 1544−38   N18   23.6 Feb. 1969

23 Feb. 0442; 24 Feb. 2305; 25 Feb. 0900; 26 Feb. 0418; 27 Feb. 1348

Age 2   return of McM 9903 and 9902, gradual decrease in the area
                                                      260/34;   D
CMP: Ca 8600/3.5;   spots 120/18;   D   (βγ);   4.0/0.55
                                                      380/36;   D
                                                      10/6 ;   C
Ref.:   210(4), 215(6), 222(6), 245(4), 246(4,6), 247(4,6), 248(4), 253(6), 271(5,6), 305(1−6), 341(6),
                   390(6), 398(4)

McM 9966   M 1545−16   N13   06.6 Mar. 1969

12 Mar. 1739

Age 3   return of McM 9911
CMP: Ca 6200/3.0;   spots 100/28; D;   (βf)   (7.8/0.67 − for 11 Mar.)
                                                       10/8 ; C;   (β)
Ref.:   247(4,6), 282(3), 340(1,2)

McM 9994	M 1545−34/38	N18	21.9 Mar. 1969

21 Mar. 0139, 1312; 24 Mar. 1449; 27 Mar. 1315; 30 Mar. 0332

Age 2	return of McM 9941

CMP: Ca 28000/3.5; spots

20/6 ;	C;	(βf)
760/40;	D;	$(\beta\gamma)$
660/38;	E;	(βp)
290/14;	C;	$(\beta\gamma)$

36.2/0.94

Ref.: 217(4,6), 222(6), 227(4), 230(6), 245(4), 246(4,6), 247(4,6), 248(4), 277(4,5), 364(4), 395(6)

McM 10035	M 1546−34/36	N17	18.3 Apr. 1969

10 Apr. 0410

Age 3−4   return of part of old McM 9994; complex region with different development of its parts (max. phase of development − 20 Apr.)

CMP: Ca (12300)/(3.5); spots

(250)/(18);	D;	$(\beta\gamma)$
(10)/(3) ;	C;	(αp)
(0)/(5) ;	B;	$(\beta\gamma)$
(180)/(14);	D;	$(\beta\gamma)$

(163/0.82 − for 11 Apr.)

Ref.: 247(4,6), 395(6)

McM 10057	M 1547−15	N13	30.4 Apr. 1969

2 May 1745; 5 May 1237

Age 2, 4+5   return of McM 10014; fast change in the area and the type of the group
CMP: Ca 9100/3.0;   spots 180/34;   H; $(\beta\gamma)$
No ref.

McM 10109	M 1548−11	N14	26.3 May 1969

28 May 1241; 29 May 0020, 1939

Age 3, 5+6   return of McM 10057; fast change in the area and the type of the group, maximal plage − 28 May than of McM 10057, maximal plage − 28 May

CMP: Ca 11000/3.5; spots

20/3 ;	G;	(αp)
40/19;	C;	(β)
10/6 ;	C;	(αp)

No ref.

McM 10134	M 1548−37	N17	10.5 June 1969

5 June 0952, < 1442; 7 June < 0945; 11 June 1615

Age 2, 5+6   return of McM 10088, 10095, 10098; 7−8 June − maximal phase of the development, spots 700/14, H, then − fast decrease in the area
CMP: Ca 13000/3.5;   spots 30/1;   J; (βp)
No ref.

MCM 10326	M 1552−23	N17	24.5 Sept. 1969

25 Sept. < 0658

Age 3   return of McM 10289
CMP: Ca 4200/2.5;   spots 0/1;   A;   (αp)
Ref.:   247(4,6)

McM 10333	M 1552−31/33	N10	27.4 Sept. 1969

27 Sept. < 0350

Age 3   return of McM 10289, fast decrease in the area after 26 Sept.

CMP: Ca 4800/3.5;   spots   30/26;   D;   (γ+δ)   6.3/0.77
                                      0/2 ;   A;   (αf)
Ref.:   247(4,6), 376(5), 379(4)

McM 10352	M 1553−07	N27	08.6 Oct. 1969

14 Oct. 0539

New   fast decrease in the area after 6 Oct.
CMP: Ca 1900/3.5;   spots 60/3;   C;   (βp)
No ref.

McM 10385	M 1553−32	N12	26.6 Oct. 1969

2 Nov. < 1102

Age 3   return of McM 10335
CMP: Ca 10000/3.5;   spots (420)/(16);   F;   (γ+δ);   (24.6/0.97 − for 31 Oct.)
Ref.:   247(4,6), 355(4)

McM 10432	M 1554−35	N10	22.1 Nov. 1969

18 Nov. 1633; 20 Nov. 1619; 23 Nov. 0958; 24 Nov. 0914

Age 4 + 5   return of McM 10381 and McM 10385

CMP: Ca 14400/3.5;   spots   250/50;   E;   (γ)   (81.4/0.91 − for 23 Nov.)
                                      600/40;   E;   (γ+δ)
Ref.:   245(4), 247(4,6), 248(4)

McM 10491	M 1555−41	S13	24.4 Dec. 1969

30 Dec. 1927

New   gradual decrease in the area
CMP: Ca 1500/3.0;   spots 490/31;   D;   (β);   (5.2/0.59 − for 29 Dec.)
Ref.:   279(5)

REFERENCES TO PART 3 AND APPENDIX B

1   Akasofu, S.-I. and Chapman, S.: 1961, Geophys. Inst. Univ. Alaska Sci. Rept. No. 7, UAG, R 112.
2   Akasofu, S.-I. and Chapman, S.: 1962, *J. Atmospheric Terrest. Phys.* **24**, 785.
3   Akasofu, S.-I. and Chapman, S: 1960, *I.U.G.G. Monogr.* No. 7, p. 93. (Symp. on the July 1959 Events and Assoc. Phenom., Helsinki, July 1960).
4   Anderson, K. A.: 1958, *Phys. Rev.* **111**, 1397.
5   Anderson, K. A., Arnoldy, R., Hoffman, R., Peterson, L., and Winckler, J. R.: 1959, *J. Geophys. Res.*, **64**, 1133.
6   Anderson, K. A. and Winckler, J. R.: 1962, *J. Geophys. Res.* **67**, 4103.
7   Anderson, K. A.: 1964, *J. Geophys. Res.* **69**, 1743.
8   Arnoldy, R. L., Kane, S. R., and Winckler, J. R.: 1967, *Solar Phys.* **2**, 171.
9   Arnoldy, R. L., Kane, S. R., and Winckler, J. R.: 1968, *Astrophys. J.* **151**, 711.
10   Arnoldy, R. L., Kane, S. R., and Winckler, J. R.: 1968, University of Minnesota Technical Report, C. R. 108, Jan. 1968.
11   Bailey, D. K.: 1964, *Planetary Space Sci.* **12**, 495.
12   Basler, P. and Owren, L.: 1964, Geophys. Inst. Univ. Alaska, Sci. Rept., UAG-R 152.
13   *Beobachtungsergebnisse, 1957–1961*, 8–12, Akademie der Wissenschaften D.D.R., Zentralinstitut für solar-terrestrische Physik, Berlin-Adlershof.
14   Bhargava, B. N. and Subrahmanyan, R. V.: 1966, *Planetary Space Sci.* **14**, 871.
15   Biswas, S., Freier, P. S., and Stein, W.: 1962, *J. Geophys. Res.* **67**, 13.
16   Boischot, A. and Warwick, J. W.: 1959, *J. Geophys. Res.* **64**, 683.
17   Boorman, J. A., McLean, D. J., Sheridan, K. V., and Wild J. P.: 1961, *Monthly Notices Roy. Astron. Soc.* **123**, 87.
18   Bouška, J.: 1968, *Ann. Geophys.* **24**, 807.
19   Brown, R. R. and D'Arcy, R. G.: 1962, *Arkiv. Geophys.* **3**, 443.
20   Bruzek, A.: 1958, *Z. Astrophys.* **44**, 183.
21   Bruzek, A.: 1964, *Astrophys. J.* **140**, 746.
22   Bruzek, A: 1964, AAS-NASA Symp. on the Physics of Solar Flares, NASA-SP 50, p. 301.
23   Bryant, D. A., Cline, T. L., Desai, U. D., and McDonald, F. B.: 1964, NASA, Goddard Space Flight Center, Greenbelt, Maryland, X-611-64-217.
24   Chapman, J. H.: 1960, *Can. J. Phys.* **38**, 1195.
25   Boischot, A. and Clavelier, B.: 1968, *Ann. Astrophys.* **31**, 445.
26   Covington, A. E. and Harvey, G. A.: 1961, *Phys. Rev. Letters* **6**, 51.
27   Davis, L. R., Fichtel, C. E., Guss, D. E., and Ogilvie, K. W.: 1961, *Phys. Rev. Letters* **6**, 492.
28   De Feiter, L. D., Fokker, A. D., Van Lohuizen, H. P., and Roosen, J.: 1960, *Planetary Space Sci.* **2**, 223.
29   De Jager, C.: 1967, *Solar Phys.* **2**, 327.
30   Dodson, H. W. and Hedeman, E. R.: 1964, AAS-NASA Symposium on the Physics of Solar Flares, NASA-SP-50, p. 15.
31   Dodson, H. W. and Hedeman, E. R.: 1966, *Astrophys. J.* **145**, 224.
32   Dorman, L. I. and Feinberg, E. L.: 1958, *Nuovo Cimento* 8, Suppl. 2, 358.
33   Ellison, M. A.: 1957, in L. d'Azambuja (ed.), *Neuvième Rapport de la Commission pour l'Etude des Relations entre les Phénomènes Solaires et Terrestres,* ICSU, Paris, p. 15.
34   Elliot, I. and Reid, J. H.: 1965, *Planetary Space Sci.* **13**, 163.
35   Elliot, I. and Reid, J. H.: 1966, *Observatory* **86**, 63.
36   Ellison, M. A. and Reid, J. H.: 1956, *J. Atmospheric Terrest. Phys.* **8**, 291.
37   Ellison, M. A., McKenna, S. M. P., and Reid, J. H.: 1960, *Dunsink Obs. Publ.* **1**, 1.
38   Ellison, M. A., McKenna, S. M. P., and Reid, J. H.: 1960, *Dunsink Obs. Publ.* **1**, 39.
39   Ellison, M. A., McKenna, S. M. P., and Reid, J. H.: 1960, *Dunsink Obs. Publ.* **1**, 53.
40   Ellison, M. A., McKenna, S. M. P., and Reid, J. H.: 1961, *Monthly Notices Roy. Astron. Soc.* **122**, 491.
41   Ellison, M. A., McKenna, S. M. P., and Reid, J. H.: 1962, *Monthly Notices Roy. Astron. Soc.* **124**, 263.

42   Falciani, R., Landini, M., Righini, A., and Rigutti, M.: 1968, in K. O. Kiepenheuer (ed.), 'Structure and Development of Solar Active Regions', *IAU Symp.* **35**, 451.
43   Fokker, A. D.: 1962, *Arkiv Geophys.* **3**, 459.
44   Fokker, A. D., Goh, T., Landré,E., and Roosen, J.: 1966, *Bull. Astron. Inst. Neth. Suppl.* **1**, 309.
45   Forbush, S. E.: 1956, *J. Geophys. Res.* **61**, 155.
46   Freier, P. S. and Weber, W. R.: 1963, *J. Geophys. Res.* **68**, 1605.
47   Fritzová-Švestková, L. and Hřebík, F.: 1964, *Bull. Astron. Inst. Czech.* **15**, 222.
48   Fritzová-Švestková, L. and Švestka, Z.: 1964, *Bull. Astron. Inst. Czech.* **17**, 249.
49   Frost, K. J.: 1964, NASA, Goddard Space Flight Center, Greenbelt, Maryland, X-610-64-60.
50   Giovanelli, R. G. and Roberts, J. A.: 1958, *Australian J. Phys.* **11**, 353.
52   Goedeke, A. D., Masley, A. J., and Adams, G. W.: 1967, *Solar Phys.* **1**, 285.
53   Gold, T. and Palmer, D. R.: 1956, *J. Atmospheric Terrest. Phys.* **8**, 287.
54   Greenstadt, E. W. and Moreton, G. E.: 1962, *J. Geophys. Res.* **67**, 3299.
55   Hachenberg, O.: 1958, *Z. Astrophys.* **46**, 67.
56   Hachenberg, O., Fürstenberg, F., Helms, B., and Krüger, A.: 1963, 'Radiostrahlungsausbrüche der Sonne im IGI Katalog der im Heinrich-Hertz-Institut beobachteten besonderen Ereignisse des cm- und unteren dm-Wellengebietes', Berlin-Adlershof.
57   Hachenberg, O. and Wallis, G.: 1961, *Z. Astrophys.* **52**, 42.
58   Hakura, Y. and Goh, T.: 1959, *J. Radio Res. Lab. Japan* **6**, 635.
59   Hakura, Y. and Goh, T.: 1961, *Rept. Ionos. Space Res. Japan* **15**, 235.
60   Hakura, Y. and Nagai, M.: 1964, *J. Radio Res. Lab. Japan* **11**, 197.
61   Hakura, Y.: 1966, *Rept. Ionos. Space Res. Japan* **20**, 128.
62   Harvey, G. A: 1965, *J. Geophys. Res.* **70**, 2961.
63   Haurwitz, M. W., Yoshida, S., and Akasofu, S. I.: 1965, *J. Geophys. Res.* **70**, 2977.
64   Hřebík, F., Kvíčala, J., Křivský, L., and Olmr, J.: 1961, *Bull. Astron. Inst. Czech.* **12**, 169.
65   Hřebík, F., Kvíčala, J., Křivský, L., and Olmr, J.: 1962, *Bull. Astron. Inst. Czech.* **13**, 199.
67   Hřebík, F., Kvíčala, J., Křivský, L., and Olmr, J.: 1965, *Bull. Astron. Inst. Czech.* **16**, 38.
68   Hřebík, F., Kvíčala, J., Křivský, L., and Olmr, J.: 1970, *Bull. Astron. Inst. Czech.* **21**, 170.
69   Hyder, C. L.: 1967, *Solar Phys.* **2**, 49.
70   Jefferies, J. T. and Orrall, F. Q.: 1961, *Astrophys. J.* **133**, 963.
71   Jelly, D. H. and Collins, C.: 1962, *Can. J. Phys.* **40**, 706.
72   WDC-A: 1957–1965, IGY Solar Activity Rep. Ser. No. 1, 2, 3, 5–11, 19, 20, 22, 26, 30, 31.
73   Kahle, A. B.: 1962, Geophys. Inst. Univ. Alaska, Sci. Rept. No. 2, UAG R129.
74   Kakinuma, T. and Tanaka, H.: 1961, *Proc. Res. Inst. Atm. Nagoya Univ.* **8**, 39.
75   Kamiya, Y.: 1961, *J. Geomagn. Geoelec.* **13**, 33.
76   Kleczek, J. and Křivský, L.: 1960, *Bull. Astron. Inst. Czech.* **11**, 165.
77   Koeckelenbergh, A.: 1959, *Ciel Terre* **75**, 28.
78   Křivský, L. and Růžičková, B.: 1959, *Bull. Astron. Inst. Czech.* **10**, 1.
79   Křivský, L. and Tlamicha, A.: 1960, *Bull. Astron. Inst. Czech.* **11**, 238.
81   Křivský, L.: 1962, *Bull. Astron. Inst. Czech.* **13**, 59.
82   Křivský, L.: 1964, *Bull. Astron. Inst. Czech.* **15**, 75.
83   Křivský, L.: 1964, *Bull. Astron. Inst. Czech.* **15**, 115.
84   Křivský, L.: 1965, *Bull. Astron. Inst. Czech.* **16**, 27.
85   Křivský, L.: 1966, *Bull. Astron. Inst. Czech.* **17**, 141.
86   Křivský, L. and Makarov, V. I.: 1966, *Bull. Astron. Inst. Czech.* **17**, 234.
87   Křivský, L. and Krüger, A.: 1966, *Bull. Astron. Inst. Czech.* **17**, 243.
88   Křivský, L.: 1969, *Bull. Astron. Inst. Czech.* **20**, 163.
89   Křivský, L.: 1969, *Bull. Astron. Inst. Czech.* **20**, 293.
90   Kundu, M. R.: 1961, *Astrophys. J.* **134**, 96.
91   Kvíčala, J., Hřebík, F., Letfus, V., Olmr, J., Švestka, Z., and Křivský, L.: 1960, *Czech. Acad. Sci. Astron. Inst. Publ.* No. 43.
92   Kvíčala, J., Hřebík, F., Olmr, J., Švestka, Z., and Křivský, L.: 1961, *Bull. Astron. Inst. Czech.* **12**, 47.
93   Leinbach, H.: 1962, Geophys. Inst. Univ. Alaska, Sci. Rept. No. 3, UAG R127.
94   Letfus, V.: 1957, *Observatory* **77**, 75.
95   Letfus, V., Růžičková, B., and Švestka, Z.: 1959, *Bull. Astron. Inst. Czech.* **10**, 136.
97   Lincoln, J. V.: 1966, WDC-A IGY Solar Activity Rept., Ser. No. 24.
98   Lincoln, J. V.: 1968, *Ann. Geophys.* **24**, 793.
99   Lockwood, J. A.: 1960, *J. Geophys. Res.* **65**, 3859.
100   Lüst, R. and Simpson, J. A.: 1957, *Phys. Rev.* **108**, 1563.

101   Maeda, K., Patel, V. L., and Singer, S. F.: 1961, *J. Geophys. Res.* **66**, 1569.
102   Malville, J. M. and Moreton, G. E.: 1963, *Publ. Astron. Soc. Pacific* **75**, 176.
103   Martres, M.-J. and Pick, M.: 1962, *Ann. Astrophys.* **25**, 293.
104   Mathews, T., Thambyahpillai, T., and Webber, W. R.: 1961, *Monthly Notices Roy. Astron. Soc.* **123**, 97.
105   Maxwell, A., Hughes, M. P., and Thompson, A. R.: 1963, *J. Geophys. Res.* **68**, 1347.
106   McCracken, K. G.: 1959, *Nuovo Cimento* **13**, 1074.
107   McKenna-Lawlor, S. M. P.: 1970, *Astrophys. J.* **159**, 51.
108   Michard, R.: 1959, *Ann. Astrophys.* **22**, 887.
109   Moreton, G. E. and Severny, A. B.: 1968, *Solar Phys.* **3**, 282.
110   Morgante, O. and Torrisi, S.: 1964, *Mem. Soc. Astron. Ital.* **35**, 217.
111   Mouradian, Z., Olivieri, G., and Soru, I.: 1963, A.G.I., participation française, Série VI, fasc.1, Activité Solaire, CNRS, Paris.
112   Müller, R.: 1956, *Observatory* **76**, 188.
113   Nagai, M. and Hakura, Y.: 1966, *Rept. Ionos. Space Res. Japan* **20**, 69.
114   Notuki, M., Hatanaka, T., and Unno, W.: 1956, *Publ. Astron. Soc. Japan* **8**, 52.
115   Obayashi, T.: 1958, *Rept. Ionos. Space Res. Japan* **12**, 3.
116   Obayashi, T. and Hakura, Y.: 1960, *Rept. Ionos. Space Res. Japan* **14**, 1.
117   Obayashi, T. and Hakura, Y.: 1960, *J. Atmospheric Terrest. Phys.* **18**, 101.
118   Obayashi, T. and Hakura, Y.: 1960, *J. Geophys. Res.* **65**, 3143.
119   Obayashi, T.: 1961, *J. Geomag. Geoelec.* **13**, 11.
120   Obayashi, T.: 1962, *J. Geophys. Res.* **67**, 2039.
121   Obayashi, T.: 1966, *Rept. Ionos. Space Res. Japan* **20**, 3.
122   Obayashi, T.: 1967, 'Table of Solar Geophysical Events (January 1957–December 1963)', Ionosphere Research Laboratory, Kyoto University.
123   Ortner, J., Leinbach, H., and Sugiura, M.: 1962, *Arkiv Geophys.* **3**, 429.
124   Pick-Gutman, M.: 1961, *Ann. Astrophys.* **24**, 183.
125   Pieper, G. F., Zmuda, A. J., Bostrom, C. O., and O'Brien, B. J.: 1962, *J. Geophys. Res.* **67**, 4959.
126   Pomerantz, M. A., Agarwal, S. P., and Potnis, V. R.: 1960, *J. Franklin Inst.* **269**, 235.
127   Reid, G. C. and Collins, C.: 1959, *J. Atmospheric Terrest. Phys.* **14**, 63.
128   Reid, G. C. and Leinbach, H.: 1959, *J. Geophys. Res.* **64**, 1801.
129   Ionosphere Research Committee, Sci. Council of Japan (ed.): 1957, *Rept. Ionos. Res. Japan* **11**, 67.
131   Ionosphere Research Committee, Sci. Council of Japan (ed.): 1960, *Rept. Ionos. Space Res. Japan* **14**, 321.
133   Russo, D. and Righini, G.: 1961, *Mem. Soc. Astron. Ital.* **32**, 193.
134   Sakurai, K.: 1967, *Rept. Ionos. Space Res. Japan* **21**, 213.
135   Sinno, G. and Hakura, Y.: 1958, *Rept. Ionos. Res. Japan* **12**, 285.
136   Sinno, K. and Hakura, Y.: 1958, *Rept. Ionos. Res. Japan* **12**, 296.
137   Sinno, K.: 1961, *J. Geomag. Geoelec.* **13**, 1.
138   *Solar Geophysical Data,* NBS Boulder, 1957–1965, CRPL-F part B 149–256.
139   Steljes, J. F., Carmichael, H., and McCracken, K. G.: 1961, *J. Geophys. Res.* **66**, 1363.
140   Mount Wilson Obs.: 1955, *Publ. Astron. Soc. Pacific* **67**, 187.
141   Švestka, Z.: 1962, *Bull. Astron. Inst. Czech.* **13**, 190.
142   Švestka, Z. and Olmr, J.: 1966, *Bull. Astron. Inst. Czech.* **17**, 4.
143   Takakura, T. and Kai, K.: 1961, *Publ. Astron. Soc. Japan* **13**, 94.
144   Tanaka, H., Kamada, T., Outsu, J., and Jwai, A.: 1961, *Proc. Res. Inst. Atm. Nagoya Univ.* **8**, 1.
145   Tlamicha, A. and Olmr, J.: 1964, *Bull. Astron. Inst. Czech.* **15**, 133.
146   Tandberg-Hanssen, E., Curtiss, W., and Watson, K.: 1959, *Astrophys. J.* **129**, 238.
147   Valníček, B.: 1961, *Bull. Astron. Inst. Czech.* **12**, 237.
148   Valníček, B.: 1962, *Bull. Astron. Inst. Czech.* **13**, 91.
150   Warwick, C. S. and Haurwitz, M. W.: 1962, *J. Geophys. Res.* **67**, 1317.
151   Warwick, J. W.: 1968, *Solar Phys.* **5**, 111.
152   Warwick, J. W.: 1963, WDC-A IGY Solar Activity Rept., Ser. No. 23.
153   Wayman, P. A.: 1957, *Observatory* **77**, 24.
154   Williams. D. J. and Bostrom, C. O.: 1967, *J. Geophys. Res.* **72**, 4497.
155   Winckler, J. R.: 1960, *J. Geophys. Res.* **65**, 1331.
156   Winckler, J. R. and Bhavsar, P. D.: 1960, *J. Geophys. Res.* **65**, 2637.
157   Winckler, J. R., Bhavsar, P. D., and Peterson, L.: 1961, *J. Geophys. Res.* **66**, 995.

404 PART 3

158 Winckler, J. R., Bhavsar, P. D., and Anderson, K. A.: 1962, *J. Geophys. Res.* **67**, 3717.
159 Winckler, J. R. and Bhavsar, P. D.: 1963, *J. Geophys. Res.* **68**, 2099.
160 Zirin, H. and Werner, S:: 1967, *Solar Phys.* **1**, 66.
161 Zirin, H. and Acton, L. W.: 1967, *Astrophys. J.* **148**, 501.
162 Abramenko, S. I., Dubov, Z. E., Ogir', M. B., Steshenko, N. E., Shaposhnikova, E. F., and Tsap, T. T.: 1960, *Izv. Krymsk. Astrofiz. Obs.* **23**, 341.
163 Akin'yan, S. T. and Dolginova, Yu. N.: 1962, *Soln. Dannye*, No. 6, 61.
164 Akin'yan, S. T. and Dolginova, Yu. N.: 1965, *Solnechnaya Aktivnost'*, Nauka, Moscow, p. 183.
165 Banin, V. G.: 1966, *Izv. Krymsk. Astrofiz. Obs.* **35**, 190.
167 Bumba, V., Gopasyuk, S. I., Dvoryashin, A. S., Eryushev, N. N., Moiseev, I. G., Savich, N. A., Severny, A. B., and Stepanov, V. E: 1957, *Soln. Dannye*, No. 1, 133.
168 Vladimirskij, B. M., Dvoryashin, A. S., Eryushev, N. N., Moiseev, I. G., Neshpor, Yu. N., Ogir', M. B., and Odintsova, I. N.: 1961, *Izv. Krymsk. Astrofiz. Obs.* **26**, 74.
169 Gopasyuk, S. I.: 1958, *Izv. Krymsk. Astrofiz. Obs.* **19**, 100.
170 Gopasyuk, S. I., Ogir', M. B., Severny, A. B., and Shaposhnikova, E. F.: 1963, *Izv. Krymsk. Astrofiz. Obs.* **29**, 15.
171 Gurtovenko, E. A. and Semenova, N. I.: 1961, *Byull. M.G.G. USSR* **3**, 47.
172 Gusejnov, R. Z., Avakova, L. M., and Gusejnov, K. I.: 1961, *Soln. Dannye*, No. 1, 59.
173 Dvoryashin, A. S.: 1962, *Izv. Krymsk. Astrofiz. Obs.* **28**, 293.
174 Dvoryashin, A. S.: 1965, *Solnechnaya Aktivnost'*, Nauka, Moscow, p. 92.
175 Dolginova, Yu. N. and Korchak, A. A.: 1968, *Soln. Dannye*, No. 5, 99.
176 Dolginova, Yu. N. and Korchak, A. A.: 1968, *Soln. Dannye*, No. 6, 81.
177 Dolginova, Yu. N. and Odintsova, I. N.: 1967, *Geomag. Aeronom.* **7**, 658.
178 Dorman, L. I., Kaminer, N. S., Konava, V. K., and Shvartsman, B. F.: 1956, *Dokl. Akad. Nauk. SSSR* **108**, 809.
179 Driatskij, V. M.: 1960, *Issled. Ionos.* No. 3, 27.
181 Koval', A. N. and Steshenko, V. N.: 1963, *Izv. Krymsk. Astrofiz. Obs.* **30**, 200.
183 Křivský, L. and Serafimov, K: 1967, *Geomag. Aeronom.* **7**, 829 and Křivský, L. and Nestorov, G.: 1967, *Bull. Astron. Inst. Czech.* **18**, 143.
184 Levitskij, L. S.: 1965, *Izv. Krymsk. Astrofiz. Obs.* **34**, 16.
185 Levitskij, L. S.: 1970, *Izv. Krymsk. Astrofiz. Obs.* **41–42**, 203.
186 Levitskij, L. S.: 1966, *Izv. Krymsk. Astrofiz. Obs.* **35**, 253.
187 Moiseev, I. G., Yurovskaya, L. I., and Yurovskij, Yu. F.: 1969, *Izv. Krymsk. Astrofiz. Obs.* **39**, 325.
188 Ogir', M. B. and Steshenko, N. E.: 1961, *Izv. Krymsk. Astrofiz. Obs.* **25**, 134.
189 Ogir', M. B. and Shaposhnikova, E. F.: 1965, *Izv. Krymsk. Astrofiz. Obs.* **34**, 272.
190 Ogir', M. B.: 1967, *Izv. Krymsk. Astrofiz. Obs.* **37**, 94.
191 Polupan, P. N.: 1968, *Problemy Kosmicheskoj Fisiki* **3**, 196.
192 Razmadze, T. S.: 1960, *Soln. Dannye*, No. 7, 68.
193 Razmadze, T. S. and Tskhovrebadze, A. S.: 1962, *Abastumansk. Obs. Byull.* **29**, 3.
194 Razmadze, T. S. and Gogosashvili, N. Z.: 1965, *Abastumansk. Obs. Byull.* **32**, 123.
195 Samojlova, L. D.: 1966, *Issledovaniya po geomagnetizmu i aeronomii*, Nauka, Moscow, p. 160.
196 Severny, A. B.: 1963, *Izv. Krymsk. Astrofiz. Obs.* **30**, 161.
197 Slonim, Yu. M.: 1956, *Soln. Dannye*, No. 3, 138.
198 Slonim, Yu. M.: 1960, *Astron. Zh.* **37**, 347.
199 Slonim, Yu. M.: 1968, *Astron. Zh.* **45**, 286.
200 Slonim, Yu. M.: 1968, *Soln. Dannye*, No. 5, 79.
202 Slonim, Yu. M.: 1969, *Astron. Zh.* **46**, 697.
203 Smith, H. J. and Smith, E. v. P.: 1966, *Solnechnye Vspyshki* (translation of *Solar Flares*), Mir, Moscow, p. 73.
204 Friedman, H., Chubb, T. A., Kupperian, J. E., Jr., Kreplin, R. W., and Lindsay, J. C.: 1958, *Ann. Geophys.* **14**, 232.
205 Khetsuriani, Ts. S. and Tskhovrebadze, A. S.: 1965, *Abastumansk. Obs. Byull.* **32**, 147.
206 Tsimakhovich, N. P.: 1968, *Bolchie Radiovspleski Solntza*, Zinatie, Riga.
207 Shaposhnikova, E. F. and Ogir', M. B.: 1959, *Izv. Krymsk. Astrofiz. Obs.* **21**, 112.
208 Yurovskaya, L. I.: 1965, *Izv. Krymsk. Astrofiz. Obs.* **34**, 9.
209 Akasofu, S.-I., Perreault, P. D., and Yoshida, S.: 1969, *Solar Phys.* **8**, 464.
210 Akin'yan, S. T., Mogilevsky, E. I., Böhme, A., and Krüger, A.: 1971, *Solar Phys.* **20**, 112.
211 Anastassiades, M. and Macris, C.: 1969, *Solar Phys.* **10**, 188.
212 Anderson, K. A.: 1969, *Solar Phys.* **6**, 111.
213 Arnoldy, R. L., Kane, S. R., and Winckler, J. R.: 1967, *Solar Phys.* **2**, 171. ₂

214 Arnoldy, R. L., Kane, S. R., and Winckler, J. R.: 1968, in K. O. Kiepenheuer (ed.), 'Structure and Development of Solar Active Regions', *IAU Symp.* **35**, 490.

215 Axisa, F., Bewick, A., Durney, A. C., Engelmann, J., Hynds, R., Koch, L., and Morfill, G.: 1971, in K. Ya. Kondratyev, M. J. Rycroft, and C. Sagan (eds.), *Space Research* **XI**, North-Holland, Amsterdam, p. 1229.

216 Badillo, V. L. and Castelli, J. P.: 1969, *Astrophys. Letters* **4**, 5.

217 Badillo, V. L. and Salcedo, J. E.: 1969, *Nature* **224**, 503.

218 Baird, G. A., Bell, G. G., Duggal, S. P., and Pomerantz, M. A.: 1967, *Solar Phys.* **2**, 491.

219 Balogh, A., Hedgecock, P. C., Hynds, R. J., and Sear, J.: 1971, *Solar Phys.* **20**, 150.

220 Barcus, J. R.: 1969, *Solar Phys.* **8**, 186.

221 Bhattacharyya, J. C. and Balakrishnan, T. K.: 1967, *J. Atmospheric Terrest. Phys.* **29**, 1573.

222 Bland, C. J., Dilworth, C., Maccagni, D., Tanzi, E. G., Mercier, J. P., Raviart, A., and Treguer, L.: 1971, in K. Ya. Kondratyev, M. J. Rycroft, and C. Sagan (eds.), *Space Research* **XI**, North-Holland, Amsterdam, p. 1205.

223 Bohlin, J. D. and Simon, M.: 1969, *Solar Phys.* **9**, 183.

224 Böhme, A., and Krüger, A.: 1968, *H.H.I. Suppl. Ser. of Solar Data* **1**, 63.

225 Böhme, A. and Krüger, A.: 1969, in V. N. Obridko (ed.), *Solnechno-zemnaya fisica 1*, Akad. Nauk S.S.S.R., Sovet Solntse-zemlya, Moscow, p. 19.

226 Blake, J. B., Paulikas, G. A., and Freden, S. C.: 1969, in C. de Jager and Z. Švestka (eds.), *Solar Flares and Space Research,* North-Holland, Amsterdam, p. 258.

227 Bougeret, J. L., Caroubalos, C., Mercier, C., and Pick, M.: 1970, *Astron. Astrophys.* **6**, 406.

228 Brückner, G. and Waldmeier, M.: 1969, *Ann. IQSY* **3**, 24.

229 Bukata, R. P., Gronstal, P. T., Palmeira, R. A. R., McCracken, K. G., and Rao, U. R.: 1969, *Solar Phys.* **10**, 198.

230 Bukata, R. P., Gronstal, P. T., and Palmeira, R. A. R.: 1970, *Solar Phys.* **14**, 419.

231 Bumba, V. and Obridko, V. N.: 1969, *Solar Phys.* **6**, 104.

232 Castelli, J. P. and Michael, G. A.: 1967, *Solar Phys.* **1**, 125.

233 Castelli, J. P., Aarons, J., and Michael, G. A.: 1967, *J. Geophys. Res.* **72**, 5491.

234 Castelli, J. P. and Aarons, J.: 1968, *Ann. Geophys.* **24**, 813.

235 Castelli, J. P., Aarons, J., Michael, G. A., Jones, C. and Ko, H. C.: 1969, in C. de Jager and Z. Švestka (eds.), *Solar Flares and Space Research,* North-Holland, Amsterdam, p. 194.

236 Castelli, J. P. and Aarons, J.: 1968, *Astrophys. J.* **153**, 267.

237 Castelli, J. P., Aarons, J., and Michael, G. A.: 1968, in K. O. Kiepenheuer (ed.), 'Structure and Development of Solar Active Regions', *IAU Symp.* **35**, 601.

238 Ionosphere Research Committee, Sci. Council of Japan (ed.): 1967, *Rept. Ionos. Space Res. Japan* **21**, 143.

239 Ionosphere Research Committee, Sci. Council of Japan (ed.): 1967, *Rept. Ionos. Space Res. Japan* **21**, 257.

240 Clavelier, B., Jarry, M. F., and Pick, M.: 1968, *Ann. Astrophys.* **31**, 523.

241 Cline, T. L., Holt, S. S., and Hones, E. W.: 1968, *J. Geophys. Res.* **73**, 434.

242 Cline, T. L. and McDonald, F. B.: 1968, *Solar Phys.* **5**, 507.

243 Croom, D. L. and Powell, R. J.: 1969, *Nature* **221**, 945.

244 Croom, D. L. and Powell, R. J.: 1970, *Solar Phys.* **14**, 221.

245 Croom, D. L.: 1970, *Solar Phys.* **15**, 414.

246 Croom, D. L.: 1970, *J. Geophys. Res.* **75**, 6940.

247 Croom, D. L.: 1971, *Solar Phys.* **19**, 152.

248 Croom, D. L. and Powell, R. J.: 1971, *Solar Phys.* **20**, 136.

249 Culhane, J. L., Sanford, P. W., Shaw, M. L., Pounds, K. A., and Smith, D.: 1968, *Astron. J.* **73**, 558.

250 Culhane, J. L. and Phillips, K. J. H.: 1970, *Solar Phys.* **11**, 117.

251 Lincoln, J. V. (Compiler): 1970, WDC-A, Upper Atmosphere Geophys. Rept. 9, ESSA, Boulder.

252 Lincoln, J. V. (Compiler): 1970, WDC-A, Upper Atmosphere Geophys. Rept. 8, Parts I, II, ESSA, Boulder.

253 Datlowe, D.: 1971, *Solar Phys.* **17**, 436.

254 De Feiter, L. D.: 1971, *Solar Phys.* **19**, 207.

255 Delys, C. and Gonze, R.: 1968, *Bull. Astron. Obs. Roy. Belg.* **6**, 237.

256 DeMastus, H. L. and Stover, R. R.: 1967, *Publ. Astron. Soc. Pacific* **79**, 615.

257 Dodson, H. W. and Hedeman, E. R.: 1968, *Solar Phys.* **4**, 229.

258 Dodson, H. W. and Hedeman, E. R.: 1968, *Astron. J.* **73**, S59.

259 Dodson, H. W.: 1969, *Ann. IQSY* **3**, 154.

260 Dodson, H. W., Hedeman, E. R., Kahler S. W. and Lin, R. P.: 1969, *Solar Phys.* **6**, 294.

261   Dodson, H. W. and Hedeman, E. R.: 1969, *Solar Phys.* **9**, 278.
262   Dodson, H. W. and Hedeman, E. R.: 1970, *Solar Phys.* **13**, 401.
263   Dollfus, A.: 1968, in K. O. Kiepenheuer (ed.), 'Structure and Development of Solar Active Regions', *IAU Symp.* **35**, 359.
264   Donnelly, R. F.: 1968, *Solar Phys.* **5**, 123.
265   Donnelly, R. F.: 1971, *Solar Phys.* **20**, 188.
266   Drago, F. G.: 1968, *Mem. Soc. Astron. Ital.* **39**, 377.
267   Duggal, S. P., Guidi, I., and Pomerantz, M. A.: 1971, *Solar Phys.* **19**, 234.
269   Durgaprasad, N., Fichtel, C. E., Guss,   D. E., and Reames, D. V.: 1968, *Astrophys. J.* **154**, 307.
270   Durgaprasad, N., Fichtel, C. E., Guss, D. E., and Reames, D. V.: 1968, *Can. J. Phys.* **46**, S749.
271   Engelmann, J., Hynds, R. J., Morfill, G., Axisa, F., Bewick, A., Durney, A. C., and Koch, L.: 1971, *J. Geophys. Res.* **76**, 4245.
272   Fan, C. Y., Pick, M., Pyle, R., Simpson, J. A., and Smith, D. R.: 1968, *J. Geophys. Res.* **73**, 1555.
273   Feit, J.: 1971, *Solar Phys.* **17**, 473.
274   Fortini, T. and Torelli, M.: 1968, in K. O. Kiepenheuer (ed.), 'Structure and Development of Solar Active Regions', *IAU Symp.* **35**, 50.
275   Fortini, T. and Torelli, M.: 1970, *Solar Phys.* **11**, 425.
276   Friedman, H. and Kreplin, R. W.: 1969, *Ann. IQSY* **3**, 144.
277   Frost, K. J. and Dennis, B. R.: 1971, *Astrophys. J.* **165**, 655.
278   Garriott, O. K., Da Rosa, A. V., Davis, M. J., Wagner, L. S., and Thome, G. D.: 1969, *Solar Phys.* **8**, 226.
279   Genesio-Elgarten, V. and Joukoff, A. A.: 1970, *Solar Phys.* **14**, 234.
280   Goedeke, A. D. and Masley, A. J.: 1969, in C. de Jager and Z. Švestka (eds.), *Solar Flares and Space Research,* North-Holland, Amsterdam, p. 284.
281   Greenstadt, E. W., Green, I. M., Inouye, G. T., and Sonnett, C. P.: 1970, *Planetary Space Sci.* **18**, 333.
282   Grossi Gallegos, H., Molnar, H., and Seibold, J. R.: 1971, *Solar Phys.* **16**, 120.
283   Harang, L.: 1968, *Planetary Space Sci.* **16**, 1081.
284   Harvey, G. A. and McNarry, L. R.: 1970, *Solar Phys.* **11**, 467.
285   Hill, H.: 1966, *J. Br. Astron. Assoc.* **77**, 28.
286   Holt, S. S. and Ramaty, R.: 1969, *Solar Phys.* **8**, 119.
287   Hsieh, K. C. and Simpson, J. A: 1970, *Astrophys. J. Letters* **162**, L191.
288   Hudson, H. S., Peterson, L. E., and Schwartz, D. A.: 1969, in C. de Jager and Z. Švestka (eds.), *Solar Flares and Space Research,* North-Holland, Amsterdam, p. 113.
289   Hultqvist, B.: 1969, in C. de Jager and Z. Švestka (eds.), *Solar Flares and Space Research,* North-Holland, Amsterdam, p. 215.
290   Hyder, C. L.: 1967, *Solar Phys.* **2**, 267.
291   Kahler, S. W.: 1968, *Astron. J.* **73**, S66.
292   Kahler, S. W. and Lin R. P.: 1969, *Ann. IQSY* **3**, 299.
293   Kahler, S. W., Meekins, J. F., Kreplin, R. W., and Bowyer, C. S.: 1970, *Astrophys. J.* **162**, 293.
294   Kai, K.: 1970, *Solar Phys.* **11**, 310.
295   Kane, S. R. and Winckler, J. R.: 1969, *Solar Phys.* **6**, 304.
296   Kahler, S. W., Primbsch, J. H., and Anderson, K. A.: 1967, *Solar Phys.* **2**, 179.
297   Kinsey, J. H. and McDonald, F. B.: 1968, in K. O. Kiepenheuer (ed.), 'Structure and Development of Solar Active Regions', *IAU Symp.* **35**, 536.
298   Knuth, R., Lauter, E. A., and Lippert, W.: 1969, in V. N. Obridko (ed.), *Solnechno-zemnaya fisika 1,* Akad. Nauk S.S.S.R., Sovet Solntze-zemlya, Moscow, p. 34.
299   Kondo, I., Nagase, F., and Yasue, H.: 1970, *Rept. Ionos. Space Research, Japan* **24**, 147.
300   Konradi, A.: 1968, Goddard Space Flight Center, Greenbelt, Maryland, X-612-321.
301   Konradi, A.: 1969, *J. Geophys. Res.* **74**, 1158.
302   Křivský, L. and Nestorov. G.: 1968, *Bull. Astron. Inst. Czech.* **19**, 197.
303   Křivský, L. and Pintér, Š.: 1969, *Bull. Astron. Inst. Czech.* **20**, 147.
304   Křivský, L.: 1970, *Bull. Astron. Inst. Czech.* **21**, 67.
305   Křivský, L. and Švestka, Z.: 1970, in T. M. Donahue, P. A. Smith, and L. Thomas (eds.), *Space Research* X, North-Holland, Amsterdam, p. 817.
306   Krimigis, S. M. and Verzariu, P.: 1971, *J. Geophys. Res.* **76**, 792.
307   Künzel, H.: 1971, *Astron. Nachr.* **293**, 105.
308   Landt, J. A. and Croft, T. A.: 1970, *J. Geophys. Res.* **75**, 4623.
309   Lange, H. and Taubenheim, J.: 1969, in V. N. Obridko (ed.), *Solnechno-zemnaya fisika 1,* Akad. Nauk S.S.S.R., Sovet Solntze-zemlya, Moscow, p. 35.

310   Lanzerotti, L. J. and Robbins, M. R.: 1969, *Solar Phys.* **10**, 212.
311   Lanzerotti, L. J.: 1970, *Solar Phys.* **11**, 145.
312   Lanzerotti, L. J.: 1969, *J. Geophys. Res.* **74**, 2851.
313   Lazarus, A. J. and Binsack, J. H.: 1969, *Ann. IQSY* **3**, 378.
314   Lin, R. P., Kahler, S. W., and Roelof, E. C.: 1968, *Solar Phys.* **4**, 338.
315   Lin, R. P. and Hudson, H. S.: 1971, *Solar Phys.* **17**, 412.
316   Lindgren, S. T.: 1968, *Solar Phys.* **5**, 382.
317   Lockwood, J. A.: 1968, *J. Geophys. Res.* **73**, 4247.
318   Macris, C. J.: 1968, in M. G. Fracastoro (ed.), *Atti dell' XI Convegno, Padova, 1967*, Tipog. Villagio del Fanciullo-opicina, Trieste, p. 79.
319   Macris, C. J. and Prokakis, T. J.: 1968, in K. O. Kiepenheuer (ed.), 'Structure and Development of Solar Active Regions', *IAU Symp.* **35**, 85.
320   Malville, J. M., Tandberg-Hanssen, E., and Zei, D.: 1969, *Solar Phys.* **7**, 253.
321   Wada, M.: 1967, *Rept. Ionos. Space Res. Japan* **21**, 223.
322   Masley, A. J.: 1968, in C. de Jager and Z. Švestka (eds.), *Solar Flares and Space Research*, North-Holland, Amsterdam, p. 279.
323   Masley, A. J. and Goedeke, A. D.: 1969, *Ann. IQSY* **3**, 353.
324   McClinton, A. T., Jr.: 1968, in NRL Space Research Seminar, April 17, 1968, NRL Washington D.C., p. 63.
325   Malville, J. M. and Tandberg-Hanssen, E.: 1969, *Solar Phys.* **6**, 278.
326   Mercer, J. B. and Wilson, B. G.: 1968, *Can. J. Phys.* **46**, S849.
327   Michard, R. and Ribes, E.: 1968, in K. O. Kiepenheuer (ed.), 'Structure and Development of Solar Active Regions', *IAU Symp.* **35**, 420.
328   Monsignori Fossi, B. C., Poletto, G., and Tagliaferri, G. L.: 1969, *Solar Phys.* **10**, 196.
329   Najita, K. and Orrall, F. Q.: 1970, *Solar Phys.* **15**, 176.
330   Ogawa, T., Tanaka, Y., Miura, T., and Owaki, M.: 1966, *Rept. Ionos. Space Res. Japan* **20**, 528.
332   Ondoh, T., Isozaki, S., and Ouchi, E.: 1967, *Rept. Ionos. Space Res. Japan* **21**, 132.
333   Palmeira, R. A. R., Bukata, R. P., and Gronstal, P. T.: 1970, *Can. J. Phys.* **48**, 419.
334   Paolini, F. R., Giacconi, R., Manley, O., Reidy, W. P., Vaiana, G. S., and Zehnpfennig, T.: 1968, *Astron. J.* **73**, S73.
335   Pfitzer, K. A. and Winckler J. R.: 1968, *J. Geophys. Res.* **73**, 5792.
336   Pintér, Š.: 1969, *Bull. Astron. Inst. Czech.* **20**, 151.
337   Pintér, Š.: 1969, *Bull. Astron. Inst. Czech.* **20**, 73.
339   Popovici, C. and Dimitriu, A.: 1969, *Ann. IQSY* **3**, 31.
340   Prata, S. W.: 1971, *Solar Phys.* **19**, 92.
341   Quenby, J. J. and Sear, J. F.: 1971, *Planetary Space Sci.* **19**, 95.
342   Rao, U. R., Allum, F. R., Bartley, W. C., Palmeira, R. A. R., Harries, J. A., and McCracken, K. G.: 1969, in C. de Jager and Z. Švestka (eds.), *Solar Flares and Space Research*, North-Holland, Amsterdam, p. 267.
343   Rao, U. R., McCracken, K. G., Allum, F. R., Palmeira, R. A. R., Bartley, W. C., and Palmer, I.: 1971, *Solar Phys.* **19**, 209.
344   Reames, D. V. and Fichtel, C. E.: 1969, in C. de Jager and Z. Švestka (eds.), *Solar Flares and Space Research*, North-Holland, Amsterdam, p. 277.
345   Ribes, E.: 1969, *Astron. Astrophys.* **2**, 316.
346   Ribes, E.: 1970, *Astron. Astrophys.* **4**, 70.
347   Sakurai, K.: 1971, *Solar Phys.* **20**, 147.
348   Sawyer, C.: 1968, in K. O. Kiepenheuer (ed.), 'Structure and Development of Solar Active Regions', *IAU Symp.* **35**, 543.
349   Severny, A.: 1969, *Ann. IQSY* **3**, 11.
350   Severny, A.: 1969, in C. de Jager and Z. Švestka (eds.), *Solar Flares and Space Research*, North-Holland, Amsterdam, p. 38.
351   Sheridan, K. V.: 1968, *Proc. Astron. Soc. Australia* **1**, 138.
352   Shimabukuro, F. I.: 1970, *Solar Phys.* **15**, 424.
353   Shimabukuro, F. I.: 1968, *Solar Phys.* **5**, 498.
354   Énomé, S., Kakinuma, T., and Tanaka, H.: 1969, *Solar Phys.* **6**, 428.
355   Smerd, S. F.: 1971, *Australian J. Phys.* **24**, 229.
356   Snijders, R.: 1969, *Solar Phys.* **6**, 290.
357   *Solar and Geophysical Data*, National Physical Laboratory, Delhi, 1966, RRC-A 130, part 2, p. 26.
358   *Solar Geophysical Data*, ESSA Boulder, 1966, CRPL-FB 264, IVd.
359   *Solar Geophysical Data*, ESSA Boulder, 1966, CRPL-FB 266, IVe.

360 *Solar Geophysical Data*, ESSA Boulder, 1966, CRPL-FB 266, IVe.
361 *Solar Geophysical Data*, ESSA Boulder, 1967, IER-FB 274, 26 and 275, 82.
362 The white-light flare of 23 May 1967 (Pictures, Sacramento Peak Obs.): 1968, *Solar Phys.* **5**, 2.
363 Steshenko, N. V.: 1968, in K. O. Kiepenheuer (ed.), 'Structure and Development of Solar
     Active Regions', *IAU Symp.* **35**, 201.
364 Stewart, R. T. and Sheridan, K. V.: 1970, *Solar Phys.* **12**, 229.
365 Stiber, G.: 1968, *Arkiv. Astron.* **4**, 571.
366 Sud, L. V.: 1968, *Australian J. Phys.* **21**, 755.
367 Sugiura, M., Skillman, T. L., Ledley, B. G., and Heppner, J. P.: 1968, Goddard Space Flight
     Center, Greenbelt, Maryland, X-612-68-235.
368 Švestka, Z.: 1968, in K. O. Kiepenheuer (ed.), 'Structure and Development of Solar Active
     Regions', *IAU Symp.* **35**, 287.
369 Švestka, Z.: 1968, in K. O. Kiepenheuer (ed.), 'Structure and Development of Solar Active
     Regions', *IAU Symp.* **35**, 513.
370 Švestka, Z.: 1968, *Solar Phys.* **4**, 361.
371 Švestka, Z.: 1969, in C. de Jager and Z. Švestka (eds.), *Solar Flares and Space Research*,
     North-Holland, Amsterdam, p. 319.
372 Švestka, Z. and Simon, P.: 1969, *Solar Phys.* **10**, 3.
373 Švestka, Z.: 1971, *Solar Phys.* **19**, 202.
374 Švestka, Z.: 1970, in T. M. Donahue, P. A. Smith, and L. Thomas (eds.), *Space Research* **X**,
     North-Holland, Amsterdam, p. 797.
375 Takakura, T.: 1969, in C. de Jager and Z. Švestka (eds.), *Solar Flares and Space Research*,
     North-Holland, Amsterdam, p. 165.
376 Takakura, T., Ohki, K., Shibauya N., Fujii, M., Matsuoka, M., Miyamoto, S., Nishimura, J.,
     Oda, M., Ogawara, Y., and Ota, S.: 1971, *Solar Phys.* **16**, 454.
377 Takao, K.: 1967, *Rept. Ionos. Space Res. Japan* **21**, 125.
378 Tandberg-Hanssen, E.: 1967, *Solar Phys.* **2**, 98.
379 Tanaka, H. and Énomé, S.: 1971, *Solar Phys.* **17**, 408.
380 Tanskanen, P. J. and Kangas, J.: 1968, *Ann. Acad. Sci. Fenn.*, **AVI**, *Physica* **287**, 3.
381 Stickland, A. C. (ed.): 1969, 'The Proton Flare Project (The July 1966 Event)', *Ann. IQSY* **3**,
     The M.I.T. Press, Cambridge (U.S.A.) and London.
382 Thomas, G. R. and Dalziel, R.: 1970, in T. M. Donahue, P. A. Smith, and L. Thomas (eds.),
     Space Research **X**, North-Holland, Amsterdam, p. 797.
383 Ulwick, J. C., Baker, K. D., and Sellers, B.: 1970, in T. M. Donahue, P. A. Smith, and L. Thomas
     (eds.), *Space Research* **X**, North-Holland, Amsterdam, p. 825.
384 Urbarz, H.: 1970, *Solar Phys.* **13**, 458.
385 Van Allen, J. A.: 1967, *Trans. Am. Geophys. Un.* **48**, 177.
386 Van Allen, J. A.: 1967, *Astron. J.* **72**, 833.
387 Van Allen, J. A.: 1968, *Astrophys. J.* **152**, L 85.
388 Van Allen, J. A. and Wende, C. D.: 1969, *J. Geophys. Res.* **74**, 3046.
389 Van Allen, J. A. and Fennell, J. F.: 1971, *J. Geophys. Res.* **76**, 4262.
390 Van Beek, H. F. and Van Gils, J. N.: 1970, in T. M. Donahue, P. A. Smith, and L. Thomas
     (eds.), *Space Research* **X**, North-Holland, Amsterdam, p. 831.
391 Vaiana, G. S., Reidy, W. P., Zehnpfennig, T., Van Speybroek, L., and Giacconi, R.: 1968,
     Science **161**, 564.
392 Yajima, S., Mizugaki, K., and Yamaguchi, K.: 1969, *Tokyo Astron. Bull.* **197**, 2283.
393 Valníček, B.: 1968, in K. O. Kiepenheuer (ed.), 'Structure and Development of Solar Active
     Regions', *IAU Symp.* **35**, 282.
394 Vernov, S. N., Chudakov, A. E., Vakulov, P. V., Gorchakov, E. V., Ifnatiev, P. P., Kontor, N. N,
     Kuznetsov, S. N., Lofachev, Yu. I., Lyubimov, G. P., Nikolaev, A. G., and Pereslegina, N. V.:
     1969, in K. S. W. Champion, P. A. Smith, and R. L. Smith-Rose (eds.), *Space Research* **IX**,
     North-Holland, Amsterdam, p. 203.
395 Vernov, S. N., Chuchkov, E. A., Kontor, N. N., Lyubimov, G. P., Nikolaev, A. G., and
     Pereslegina, N. V.: 1971, in K. Ya. Kondratyev, M. J. Rycroff, and C. Sagan (eds.), *Space
     Research* **XI**, North-Holland, Amsterdam, p. 1213.
396 Walker, A. B. C., Jr. and Rugge, H. R.: 1969, in C. de Jager and Z. Švestka (eds.), *Solar Flares
     and Space Research*, North-Holland, Amsterdam, p. 102.
397 Warwick, J. W.: 1966, *Solar Phys.* **4**, 446.
398 Wassenberg, W.: 1971, *Solar Phys.* **20**, 130.
399 Webb, H. D.: 1969, *J. Geophys. Res.* **74**, 1880.
400 White, III, K. P. and Janssens, T. J.: 1970, *Solar Phys.* **11**, 291.

401   White-light solar flare and the May 25th Aurora: 1967, *Sky Telesc.* **34**, 57.
402   Wilcox, J. M.: 1969, in C. de Jager and Z. Švestka (eds.), *Solar Flares and Space Research*, North-Holland, Amsterdam, p. 294.
403   Wilcox, J. M.: 1968, Space Sciences Laboratory, Univ. of California, Berkeley, Ser. 9, No. 24.
404   Williams, D. J. and Bostrom, C. O.: 1969, NASA, Goddard Space Flight Center, Greenbelt, Maryland, X-162-69-141.
405   Williams, D. J.: 1969, NASA, Goddard Space Flight Center, Greenbelt, Maryland, X-612-69-258.
406   Zirin, H. and Lackner, D. R.: 1969, *Solar Phys.* **6**, 86.
407   Ageshin, P. N., Bayarevich, V. V., Stozhkov, Yu. I., and Charakhchyan, T. N.: 1969, *Geomag. Aeronom.* **9**, 538.
408   Akin'yan, S. T.: 1968, *Soln. Dannye*, No. 7, 83.
409   Akin'yan, S. T., Korolev, O. S., and Mogilevsky, E. I.: 1969, in V. N. Obridko (ed.), *Solnechno-zemnaya fizika 1*, Akad. Nauk. S.S.S.R., Sovet Solntse-zemlya, Moscow, p. 10.
410   Babin, A. N.: 1970, *Izv. Krymsk. Astrofiz. Obs.* **41–42**, 45.
411   Belikovich, V. V., Benediktov, E. A., and Grishkevich, L. V.: 1969, *Geomag. Aeronom.* **9**, 172.
412   Belikovich, V. V., Benediktov, E. A., and Rapoport, Z. Ts.: 1969, *Geomag. Aeronom.* **9**, 666.
413   Best, A., Bernik, A., Gal'perin, Yu. I., Grafe, A., Knut, R., Lashtovichka, Ya, Lippert, V., Mulyarchik, T. M., Nestorov, G., Samordzhiev, D., Serafimov, K. B., Trzhiska, P., Shujskaya, F. K., and Ehnttsian, G.: 1971, *Geomag. Aeronom.* **11**, 29.
414   Vernov, S. N., Senguro, I. N., Tel'tsov, M. V., and Shavrin, P. I.: 1969, *Geomag. Aeronom.* **9**, 968.
415   Vernov, S. N., Kontor, N. N., Kuznetsov, S. N., Logachev, Yu. I., Lyubimov, G. P., and Pereslegina, N. V.: 1969, in V. N. Obridko (ed.), *Solnechno-zemnaya fizika 1*, Akad. Nauk S.S.S.R., Sovet Solntse-zemlya, Moscow, p. 85.
416   Volodichev, N. N. and Madeeva, M. O.: 1970, *Vestn. Mosk. Univ., ser. 3*, **11**, 582.
417   Gajnova, L. E., Ivanov, V. I., Kolomeets, E. V., and Kobzev, V. A.: 1971, *Tr. mezhdunar. seminara po problemi generatsiya kosmich. luchej na Solntse, 1970*, Moscow, p. 379.
418   Goncharova, E. E., Zevakina, R. A., Lavrova, E. V., and Yudovich, L. A.: 1970, *Geomag. Aeronom.* **10**, 67.
419   Grigorov, N. L., Lutsenko, V. N., Madnev, V. L., Pisarenko, I. F., and Savenko, I.A.: 1967, *Kosm. Issled* **5**, 946.
420   Grigorov, N. L., Kurt, V. G., Lutsenko, V. N., Pisarenko, I. F., and Savenko, I. A.: 1969, in V. N. Obridko (ed.), *Solnechno-zemnaya fizika 1*, Akad. Nauk. S.S.S.R., Sovet Solntse-zemlya, Moscow, p. 107.
421   Degtyarev, V. I. and Kurchenko, Yu. A.: 1970, *Issledovaniya po geomagnetizmu aeronomii i fisike solntsa*, Nauka, Moscow **10**, 286.
422   Driatskij, V. M.: 1969, in V. N. Obridko (ed.), *Solnechno-zemnaya fizika 1*, Akad. Nauk. S.S.S.R., Sovet Solntse-zemlya, Moscow, p. 28.
423   Driatskij, V. M: 1969, *Geomag. Aeronom.* **9**, 56.
424   Driatskij, V. M.: 1970, *Tr. Arkt. Antarkt. in-ta* **288**, 74.
425   Driatskij, V. M., Lazutin, L. L., and Borovkov, L. P.: 1969, *Tr. mezhdunar. seminara po izucheniyu fiziki mezhplanetnogo prostranstva s pomoshch' yu kosmich. luchej*, Leningrad, p. 255.
427   Eryushev, I. N. and Tsvetkov, L. I.: 1969, *Soln. Dannye*, No. 12, 95.
428   Zvereva, A. M., Severny, A. B.: 1970, *Izv. Krymsk. Astrofiz. Obs.* **41–42**, 97.
429   Zel'dovich, M. A., Kovrizhnykh, O. M., Madeev, M. O., and Savenko, I. A.: 1971, *Geomag. Aeronom.* **11**, 145.
430   Ivanov, K. G. and Mikerina, I. V: 1969, *Geomag. Aeronom.* **9**, 359.
431   Ivanov, V. I., Kobzev, V. A., and Kolomeets, E. V.: 1970, *Fizika Alma-Ata* **1**, 62.
432   Kaminer, I. S. and Miroshnichenko, L. I.: 1969, *Geomag. Aeronom.* **9**, 336.
433   Konstantinov, V. A.: 1967, *Geomag. Aeronom.* **7**, 876.
434   Matveeva, Eh. T. and Troitskaya, V. A.: 1968, *Geomag. Aeronom.* **8**, 598.
435   Moiseev, I. G. and Efanov, V. A.: 1967, *Izv. Krymsk. Astrofiz. Obs.* **37**, 128.
436   Nestorov, G. and Letfus, V.: 1969, in V. N. Obridko (ed.), *Solnechno-zemnaya fizika 1*, Akad. Nauk. S.S.S.R., Sovet Solntse-zemlya, Moscow, p. 44.
437   Nefed'ev, V. P. and Turchina, V. D.: 1970, *Issledovaniya po geomagnetizmu aeronomii i fizike solntsa*, Nauka, Moscow, **6**, 87.
438   Popov, N. P., Fedchenko, Z. A., Starkova, I. P., and Plotsenko, O. S.: 1969, *Issledovaniya po geomagnetizmu aeronomii i fizike solntsa*, Nauka, Moscow **5**, 146.
439   Rudina, M. P. and Solonitsyna, N. F.: 1970, *Tr. Sekt. Ionos. A. N. Kaz. S.S.R.,* **1**, 20.
440   Stepanyan, N. N.: 1969, *Astron. Zh.* **46**, 580.

441 Struin, O. N. and Shirochkov, A. V.: 1970, *Inform. Byull. Sov. Antarkt. Exsped.* 78, 54.
442 Troitskaya, V. A. and Matveeva, Eh. T.: 1969, in V. N. Obridko (ed.), *Solnechno-zemnaya fizika 1*, Akad. Nauk. S.S.S.R., Sovet Solntse-zemlya, Moscow, p. 41.
443 Fedyakina, N. I.: 1969, in V. N. Obridko (ed.), *Solnechno-zemnaya fizika 1*, Akad. Nauk. S.S.S.R., Sovet Solntse-zemlya, Moscow, p. 51.
444 Khetsuriani, Ts. S. and Tskhovberidze, A. S.: 1969, *Byull. Abastumansk. Obs.* 37, 147.
445 Khetsuriani, Ts. S. and Tetriashvili, Eh. I.: 1971, *Byull. Abastumansk. Obs.* 40, 93.

27-DAY INDEX OF SOLAR PARTICLE EVENTS, 1955–1969

	1955	Jan/Feb	Feb/Mar	Mar/Apr	Apr/May	May	May/Jun	Jun/Jul	Jul/Aug	Aug/Sep	Sep/Oct	Oct/Nov	Nov/Dec	Dec	
1		15	11	10	6	3	30	26	23	19	15	12	8	5	1
2		16 X20												6 UN	2
3															3
4															4
5															5
6															6
7															7
8															8
9															9
10															10
11													19 UN		11
12															12
13															13
14															14
15															15
16															16
17															17
18		1 UN													18
19															19
20															20
21															21
22															22
23															23
24															24
25															25
26															26
27															27

1956	Jan	Jan/Feb	Feb/Mar	Mar/Apr	Apr/May	May/Jun	Jun/Jul	Jul/Aug	Aug	Aug/Sep	Sep/Oct	Oct/Nov	Nov/Dec
1	1	28	24	22	18	15	11	8	4	31 X21	27	24	20
2													
3													
4													
5													
6													
7													
8													
9													
10					27 X(1)0								
11													
12													
13													
14													
15													
16			10 X20										
17												8 UN	
18													
19													
20													
21												13 X30	
22													
23													
24													
25				15 UN					28 UN				
26													
27		23 X34			14 UN								

	1956/57	1957											
	Dec/Jan	Jan/Feb	Feb/Mar	Mar/Apr	Apr	May	May/Jun	Jun/Jul	Jul/Aug	Aug/Sep	Sep/Oct	Oct/Nov	Nov/Dec
1	17	13	9	8	4	1	28	24	21	17	13	10	6
2													
3													
4					6 UN		30 UN		24 X(1)0				
5						5 UN		28 UN					
6											18 X(1)0		
7													
8		20 X20									20 UN		
9	25 UN				11 X(1)0	8 X(1)0		1 X(1)0	28 X(1)0		21 X30		
10								3 X30					
11									9 X20	27 X(1)0	20 X20	20 ⌈UN / X30	
12										28 X20			
13			21 X(1)0							29 X30			
14											26 X20		
15										31 X30			
16					19 X(1)0								
17										2 X30			
18										3 UN			
19						19 UN							24 UN
20													
21				28 UN									
22													26 UN
23							19 X(1)0				5 UN		
24										10 UN			
25													
26				2 UN			22 X30	19 UN				4 X20	
27				3 X20							12 X10		

	1957	1957/58	1958											
	Dec	Dec/Jan	Jan/Feb	Feb/Mar	Mar/Apr	Apr/May	May/Jun	Jun/Jul	Jul/Aug	Aug	Aug/Sep	Sep/Oct	Oct/Nov	
1	3	30	26	22	21 UN	17	14	10	7 X40	3	30	26	23	1
2														2
3														3
4					23 X20									4
5														5
6					25 X30									6
7														7
8												3 UN		8
9														9
10				3 UN										10
11					31 X(1)0									11
12														12
13														13
14										16 X30				14
15	17 X(1)0													15
16			10 X20								14 UN			16
17														17
18				11 X(1)0										18
19										21 X20				19
20										22 X30				20
21				14 X(1)0	10 X20		4 X(1)0							21
22														22
23									29 X20					23
24				17 XXO			6 X(1)0			26 X40	22 UN X30			24
25														25
26	28 X(1)0													26
27		25 UN												27

	Nov/Dec 1958	Dec/Jan 1958/59	Jan/Feb 1959	Feb/Mar	Mar/Apr	Apr	Apr/May	May/Jun	Jun/Jul	Jul/Aug	Aug/Sep	Sep/Oct	Oct/Nov	
1	19	16	12	8	7	3	30	27	23	20	16	12 UN	9	1
2														2
3											18 X20			3
4											19 X(1)0			4
5														5
6				13 X20										6
7														7
8														8
9														9
10														10
11							10 X40							11
12														12
13														13
14								9 X(1)0		2 UN				14
15			26 X(1)0											15
16														16
17									9 [UN UN					17
18								13 X10	10 X40		2 X(1)0			18
19														19
20														20
21														21
22									14 X40					22
23														23
24									16 X42					24
25														25
26														26
27														27

Day	Nov/Dec 1959	Dec	Dec/Jan 1959/60	Jan/Feb 1960	Feb/Mar	Mar/Apr	Apr/May	May/Jun	Jun/Jul	Jul	Aug	Aug/Sep	Sep/Oct	Day
1	5	2 UN	29	25	21	19	15 UN	13 (1)20	8	5	1	28	24	1
2								12 [UN / UN				29 UN	25 UN	2
3													26 (1)(2)0	3
4														4
5	9 UN													5
6								17 (0)(1)0				1 UN		6
7												3 221		7
8														8
9					29 UN				15 UN					9
10										14 UN			3 X(1)0	10
11						28 [UN / UN	28 (2)20	26 (0)(1)0			12 (0)(1)0			11
12						29 (-1)20	29 (1)30							12
13						30 [X(1)0 / X30								13
14			11 (0)(2)0	7 UN		1 120	4 (1)24		25 X(1)0		14 (0)(1)0			14
15														15
16			13 X(1)0			3 UN	6 (1)30							16
17						4 UN		28 UN						17
18						5 120							11 UN	18
19			16 UN											19
20		21 UN			10 UN				27 X(1)0					20
21								1 (1)(1)0	28 UN					21
22				15 UN			9 X(1)0							22
23														23
24														24
25														25
26	30 UN				17 (-1)(1)0							26 X(1)0		26
27														27

Day	1960		1960/61	1961										Day
---	Oct/Nov	Nov/Dec	Dec/Jan	Jan/Feb	Feb/Mar	Mar	Apr	Apr/May	May/Jun	Jun/Jul	Jul/Aug	Aug/Sep	Sep/Oct	
1	21	17	14	10	6	5	1	28	25	21	18 333	14	10 (2)20	1
2														2
3		19 X(1)0									20 (1)22			3
4		20 (3)32												4
5														5
6														6
7											24 (0)(1)0			7
8					13 UN									8
9	29 X(1)0												18 L	9
10														10
11											28 UN			11
12														12
13					18 UN	17 UN	13 X(1)0							13
14														14
15														15
16														16
17														17
18														18
19													28 220	19
20		6 X(1)0												20
21	10 X(1)0									11 X10				21
22	11 X(1)0									12 (2)40				22
23	12 444													23
24	13 444										10 UN			24
25	14 X(1)0									15 X(2)0		7 (0)10		25
26	15 443													26
27										17 UN				27

	1961 Oct/Nov	Nov	Nov/Dec	1961/62 Dec/Jan	1962 Jan/Feb	Feb/Mar	Mar/Apr	Apr/May	May/Jun	Jun/Jul	Jul	Jul/Aug	Aug/Sep
1	7	3	30	27	23	19	18	14	11	7	4	31	27
2			1 UN										
3													
4													
5													
6													
7													
8		10 110											
9						20 (0)10							
10					1 (1)(2)0								
11													
12													
13					4 (0)(1)0								
14													
15													
16													
17													
18													
19													
20	26 (-1)(1)0												
21													
22													
23													
24													
25													
26													
27													

	1962			1962/63	1963								
	Sep/Oct	Oct/Nov	Nov/Dec	Dec/Jan	Jan/Feb	Feb/Mar	Mar	Mar/Apr	Apr/May	May/Jun	Jun/Jul	Jul/Aug	Aug/Sep
1	23	20	16	13	10	5	4	31	27	24	20	17	13
2													
3													
4		23 010								27 L			
5													
6										29 UN	25 L		
7													
8													
9						9 -110	8 L	4 L	1 -1(1)0				
10													
11													
12						15 L							
13.													
14	6 UN												
15													
16								15 110					
17													
18													
19													
20												6 (0)(1)0	
21													
22										14 (-1)(0)0			
23												9 0(1)0	
24								24 L					
25													
26													
27													

| | 1963 | | | | 1963/64 | 1964 | | | | | | | | |
#	Sep/Oct	Oct/Nov	Nov	Nov/Dec	Dec/Jan	Jan/Feb	Feb/Mar	Mar/Apr	Apr/May	May/Jun	Jun/Jul	Jul	Jul/Aug	#
1	9	6	2	29	26	22	18 UN	16 [UN UN 010	12	9	5	2	29	1
2						23 UN				10 UN				2
3														3
4								19 UN	15 L					4
5				3 UN										5
6	14 UN													6
7	15 X10	12 UN				28 [UN UN		22 L						7
8	16 X10				2 UN									8
9														9
10														10
11	19 UN													11
12								27 L						12
13	21 120													13
14														14
15							3 UN							15
16								31 L						16
17														17
18	26 (1)20													18
19														19
20														20
21														21
22														22
23		28 (-1)(1)0												23
24														24
25														25
26														26
27														27

	1964				1964/65	1965							
	Aug/Sep	Sep/Oct	Oct/Nov	Nov/Dec	Dec/Jan	Jan/Feb	Feb/Mar	Mar	Mar/Apr	Apr/May	May/Jun	Jun/Jul	Jul/Aug
1	25	21	18	14	11	7	3	2	29	25	22	18	15
2						8 L							
3							5 120						
4						10 UN							
5											25 L		
6													
7													
8													
9													
10											31 UN		
11											1 UN	28 UN	
12												29 L	
13										7 UN			
14		4 UN											28 UN
15											5 L	2 UN	
16													
17												4 L	
18													
19													
20									17 UN				
21													
22										16 UN	12 UN		
23									20 UN		13 L	10 UN	
24													
25											15 L	13 L	
26						2 PE				20 UN			
27											17 UN		

	1965 Aug/Sep	Sep/Oct	Oct	Oct/Nov	Nov/Dec	1965/66 Dec/Jan	1966 Jan/Feb	Feb/Mar	Mar/Apr	Apr/May	May/Jun	Jun	Jul	
1	11	7	4 0(1)0	31	27	24 UN	20	16	15	11	8 -(2)00	4	1	1
2													2	2
3										12 [UN L			3 [PE UN.	3
4						27 L							4 -200	4
5	16 L		8 L					19 UN	17 UN					5
6						29 L	19 UN	18 UN	19 [L PE -200					6
7						30[-1(0)0 -200	19 UN		20 -200 21[UN -200	16 -(2)00			7 121	7
8								22 UN	22 -200					8
9													9 000	9
10														10
11					7 L	2 L		27 L	24 [-100 120 PE -200 25 -200					11
12									26 PE					12
13								28 PE	27 -200				14 -100	13
14									28 -200				14 -100	14
15							3 L			25 UN				15
16									31 UN	26 UN	26 UN		16 -100	16
17					17 UN								25 UN	17
18									1 UN					18
19							7 UN			29 -100				19
20											27 PE			20
21							9 UN							21
22		1 L								2 0(1)0	28 -(2)00	25 -200		22
23		29 L						10 UN						23
24				23 UN		17 -200	12 UN			4 -200			25 UN	24
25						18 -200					2 PE		25	25
26				25 L		18 -200				6 -100		28 PE	26	26
27						19 -200				7 PE			27	27

Year groupings: **1966** spans Jul/Aug–Nov/Dec; **1966/67** = Dec/Jan; **1967** spans Jan/Feb–Jun/Jul.

Day	Jul/Aug	Aug/Sep	Sep/Oct	Oct/Nov	Nov/Dec	Dec/Jan	Jan/Feb	Feb	Mar	Mar/Apr	Apr/May	May/Jun	Jun/Jul	Day
1	28	24	20 -100	17	13	10	6	2 120	1	28	24	21 -200	17	1
2	29 -100					11 -(2)00	7 -200							2
3				19 UN		12 UN	8 PE		3 L			23 330		3
4		27 UN		20 [PE/UN		13 -200			4 UN					4
5		28 120								1 [PE/L/PE				5
6			25 -200					7 UN						6
7										3 L				7
8		31 PE	27 -100	24 PE	20 UN					4 UN		28 220		8
9							14 PE		9 [-200/L/PE-				25 -100	9
10	6 UN	2 230				19 [UN/L		11 010						10
11							16 -100		11 010	7 L				11
12		4 [(1)(00/(1)(00					17 L	13 -110		8 UN		1 L		12
13						22 UN	18 -200	14 PE	13 [PE/L					13
14				30 L					14 PE			3 0(1)0		14
15			4 UN				20 -200					4 L		15
16				1 PE								5 -100		16
17				2 PE								6 120		17
18				3 PE	30 UN					14 -200				18
19		11 PE		4 PE									5 -110	19
20		12 -200				29 L								20
21				6 PE									7 [X00/-100	21
22	18 UN	14 [000/010				31 [PE/PE/PE							9 L	22
23							28 [011/133					12 010		23
24									24 L					24
25		17 -100				3 L							11 L	25
26							31 UN	27 000			19 UN			26
27	23 UN					5 -200			27 [UN/(-1)00					27

1967	Jul/Aug	Aug/Sep	Sep/Oct	Oct	Oct/Nov	Nov/Dec	1967/68 Dec/Jan	1968 Jan/Feb	Feb/Mar	Mar/Apr	Apr/May	May/Jun	Jun	
1	14	10	6	3 L	30 -100	26	23 PE	19	15 -100	13	9 L	6	2	1
2		12 PE	9 PE	4 [PE / PE	2 010	27 -200				14 -200	8 L		3 -200	2
3	16 PE	13 [PE / PE	10 [PE / PE											3
4			7 [L / -200	7 [3L / -200		1 3PE								4
5		15 L	11 L	8 [L		30 2PE	28 PE			31 -100	27 [2PE / L	8 L	5	5
6			12 [L / PE				26	24 L		26 -200	10 PE		6	6
7		17 UN				3 120	29 [-100 / PE	21 -100			11 -100	11 -100	9 [L / 230	7
8				12 [L / PE		2 PE	30 [-200 / (0)00		21 -100				7 -100	8
9	23 L	18 L			7 [2PE / -110	3 120	31 L		21 -200					9
10				12 L	8 L	9 UN		30 L	26 -100	23 L	19 -200			10
11					10 -100	11 -200								11
12			17 -210		11 [-100 / 000	12 -100	30 L					12 -200		12
13			18 PE	11 [-100 / 000										13
14			19 [-2(1)0 / -100			9 UN	5 -100	2 -100		27 -200	22 -200			14
15	28 3PE	24 L	20 -110	13 010							23 -100		17 UN	15
16	29 [3PE / -100		26 L						26 -100	24 -200		22 -200		16
17	30 [3PE / -100	26 L		15 000		11 -200	8 UN			28 L			17 UN	17
18	31 [PE / -200				12 -100	16 010	11 [-1(1)0 / L	8 L			26 010			18
19	1 [000 / -100			19 L	9 UN	17 000 / 010	12 [000 / 0(1)0 / [PE	6 -100		31 -100	27 [L / 2PE			19
20	2 [000 / -100				5 -100	18 000			10 [-100 / -100 / -100					20
21	3 [000 / -100		26 L			16 010					29 -100			21
22	4 [L / 000		27 PE			17 000 / 010	12 L				30 -100			22
23				25 [3PE / L		18 000	14 -200			4 (-1)00				23
24		29 UN	29 UN	26 [L / -200			15 L		8 L		26 010			24
25		3 PE		27 [L / -200				12 -100	10 [-100 / -100 / -100	6 -200		26 [L / -100		25
26	8 -100			23 L			17 -100				4 [-200 / (2)00			26
27	9 000		29 -100											27

Day	1968 Jun/Jul	Jul/Aug	Aug/Sep	Sep/Oct	Oct/Nov	Nov/Dec	1968/69 Dec/Jan	1969 Jan	Jan/Feb	Feb/Mar	Mar/Apr	Apr/May	May/Jun
1	29 UN	26 -100	22	18	15	11	8	4	31	27 110	26	22	19
2										28 110	27 -100	23 UN	
3					17 L	13 -1(1)0			2 L			24 000	
4													
5											30 112		23 UN
6				23 UN				9 L					
7	5 PE		28 -100										
8	6 010					18 233		11 L				29 -100	26 -100
9		3 -200		26 [110 / UN			16 PE						
10	8 110				24 000	20 100			9 UN	8 L			28 [4PE / -200
11		5 -100	1 PE	28 [UN / 110		21 100		14 L				2 -200	29 [-200 / 4PE
12		6 -100	2 L	29 [PE / 121	26 -100				11 L				30 4PE
13			3 UN		27 -100				12 L				31 2PE
14	12 [010 / **120**		4 UN	1 UN				17 000		12 010		5 -200	1 PE
15					29 000	25 UN						6 PE	2 PE
16				3 PE	30 000		23 UN						3 [L / UN
17				4 120	31 230		24 -100				11 330	8 L	
18					1 230					16 -100			5 L
19							26 UN			17 000			
20		14 -100	10 UN		3 120		27 -100						7 [L / 110
21					4 120		28 2PE	24 [L / 010					
22					5 230							13 110	
23	21 UN	17 L					30 2UN			21 010			
24													11 L
25		19 -100							24 [PE / -100	23 -100			
26					9 -100				25 123				
27		21 UN							26 110				

No.	1969	Jun/Jul	Jul/Aug	Aug/Sep	Sep	Oct	Oct/Nov	Nov/Dec	Dec/Jan	1969/70
1		15	12	8	4	1	28	24 010	21	
2		16 L	13 PE					25 UN	23 L	
3				10 L						
4										
5										
6					7 L		2 330	30 -100		
7						7 UN				
8										
9			20 L	16 L		10 L				
10				17 L					30 010	
11					14 UN	11 UN	7 010			
12				19 UN						
13			24 L				10 -100			
14		28 UN			17 UN	14 [-200 / -100 / PE / 010 / 000		7 UN		
15										
16								9 L		
17							13 -100			
18		2 UN				18 UN				
19										
20						20 UN		13 UN		
21										
22					25 [110 / PE		18 [-100 / -100			
23					27 [-110 / **120**					
24			4 L				20 -200	18 010		
25					30 UN		21 -100			
26										
27			7 L	3 UN			23 UN	20 010		

ASTROPHYSICS AND SPACE SCIENCE LIBRARY

Edited by

J. E. Blamont, R. L. F. Boyd, L. Goldberg, C. de Jager, Z. Kopal, G. H. Ludwig, R. Lüst,
B. M. McCormac, H. E. Newell, L. I. Sedov, Z. Švestka, and W. de Graaff

1. C. de Jager (ed.), *The Solar Spectrum, Proceedings of the Symposium held at the University of Utrecht, 26–31 August, 1963.* 1965, XIV + 417 pp.
2. J. Ortner and H. Maseland (eds.), *Introduction to Solar Terrestrial Relations, Proceedings of the Summer School in Space Physics held in Alpbach, Austria, July 15–August 10, 1963 and Organized by the European Preparatory Commision for Space Research.* 1965, IX + 506 pp.
3. C. C. Chang and S. S. Huang (eds.), *Proceedings of the Plasma Space Science Symposium, held at the Catholic University of America, Washington, D.C., June 11–14, 1963.* 1965, IX + 377 pp.
4. Zdeněk Kopal, *An Introduction to the Study of the Moon.* 1966, XII + 464 pp.
5. B. M. McCormac (ed.), *Radiation Trapped in the Earth's Magnetic Field. Proceedings of the Advanced Study Institute, held at the Chr. Michelsen Institute, Bergen, Norway, August 16–September 3, 1965.* 1966, XII + 901 pp.
6. A. B. Underhill, *The Early Type Stars.* 1966, XII + 282 pp.
7. Jean Kovalevsky, *Introduction to Celestial Mechanics,* 1967, VIII + 427 pp.
8. Zdeněk Kopal and Constantine L. Goudas (eds.), *Measure of the Moon. Proceedings of the 2nd International Conference on Selenodesy and Lunar Topography, held in the University of Manchester, England, May 30–June 4, 1966.* 1967, XVIII + 479 pp.
9. J. G. Emming (ed.), *Electromagnetic Radiation in Space. Proceedings of the 3rd ESRO Summer School in Space Physics, held in Alpbach, Austria, from 19 July to 13 August, 1965.* 1968, VIII + 307 pp.
10. R. L. Carovillano, John F. McClay, and Henry R. Radoski (eds.), *Physics of the Magnetosphere, Based upon the Proceedings of the Conference held at Boston College, June 19–28, 1967.* 1968, X + 686 pp.
11. Syun-Ichi Akasofu, *Polar and Magnetospheric Substorms.* 1968, XVIII + 280 pp.
12. Peter M. Millman (ed.), *Meteorite Research. Proceedings of a Symposium on Meteorite Research, held in Vienna, Austria, 7–13 August, 1968.* 1969, XV + 941 pp.
13. Margherita Hack (ed.), *Mass Loss from Stars. Proceedings of the 2nd Trieste Colloquium on Astrophysics, 12–17 September, 1968.* 1969, XII + 345 pp.
14. N. D'Angelo (ed.), *Low-Frequency Waves and Irregularities in the Ionosphere. Proceedings of the 2nd ESRIN-ESLAB Symposium, held in Frascati, Italy, 23–27 September, 1968.* 1969, VII + 218 pp.
15. G. A. Partel (ed.), *Space Engineering. Proceedings of the 2nd International Conference on Space Engineering, held at the Fondazione Giorgio Cini, Isola di San Giorgio, Venice, Italy, May 7–10, 1969.* 1970, XI + 728 pp.
16. S. Fred Singer (ed.), *Manned Laboratories in Space. Second International Orbital Laboratory Symposium.* 1969, XIII + 133 pp.
17. B. M. McCormac (ed.), *Particles and Fields in the Magnetosphere. Symposium Organized by the Summer Advanced Study Institute, held at the University of California, Santa Barbara, Calif., August 4–15, 1969.* 1970, XI + 450 pp.
18. Jean-Claude Pecker, *Experimental Astronomy.* 1970, X + 105 pp.
19. V. Manno and D. E. Page (eds.), *Intercorrelated Satellite Observations related to Solar Events. Proceedings of the 3rd ESLAB/ESRIN Symposium held in Noordwijk, The Netherlands, September 16–19, 1969.* 1970, XVI + 627 pp.
20. L. Mansinha, D. E. Smylie, and A. E. Beck, *Earthquake Displacement Fields and the Rotation of the Earth. A NATO Advanced Study Institute Conference Organized by the Department of Geophysics, University of Western Ontario, London, Canada, June 22–28, 1969.* 1970, XI + 308 pp.
21. Jean-Claude Pecker, *Space Observatories.* 1970, XI + 120 pp.
22. L. N. Mavridis (ed.), *Structure and Evolution of the Galaxy, Proceedings of the NATO Advanced Study Institute, held in Athens, September 8–19, 1969.* 1971, VII + 312 pp.

23. A. Muller (ed.), *The Magellanic Clouds. A European Southern Observatory Presentation: Principal Prospects, Current Observational and Theoretical Approaches, and Prospects for Future Research. Based on the Symposium on the Magellanic Clouds, held in Santiago de Chile, March 1969, on the Occasion of the Dedication of the European Southern Observatory*. 1971, XII + 189 pp.

24. B. M. McCormac (ed.), *The Radiating Atmosphere. Proceedings of a Symposium Organized by the Summer Advanced Study Institute, held at Queen's University, Kingston, Ontario, August 3–14, 1970*. 1971, XI + 455 pp.

25. G. Fiocco (ed.), *Mesospheric Models and Related Experiments. Proceedings of the 4th ESRIN-ESLAB Symposium, held at Frascati, Italy, July 6–10, 1970*. 1971, VIII + 298 pp.

26. I. Atanasijević, *Selected Exercises in Galactic Astronomy*. 1971, XII + 144 pp.

27. C. J. Macris (ed.), *Physics of the Solar Corona. Proceedings of the NATO Advanced Study Institute on Physics of the Solar Corona, held at Cavouri-Vouliagmeni, Athens, Greece, 6–17 September 1970*. 1971, XII + 345 pp.

28. F. Delobeau, *The Environment of the Earth*. 1971, IX + 113 pp.

29. E. R. Dyer (general ed.), *Solar-Terrestrial Physics/1970. Proceedings of the International Symposium on Solar-Terrestrial Physics, held in Leningrad, U.S.S.R., 12–19 May 1970*. 1972, VIII + 938 pp.

30. V. Manno and J. Ring (eds.), *Infrared Detection Techniques for Space Research, Proceedings of the 5th ESLAB-ESRIN Symposium, held in Noordwijk, The Netherlands, June 8–11, 1971*. 1972, XII + 344 pp.

31. M. Lecar (ed.), *Gravitational N-Body Problem, Proceedings of IAU Colloquium No. 10, held in Cambridge, England, August 12–15, 1970*. 1972, XI + 441 pp.

32. B. M. McCormac (ed.), *Earth's Magnetospheric Processes. Proceedings of a Symposium Organized by the Summer Advanced Study Institute and Ninth ESRO Summer School, held in Cortina, Italy, August 30–September 10, 1971*. 1972, VIII + 417 pp.

33. Antonin Rükl, *Maps of Lunar Hemispheres*. 1972, V + 24 pp.

34. V. Kourganoff, *Introduction to the Physics of Stellar Interiors*. 1973, XI + 115 pp.

35. B. M. McCormac (ed.), *Physics and Chemistry of Upper Atmospheres. Proceedings of a Symposium Organized by the Summer Advanced Study Institute, held at the University of Orléans, France, July 31–August 11, 1972*. 1973, VIII + 389 pp.

36. J. D. Fernie (ed.), *Variable Stars in Globular Clusters and in Related Systems. Proceedings of the IAU Colloquim No. 21, held at the University of Toronto, Toronto, Canada, August 29–31, 1972*. 1973, IX + 234 pp.

37. R. J. L. Grard (ed.), *Photon and Particle Interaction with Surfaces in Space. Proceedings of the 6th ESLAB Symposium, held at Noordwijk, the Netherlands, 26–29 September, 1972*. 1973, XV + 577 pp.

38. Werner Israel (ed.), *Relativity, Astrophysics and Cosmology. Proceedings of the Summer School, held 14–26 August, 1972, at the BANFF Centre, BANFF, Alberta, Canada*. 1973, IX + 323 pp.

39. B. D. Tapley and V. Szebehely (eds.), *Recent Advances in Dynamical Astronomy, Proceedings of the NATO Advanced Study Institute in Dynamical Astronomy, held in Cortina d'Ampezzo, Italy, August 9–12, 1972*. 1973, XIII + 468 pp.

40. A. G. W. Cameron (ed.), *Cosmochemistry. Proceedings of the Symposium on Cosmochemistry, held at the Smithsonian Astrophysical Observatory, Cambridge, Mass., August 14–16, 1972*. 1973, X + 173 pp.

41. M. Golay, *Introduction to Astronomical Photometry*. 1974, IX + 364 pp.

42. D. E. Page (ed.), *Correlated Interplanetary and Magnetospheric Observations. Proceedings of the 7th ESLAB Symposium, held at Saulgau, W. Germany, 22–25 May, 1973*. 1974, XIV + 662 pp.

43. Riccardo Giacconi and Herbert Gursky (eds.), *X-Ray Astronomy*. 1974, X + 450 pp.

44. B. M. McCormac (ed.), *Magnetospheric Physics. Proceedings of the Advanced Summer Institute, held in Sheffield, U.K., August 1973*. 1974, VII + 399 pp.

45. C. B. Cosmovici (ed.), *Supernovae and Supernova Remnants. Proceedings of the International Conference on Supernovae, held in Lecce, Italy, May 7–11, 1973*. 1974, XVII + 387 pp.

46. A. P. Mitra, *Ionospheric Effects of Solar Flares*. 1974, XI + 294 pp.